Cytokinins

Chemistry, Activity, and Function

Cytokinins

Chemistry, Activity, and Function

Edited by

David W. S. Mok, Ph.D.

and

Machteld C. Mok, Ph.D.

Department of Horticulture

and

Center for Gene Research and Biotechnology

Oregon State University

Corvallis, Oregon

CRC Press
Taylor & Francis Group
Boca Raton London New York

CRC Press is an imprint of the
Taylor & Francis Group, an **informa** business

First published 1994 by CRC Press
Taylor & Francis Group
6000 Broken Sound Parkway NW, Suite 300
Boca Raton, FL 33487-2742

Reissued 2018 by CRC Press

Library of Congress Cataloging-in-Publication Data

Cytokinins: chemistry, activity, and function / edited by David W. S. Mok and Machteld C. Mok
p. cm.
Includes bibliographical references and index.
ISBN 0-8493-6252-0
1. Cytokinins. I. Mok, David W. S. II. Mok, Machteld C.
QK898.C94C97 1994
581.19'27—dc20 93-6461

A Library of Congress record exists under LC control number: 93006461

Publisher's Note
The publisher has gone to great lengths to ensure the quality of this reprint but points out that some imperfections in the original copies may be apparent.

Disclaimer
The publisher has made every effort to trace copyright holders and welcomes correspondence from those they have been unable to contact.

ISBN 13: 978-1-315-89218-4 (hbk)
ISBN 13: 978-1-351-07128-4 (ebk)

Visit the Taylor & Francis Web site at http://www.taylorandfrancis.com and the
CRC Press Web site at http://www.crcpress.com

Preface

The discovery of cytokinins by Skoog, Miller, and associates in 1955 represented a milestone in plant hormone biology and ushered in several new areas of plant research. Subsequent studies demonstrated the extensive involvement of cytokinins in plant development, from seed germination to flowering and senescence. This knowledge of cytokinins led to the understanding of some important plant–microbial interactions, as illustrated by the unraveling of the tumor-inducing activities and genes of *Agrobacterium*. Furthermore, cytokinins have been essential for *in vitro* manipulations of plant cells and tissues, thereby directly contributing to advances in plant biotechnology.

Much information has been gathered regarding the chemistry and biological effects of cytokinins; however, many critical issues, such as the mode of action of cytokinins and the mechanisms regulating the levels of cytokinins, remain unresolved. Increasing application of molecular biology techniques to the study of cytokinins should bring more rapid progress and shift cytokinin research from predominantly descriptive studies to analyses of cytokinin-related genes and cytokinin effects on gene expression. A volume summarizing previous findings may be appropriate at this transition phase in research emphasis.

We are grateful to all the contributors for assembling and reviewing key findings in their respective areas of expertise. Without their involvement this volume would have remained just a good idea. We wish to pay special tribute to Professor Skoog for his prominent role in cytokinin research and for his great effort in documenting his personal involvement in hormone research (Chapter 1).

The chapters are arranged into sections focusing on chemistry, biosynthesis, metabolism, activity, function, genetics, and analysis. The three chapters on chemistry, by Shaw, Shudo, and Iwamura, describe the synthesis and biological activities of two main classes of cytokinin-active compounds: the adenine and phenylurea derivatives, and their antagonists. Organic and analytical chemistry have played a crucial role in the identification and authentication of cytokinins and will continue to be important for the detection and synthesis of new compounds. Research on antagonists has contributed to the determination of the structural requirements for cytokinin activity and may help resolve the paradox of chemically distinct compounds exerting similar biological effects.

The biosynthesis of cytokinins in plants is summarized in Chapters 5 to 8, starting with an overview of the sites of cytokinin synthesis and the patterns of cytokinin translocation in plants, presented by Letham. Cytokinin biosynthesis in plants may occur by *de novo* synthesis and possibly via metabolism of cytokinin-containing RNAs. Enzyme activities involved in *de novo* synthesis are described by Chen and Ertl (Chapter 6), and cytokinins in tRNAs and RNAs are discussed by Murai (Chapter 7) and Taller (Chapter 8). Although the biosynthetic pathways in higher plants and lower organisms are believed to be the same, no homologous sequence to the bacterial isopentenyltransferase (*ipt*) gene has been detected in plants thus far. The recent report of a zeatin *cis-trans* isomerase (Chapter 10) may be relevant to the RNA route of cytokinin biosynthesis. It is apparent that much remains to be learned regarding the synthesis of cytokinins in higher plants.

Key findings on interconversions between the various forms of cytokinins and the possible significance of the metabolites are summarized by Jameson in Chapter 9. While many enzymes mediating the metabolism of adenine-type cytokinins are also active in converting adenine, adenosine, or AMP, others are clearly specific to cytokinins. Most critical is the characterization of cytokinin-specific enzymes and the cloning of genes encoding these enzymes. Recent studies on enzymes modifying or removing the isoprenoid side chain are summarized, respectively, in Chapters 10 (D. Mok and Martin) and 11 (Armstrong).

The diversity in forms of cytokinins is matched by their wide-ranging effects on plant growth and development. An overview is provided in Chapter 12 (M. Mok), followed by expanded discussions on selected topics. Cytokinin effects on oxidative processes, described in Chapter 13 by Musgrave, are potentially important, but remain relatively unexplored. The effects of cytokinins on greening and other light-regulated processes have been known for many years; recent studies on plastid genes are described in Chapter 14 by Reski. In spite of the fact that cytokinins were identified as cell division factors, the precise target of cytokinin action in cell division is still unknown. Jacqmard, Houssa, and Bernier (Chapter 15) review recent studies on this topic and suggest potential areas of future research.

The mechanisms of cytokinin action remain elusive. Based on models in animal systems, protein receptors were the primary targets, leading to the isolation of several cytokinin-binding proteins, as discussed in Chapter 16 by Brinegar. The advent of molecular biology led to the search for cytokinin-responsive genes (described by Crowell and Amasino in Chapter 17). It is expected that continuing efforts will uncover more cytokinin-responsive genes and lead to the characterization of their function. Signal transduction and the possible role of calcium are discussed in Chapter 18 by Saunders.

Genetic analyses have contributed to the understanding of the action of gibberellins, abscisic acid, and, to some extent, auxin. Genetic studies involving cytokinins have been somewhat limited, as discussed by Wang in Chapter 19, perhaps due to the potential lethality of cytokinin mutants and the difficulty of screening for less severe, but true, cytokinin mutants in higher plants. One interesting trait is the cytokinin autonomy of cell cultures, caused by genetic and epigenetic changes, as summarized by Meins in Chapter 20. Transformation of plants with the *ipt* gene of *Agrobacterium,* and the accompanying changes in plant development, is the subject of Chapter 21, by Klee. The findings with transgenic plants show that even relatively small increases in cytokinin concentrations can bring about significant modifications in plant stature, supporting the contention that cytokinin levels must be carefully regulated in order for normal plant development to take place. Although quantitative effects are often thought to be paramount, qualitative differences may have pronounced effects on plant development as well, as illustrated by Durand and Durand in Chapter 22, by the presence of different cytokinins in male vs. female, and fertile vs. sterile male flowers of *Mercurialis.*

The methods for the separation, identification, and quantitation of cytokinins have been refined in recent years through advances in liquid chromatography, immunology, and mass spectrometry. In Chapters 23 and 24, Banowetz and Teller describe the immunological and GC-MS analyses crucial for accurate measurements of endogenous cytokinin.

Due to space limitations, only selected areas of cytokinin research are included. Moreover, the treatment of each subject area was left to the judgment of the authors, leading to diversity in style and approach. Nevertheless, the chapters represent the current state of knowledge and will hopefully serve to identify future research directions needed to fill the many gaps in our understanding of this important group of growth regulators.

David W. S. Mok, Ph.D.
Machteld C. Mok, Ph.D.

The Editors

David Mok, Ph.D., is Professor of Genetics in the Department of Horticulture and a member of the Center for Gene Research and Biotechnology at Oregon State University, Corvallis, Oregon.

Dr. Mok received his Bachelor of Science degree from National Taiwan University in 1967; his Master of Science degree, in Genetics, from the University of Guelph in 1970; and his Ph.D., also in Genetics, from the University of Wisconsin–Madison in 1975. Dr. Mok was a Pre-Doctoral Fellow of the Donald Jones Funds of the Research Corporation, and has been associated with Oregon State University since his appointment as Assistant Professor in 1975.

Dr. Mok is a member of the American Society of Plant Physiologists, American Society of Genetics, Genetics Association of America, American Association for the Advancement of Science, International Plant Growth Substances Association, and American Society of Horticultural Science.

Dr. Mok's research is centered on the genetics of food legumes, with emphasis on the genetic regulation of cytokinin metabolism, interspecific hybridization, and developmental mutants. He has written over 60 refereed papers and is currently active in studying the molecular biology and enzymology of zeatin metabolic enzymes.

Machteld Mok, Ph.D., is Professor of Horticulture and Plant Physiology, and is associated with the Center for Gene Research and Biotechnology at Oregon State University, Corvallis, Oregon.

Dr. Mok received her undergraduate degree from the University of Wageningen, The Netherlands, in 1969; and her Master of Science and Ph.D. from the University of Wisconsin–Madison, in 1973 and 1975, respectively. She has been at Oregon State University since 1975, when she was appointed Assistant Professor.

Dr. Mok is a member of the American Society of Plant Physiologists, Genetics Society of America, Tissue Culture Association, American Society of Horticultural Science (presently serving as Associate Editor for the journal), International Plant Growth Substances Association, and International Association for Plant Tissue Culture.

Dr. Mok's major research interests focus on cytokinin metabolism, enzymology, gene regulation, and gene transfer. She has written over 50 research publications and received grants from the National Science Foundation and the USDA Competitive Research Grants Office.

The Contributors

Richard M. Amasino, Ph.D.
Department of Biochemistry
University of Wisconsin–Madison
Madison, Wisconsin

Donald J. Armstrong, Ph.D.
Professor
Department of Botany and Plant Pathology
Oregon State University
Corvallis, Oregon

Gary M. Banowetz, Ph.D.
Research Microbiologist/Plant Physiologist
Agricultural Research Service
United States Department of Agriculture
Corvallis, Oregon

Dr. Georges A. E. J. Bernier
Professor
Département de Botanique
Université de Liège
Liège, Belgium

A. Chris Brinegar, Ph.D.
Associate Professor
Department of Biological Sciences
San Jose State University
San Jose, California

Chong-maw Chen, Ph.D.
Professor of Biological Sciences
Department of Biological Sciences
Biomedical Research Institute
University of Wisconsin–Parkside
Kenosha, Wisconsin

Dring N. Crowell, Ph.D.
Assistant Professor
Department of Biology
Indiana University–Purdue University at Indianapolis
Indianapolis, Indiana

Bernard Durand, Ph.D.
Professor
Laboratoire de Biologie et Biochimie Végétale
Université d'Orléans
Orléans, France

Raymonde Durand, Ph.D.
Professor
Laboratoire de Biologie et Biochimie Végétale
Université d'Orléans
Orléans, France

John R. Ertl, M.S.
Research Specialist
Department of Biological Sciences
Biomedical Research Institute
University of Wisconsin–Parkside
Kenosha, Wisconsin

Claude Houssa
Département de Botanique
Université de Liège
Liège, Belgium

Hajime Iwamura, Ph.D.
Department of Agricultural Chemistry
Faculty of Agriculture
Kyoto University
Kyoto, Japan

Dr. Annie Jacqmard
Senior Scientist
Département de Botanique
Université de Liège
Liège, Belgium

Paula E. Jameson, Ph.D.
Professor of Plant Biology and Biotechnology
Department of Plant Biology and Biotechnology
Massey University
Palmerston North, New Zealand

Harry J. Klee, Ph.D.
Fellow
Agricultural Group of Monsanto
St. Louis, Missouri

David S. Letham, Ph.D.
Professor Emeritus
Research School of Biological Sciences
Australian National University
Canberra, Australia

Ruth C. Martin, M.S.
Senior Research Assistant
Department of Horticulture
Center for Gene Research and Biotechnology
Oregon State University
Corvallis, Oregon

Dr. Frederick Meins, Jr.
Friedrich Miescher–Institut
Basel, Switzerland

David W. S. Mok, Ph.D.
Professor
Department of Horticulture
Center for Gene Research and Biotechnology
Oregon State University
Corvallis, Oregon

Machteld C. Mok, Ph.D.
Professor
Department of Horticulture
Center for Gene Research and Biotechnology
Oregon State University
Corvallis, Oregon

Norimoto Murai, Ph.D.
Professor
Department of Plant Pathology and Crop Physiology
Louisiana State University and LSU Agricultural Center
Baton Rouge, Louisiana

Mary E. Musgrave, Ph.D.
Associate Professor
Department of Plant Pathology and Crop Physiology
Louisiana Agricultural Experiment Station
Louisiana State University Agricultural Center
Baton Rouge, Louisiana

Dr. Ralf Reski
Universität Hamburg
Institut für Allgemeine Botanik
Hamburg, Germany

Mary Jane Saunders, Ph.D.
Associate Professor
Biology Department
University of South Florida
Tampa, Florida

Gordon Shaw, D.Sc., Ph.D.
Professor
Department of Chemistry and Chemical Technology
University of Bradford
Bradford, West Yorkshire
United Kingdom

Koichi Shudo, Ph.D.
Faculty of Pharmaceutical Sciences
University of Tokyo
Bunkyo-ku, Hongo
Tokyo, Japan

Folke Skoog, Ph.D.
Professor Emeritus
Department of Botany
University of Wisconsin–Madison
Madison, Wisconsin

Barbara J. Taller, Ph.D.
Assistant Professor
Department of Biology
Memphis State University
Memphis, Tennessee

Dr. Gerard Teller
Chemical Engineer
Laboratoire de Spectrométrie de Masse
Institut de Chimie
Centre National de la Recherche Scientifique
Université de Strasbourg
Strasbourg, France

Trevor L. Wang, Ph.D.
Applied Genetics
John Innes Institute
Colney, Norwich
United Kingdom

Contents

Genetics

Analyses

Cytokinins

Chemistry, Activity, and Function

Professor Folke Skoog, 1993
Photographed by M. T. Rosenbaum, Oregon State University

Chapter 1

A Personal History of Cytokinin and Plant Hormone Research

Folke Skoog

TABLE OF CONTENTS

PROLOGUE

I was asked to submit an account of the early research on plant hormones and especially on the work on cytokinins that took place in my laboratory at the University of Wisconsin. Considering my limited capacity to recall, to concentrate, and to write, and having seen the results of similar tasks by other octogenarians, I am convinced it would be in the best interest of all concerned (eventual readers included) for me to refrain. Instead, I am pleased to discuss and answer a set of questions, prepared by Professor Donald Armstrong, on my experience with plant hormone research. Professor Armstrong has been engaged in plant hormone research ever since he started as a graduate student with me in 1961. He continued as a research associate with me and others and is now a faculty member of the Department of Botany and Plant Pathology at Oregon State University. He has made outstanding contributions to our knowledge of cytokinins, especially those he has isolated and studied in specific tRNA species from a wide variety of organisms. He has studied the action of different types of cytokinins, their metabolism, and their degradation. He has shared in much of my work and is thoroughly familiar with all of it, and he is uniquely qualified to review it. I am deeply grateful to him for his interest in this matter.

I. BEGINNINGS OF PLANT GROWTH SUBSTANCE RESEARCH IN THE U.S.

Dr. Skoog, you received an undergraduate degree in chemistry from the California Institute of Technology in 1932 and a Ph.D. in biology from the same institution in 1936. During this period, the Division of Biology at Cal Tech became an important initial center of plant hormone research in this country. How did this come about?

It was introduced by T. H. Morgan, the eminent geneticist, who in 1928, when he was facing mandatory retirement at Columbia University, was persuaded by Cal Tech physical scientists (Millikan and others) to come to Cal Tech to start a division of biology. Morgan was interested in establishing an institute of quantitative experimental biology. He collected a staff of about ten investigators. He brought his former students and current research associates from Columbia: A. Sturtevant, C. Bridges, Jack Schultz, two plant geneticists (Sterling Emerson and Ernest Anderson), and two biochemists, one of whom was Kenneth Thimann. He also brought two plant physiologists, Robert Emerson (fresh from Otto Warburg's laboratory) and one plant growth specialist, Herman Dolk, from F.A.F.C. Went's laboratory in Utrecht. He brought these men together and looked after their research

0-8493-6252-0/94/$0.00+$.50

interests and personal needs, with truly paternal care.

Why Dolk? Morgan at heart was a classical embryologist or developmental biologist. He had spent much time slicing earthworms to study their regeneration. He was interested in the polar regeneration of head and tail regions, and it is said he had the notion of a tail-forming and a head-forming substance. In any case he was greatly interested in polarity. Around 1910 he wrote an article on polarity in plants. When he heard that two young students of Went, his son (Frits) and Dolk, had demonstrated that a substance that promoted growth moved polarly down oat coleoptiles, he arranged for Dolk to join his group at Cal Tech. A small building with two underground rooms, superficially environmentally controlled, was constructed for Dolk's work on the plant growth substance.

How did you personally get involved in plant hormone research?

Well, in 1929 when I was a sophomore at Cal Tech, taking physics, I met Carl Lindegren. Lindegren was a graduate student with Morgan, but he also was taking sophomore physics. Since he was an old Swede from Rhinelander (Wisconsin), I told him that I didn't have to take German, but I had to take something else. He said, "Fine, you come to work with me." Lindegren was working on the genetics of *Neurospora*, which eventually, largely through Lindegren's influence, became a very important organism in microorganism genetics, especially as developed by Beadle, who was a young instructor at Cal Tech at that time. I worked with Lindegren while I was doing my regular undergraduate work in chemistry. My task was to study the mating types of *Neurospora*. This was fine, but the study of *Neurospora* at that time, before the availability of B-vitamins, was more or less a kitchen science. You had to cook the oatmeal and grow them, and it became routine and pretty monotonous. So, I told Lindegren that I was going to see what I could do with the biochemists. So he said, "Well, those biochemists..., I tell you, I wouldn't go there. There's a young Dutchman just come here to work, and he has a new thing, something that stimulates growth. I'll talk to him, and you can work with him." So I started to work with Dolk, which was the beginning of my work with plant hormones.

At that time the growth substance, later called auxin, had been shown to promote elongation of cells, especially in regions some distance below the tip of the coleoptile where it was formed. Frits Went suspected that a second growth factor moving from the seed was required for auxin to be effective in promoting growth. My project was to obtain evidence for this second factor. Using time-lapse photography, I studied the distribution of growth rates along the coleoptile, to see how it might be influenced by factors coming from the seed. Unfortunately, Dolk had an accident at the Grand Canyon, in which his car went off the road. He came back from this, but his car might have been damaged. Two weeks later he took the same car out in the desert, and he blew a tire and was killed. He was a fine man and a keen, exceptionally talented experimenter.

When Dolk died, Thimann, who had started to collaborate with Dolk, wanted us to continue the work begun with Dolk. I worked with Thimann as an undergraduate (but on a different problem) and continued to work with him as a graduate student at Cal Tech, starting in 1932, until he left for Harvard University in 1935. His early guidance and my close contact with him were most beneficial. Our friendship has continued for more than 60 years.

I finished my graduate work in 1936 under the direction of Professor Frits Went, who was brought to Cal Tech by Morgan soon after Dolk's death. Went's broad botanical knowledge and his imaginative, enthusiastic approach to research were both beneficial and inspiring. His deep concern for my personal well-being as a lone student during the economic stress of the Depression is something I shall always remember with deep gratitude.

II. AUXIN AND THE REGULATION OF BUD GROWTH

Much of your early career in plant hormone research appears to have centered on the role of auxin in regulating the growth of buds. Would you trace how this work developed and comment on the ideas, events, and people that influenced your thinking during this period?

In 1925 Keeble and Snow proposed that the inhibition of lateral buds by the terminal bud in the broad bean *Vicia faba* was due to a factor moving polarly from the terminal bud, down the stem, to the lateral buds. The reason they concluded that the movement was polar is still not clear to me. When I began to work with Thimann, he suggested that we attempt to find evidence for this factor by procedures that had been used to demonstrate the growth substance in *Avena* coleoptiles. We found that material promoting growth in *Avena* coleoptiles could be obtained from terminal buds of beans, by diffusion into agar blocks. Furthermore, when the agar blocks or excised terminal buds of beans were placed directly on the cut surfaces of decapitated plants and replaced every 3 h, inhibition of the laterals could be maintained. *Avena* coleoptile tips could be substituted for terminal buds. Small amounts of active material could also be obtained from leaves. We found that the buds could be inhibited with extracts from *Rhizopus* (the bread mold) and, eventually, with a synthetic preparation of the auxin indoleacetic acid. We also showed that auxin moved polarly in beans, just as it did in coleoptiles, and promoted elongation of the stems of defoliated bean plants. Thus, we concluded that auxin was not confined to coleoptiles, but occurred generally as a growth-promoting factor in plants. This work was published in 1933 and 1934.

The fact that a growth-promoting substance from the terminal bud could also inhibit the growth of laterals was not readily accepted. We argued that the effect was exerted *directly* in the buds by relatively high concentrations of the hormone being supplied continuously from the terminal. Others proposed that the inhibition resulted *indirectly* from depletion of nutrients (Went said specific growth factors) being utilized for the growth of the main stem in the presence of the hormone.

Soon after I began work as a graduate student, I wanted to test the effect of the growth substance on buds in isolation from the plant. I went to Morgan to tell him that I wanted to do tissue culture with plant material and needed equipment (forceps, scissors, etc.) for this purpose. He told me that "you don't need all this equipment. There is a Frenchman coming in a few months. He is bringing instruments, and I will put you to work with him." This was Boris Ephrussi. He was an expert in animal tissue culture. He did bring beautiful equipment, and I learned a great deal about tissue culture techniques (bleeding the rooster, making embryo extracts, and growing animal tissues) from him, but this did not accomplish much with plants. Not until some years later did I grow excised pea epicotyls and show that their growth could be promoted or inhibited as a function of the applied concentration of indoleacetic acid.

I was still interested in the possibility of a second factor, and as a graduate student, I also studied the effect of deseeding on the growth of *Avena* coleoptiles. I was able to demonstrate that a factor from the seed was moving up the coleoptile and was converted to auxin in the tip. I also found that deseeded oat seedlings in which the synthesis of auxin was reduced were more sensitive to applied auxin and could be used not only to test for five to ten times lower auxin concentrations than the standard *Avena* test, but could also be used for the detection and quantitative assay of auxin precursors. For example, tryptophan and tryptamine were both shown to be slowly converted to indoleacetic acid, when applied to coleoptiles. However, this work did not reveal a second factor.

At the time of my work on bud inhibition, I saw a report by an Irish botanist, Hudson, on gemma cup formation in the liverwort *Marchantia*. He showed that, normally, gemma cups develop only on the basal portion of the thallus, but when the thallus is cut into segments, gemma cups are formed on the basal edge of all segments except on the segment containing the apical cell. Furthermore, if the thallus is exposed to X-rays, gemma cups appear on all parts of the thallus. It seemed possible that gemma cup formation might be inhibited by auxin in the same way that lateral buds were inhibited by auxin from the terminal bud. Also, it seemed possible that auxin might be inactivated by X-rays. I found that terminal buds of young *Vicia* or *Pisum* plants, also coleoptile tips of *Avena* seedlings, exposed to moderate doses of high-voltage X-rays quickly lost about 30 to 40% of their auxin content as measured by diffusion into agar blocks, and in the former two, auxin production was entirely lost in a period of about 4 to 5 days. In parallel with this, there was outgrowth of lateral buds. Similarly, indoleacetic acid in aqueous solution exposed to moderate dosages of X-rays was inactivated by 30 to 40% in the presence of air. If kept under nitrogen, the inactivation was greatly reduced. It was deduced that the inactivation was due to peroxide, which is formed on X-irradiation of water in the presence of air and to which auxin had been shown to be very sensitive. I could show that the immediate effect of X-rays on the growth could be reversed, to a slight extent, by the application of indoleacetic acid to the plants. Furthermore, X-irradiation of internodes between the terminal and lateral buds led to the development of lateral buds in the portion of the plant below the irradiated region. This outgrowth of buds could be counteracted by indoleacetic acid applied above the buds, thus providing further evidence that the action of auxin in bud inhibition is exerted directly in the buds.

On completion of my graduate work at Cal Tech, and no doubt on the basis of Morgan's recommendation, I was awarded a National Research Council Fellowship to work with Professor Dennis Hoagland in the Division of Plant Nutrition at the University of California-Berkeley. Investigations in his laboratory indicated that incipient effects of zinc deficiency of plants grown in nutrient solution could be counteracted by addition of the auxin indoleacetic acid. It was therefore of interest to determine to what extent zinc might affect the synthesis or action of the endogenous growth substance. The results of that work demonstrated that even 1 day before visible symptoms of zinc deficiency appeared, auxin production in the terminal shoots of tomato plants was reduced 50%, and within 2 days, to practically zero, leading to complete secession of growth. This effect could be prevented by growing the plants under red light, and it was shown that deficiency of zinc was associated with striking increases in peroxidase and oxidase activity in the tissue. Although the effect of zinc deficiency on auxin was both striking and specific, the essential requirement for zinc was not due exclusively to auxin inactivation and the loss of auxin synthesis.

My experience in the Division of Plant Nutrition was exceptionally valuable to me. Hoagland was a kind, wise, and practical scientist and administrator. It is of interest that he and Truog, the two outstanding names in plant nutrition in this country, worked together under E. V. McCollum at the University of Wisconsin, where they both terminated their formal studies with the M.S. degree. Hoagland's group of investigators included faculty members and research associates. Each member had his own area of specialty, and each one was responsible also for helping all the others with work in his area. It was a very sound and effective team. In my opinion, it was the most active and efficient organization in experimental plant biology anywhere at that time. I had the special favor of often accompanying Hoagland on his daily inspection tours of the U.C.–Berkeley greenhouses where a large number of nutrition experiments were in progress. He took time to discuss the special features and the purpose and nature of each one as it was started. I especially recall his pleasure when he observed (I believe, for the first time) and described symptoms of molybdenum deficiency. His instructions on plant behavior and his emphasis on improving methodology have been of lasting value in my work.

While at U.C.–Berkeley, I also had an opportunity to work with Professor J. P. Bennett, who had a long-term interest in the dormancy of fruit trees. He had a greenhouse full of 6-ft-tall pear trees that had been kept dormant by high temperature for 4 years. I jokingly told him that I thought I could break the dormancy of these trees with an extract, prepared from yeast, that I had used to germinate dormant bacterial spores. Two ml of an aqueous solution of this preparation, injected through the tip of a branch near the top of a tree, moved down the stem and caused the leaves to come out all the way to the base of the tree. This led to further collaboration with him, showing the presence of auxin-inhibiting substances in dormant buds in trees and their gradual disappearance and the development of increasing auxin content in buds of dormant trees kept in cold storage to break dormancy. The chemical nature of the inhibitors, or of the auxin-active substances, was not established.

In 1937 I accepted an appointment as instructor and research associate to work with Thimann, who had moved to Harvard University. At Harvard my work was mainly on the analysis and extraction of free and bound auxins, but there I also carried out some experiments with dwarf apple trees, showing that not only breaking of dormancy, but also profuse production of new buds, was obtained by treatment with some preparations from yeast, suggesting, in retrospect, that cytokinins may well have been part of the active material obtained from yeast.

III. REGULATION OF BUD FORMATION

Your work on the role of auxin in regulating the growth of buds was followed by investigations of the regulation of bud formation and your first attempts to utilize plant tissue cultures as experimental systems. How did this come about?

In 1939, while I was at Harvard, Philip White, in a paper demonstrating continuous growth of tobacco genetic tumor tissue (derived from the tumor-forming hybrid *Nicotiana langsdorffii* x *N. glauca*), emphasized that this was growth without auxin. At this time, White was still one of the most active opponents of the idea that auxin played a significant role in plant growth. In fact, skepticism about the existence of a growth-promoting hormone (auxin) was very prevalent in this country. In 1935 Frits Went had been invited to an AAAS symposium in Pittsburgh, to demonstrate the *Avena* curvature test for auxin. For some reason, Went was unable to obtain curvatures at this meeting. White, representing a group of skeptics, stood up to declare that the auxin business was obviously a hoax and that it was high time that these charlatans from Europe were sent back to where they belonged. Went was greatly disturbed by this experience and deduced that it must have been the high degree of air pollution in Pittsburgh that was responsible for his failure. On return to Pasadena, he set Haagen-Smit, at that time a research associate with Went, to investigate the effect of air pollution on the *Avena* test and the growth of plants. This eventually led to Haagen-Smit's identification of the major toxic components in smog, and his outstanding contributions to both technical and legal means of preventing air pollution, for which he was awarded the National Medal of Science.

White had an intense interest in achieving a plant tissue culture medium of completely known composition. This may well have been a reason for his early distrust of auxins and organic growth factors in general. In fact, he was so engrossed in this project that I used to tease him by telling him that he was a completely frustrated fellow going through life trying to prove carbon was a nonessential element. This did not deter him, but gradually he became more tolerant. In fact, in the case of auxins, he finally reached the point of believing that auxin was the principal causal agent in pathogenic overgrowths in plants, and he published a report, more or less, to this effect.

In spite of White's earlier disbelief in auxin, by 1939 he must have harbored some doubts, because he sent a culture of his genetic tumor tissue to Harvard and requested that I test it for the presence of auxin. I found that very high concentrations of auxin were present in ethyl ether extracts of the tissue, even though high levels of inhibitors masked the auxin activity in the extracts and had to be removed by diffusion through stacks of agar blocks. White was correct, however, in claiming that added indoleacetic acid did not improve the growth of these cultures and, in fact, tended to inhibit their growth.

After I moved to Johns Hopkins University in 1941, I became interested in what might be done to counteract auxin's inhibiting effect on the growth of the tumor cultures. I was especially interested in this because White had shown that, occasionally, buds were formed in submerged cultures of the tumor tissue. I found that formation of the buds could be completely prevented by addition of auxin (indoleacetic acid) to the medium. If the buds were allowed to develop, occasionally root formation would occur, and I was able to recover small tobacco plantlets from these cultures (Figure 1). These plantlets were the first complete plants to be regenerated from plant cultures.

I recognized that the tobacco tumor system provided opportunities to test a role of auxin in initiating the formation of buds as well as in promoting their subsequent development. On the assumption, at that time, that auxin, in some manner, acted through stimulating respiration, I started to test factors that might enhance respiration and thus act as cofactors in the auxin response. I wanted to test ATP. As it was unavailable, I used adenosine in combination with increased sucrose and phosphate concentrations. In the presence of high phosphate, I also used increased iron concentrations. By this procedure I was able to prevent completely the

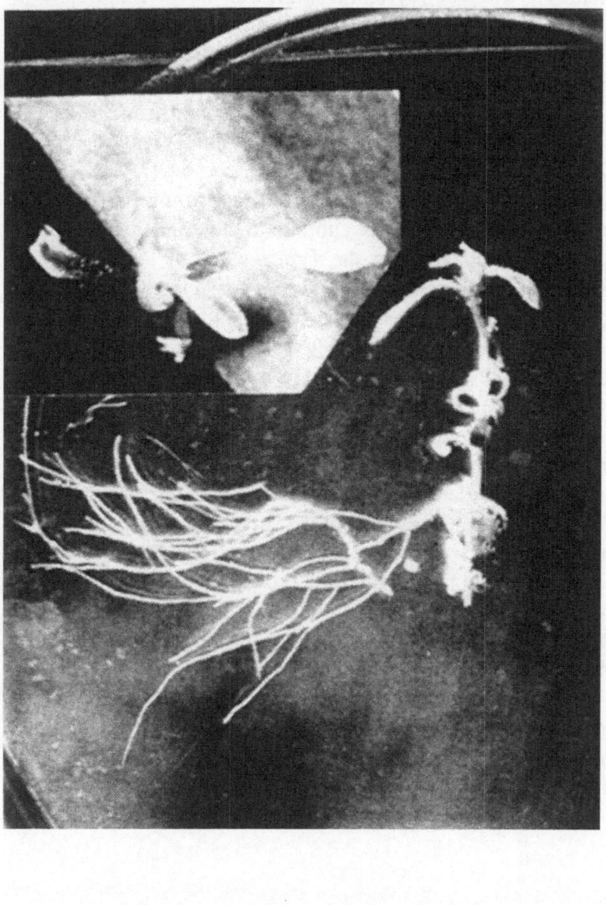

Figure 1-1 The first complete plant obtained from a callus tissue culture (*Nicotiana glauca* x *N. langsdorffi* callus cultured *in vitro* for more than ten transfers in 1939–1940). (Reprinted from Skoog, F., *Am. J. Bot.*, 31:19–24, 1944. With permission.)

inhibitory effect, and actually to increase the growth, of the tumor tissue in the presence of relatively high auxin concentrations, but I did not restore bud formation in submerged cultures treated with auxin.

At this point my work on plants was interrupted by other activities during the war, and I did not return to it until coming to the University of Wisconsin in January 1947. At that time, my cultures of hybrid tumor tissue had been lost, and a shipment of new cultures from White was so seriously contaminated with microorganisms that I decided to start from scratch with a new tobacco cultivar, Wisconsin 38, that I obtained from the Department of Horticulture. Cheng Tsui (who had come from Michigan as a postdoctoral research associate) and I prepared stem-segment cultures of this tobacco cultivar. The behavior of these cultures differed from that of the genetic tumor tissue. Even when cultured on agar, the stem segments formed occasional buds as well as callus. With increasing auxin concentration, growth of callus extended from the base to the more apical portions of the segments; formation of buds was inhibited, and some roots were formed. Adenine supplied to cultures with low auxin concentrations markedly stimulated bud formation. In fact, the amount of bud and root formation depended on the proportions of auxin and adenine supplied in the medium. Approximately 15,000 molecules of adenine were required to counteract one molecule of indoleacetic acid.

IV. THE DISCOVERY OF KINETIN

Your work on the regulation of growth and morphogenesis in tobacco stem segments led you to search for factors that might stimulate cell division in plant cells. Would you describe the events that culminated in the isolation and identification of kinetin from partially degraded samples of DNA?

As the tobacco stem segments were anatomically and morphologically too complex to determine composition changes associated with treatments affecting growth, an attempt was made to culture pith tissue, which contained parenchyma cells and no vascular tissue. Slabs of pith tissue placed on the basal medium exhibited no growth. In response to auxin, the cells enlarged and formed masses of swollen parenchyma cells. Some mitoses occurred, but no cell divisions were observed. If these pith slabs were placed in contact with segments containing vascular tissue, a callus would form at the basal ends of the slabs. Also, if segments of vascular tissue were placed on the medium next to the pith slabs, some cell division occurred in the slabs.

A graduate student doing these experiments, John Jablonski, showed that cell divisions could also be obtained in tobacco pith tissue by supplying coconut milk (which had been used by experimenters in growing plant cells) and also by supplying yeast extract or malt extract to the medium. Therefore, an attempt was made to isolate the active material in coconut milk and in yeast extract. Work on coconut milk resulted in preparations approximately 4000 times more active than the starting material, but no active compound was identified. Work on the yeast extract, carried out mainly by Carlos Miller, a postdoctoral fellow, did result in preparations that were tentatively characterized as containing purines, on the basis of their UV spectrum and other properties.

Various purines and purine-containing materials were tested by Miller, including an old commercial preparation of DNA, which was highly active. However, a large new sample of commercial herring sperm DNA was practically inactive. But on standing, and especially on heating in weakly acid solution in an autoclave, this material became highly active. From this he eventually isolated, by chromatographic procedures, a crystalline compound. Before the structure was determined, because of its action in promoting cell division (cytokinesis), it was named kinetin. With the aid of Professor Frank Strong and associates in the Department of Biochemistry, the elemental composition and structure of this material was deduced and confirmed by synthesis to be 6-furfurylaminopurine. The synthesis was performed in March 1955.

The initial manuscript describing the isolation of kinetin was submitted, but not accepted, for publication in the *Proceedings of the Society of Experimental Biology and Medicine*. One reviewer stated (among other complaints), "These authors are obviously trying to put something on the record." This prompted me to terminate my membership in the Society. This manuscript, and two later ones, were accepted by the *Journal of the American Chemical Society*. The last manuscript reported on the identification of kinetin. One reviewer said only, "Publish as is," but the second reviewer, in several pages of condemnation, stated that "this report has no place in either the biological or chemical literature." The editor noted that the two curves in one figure (which compared the infrared spectrum of the isolated compound with that of the authentic synthesized compound) were identical, and he suggested that "in deference to the second reviewer," one curve might be removed. I was unhappy, but Strong was highly amused, and we complied with the editor's instructions.

Strong and co-workers undertook the synthesis of analogs of kinetin. Three days after obtaining synthetic kinetin, he and F. S. Okumura synthesized 6-benzylaminopurine, which was even slightly more active than kinetin in promoting the growth of tobacco cultures. Of approximately 40 compounds that they synthesized and we tested for kinetin-like activity, 21 were active (but less so than kinetin itself) in promoting growth. Thirty compounds, including some available from other sources, were inactive. None of the compounds tested showed marked activity as kinetin antagonists.

Some time after the discovery of kinetin, F. C. Steward reported that an aza derivative of kinetin was much more active than kinetin itself and concluded that kinetin was an artifact without biological significance. In tobacco bioassays of the two compounds, we found that the concentrations tested by Steward were optimal for the aza derivative, but that for kinetin they were ten times higher than optimal and, therefore, had only a slight growth-promoting effect. In his report Steward stated that optimal concentrations of both compounds were used, but in a discussion years later, he admitted that "more tests might have been done."

The notion that kinetin was an artifact was widely circulated. It seriously hindered me and others from obtaining cytokinin research funds from granting agencies. It has persisted even though data reported by Nelson Leonard and co-workers have shown that deoxyadenosine is converted to kinetin and that on the basis of the kinetics of the reactions involved, the formation of kinetin is expected to occur spontaneously where DNA is present. Therefore, it is expected to occur naturally, especially in wounded tissue in which there are highly elevated levels of nuclease activity.

You encountered a number of interesting difficulties in attempting to patent kinetin and other cytokinins. Would you describe some of your personal experiences with the patent process?

I have held one patent on the use of adenine and adenine derivatives in the growth regulation and propagation of plants and have been a co-holder with others on three patents on specific cytokinins or antagonists and their use in plant growth regulation. All these patents were obtained through, and assigned to, the Wisconsin Alumni Research Foundation (WARF).

When an application was filed on kinetin and its use, it was ruled that my patent on adenine and derivatives stood in its way. The lawyer representing WARF, in his wisdom, suggested, and we (Strong, Miller, and I), being equally enlightened, agreed to waive the claim on adenine, thus obtaining the kinetin patent and making both patents useless for practical purposes.

When WARF applied for a patent on N^6-benzyladenine, three other organizations had made prior applications. On the basis of the data submitted, we were judged to have 2 months priority. Two organizations withdrew their applications, the third (investigators of the Lederle Laboratories) claimed priority in "reduction to practice." Our claim for reduction to practice was based on the original bioassays in Miller's notebook and could only be used if he were not an applicant. Our lawyer stated that Strong and I would be granted the patent if Miller's name were removed. Considering that Lederle's claim was based on information given to them by Strong, I insisted that Miller's name be kept on, and as a result The American Cyanamid Co., parent company of Lederle Laboratories, was assigned the patent.

We were assigned a patent on cytokinin antagonists, but no license on the use of this patent has been assigned to anyone by WARF. To the best of my information, no income has accrued to WARF, or anyone else, on this patent or on any of the cytokinin patents we assigned to WARF. In short, on the basis of my experience, I doubt that any individuals without exceptional financial resources would be able to defend successfully patent rights in competition with large industrial corporations. As far as I know, no original investigator of plant hormones has profited financially from practical applications.

V. TRIACANTHINE AND THE CYTOKININ ACTIVITY OF ISOTRIACANTHINE

The chemical identification of kinetin, which was carried out in collaboration with Frank Strong's laboratory at the University of Wisconsin, led to interactions with Nelson Leonard of the Department of Chemistry at the University of Illinois. Would you describe your initial work with Leonard on the biological activity of the alkaloid triacanthine, and comment on the significance of the synthesis of the compound that was then called isotriacanthine?

Leonard and Strong were members of the same NIH biochemistry panel, and Leonard informed Strong that he and his group had isolated and identified an alkaloid (triacanthine) from young leaves of *Gleditsia triacanthos* (honey locust). Because triacanthine, which they had characterized as 3-(γ, γ-dimethylallyl)-6-aminopurine [i.e.,

3-(Δ^2-isopentenyl)adenine], had a structure similar to that of kinetin, he asked that we test this compound for cytokinin activity. It was found that cold-sterilized samples of the compound were inactive, but a sample added to the medium and then autoclaved was active in promoting the growth of tobacco callus. At about the same time, French investigators (Cavé, Chouard, and Beauchesne) identified and tested triacanthine from another source and reported that it had cytokinin activity. We realized that the difference in our results must have been due to the fact that they used only autoclaved material, and we suspected that the mechanism by which the material became active must have involved some intramolecular rearrangement. At this time, Leonard and co-workers had already synthesized the N^6-isomer of triacanthine, which we referred to as isotriacanthine or 6-γ, γ-dimethylallylaminopurine and which was later commonly called N^6-(Δ^2-isopentenyl)adenine. We had found this compound to be ten times more active than kinetin in the tobacco callus bioassay and even more effective than kinetin or N^6-benzyladenine in promoting the formation and growth of buds in tobacco cultures. Credit should be given to Leonard and co-workers for making this compound available for use for the identification of naturally occurring cytokinin-active N^6-substituted adenine derivatives. It was available to Letham for use in identifying zeatin, and to other investigators.

VI. THE RACE TO ISOLATE A NATURALLY OCCURRING CYTOKININ

The discovery of kinetin triggered an intensive effort to isolate and identify a naturally occurring compound with similar activity. The main players were Miller at Indiana University, Letham in New Zealand, Steward at Cornell, and yourself at the University of Wisconsin. First of all, could you comment on the early work on the isolation of growth-promoting factors from coconut milk and on the work in Steward's laboratory that led to the isolation of diphenylurea?

The use of coconut milk for culturing plant tissues has a long history, starting in the laboratory of A. F. Blakeslee. He engaged Johannes Van Overbeek to grow haploid plants from *Datura* egg cells and suggested that coconut milk (a liquid endosperm) might be a suitable medium for this purpose. Van Overbeek and Conklin found that the coconut milk enabled very young embryos to develop into mature embryos *in vitro*, but they were unable to culture the egg cells. Haagen-Smit and Siu attempted to isolate the so-called embryo factor. In a meeting organized by Blakeslee in 1943, Van Overbeek repeatedly stated that this had been achieved and would be reported later by Haagen-Smit. But later at the same meeting, Haagen-Smit explained that the bioassays provided by Van Overbeek were inadequate for this purpose. No specific growth factor, let alone embryo factor, had been found.

At about this time, Steward was interested in the interaction of growth factors found in coconut milk with auxins, especially 2,4-D, in promoting the growth of carrot cultures. He and co-workers carried out a systematic fractionation of the coconut milk. They obtained large quantities of coconuts from trees damaged by a storm in Florida, and from this material they isolated a number of growth-promoting factors, including *myo*-inositol and some other known organic substances that even if not essential from an external source, markedly increase the growth of carrot tissues. They also isolated a compound with cell division activity, which they identified as *N,N′*-diphenylurea. This compound has been shown to be weakly active as a cytokinin, and many synthetic phenylurea derivatives have subsequently been found to be highly active cytokinins. Diphenylurea itself has not since been found in either coconut milk or any other plant tissue by subsequent investigators. On the contrary, Letham and co-workers later found the principal cytokinin-active compounds in coconut milk to be zeatin and its derivatives.

Your approach to the isolation of a naturally occurring cytokinin was rather novel and may serve to illustrate the nature of some of the problems involved in such an undertaking. Would you describe the efforts in your laboratory to isolate a naturally occurring cytokinin from pea seeds?

In our laboratory, in addition to work on coconut milk, an effort was made to isolate cytokinins from pea blanching water. A milk truck was used to transport 2000 gal of blanching water (representing one day's processing of 25 tons of peas at the Green Giant Cannery at Fox Lake, WI) to the University of Wisconsin Dairy Department, where it was concentrated to 400 gal and evaporated to dryness in the facilities used for preparing dry milk. Thirteen kilograms of the more than 100 kg of dry material were used for purification of cytokinins by a series of extractions and chromatographic procedures. Several active fractions were obtained. One of these yielded about 2 mg of highly active crystals, and another, about 5 mg, similarly highly active. By misadventure, these two crystalline fractions were pooled, and when again separated by chromatography, only 10% of the original activity was recovered, too little for rigorous chemical identification at that time. But on the basis of their properties, they likely were adenine derivatives.

While your attempt to isolate cell-division factors from pea seeds was in progress, both Miller and Letham were trying to isolate naturally occurring cytokinins from other plant sources. Their work culminated in the isolation and characterization of zeatin from corn endosperm in 1964. What are your recollections of the work that led to the identification of zeatin?

Already in 1959, Miller reported the isolation of cytokinin-active material from corn endosperm, and in 1961 he reported some of the properties of this material.[1,2] At the International Plant Growth Substances Association symposium in 1963 at Gif in France, he reported the isolation of a crystalline compound (from corn endosperm) that had high cytokinin activity and which he identified as an adenine derivative containing a substituent with a hydroxyl group and a double bond between the β- and γ carbons.[3] Letham was present at this meeting, and on his way home via the U.S., he visited my laboratory, telling me that he had a cytokinin-active material from plums. He requested, and was sent a sample of synthetic N^6-(Δ^2-isopentenyl)adenine, synthesized by Leonard and coworkers, to compare its properties with those of his material. Shortly thereafter, he sent me a letter informing me that the N^6-(Δ^2-isopentenyl)adenine and his compound from plums had different chromatographic properties. In 1964 the structure of a compound (called zeatin) isolated by Letham from corn endosperm was deduced to be N^6-(*trans*-4-hydroxy-3-methyl-2-butenyl)adenine. This compound was synthesized by G. Shaw and D. V. Wilson. Shaw, in a discussion with a group of investigators, told us that Letham provided him with two spectra: one of Letham's compound and one of a known purine derivative. Shaw also told us that he did not see the crystalline compound, but did his synthesis on the basis of data obtained by comparing the properties of the two spectra. The identity of the known 6-substituted purine, the spectrum of which Letham had given Shaw, was not reported. Presumably, this same compound was one of the 6-aminopurines used by Letham and coworkers in 1964 to deduce the structure of zeatin.

The data reported by Miller in 1961 and 1963 contradicted both Letham's claims and common statements in textbooks that zeatin was first isolated and identified by Letham. Letham's statement (*Annu. Rev. Plant Physiol.*, 1967) that Miller's factor was "*later* isolated in crystalline form" may be true, but is highly misleading. Miller also isolated the compound *earlier*. On the basis of the available information, my conclusion is that Miller should be given priority for the isolation and composition of zeatin. The work of Letham and coworkers, together with that of Shaw and Wilson, established the precise location of the hydroxyl group in the side chain. This is not said to detract from valuable contributions that have since come from Letham's laboratory.

VII. CHEMICAL STUDIES OF CYTOKININS

An important result of the early work with triacanthine was the establishment of a long-term collaboration between your laboratory and the laboratory of Nelson Leonard of the Department of Chemistry at the University of Illinois. Would you describe something of the nature of the collaboration and identify some of the important outcomes of the work with Leonard's laboratory?

The work on triacanthine started 20 years of close collaboration between Leonard and co-workers at the University of Illinois and my laboratory, on isolation, identification, structure–activity relations, and the biological actions of cytokinins. He was an exceptional organic chemist, but he also took a very active interest in and contributed greatly to, all phases of our joint investigations. During our collaboration, more than 450 compounds were synthesized, purified, structurally characterized, and bioassayed.

A number of the compounds synthesized by Leonard et al., including N^6-(Δ^2-isopentenyl)adenine and its derivatives, have since been found to be naturally occurring cytokinins. Among the most highly cytokinin-active synthetic compounds was 8-azidozeatin, which was active down to about 10^{-13}- or 10^{-14}-molar concentrations in tobacco cultures grown in darkness. It was to be used for determining the sites of hormone action, by photoaffinity labeling procedures. Also, the N^6-substituted isopentenyl and benzyl derivatives of 8-azidoadenine were appreciably more active than their corresponding parent compounds. The 2-azidoadenine derivatives were also active. The work on structure–activity relationships also included elucidation of the steps whereby 1-, 3-, or 9-substituted adenines, inactive as cytokinins, are converted to the more stable and cytokinin-active N^6-isomer, and also investigations of the limits to which the ring structure of the purine moiety can be modified without completely removing cytokinin activity. In this connection, some highly effective and specific cytokinin antagonists were synthesized, especially in extensive studies by S. M. Hecht, who began this work as a graduate student in Leonard's laboratory and continued it as a postdoctoral fellow working in Professor R. M. Bock's laboratory on our campus and as a faculty member with an active group in the Department of Chemistry at the Massachusetts Institute of Technology. All this chemistry has been more adequately described in articles by Leonard and Hecht.

VIII. ISOLATION OF CYTOKININS FROM *CORYNEBACTERIUM FASCIANS*

In the early 1960s your laboratory began work with the bacterial plant pathogen, Corynebacterium fascians. *This work resulted in the first isolation of a cytokinin from a bacterial source. What caused you to turn your attention to this organism?*

Already around 1960, R. Samuels, a graduate student at the University of Indiana, found that *Corynebacterium fascians* applied to peas led to fasciation and release of apical dominance. Later work in Thimann's laboratory demonstrated that the effect of the organism could be duplicated by application of synthetic cytokinins. In 1964 we obtained a culture of this strain, which we grew in liter batches, and from the medium, Dieter Klämbt and Gail Thies, in my laboratory, extracted three active compounds. One of these was obtained in crystalline form in sufficient quantity to be identified by Leonard and Helgeson as N^6-(Δ^2-isopentenyl)adenine, the identical compound that Leonard and co-workers had already synthesized a few years earlier. Later, in collaboration with Leonard's laboratory, the other two were also purified and identified: one as *cis*-zeatin and the other as 2-methylthiozeatin.

The action of *Corynebacterium* in the release of apical dominance and the induction of bud formation was studied by analysis of the cytokinins present in the organism and excreted into the culture medium. Murai et al. in our laboratory and Hanson et al. in the Department of Bacteriology, utilizing mutant strains of the organism with very different virulence and using the release of lateral buds in pea plants as a bioassay, showed that the virulence of the strains was correlated with the quantities of cytokinins produced and secreted into the culture medium and with the presence of plasmids. The major cytokinin found in culture filtrates from the highly virulent strains was N^6-(Δ^2-isopentenyl)adenine, not zeatin.

IX. CYTOKININS AS CONSTITUENTS OF RNA MOLECULES

In the late 1960s, your laboratory was one of several that were actively involved in characterizing the cytokinin-active constituents of transfer RNA. Would you comment on some of the accomplishments and results of this effort?

Bock told me in 1966 that Zachau and co-workers had reported that the cytokinin-active nucleoside N^6-(Δ^2-isopentenyl)adenosine (which we had been using) was present at a locus adjacent to the anticodon in serine-specific tRNA from yeast. This finding prompted intensive study of the identity and distribution of cytokinin-active nucleosides in tRNA. In our laboratory, Don Armstrong started, and John Burrows, Norimoto Murai, and others participated in, a systematic investigation of the cytokinin-active nucleosides associated with tRNA species, including an examination of the distribution of cytokinin activity in tRNA preparations from a large number of different biological sources, the isolation and characterization of the cytokinin-active nucleosides present in tRNA from different sources, and the identification of tRNA species that contained cytokinin-active nucleosides in various organisms. These studies demonstrated that cytokinins are ubiquitous constituents of tRNA preparations, occurring in all types of organisms, from *Escherichia coli* cells to human brain cells. The natural occurrence of 2-methylthio-N^6-(Δ^2-isopentenyl)adenosine and 2-methylthioribosylzeatin as constituents of tRNA was also established in these studies, and work in our laboratory, carried out in collaboration with the laboratories of Bock, G. M. Tener, and D. Söll, contributed significantly to the determination that the distribution of cytokinin moieties in tRNA was restricted to tRNA species that responded to codons beginning with U. Experiments by John Einset, in my laboratory, showed that a $tRNA^{Leu}$ from *Pisum* root tips contained ribosyl-*cis*-zeatin in the same triple A sequence in which cytokinin nucleosides occur in microbial tRNA species for which sequence data are available.

The discovery of cytokinin-active nucleosides in tRNA raised the question of whether this observation was related to the hormonal action of cytokinins in the regulation of plant growth. Initially (1966), this possibility appeared to be strengthened by a report from the laboratory of Gene Fox, that radioactive N^6-benzyladenine was incorporated in small amounts into the low-molecular-weight RNA fraction of cytokinin-dependent tobacco callus tissue. By the use of double-labeled N^6-benzyladenine synthesized by Leonard and co-workers, we determined that N^6-benzyladenine was incorporated intact (as the entire base) into both tRNA and rRNA of tobacco callus. Furthermore, the cytokinin composition of tRNA from 70 kg of cytokinin-dependent tobacco callus grown on medium containing N^6-benzyladenine was analyzed. As expected on the basis of the labeling experiments, a small amount of N^6-benzyladenine was recovered from the tobacco callus tRNA, and it was identified by mass spectrometry. However, most of the cytokinin moieties present in the tobacco callus tRNA were the isoprenoid cytokinins (e.g., *cis*-ribosylzeatin) normally present in plant tRNA preparations. Later, Taller et al. reported the isolation of naturally occurring cytokinin-active nucleosides from ribosomal RNA preparations, but the significance of this observation with reference to growth promotion has yet to be established.

Although our efforts, and the efforts of several other laboratories, failed to establish a connection between the growth-promoting activity of cytokinins in plant systems and the occurrence of cytokinin-active moieties in tRNA, these studies did contribute a large body of new information concerning the structure and function of modified nucleosides in tRNA.

X. CONCEPTS OF HORMONE ACTION

You appear to have been the first to recognize and emphasize the importance of hormonal interactions in regulating plant differentiation and morphogenesis. How did your ideas differ from earlier concepts, and how were your ideas received by others in the field?

In 1890 Wiesner proposed that cell enlargement and cell division were controlled by separate and specific factors. Because of the fact that no cell division is associated with the curvature response of the *Avena* coleoptile, auxin at first was assumed to be a specific factor promoting cell elongation. The work of Haberlandt in the early decades of this century provided evidence for two factors that were considered specific for cell division. Sachs, already in 1880, had proposed the existence of specific organ-forming substances. This concept was further promoted by Went. He proposed that separate factors (calines) were required for the initiation and development of roots, buds, and leaves. Also, florigen is a well-known example of a proposed specific organ-forming substance.

I first reported my data on the regulation of the formation of buds and roots in terms of the quantitative interaction of indoleacetic acid and adenine, at a symposium in Strasbourg in 1949. I met with vehement opposition. J. Gautheret insisted my results were due to the presence of preformed buds, specifically in the species that I used, and could never be duplicated with tissues that did not form buds spontaneously. He claimed that the formation of organs was due to specific endogenous substances and could not be achieved with exogenous factors *in vitro*. Gautheret's remarks were repeated in a lengthy denunciation of my report by Bouillenne, who claimed to have evidence that root formation depended on the auxin response of special cells with unique genetic composition. His attack was then augmented by lengthy remarks by his wife, in French, which I could not understand.

Kinetin and other highly active cytokinins were subsequently demonstrated to interact with auxin, even much more effectively than adenine, in regulating organ formation. Auxins and cytokinins interact quantitatively in the regulation of roots, shoots, and even flowers in excised tissues and callus cultures of many species. Gibberellins and cytokinins interact quantitatively in regulating leaf formation. Gibberellins markedly promote polar growth of the vascular tissue of leaves and tend to hold back laminar growth. Cytokinins greatly stimulate expansion of the interveinal laminar tissue. And it is now evident from the manipulation of hormone levels by the introduction of genes for their synthesis that the same quantitative regulation of growth and morphogenesis by hormones must occur in intact plants. Many growth factors interact in concert in every stage of development. Even though one factor may be quantitatively dominant in any one stage, there is no valid evidence to support the concept advocated by Van Overbeek and by Steward that hormones act in sequence in successive stages of development. The above concept of quantitative interactions of growth factors in the regulation of growth and morphogenesis met with persistent skepticism and opposition by botanists, and others, who refused to accept the chemical law of mass action and liked to think of developmental processes as predetermined, qualitative, separate events, rather than in quantitative terms.

You appear to have been one of the first to propose and strongly advocate the idea that plant hormones might act by influencing events in nucleic acid metabolism. What led you to this conclusion?

The early finding that indoleacetic acid interacted with nucleic acid components (purines and pentoses) in organ formation suggested an effect of auxins on nucleic acid metabolism. In 1953 auxin was shown (in experiments by J. Silberger, a graduate student in my laboratory) to stimulate both

DNA and RNA synthesis in tobacco pith tissue. And then we found that auxin and cytokinin induced mitosis and cell division in this tissue. In collaboration with cytogeneticists K. Patau and N. Das, the interaction of kinetin and indoleacetic acid in DNA replication, cell enlargement, and cell division was examined in detail. Both hormones were required in all of these processes. Patau stressed evidence that auxin might be more primarily involved, but that was his personal view. Hormone effects on nucleic acid metabolism in the regulation of growth were further substantiated by later investigations, notably by the the work of Joe Key and co-workers.

XI. DEVELOPMENT OF THE REVISED MEDIUM FOR PLANT TISSUE CULTURE

Your work includes major contributions to plant tissue culture as a technique. How did the revised medium come to be developed?

With the increase in growth obtained with the plant hormones and to improve bioassaying, it became important to modify the requirements for other factors, including the mineral nutrients. We wished to develop a nutrient medium in which all essential elements would be present in sufficiently high concentrations to not limit the growth in bioassays for cytokinins. Considering what might be the specific requirements of tobacco cultures, a graduate student, Toshio Murashige, started this project by obtaining expressed sap from vigorously growing tobacco plants. In testing its effect on tobacco cultures, such a remarkable increase in the growth of the tissue was observed that attention was first given to the possibility of its containing important organic growth-promoting factors. Several years of work by Murashige did not lead to the isolation of any new organic factor with significant effect on growth. When Murashige became a postdoctoral research associate, we started over by ashing the tissue and found that a large portion of the activity was due to the mineral content. With analysis of the ash as a basis, the mineral salts, in various combinations and increasing concentrations, were tested until we obtained a medium we felt would provide slightly more than the optimal amounts of all essential mineral nutrients, so that further addition of minerals present in test materials would not interfere in assays of cytokinin activity.

As there has been persistent, including public, bickering as to who did what in the development of this medium, I now provide the following information. I prepared the manuscript in an attic room at my disposal in the Department of Food Chemistry at the Royal Institute of Technology at Stockholm during the cold winter days of 1962. The manuscript was based, in large part, on the results of the extensive quantitative testing of mineral nutrient combinations by Murashige. Data on sucrose from tests of carbohydrate requirements done by Elfriede Linsmaier, also a postdoctoral associate in my laboratory, were included, as was information from Ruth Schmitz's thesis on the uptake and availability of ferrous iron salts in excised tobacco tissues. In addition, we made use of the experience I had gained in work on minor elements with Perry Stout in Hoagland's laboratory. The manuscript was graciously typed by Department of Food Chemistry secretaries who had no previous experience with either the English language or my penmanship. On Easter eve I carried the manuscript by hand to my friend, Professor Hans Burström, editor of *Physiologia Plantarum*, at his vacation residence on the outskirts of the city. He accepted it for publication, practically sight unseen, and, fortunately for me, without review by experts, or else it might not have been published, certainly not in the detailed form submitted.

The organic growth factor requirements of tobacco callus tissue were further investigated in detail, mainly by Linsmaier, who found that the beneficial effect of the organic content of the leaf extract could be largely duplicated by addition of known compounds, including vitamins, amino acids, and nucleotides. With their inclusion, yields of up to 700 g fresh weight of tissue per liter of medium were achieved. Most of the growth increase obtained by the addition of organic factors other than plant hormones was obtained with the addition of thiamine (vitamin B_1) and *myo*-inositol alone. These results were included in the report on the revised medium published by Linsmaier and Skoog in 1965. This medium has found wide application for the culture of all types of plant tissues, including gymnosperms and monocots as well as many kinds of dicots.

XII. PERSPECTIVES

As you look to the future of plant hormone research, do you have any thoughts or advice about the directions that research in this field should take?

I would certainly hesitate to make suggestions concerning the specific directions of future research. However, there are certain areas where feasible experimentation could give information on the mode of action of growth regulators, in more strict chemical terms than we now have. The association of hormone systems with other metabolic systems in the synthesis and assembly of cell components is now ripe for investigation. The hormonal interactions we have dealt with on a biochemical level must reflect definite integrated molecular reactions that might be elucidated. In addition, from various evidence far from conclusive, I have the impression that regulation of plant growth may involve factors yet to be discovered that are active at concentrations several orders of magnitude below the lowest concentrations at which the most active known plant growth regulators are effective.

On the practical side, we now have means of changing endogenous hormone levels by inserting appropriate genes and regulatory factors required for the synthesis and metabolism of plant hormones instead of treating the plants with exogenous plant growth regulators. The modification of the plant form and the improvement of the yield and nutritional quality of agricultural plant products are within reach, on a broad scale, in the foreseeable future.

Future training of experimental plant scientists will require more advanced training in the physical sciences and should give greater emphasis to earlier, hands-on participation in physiological experimentation. To go forward, instrumentation for revealing molecular structure and for observing detailed cell structure, at the submicroscopic level, also must be developed. For culturing of plants, more sophisticated environmental control facilities will be needed. Present instrumentation and facilities for plant research are far below those used in biomedical studies.

As you look back over your career, were there any special obstacles that you encountered?

I expect I encountered many of the same difficulties all others in my line of work would encounter. In addition, I think my work was hampered at the University of Wisconsin by less-than-adequate means for experimentation on a desirably large scale. Some of these needs were met by collaboration with individuals and laboratories in the College of Agriculture, but much more could have been accomplished. Reasonable understanding of the purpose and needs of my work often was not to be found in administrators in the Department of Botany and in the College of Letters and Sciences, but through the benevolent intervention of President Conrad Elvehjem, I was able to maintain an independent and active research group. I had excellent direction and support from the Graduate School and the Wisconsin Alumni Research Foundation, and my research was generously funded by the National Science Foundation.

As you look back over your career, what do you think were the keys to your success?

My success, to the extent I may have had some, and considering only professional aspects, must be ascribed mainly to my opportunities to have a large number of very capable people in different disciplines helping me. First of all, I have had exceptionally knowledgeable and patient teachers, including my chemistry teacher in high school, Professors Don Yost and Linus Pauling in chemistry, Professors H. Dolk, T. H. Morgan, K. V. Thimann, F. Went, Boris Ephrussi, and Robert Emerson in biology at Cal Tech, and Professor D. Hoagland at U.C.–Berkeley. Also, I have had a large number of enthusiastic, talented students and colleagues, devoted to their research. I have made a list of some of the laboratories and people who have participated in my work on cytokinins (Table 1). I have included only those graduate students and postdoctoral associates who spent

Table 1-1 **Long-Term Participants in Cytokinin Studies at Wisconsin**

Physiology	Chemistry	Biochemistry
Ruth Y. Schmitz[a]	Frank M. Strong and co-workers	Robert M. Bock and co-workers
Cheng Tsui		
Carlos O. Miller	Francis S. Okumura	
Toshio Murashige[a]		
Elfriede Linsmaier	Nelson J. Leonard and co-workers	
Donald J. Armstrong[a]	(U. of Illinois)	**Cytology**
William Wardell[a]		Klaus Patau and co-workers
John W. Einset[a]		
Norimoto Murai[a]	Sidney M. Hecht and co-workers	
	(M.I.T.)	Gerhard Sanders

[a] Graduate students who also spent several years as postdoctoral associates in my laboratory.

many years with me. In total, well over 100 pre- and postdoctoral students have generously contributed to the work. I am deeply grateful and in debt to all these people. Much credit must go to Dr. Ruth Schmitz, who as a graduate student started to work with me when I came to Wisconsin in 1947. Except for a few years devoted to family needs, she continued as a research associate until my retirement in 1979. She participated actively in all phases of our work, from planning to reporting of experiments. She was a careful, skillful experimenter. She supervised bioassaying and served as a much-valued guide to successive groups of young investigators coming through our laboratory.

In addition, I believe I have had a fairly long nose in smelling out problems and blind perseverance in trying to bring matters to a conclusion.

REFERENCES

1. **Miller, C. O.,** A kinetin-like compound in maize, *Proc. Natl. Acad. Sci. U.S.A.*, 47, 170, 1961.
2. **Miller, C. O.,** Additional information on the kinetin-like factor in maize, *Plant Physiol.*, 37 (suppl.), xxxv, 1962.
3. **Miller, C. O. and Witham, F. H.,** a kinetin-like factor from maize and other sources, in *Addendum to Regulateurs de la Croissance Végétale,* Nitsch, J. P., Ed., Coll. Int. Centre Nat. Recherche Sci., Gif-sur-Yvette, France, 1964, i–vi. (Editorial note: By error, the manuscript cited here was omitted from the published proceedings of the Gif symposium and was published later as a separate addendum to the original volume.)

Chapter 2

Chemistry of Adenine Cytokinins

Gordon Shaw

TABLE OF CONTENTS

I. INTRODUCTION

Suggestions that plants contain chemical factors that are capable of greatly promoting cell division were made as long ago as 1892 (Wiesner)[1] and received partial confirmation from the research of Haberlandt,[2] who, in 1913, showed that diffusates from phloem could cause cell division in potato parenchyma. In 1921 he also showed that cell division induced by wounding in several plants could be prevented by rinsing the surface of the wound, and that cell division could be reinitiated by adding crushed tissue to the wound.[3] He called the material responsible for cell induction "Wundhormone." In 1956 Miller and Skoog et al. isolated a compound from autoclaved herring sperm DNA that had potent cell-division-promoting activity and was shown to be 6-furfurylaminopurine (**1**).[4] It was given the name kinetin because of its ability to induce cell division, provided that IAA was present in the medium. Later, however, it was shown that kinetin is an artifact produced from the deoxyadenosine residues in DNA, which degrade on standing for long periods or when heated during the isolation procedure.[5]

The original name "kinin" for plant cell-division promoters conflicted with the same name used in animal biochemistry. It was replaced by the term cytokinin,[6,7] which is now accepted for compounds that promote cytokinesis in plant cells. In practice cytokinin may be defined as a compound that in the presence of auxin, induces cell division in a suitable assay material grown on a defined medium.[8] Following the discovery of kinetin, large numbers of analogous 6-*N*-alkylaminopurines were synthesized, and the availability of such compounds facilitated the collection of a large pool of information concerning the effects of cytokinins on different biological systems, their interaction with other growth factors, and the relationship between chemical structure and cytokinin activity. However, in a review of kinetin and related substances, Miller pointed out that the work would have more meaning if a naturally occurring compound of the same type could be shown to exist.[9] It had been shown from work over some years that biologically active compounds similar to kinetin did occur, especially in immature grains of maize, long recognized as a good source of plant growth substances.[10] A very active kinetin-like factor was eventually isolated from maize,[11] and its properties suggested that it was a purine derivative. In particular, it exhibited behavior similar to that of kinetin in the purification steps, and had similar UV absorption spectra in neutral, acid, or alkaline solutions. Oxidation of the compound with neutral potassium permanganate gave substances tentatively identified as *N*(purin-6-yl)glycine (**2**) and adenine. The compound was also found to react with acetic anhydride to give a new product.

Since 6-*N*-allylaminopurine did not readily produce purinyl glycine, it was assumed that the

0-8493-6252-0/94/$0.00+$.50

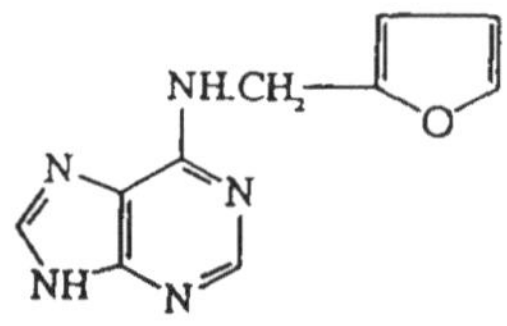

(1)

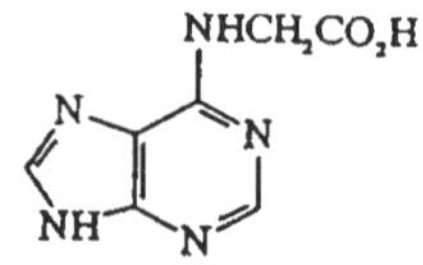

(2)

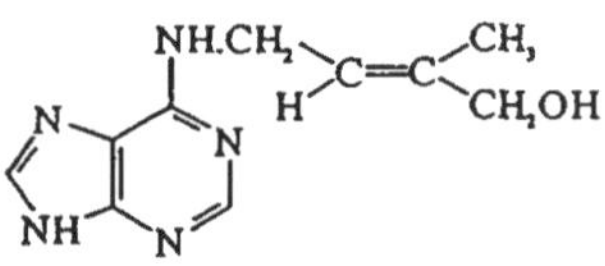

(3)

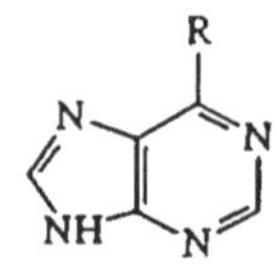

(4) $R = SCH_3$

(5) R = Cl

unknown compound might have a double bond similar to the latter, but with additional substitution on the second carbon atom. Similarly, the reaction with acetic anhydride indicated the possible presence of a hydroxyl group. The authors also reported the presence of the same compound in sunflower fruits and soybeans. However, they were unable to propose a precise structure for the compound.

At about the same time, the first naturally occurring pure crystalline cytokinin was isolated in very small amounts from *Zea mays* (0.7 mg from 60 kg) and given the name zeatin.[12] In the following year, tentative structures for zeatin, including the correct one, namely (E)-4-(hydroxy-3-methylbut-2-enyl)aminopurine **(3)**, were proposed,[13] and this was immediately confirmed by a chemical synthesis.[14,15] The natural occurrence of zeatin was soon confirmed both in *Z. mays* and in many other sources. Following this early work, several new cytokinins have been isolated from natural sources and many have been detected by chromatographic examination. The naturally occurring cytokinins are all N^6-substituted adenine or substituted (in the ring) derivatives, and many occur as *N*- or (in the case of zeatin and derivatives) *O*-glycosides and their phosphorylated derivatives.

Following the initial discoveries that kinetin had high cytokinin activity and the isolation of endogenous cytokinins such as zeatin and its derivatives from higher plants, large numbers of N^6-substituted adenines have been prepared and tested for cytokinin activity usually as part of structure–activity studies. Several of these synthetic compounds have been applied exogenously to plant cell tissue cultures and, invariably and not unexpectedly, found to produce various derivatives, including glycosides, glycoside phosphates, and amino acid conjugates.

In addition to their widespread occurrence in plant tissues, numerous cytokinins have also been isolated from tRNAs of plant and animal origin and from fungi, bacteria, algae, and marine organisms such as blue coral. Tables 1 and 2 list naturally occurring endogenous cytokinins from plant and other sources, although whether substances such as 6-methylamino-2-hydroxy purine, with weak cytokinin activity, should be considered as true cytokinins is, no doubt, open to debate.

In addition to the adenine cytokinins, another group of substances, namely diarylureas, have also been found to possess very high cytokinin activity, often in excess of that of the adenine derivatives. These compounds will be reviewed by Shudo (Chapter 3), and this chapter will be largely concerned with the naturally occurring adenine cytokinins.

II. GENERAL SYNTHESIS OF 6-ALKYLAMINOPURINES

The preparation of a purine derivative with a 6-substituent reactive enough to condense with primary and secondary amines was achieved by Elion et al.,[16] who described the preparation of 6-methylthiopurine **(4)** from hypoxanthine and also recorded the subsequent reactions of this compound with various amines. The first synthesis of kinetin employed this method and involved refluxing a mixture of **(4)** with furfurylamine overnight.[17] The reaction normally requires the use of sealed tubes, but the high boiling point of furfurylamine (150°C) made this unnecessary. An improvement in the synthesis came from the use of 6-chloropurine (**5**) prepared from hypoxanthine and phosphoryl chloride in the presence of *N,N*-dimethylaniline.[18,19] This produces a complex that is decomposed by cold alkali prior to acidification and ether extraction. 6-Chloropurine readily reacts with amines and other nucleophiles, at modest temperatures, in solvents such as n-butanol or 2-methoxy ethanol and generally in the presence of a base such as triethylamine, which reduces the amount of amine that would otherwise be required to absorb liberated hydrogen chloride. Although the chlorine atom is fairly readily removed by hydrolysis with acid or alkali, 6-chloropurine may also be reacted with amino acids in a hot aqueous solution, over 3 to 5 h, in which a pH of 9.5 is maintained by the addition of alkali.[20]

III. ADENINE CYTOKININ AGLYCONES

A. (E)-ZEATIN

Cytokinins are generally extracted from plant tissues with aqueous methanol or ethanol or with ethyl acetate in the presence of aqueous potassium phosphate.[21] The compounds may then be purified by paper or thin-layer chromatography, using cellulose and silica gel layers. Column chromatography has also been extensively used. Cytokinin bases and nucleosides are retained on cation-exchange resins (e.g., Dowex 50, cellulose phosphate) and may be eluted from them with aqueous ammonia.

Further fractionation of the compounds required the use of gel or ion-exchange gels such as Sephadex LH-20, polyvinylpyrrolidone, and silica gel. High-pressure liquid chromatography is increasingly used to give excellent separations of cytokinin mixtures. The molecular formulas of isolated cytokinins are usually determined by initially using mass spectrometry as the key technique, and this method is reviewed in Chapter 24. When sufficient material is available NMR spectroscopy is universally employed to determine precise structures. Structural analysis of the first samples of zeatin was hindered by the small amount of pure material available. However, enzyme-catalyzed oxidation to purin-6-yl glycine and mass spectrometric analysis suggested the structure (**3**) or an isomer.[13] The (E)-structure (**3**) was, however, soon confirmed by an unambiguous chemical synthesis[14,15] (Scheme 1) from 6-methylthiopurine or 6-chloropurine and (E)-4-hydroxy-3-methylbut-2-enylamine (**6**) in the

Table 2-1 **Naturally Occurring Cytokinins Identified in Plant Tissues**

Rf = β-D-ribofuranosyl, Gp α- or β-D-glucopyranosyl
Xp = β-D-xylopyranosyl

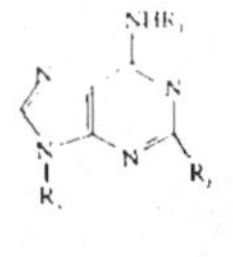

R_1	R_2	R_3	References
CH_2(H)C=C(CH_3)CH_2OH	H	H	12-15, 96-100
CH_2(H)C=C(CH_3)CH_2OH	H	Rf	45, 97, 99 - 103
CH_2(H)C=C(CH_3)CH_2OH	H	Gp	54, 55, 104
CH_2(H)C=C(CH_3)CH_2OH	H	Rf-5′-O-phosphate	51, 89, 90
CH_2(H)C=C(CH_3)CH_2OH	OH	H	45
CH_2(H)C=C(CH_3)CH_2OAc	H	Rf	60
CH_2(H)C=C(CH_3)CH_2OGp	H	H	54, 81 - 3, 87, 88
CH_2(H)C=C(CH_3)CH_2OGp	H	Rf	54, 87, 88, 104
CH_2(H)C=C(CH_3)CH_2O Xp	H	H	85, 86
$CH_2CH_2CH(CH_3)CH_2OH$	H	H	35, 36, 58, 98, 104
$CH_2CH_2CH(CH_3)CH_2OH$	H	Rf	57, 58
$CH_2CH_2CH(CH_3)CH_2OGp$	H	H	54, 84, 88
$CH_2CH_2CH(CH_3)CH_2OGp$	H	Rf	54, 88
$-CH_2CH{=}C(CH_3)_2$	H	H	41 - 43
$-CH_2CH{=}C(CH_3)_2$	H	Rf	41, 42, 67
$-CH_2CH{=}C(CH_3)_2$	H	Gp	56
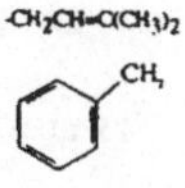	H	Rf	78
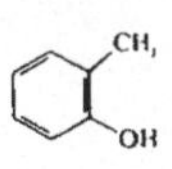	H	Rf	79
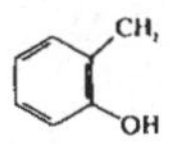	SCH_3	Gp	80

Table 2-2 **Naturally Occurring Cytokinins Identified in Material Other Than Plant Tissue**

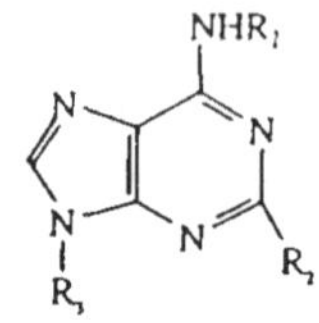

R_1	R_2	R_3	References
$-CH_2CH=C(CH_3)_2$	SCH_3	Rf	46, 47
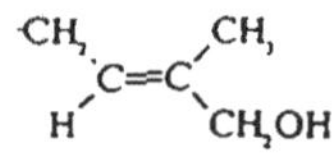	SCH_3	Rf	41, 42, 44, 67
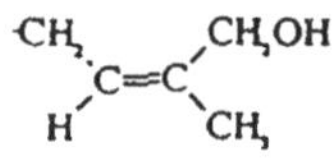	SCH_3	H	106
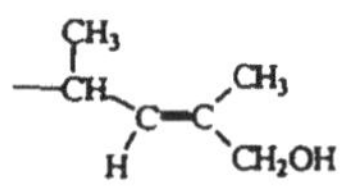	OH	H	52, 105
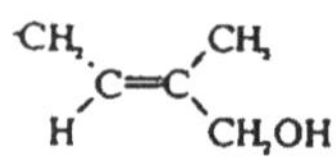	H	H	49, 50, 52
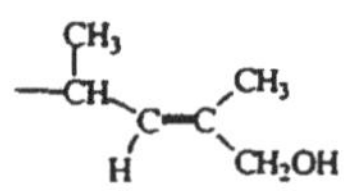	H	H	62
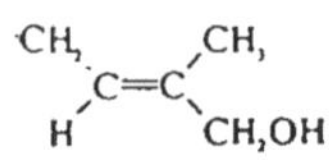	H	Rf	50, 62
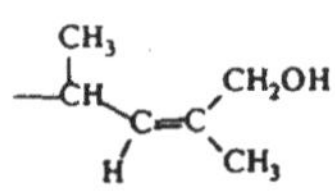	H	Rf	62
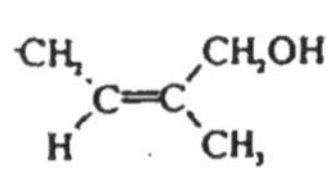	H	Rf	41, 66 - 9
$-CH_2CH=C(CH_3)_2$	$S.CH_3$	Rf	41, 48
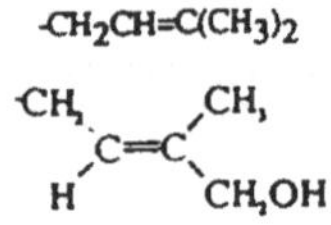	H	2'-deoxy-Rf	25, 61
CH_3	OH	H	52

Notes: Rf = β-D-ribofuranosyl

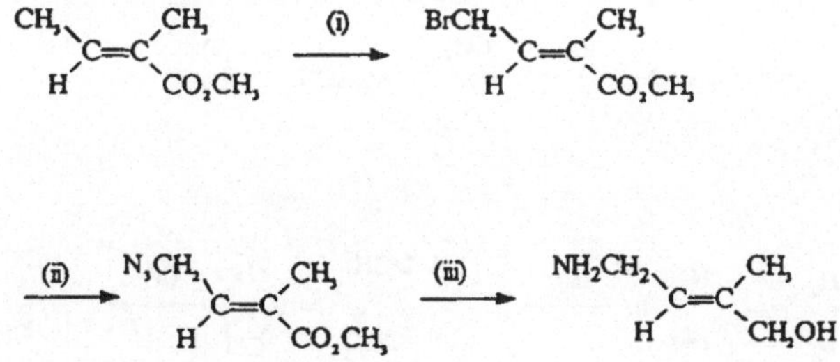

(6)

Scheme 1

(i) N-Bromosuccinimide; (ii) NaN_3; (iii) $LiAlH_4$, Et_2O.

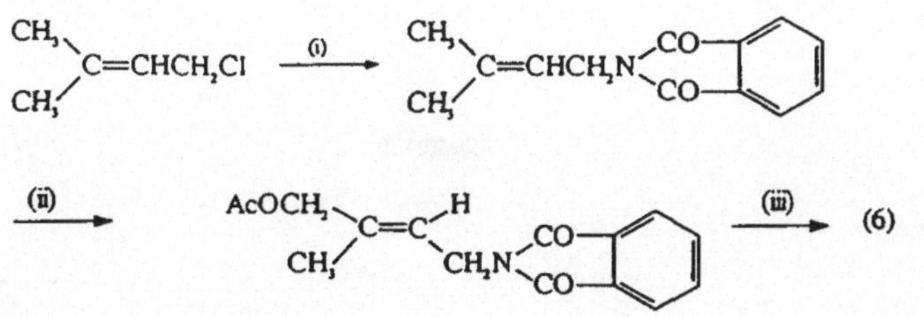

(i) Potassium Phthalimide, (ii) SeO_2, HOAc, Ac_2O; (iii)$Ba(OH)_2$

Scheme 2[22]

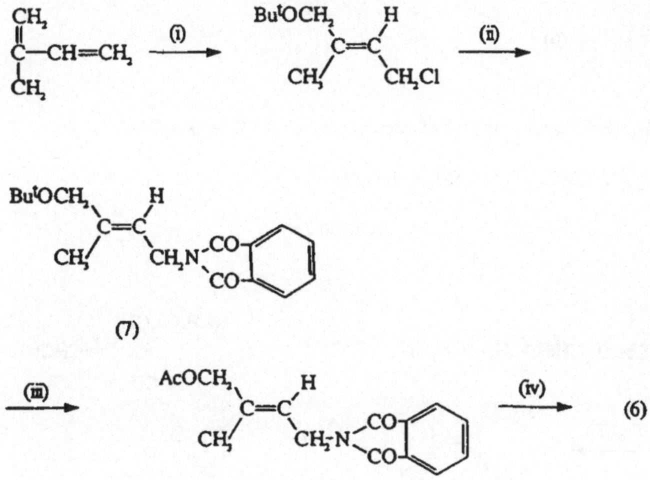

(i) Bu^tOCl; (ii) Potassium phthalimide; (iii) TsOAc; (iv) OH^-

Scheme 3[23]

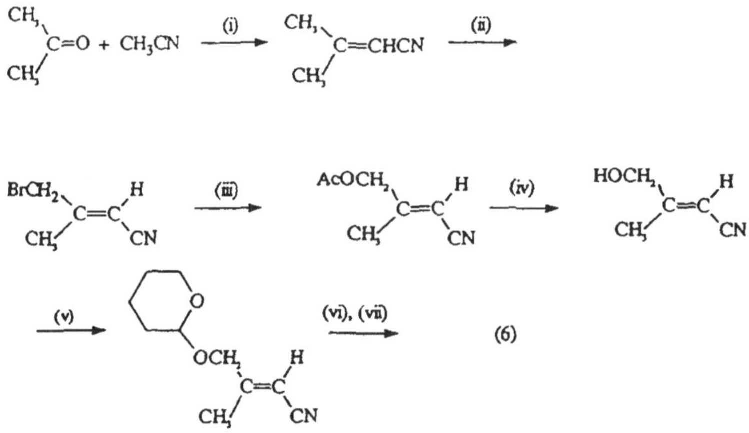

(i) $C_6H_{11}NH_2$; (ii) N-Bromosuccinimide; (iii) KOAc, HOAc, (iv) KOH

(v) Dihydropyran, H^+, (vi) $LiAlH_4$, (vii) H_3O^+

Scheme 4[24]

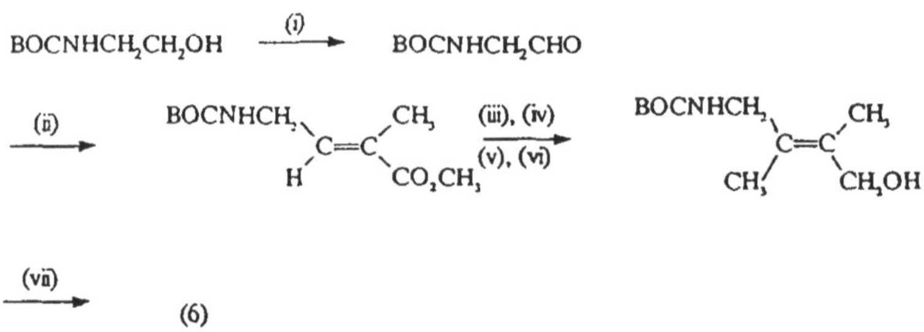

(i) $C_5H_5NH^+CrO_3Cl^-$, (ii) $Ph_3P{=}CMeCO_2Me$, (iii) OH^-, (iv) H_3O^+

(v) EtO. $COCl/NEt_3$, (vi) $NaBH_4$, (vii) H_3O^+

Scheme 5[25]

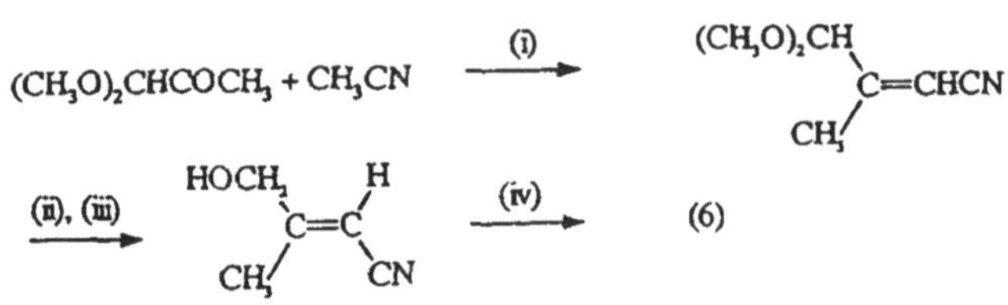

(i) OCH_3^-, (ii) H_3O^+, (iii) $NaBH_4$, (iv) $LiAlH_4$

Scheme 6[26]

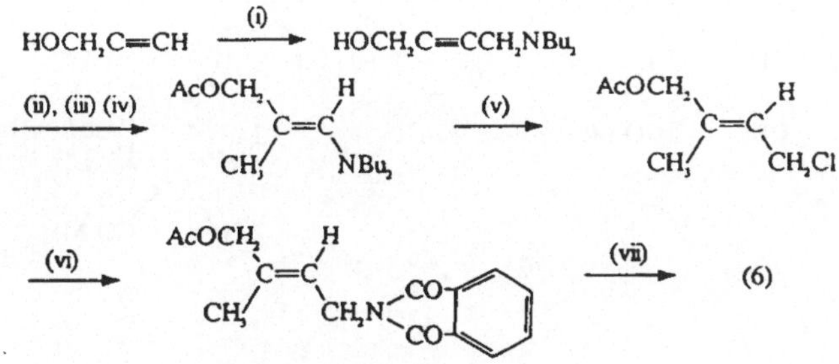

(i) CH_2O, Bu_2NH, (ii) CH_3MgCl, (iii) H_2O, (iv) Ac_2O, (v) $ClCO_2Et$,
(vi) Potassium Phthalimide, (vii) NH_2NH_2

Scheme 7[27]

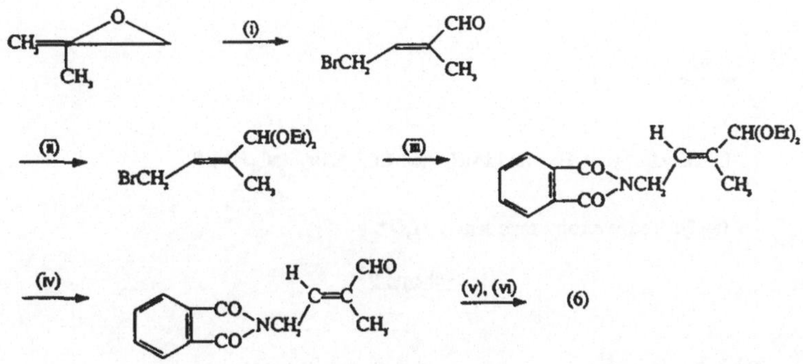

(i) $CuBr_2$, Li_2CO_3, $CHCl_3$/EtOAc, 90°C, (ii) $(EtO)_3CH$, EtOH, Me_2CO,
(iii) potassium phthalimide, (iv) H_3O^+, (v) $NaBH_3CN$, pH 3.3, (vi) NH_2NH_2

Scheme 8[28]

presence of triethylamine. The enylamine was prepared by allylic bromination of methyl tiglate with *N*-bromosuccinimide to give a bromo derivative [the (E) structure of which had been confirmed by its use in a synthesis of crocetin] that, with sodium azide, produced the corresponding azido derivative, reduction of which, with lithium aluminium hydride in ether, then gave the enylamine (**6**), which was isolated as a crystalline hemi sulfate salt.

Following the original synthesis several alternative syntheses of zeatin and derivatives (usually via the enylamine, **6**) have been described (Schemes 2–9).

A synthesis of (E)-zeatin, however, that avoids formation of the water-soluble enylamine (**6**), which is difficult to purify and has frequently been used in the final condensation step without purification, involves reaction of (E)-3-methyl-4-t-butoxybut-2-enylphthalimide (**7**, Scheme 3) with hydrazine to give a high yield of (E)-3-methyl-4-t-butoxybut-2-enylamine (**8**), which is readily purified by distillation (bp 73–74°C/3 mm Hg).[30] This compound, with 6-chloropurine and triethylamine in hot n-butanol, gave an excellent yield of (E)-*O*-t-butylzeatin (**9**), which, with dilute hydrochloric acid, gave a high yield of (E)-zeatin (Scheme 10).

The method was readily adapted to the preparation of (E)-[8-^{14}C]zeatin from (**6**) and 6-chloro [8-^{14}C]purine.

B. (E)-ZEATIN DERIVATIVES

(E)-zeatin readily gives an *O*-acetate[15] (**10**), mp 168–169°C, and an *O*-p-toluene sulfonyl derivative, mp 193–194°C, but with methanesulfonyl chloride in dimethylformamide, it produced the allylic chloride (**11**), mp 194°C, but in pyridine, the quaternary derivative (**12**) was obtained.[31] The quaternary salt was also produced from (E)-zeatin and phosphorylating agents in pyridine. Other esters, including the interesting indole acetate (**13**), have also been prepared.[32]

Methylation of zeatin with dimethyl sulfate in aqueous alkali gave 3-methylzeatin, whereas the

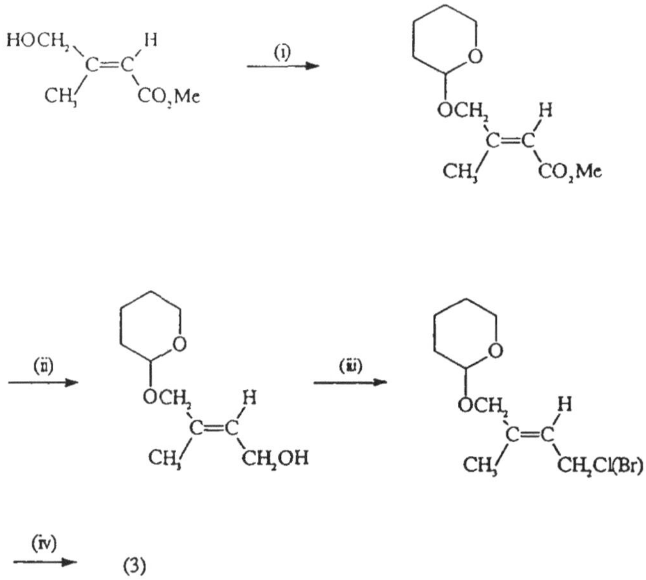

(i) Dihydropyran, H^+, (ii) $LiAlH_4$, (iii) CCl_4/CBr_4, $(nC_8H_{17})_3P$

(iv) Na Tetrahydropyranyladenine, H_3O^+.

Scheme 9[29]

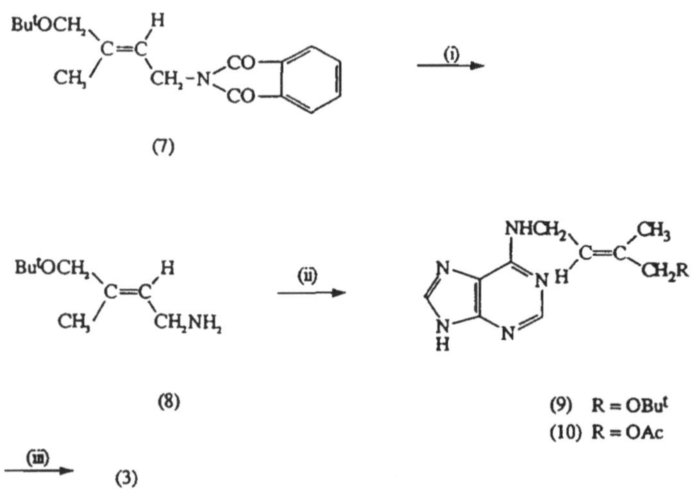

(i) NH_2NH_2, (ii) 6-chloropurine, (iii) MHCl, MNaOH

Scheme 10[30]

7- and 9-monomethyl zeatins were prepared from 7- and 9-methyl-6-chloropurine and the amine **(6)**, respectively.[33] Reaction of the allylic chloride **(11)** with potassium thioacetate produced thioacetylzeatin **(14)**, which with methanolic ammonia soon gave thiozeatin **(15)** (Scheme 11).[34] In *Phaseolus vulgaris* tissues, thiozeatin displayed much higher activity than zeatin, whereas in *P. lunatus* it was about 30 times less active. The corresponding disulfide **(16)**, prepared from an alkaline solution of thiozeatin with iodine, was less active than thiozeatin.

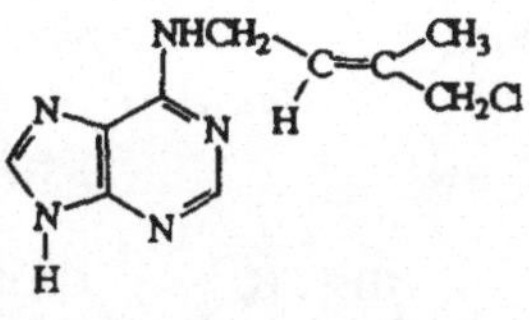

(11)

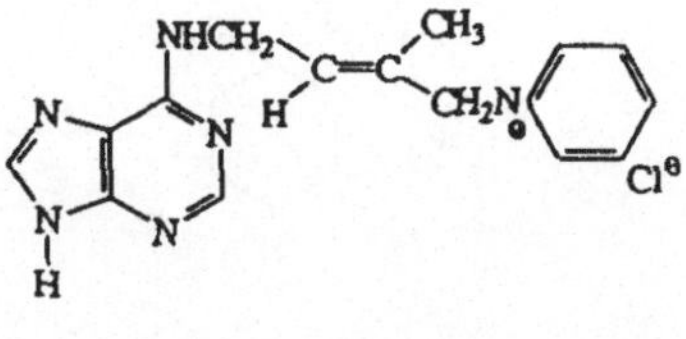

(12)

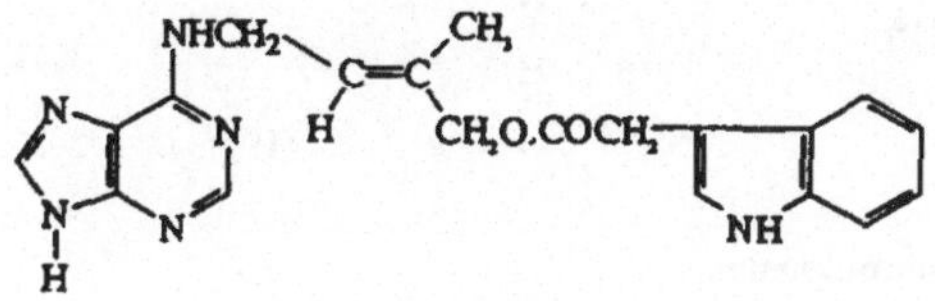

(13)

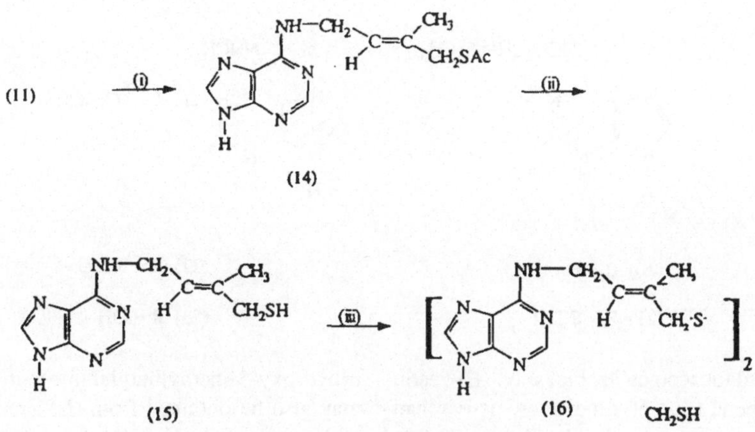

(i) KSAc, (ii) MeOH/NH$_3$, (iii) OH$^-$, I$_2$.

Scheme 11[34]

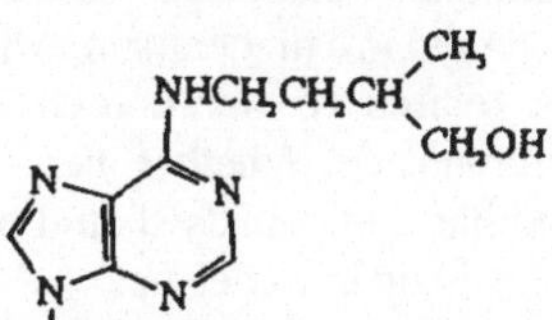

(17)

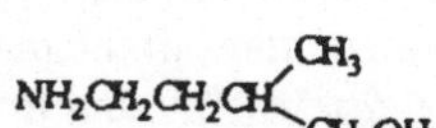

(18)

(–)Dihydrozeatin (**17**), unique among cytokinins in possessing a chiral center, was isolated[35] from immature seeds of *Lupinus luteus* and was assigned the (S) configuration from a study of its ORD spectrum.[36] The (–) form was prepared from the known (–)4-amino-2-methylbutan-1-ol (**18**) and 6-chloropurine. The (R)(+) and (RS) forms were similarly prepared, and all forms were found to have very similar NMR, UV, and IR spectra. The RS dihydro form also results from the hydrogenation of zeatin.

C. (Z)-ZEATIN AND ZEATIN ANALOGS

The geometrical isomer (**19**) of (E)-zeatin was isolated as a riboside from peas, spinach, and corn,[37,38] and the aglycone, which has been detected in several plants, was synthesized in a stereospecific manner (Scheme 12).[39,40]

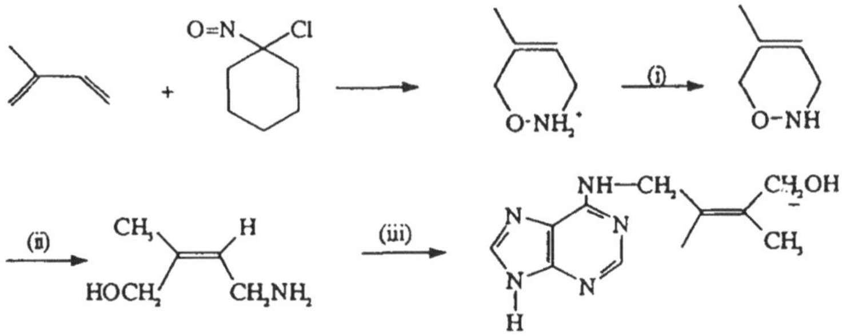

(19)

(i) OH^-, (ii) Zn / HOAc (iii) 6-chloropurine

Scheme 12[39]

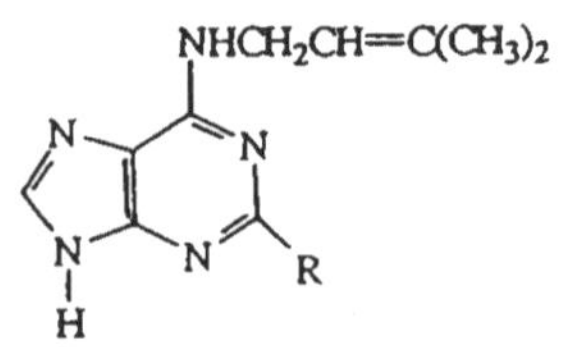

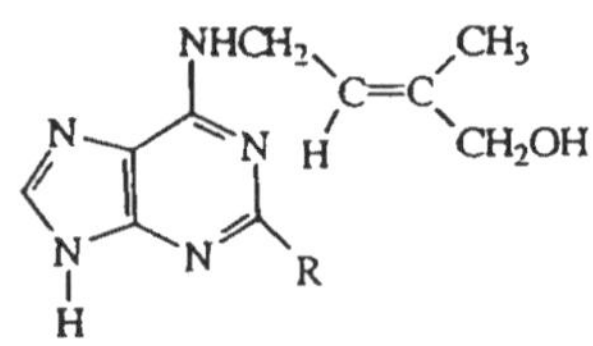

(20) R = H

(21) R = SCH_3

(22) R = SCH3

(23) R = OH

In a standard tobacco callus bioassay, (E)-zeatin was found to be at least 50 times more active than (Z)-zeatin.[39] Other compounds related to zeatin, but sometimes isolated as 9-β-D-ribofuranosides (see later), include 6-(3-methyl-2-butenyl)aminopurine (**20**),[41-43] 6-(3-methyl-2-butenylamino)-2-methylthiopurine(**21**),[41] 2-methylthio-(E)-zeatin (**22**),[41,42,44] and 2-hydroxy (E)-zeatin (**23**).[45] They are all readily synthesized by reaction of the corresponding 6-chloro or 6-methylthiopurine with dimethylallylamine or the hydroxylated derivative (**6**).

The isopentenyl derivative (**20**) has also been isolated as the free base from cultures of the plant pathogen *Corynebacterium fascians*[46,47] and from serine-specific yeast tRNA.[48] Other bacteria, generally plant pathogens, also produce cytokinins in culture. Thus *Pseudomonas amygdali* produces (E)-zeatin, isopentenyladenine (**20**), and (−) dihydrozeatin (**17**),[49] while (E)-zeatin and its 9-β-D-ribofuranoside have been isolated from liquid cultures of the puffball fungus *Rhizopogon roseolus*.[50] The production of cytokinins by plant pathogens presumably enhances plant growth to the advantage of the pathogen.

The presence, in *Zea mays*, of a series of adenines related to zeatin and, in most cases, possibly produced from zeatin by enzyme-catalyzed oxidation has also been reported.[45] They include 6-(2,3,4-trihydroxy-3-methylbutylamino)purine (**24**), which may also be obtained from (E)-zeatin by reaction with ammoniacal potassium permanganate,6-(3,4-dihydroxy-3-methylbutylamino)purine (**25**), the hydroxy acid derivatives (**26**) and (**27**), and 2-hydroxy-(E)-zeatin (**23**). The aspartic acid derivative (**27**) is a well-known intermediate (as a ribonucleotide) in purine nucleotide *de novo* biosynthesis, and this is claimed to be the first time it has been reported in a higher plant, whereas 2-hydroxyzeatin is related to the known 2-hydroxyadenosine, crotonoside. Whether the acids (**26**) and (**27**) and similar compounds should be regarded as cytokinins is open to debate.

IV. CYTOKININ GLYCOSIDES

The cytokinins are widely distributed as *N*- or [in the case of (E)-zeatin] *O*-glycosides and as corresponding 5′-*O*-phosphorylated derivatives.

A. CYTOKININ NUCLEOSIDES

1. Isoprenoid Derivatives

The original investigations of *Z. mays* extracts indicated that only a small proportion of the cytokinin activity is due to zeatin; moreover, bioassays of chromatogram zones generally reveal the presence of at least three active materials. One of these,

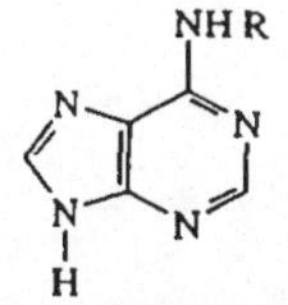

(24) R = $CH_2CH(OH)CMe(OH)CH_2OH$

(25) R = $CH_2CH_2C(OH)MeCH_2OH$

(26) R = $CH_2CHOHCH_2CO_2H$

(27) R = $CH(COOH)CH_2COOH$

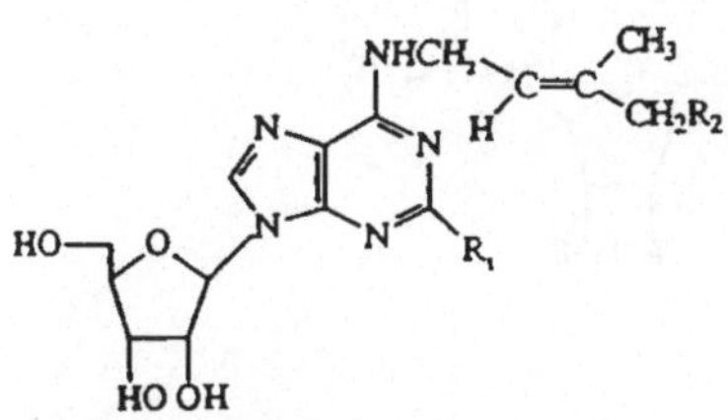

(28) $R_1 = H$, $R_2 = OH$
(29) $R_1 = R_2 = H$
(30) $R_1 = SCH_3$, $R_2 = R_3 = H$

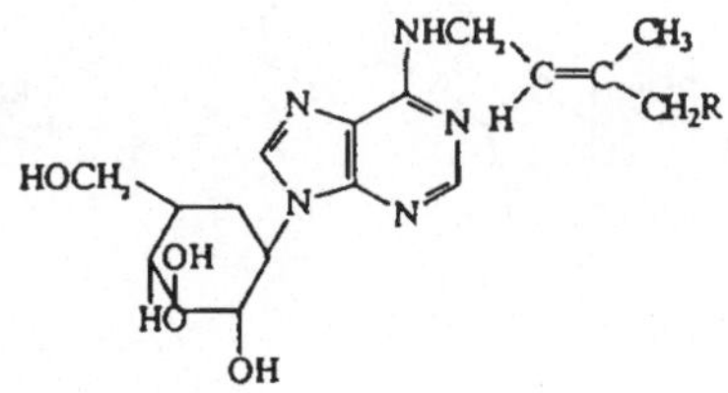

(31) R = OH
(32) R = H

namely 9-β-D-ribofuranosyl zeatin (**28**), was first detected[12] in *Z. mays* var. Golden Bantam and synthesized[14,15] by reaction of the amine (**6**) with 9-β-D-ribofuranosyl-6-chloropurine. The same method has been universally adopted for the synthesis of many other analogous nucleosides. The riboside is of widespread occurrence in plants and has also been isolated from liquid cultures of the puffball fungus *Rhizopogon roseolus*.[50]

It has recently been shown to be present in a green alga, *Chara globularis*,[51] and a blue coral.[52] Similarly, N^6-isopentenylaminopurine occurs as the 9-β-D-ribonucleoside (**29**) in spinach, garden peas,[37] and tobacco plants[52] and exists widely in tRNAs, including those of yeast and calf liver. It, too, has recently been identified as a constituent of the green algae, *Chara globularis*, the first recorded presence of a cytokinin in an algae.[53] 6-(3-Methyl-2-butenylamino)-2-methylthio-9-β-D-ribo-furanoylpurine (**30**) also occurs in the tRNA of *Escherichia coli*.[48]

The 9-glucopyranoside of (E)-zeatin (**31**) also occurs in *Z. mays* kernels,[54] *Vinca rosea* crown gall tissue, and leaf and root exudates of the short-day plant, *Perilla frutescens*,[56] where it is accompanied by the 9-glucopyranoside (**32**) of isopentenyladenine together with other zeatin derivatives.

Dihydro-(E)-zeatin riboside (**33**) has been isolated from leaves of *Phaseolus vulgaris*[53] and from the culture medium of *Vinca rosea* crown gall tissue.[58] In *Lupinus angustifolia*, the nucleoside and (E)-zeatin riboside are metabolized to the *O*-acetyl-5′-phosphates (**34**) and (**35**), respectively.[59] Recently, however, *O*-acetyl (E)-zeatin riboside (**36**) was isolated from tobacco plant cells and synthesized from the silylated 6-chloropurine (**37**) and the enylamine (**6**) (Scheme 13).[60]

(E)-zeatin-2′-deoxyriboside (**38**) has recently been isolated from a culture of the plant pathogen *Pseudomonas amygdali*,[25,61] which causes bacterial canker of almonds. The deoxyriboside has been synthesized by reaction of 6-chloro-9-β-D-2′-deoxy ribofuranosyl purine and the enylamine (**6**). A similar organism, *Pseudomonas syringae*, was found to produce 1′-methyl (E)-zeatin riboside (**39**) in addition to other cytokinins, including zeatin and its riboside.[62]

Other nucleoside derivatives of zeatin and isopentenyl adenine, including the 3- and 7-glucopyranosyl derivatives, have been recorded as metabolites of exogenously added cytokinin, but probably occur endogenously.[63-65]

(Z)-zeatin 9-β-D-ribofuranoside (**40**) occurs in plant tRNAs, including those from wheat germ, pea roots,[11] and tobacco callus tissue,[67] and has been synthesized by standard methods. It also occurs in a tRNA of *Corynebacterium fascians*[68] and in cones of the hop plant.[69]

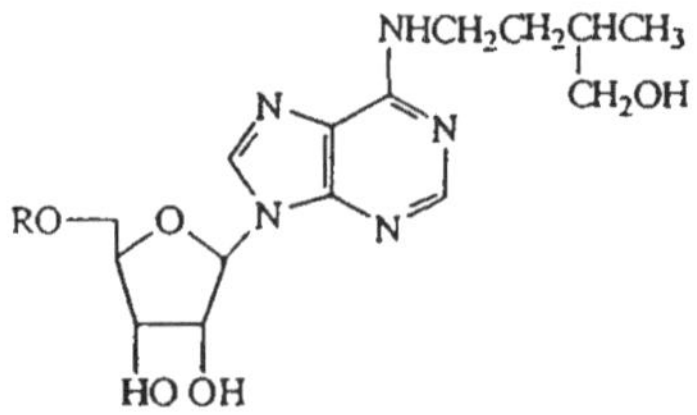

(33) R = H
(34) R = PO(OH)$_2$

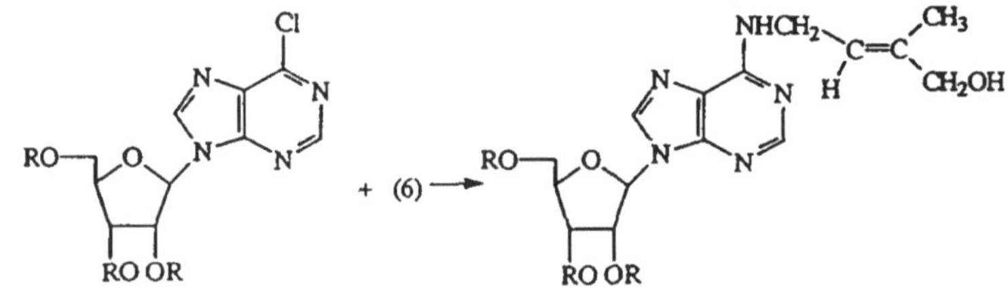

(37) R = Bu^{t}SiMe$_2$

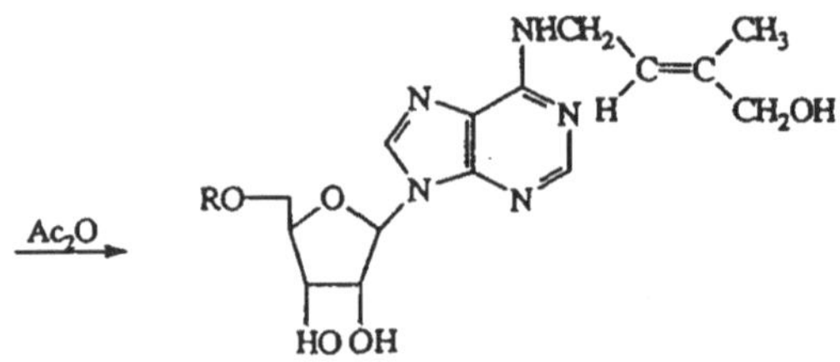

(35) R = PO(OH)$_2$
(36) R = Ac

Scheme 13

2. Aralkyladenine Derivatives

The ready availability of 6-*N*-benzylaminopurine **(41)**, and its cytokinin activity, has long made it a prime candidate for metabolic studies. Indeed, the first reported metabolic studies on a cytokinin (in *Xanthium* leaves) were carried out using **(41)**,[70] which was shown to be converted into the 9-β-D-ribofuranoside **(42)** and corresponding 5′-*O*-phosphate **(43)**. However, in *Phaseolus vulgaris* seedlings, the principal metabolite appears to be the alanine derivative **(44)**,[71] and a similar compound, lupinic acid **(45)**, was later reported as a novel metabolite of (E)-zeatin in *Lupinus angustifolia* seedlings.[72] On the other hand, intact seedlings, organ slices, and tissue cultures from several other plants were found to produce the 9-, 7-, and 3-glucopyranosides **(46)**, **(47)**, and **(48)**, respectively.[64,73-75]

The 9-glucopyranoside is also obtained from oat leaves,[76] and recently, a disaccharide derivative of **(41)** probably containing a glucosyl-ribose unit was found in 24-day-old shoot cultures of *Gerbera jamesonii*.[77]

Recently, much of this work achieved added significance by the isolation of the ribonucleoside **(42)** from a cell culture of *Pimpinella anisum*.[78] Similarly, the closely related 6-(*O*-hydroxy-benzyl)-aminopurine riboside **(49)** was obtained from leaves of *Populus robusta*,[79] and the analogous 2-methylthio derivative was isolated as a glucoside, probably the pyranoside **(50)**, from fruits of *Zantedeschia aethiopica*.[80] The benzyladenines are readily synthesized by standard processes from 6-chloropurine and the appropriate benzylamine. It seems more than likely that other aralkyladenines with cytokinin activity will eventually be found to occur in plant and other tissues.

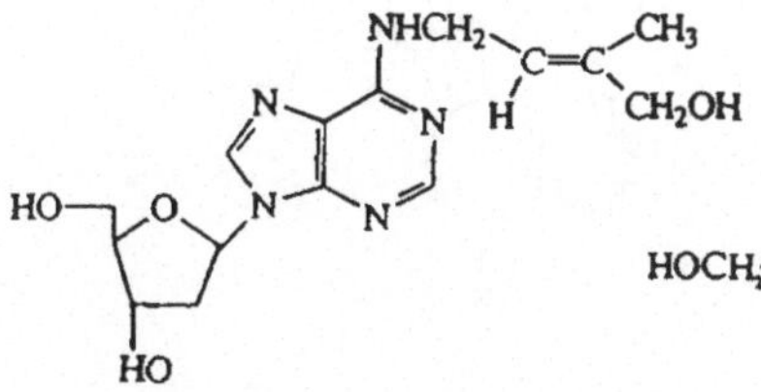

(38)

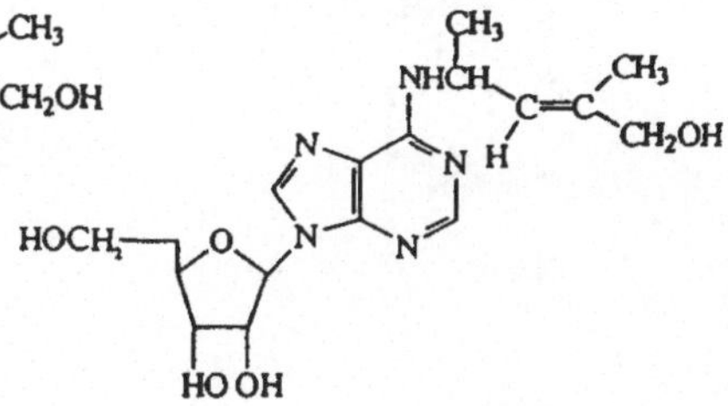

(39)

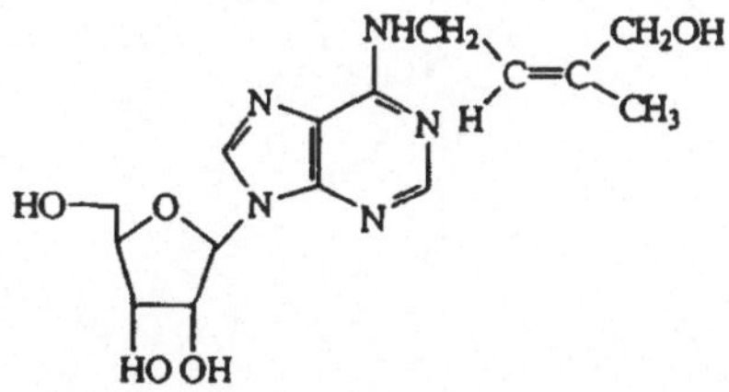

(40)

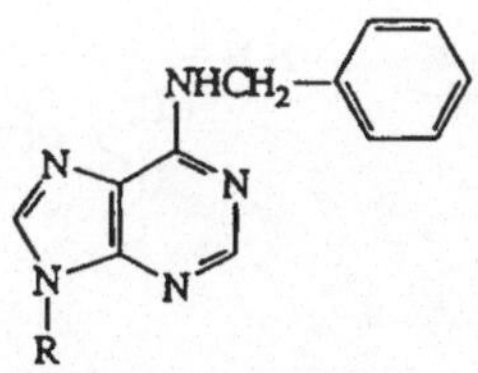

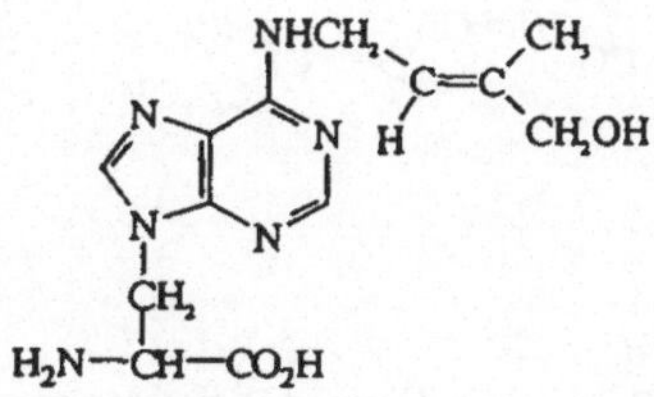

(41) R = H
(42) R = β-D-ribofuranosyl
(43) R = β-D-ribofuranosyl-5-O-phosphate
(44) R = $CH_2CH(NH_2)CO_2H$
(46) R = D-glucopyranosyl

(45)

B. CYTOKININ *O*-GLYCOSIDES

O-D-glucopyranosyl (E)-zeatin (**51**) appears to be of widespread natural occurrence. It was first reported from soybean callus tissue[81] and also as a metabolite of exogenously supplied zeatin from *Lupinus angustifolius* and, later, from *Phaseolus vulgaris*.[82] The glucoside has been synthesized by reaction of 2,3,4,6-tetra-*O*-acetyl-D-glucopyranosyl bromide with (E)-2-methyl-4-phthalimidobut-2-en-1-ol (**52**) in the presence of silver carbonate over 4 days to produce (**53**), which, after deblocking, gave the amine (**54**), which afforded the *O*-glucoside (**51**) when condensed with 6-chloropurine (Scheme 14).[83] The corresponding dihydrozeatin-*O*-glucoside has also been isolated from leaves of *Phaseolus vulgaris*.[84]

Recently, *O*-xylopyranosyl-(E)-zeatin (**57**) was isolated from *Phaseolus vulgaris* cv. Great Northern[85] and has been synthesized[86] by reaction of 2,3,4-tri-*O*-benzoyl-α-D-xylosyl bromide with the phthalimide derivative (**52**) in the presence of silver trifluoromethanesulfonate and a molecular sieve over 3 h at room temperature to give the tribenzoyl derivative (**55**), which, when deblocked with methanolic ammonia, afforded the *O*-xylosylamine (**56**), which, with 6-chloropurine and triethylamine, gave (E)-zeatin *O*-xyloside (**57**) (Scheme 14).

O-glucosyl-(E)-zeatin also occurs[87] as the ribonucleoside (**58**) in *Vinca rosea* crown gall tissue, *Lupinus luteus*, and *Z. mays* kernels,[54] and as the related *O*-glucosyldihydro-(E)-zeatin in *Lupinus luteus*,[88] *Zea mays*,[54] and probably many other plant tissues.

V. CYTOKININ RIBONUCLEOTIDES

A compound isolated from sweet corn kernels as a barium salt was tentatively identified as ribosyl-(E)-zeatin 5-phosphate (**61**).[89] The nucleotide occurs

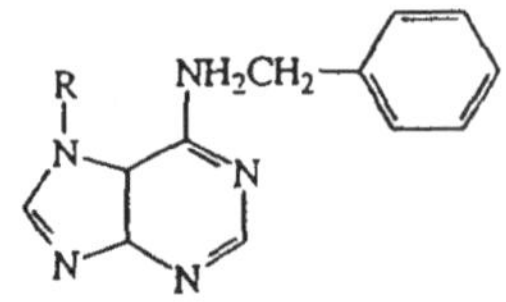

(47) R = D-glucopyranosyl

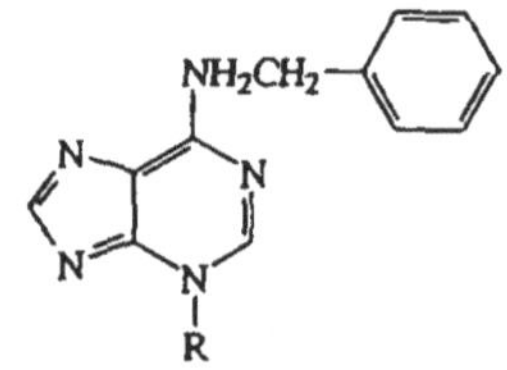

(48) R = D-glucopyranosyl

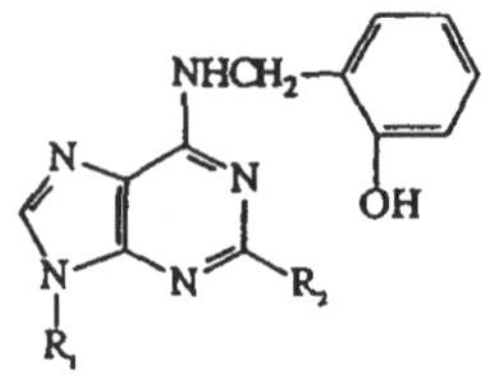

(49) R_1 = β-D-ribofuranosyl, R_2 = H

(50) R_1 = D-glucopyranosyl, R_2 = SCH_3

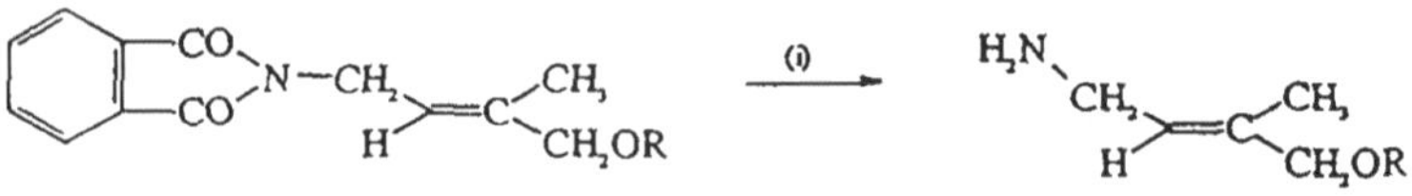

(52) R = H

(53) R = 2,3,4,6-tetra-O-acetyl-D-glucopyranosyl (54) R = D-glucopyranosyl

(55) R = 2,3,4-tri-O-benzoyl-D-xylopyranosyl (56) R = D-xylopyranosyl

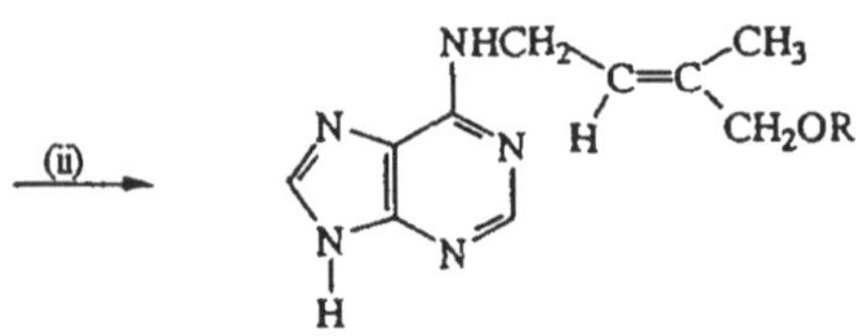

(51) R = D-glucopyranosyl

(57) R = β-D-xylopyranosyl

(i) NH_3/MeOH, (ii) 6-chloropurine

Scheme 14[83,86]

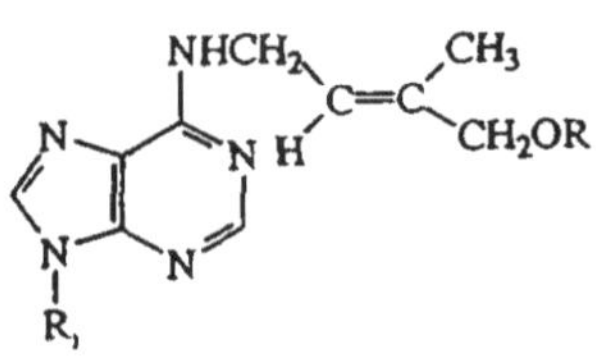

(58) R_1 = β-D-ribofuranosyl

R_2 = D-glucopyranosyl

in much greater quantity than the riboside or aglycone and has also been isolated from the puffball fungus *Rhizopogon roseolus*.[51] The structure of the nucleotide was confirmed by two syntheses.[90] Thus, reaction of the amine **(6)** with 6-chloro-9-(2,3-*O*-isopropylidene-β-D-ribofuranosyl)purine **(59)** and triethylamine in hot butanol gave isopropylidene zeatin riboside **(60)**, which was phosphorylated with pyrophosphoryl chloride and deblocked with aqueous acetic acid to give zeatin ribonucleotide **(61)** (Scheme 15). An alternative synthesis involved reaction of the amine **(6)** with 6-chloropurine ribotide **(62)** (prepared by a new improved method) in aqueous 3% potassium hydrogen carbonate at 38°C for 70 h (Scheme 16). Other nucleotides, including 5′-mono-, di-, and triphosphates, have been obtained as metabolites of exogenously added cytokinins.[91]

VI. STRUCTURE–ACTIVITY RELATIONSHIPS

Large numbers of 6-*N*-substituted aminopurine derivatives have been prepared, and their cytokinin activities have been compared, although assay results tend to vary substantially according to the specific method used. More detailed accounts (than are possible here) of the effect on cytokinin activities, of a variation of the molecular structures of large numbers of purine derivatives, have been recorded elsewhere.[92,93] However, the following general comments are relevant:

1. The active naturally occurring cytokinins are all 6-*N*-substituted adenines.
2. Substitution in the ring or at N^6 tends to lower cytokinin activity, although there is some confusion about 9-substituted derivatives.

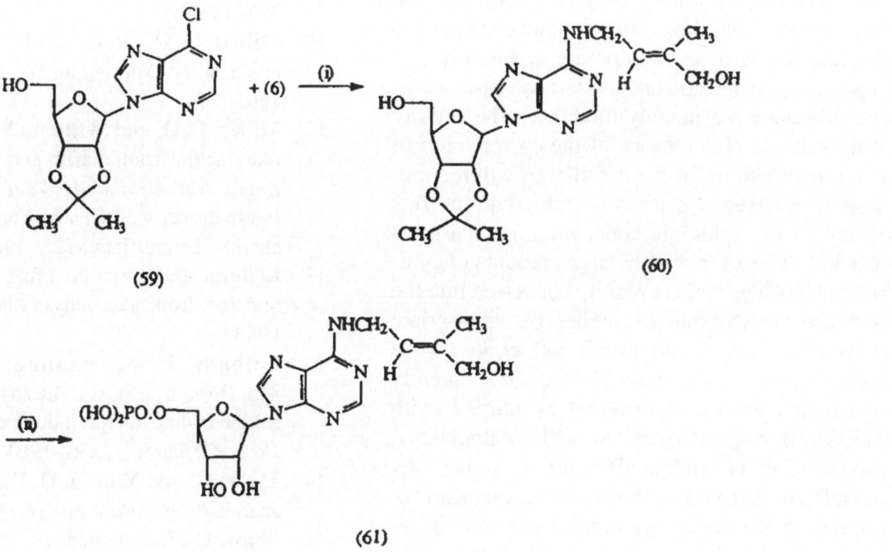

(i) NEt_3, n-BuOH, (ii) $P_2O_3Cl_4$ HOAc

Scheme 15[90]

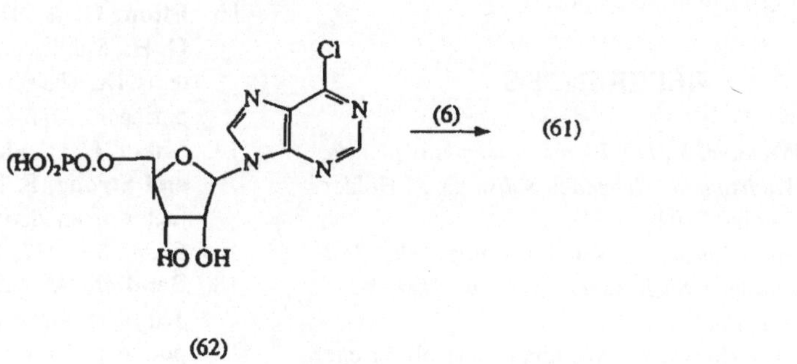

(62)

Scheme 16[90]

3. The nucleosides, and even more so the nucleotides, tend to have lower activity as do the *O*-glycosides.
4. In the case of analogous synthetic N^6-substituted adenines, the size of the substituent is critical for high activity. Thus, in a series of 6-*N*-alkylaminopurines, maximum activity is reached at about C_5. Longer chains result in less activity, and at C_{10} the compound is almost inactive.[94]
5. The presence of unsaturation in the side chain markedly enhances activity, making isopentenylamino purine and zeatin, especially, the most active of all the cytokinins. Saturation of the double bond leads to dihydroderivatives that are much less active.

Apart from the strictures imposed by the size of the substituent group, there is little or nothing known about the mechanism of action of cytokinins at the molecular level, although they are reported to be involved in gene expression reactions where the absolute levels of cytokinins are apparently less important than the ratio of cytokinin to auxin, where one substance presumably modulates the activity of the other.[95] Maintenance of the correct ratio of auxin to cytokinin for the specific growth requirement in a given species no doubt requires fine control of the cytokinin concentration. The presence in tissues of relatively large amounts of cytokinin glycosides that are readily converted into the very active cytokinin aglycones by appropriate enzyme catalysis would permit such control to be exercised. It is equally possible that the apparently confusing presence of different cytokinins (with widely differing activities and with, undoubtedly, varying receptor-binding affinities) in a given tissue reflects part of a sophisticated mechanism for controlling the active cytokinin concentration requirements. Similar comments could apply equally to indole acetic acid, which also occurs as various conjugates, mainly esters, with polysaccharides, *myo*-inositol, and other related carbohydrates. In some cases at least, these are hydrolyzed to release IAA during growth processes.[107]

REFERENCES

1. **Wiesner, J.,** *Die Elementarstruktur und das Wachstum der lebenden Substanz*, A. Holder, Vienna, 1892.
2. **Haberlandt, G.,** Zur Physiologie der Zellteilungen, *Sitzungsber. K. Preuss Akad. Wiss.*, 318, 1913.
3. **Haberlandt, G.,** Wundhormone als Erreger von Zellteilungen, *Beitr. Allg. Bot.*, 2, 1, 1921.
4. **Miller, C. O., Skoog, F., Okomura, F. S., von Salta, M. H., and Strong, F. M.,** Isolation, structure and synthesis of kinetin, a substance promoting cell division, *J. Am. Chem. Soc.*, 78, 1345, 1956.
5. **Hall, R. H. and de Rapp, R. S.,** Formation of 6-furfurylaminopurine from DNA breakdown products, *J. Am. Chem. Soc.*, 77, 6400, 1955.
6. **Letham, D. S.,** Regulations of cell division in plant tissues. I. Inhibitors and stimulants of cell division in developing fruits, their properties and activity in relation to the cell division period, *N. Z. J. Bot.*, 1, 336, 1963.
7. **Skoog, F., Strong, F. M., and Miller, C. O.,** Cytokinins, *Science*, 148, 532, 1965.
8. **Horgan, R.,** Cytokinins, in *Advanced Plant Physiology*, Wilkins, M. B., Ed., Pitman Medical, London, 1984.
9. **Miller, C. O.,** Kinetin and related compounds in plant growth, *Ann. Rev. Plant Physiol.*, 12, 395, 1961.
10. **Miller, C. O.,** A kinetin-like compound in maize, *Proc. Natl. Acad. Sci. U.S.A.*, 47, 170, 1961.
11. **Miller, C. O. and Witham, F. H.,** A kinetin-like factor from maize and other sources, *Regul. Nat. Croissance Vegetale*, Colloques Internationaux du Centre National de la Recherche Scientifique, 123, 1963.
12. **Letham, D. S.,** Zeatin, a factor inducing cell division from *Zea mays*, *Life Sci.*, 8, 596, 1963.
13. **Letham, D. S., Shannon, J. C., and MacDonald, I. R. C.,** The structure of zeatin, a kinetin-like factor inducing cell division, *Proc. Chem. Soc.*, 230, 1964.
14. **Shaw, G. and Wilson, D. V.,** A synthesis of zeatin, *Proc. Chem. Soc.*, 231, 1964.
15. **Shaw, G., Smallwood, B. M., and Wilson, D. V.,** Purines, Pyrimidines and Imidazoles. XXIV. Syntheses of zeatin, a naturally occurring adenine derivative with plant cell-division promoting activity and its 9-β-D-ribofuranoside, *J. Chem. Soc.*, (C), 921, 1966.
16. **Elion, G. B., Burgi, E., and Hitchings, G. H.,** Studies on condensed pyrimidine systems. IX. The synthesis of some 6-substituted purines, *J. Am. Chem. Soc.*, 74, 411, 1952.
17. **Miller, C. O., Skoog, F., von Saltza, M. H., and Strong, F. M.,** Kinetin, a cell division factor from deoxyribonucleic acid, *J. Am. Chem. Soc.*, 77, 1392, 1955.
18. **Bendich, A. A., Russell, P. J., and Fox, J. J.,** The synthesis and properties of 6-chloropurine and purine, *J. Am. Chem. Soc.*, 76, 6073, 1954.

19. **Albert, A. and Brown, D. J.,** Purine studies part 1. Stability to acid and alkali. Solubility ionization comparison with pteridines, *J. Chem. Soc.*, 2060, 1954.
20. **Carter, C. E.,** Synthesis of 6-succino-aminopurine, *J. Biol. Chem.*, 223, 139, 1956.
21. **Letham, D. S.,** Cytokinins, in *Phytohormones and Related Compounds*, Vol. 1, Letham, D. S., Goodwin, P. B., and Higgins, J. V., Eds., Elsevier, New York, 1978.
22. **Olomucki, M., Desvages, G., Thoai, N. V., and Roche, J.,** Étude de la configuration de l'hydroxygalegine naturelle et synthèse de son stereoisomere, *Compt. Rendu.*, 260, 4519, 1965.
23. **Corse, J. and Kuhnle, J.,** Improved synthesis of *trans*-zeatin, *Synthesis*, 11, 618, 1972.
24. **Cebalo, T. and Letham, D. S.,** Synthesis of zeatin, a factor influencing cell division, *Nature*, 86, 177, 1967.
25. **Napoli, L., de Evidente, A., Piccialli, G., Santacroce, C., and Vellone, R.,** Synthesis of the naturally occurring cytokinin, 2′-deoxyzeatin riboside, *Phytochemistry*, 29, 701, 1990.
26. **Chen, S. C.,** A simple synthesis of (E)-3-formylbut-2-enenitrile and its use as a precursor of isotope labelled zeatin and (±)dihydrozeatin, *Experientia*, 37, 543, 1981.
27. **Mornet, R. and Gouin, L.,** Synthese de l'acetate du chloro-4-methyl-2-butene-2-ol (E) Nouvelle voie d'accès à la trans zeatin, *Tet. Letts.*, 2, 167, 1977.
28. **Gray, G. M.,** An improved synthesis of (E)-2-(4-hydroxy-3-methyl-2-butenyl) 1H-isoindole-1,3(2H)-dione, *Synthesis*, 488, 1983.
29. **Letham, D. S., Hocart, C. H., and Eicholzer, J. V.,** Synthesis of radioactive zeatin riboside and related compounds by alkylation of purine moieties, *Phytochemistry*, 19, 385, 1990.
30. **Kadir, K., Gregson, S., Shaw, G., Mok, M. C., and Mok, D. W. S.,** Purines, pyrimidines and imidazoles. Part 59. A convenient synthesis of (E)-zeatin and (E)-[8-^{14}C]-zeatin, *J. Chem. Res.*, 299, 1984.
31. **Shaw, G., Smallwood, B. M., and Wilson, D. V.,** Purines, pyrimidines and imidazoles. Part 31. Some reactions of the allylic group in the plant cell-division promoting factor, zeatin and its nucleoside and nucleotide derivatives, *J. Chem. Soc.*, (C), 2999, 1968.
32. **Schmitz, R. Y., Skoog, F., Hecht, S. M., and Leonard, N. J.,** Cytokinins: synthesis and biological activity of zeatin esters and related compounds, *Phytochemistry*, 10, 275, 1971.
33. **Shaw, G., Smallwood, B. M., and Steward, F. C.,** Synthesis and cytokinin activity of the 3-, 7-, and 9-methyl derivatives of zeatin, *Experientia*, 24, 1089, 1968.
34. **Gregson, S., Shaw, G., Mok, M. C., and Mok, D. W. S.,** Cytokinin activities of thiozeatin and related compounds in *Phaseolus* callus culture bioassays, *Plant Sci.*, 38, 111, 1985.
35. **Koshimizu, K., Matsubara, S., Kasaki, T., and Mitsui, T.,** Isolation of new cytokinin from immature yellow lupin seeds, *Agric. Biol. Chem.*, 31, 795, 1967.
36. **Fujii, T. and Ogawa, N.,** Absolute configuration of (–) dihydrozeatin, *Tet. Letts.*, 30, 3075, 1972.
37. **Hall, R. H., Csonka, L., David, H., and McLennan, D. B.,** Cytokinins in the soluble RNA of plant tissues, *Science*, 156, 69, 1967.
38. **Babcock, D. F. and Morris, R. O.,** Quantitative measurement of isoprenoid nucleosides in transfer RNA, *Biochemistry*, 9, 3701, 1970.
39. **Leonard, N. J., Playtis, A. J., Skoog, F., and Schmitz, R. Y.,** *J. Am. Chem. Soc.*, 93, 3056, 1971.
40. **Leonard, N. J.,** Chemistry of the cytokinins, *Recent Adv. Phytochem.*, 7, 21, 1974.
41. **Burrows, W. J., Armstrong, D. J., Kamínek, M., Skoog, F., Bock, R. M., Hecht, S. M., Dammann, L. G., Leonard, N. J., and Occolowitz, J.,** Isolation and identification of four cytokinins from wheat germ transfer RNA, *Biochemistry*, 9, 1867, 1970.
42. **Vreman, H. J., Thomas, R., Corse, J., Swaminathan, S., and Murai, N.,** Cytokinins in tRNA obtained from *Spinacia oleracia* L. leaves and isolated chloroplasts, *Plant Physiol.*, 61, 296, 1978.
43. **Letham, D. S.,** Cytokinins, in *Phytohormones and Related Compounds: A Comprehensive Treatise*, Vol. 1, Letham, D. A., Goodwin, P. B., and Higgins, T. J. V., Eds., Elsevier, Amsterdam, 1978, 205.
44. **Vreman, H. J., Schmitz, R. Y., Skoog, F., Playtis, A. J., Frihart, C. R., and Leonard, N. J.,** Synthesis of methylthio-*cis*- and *trans*-ribosylzeatin and their isolation from *Pisum* tRNA, *Phytochemistry*, 13, 31, 1974.
45. **Letham, D. S.,** Cytokinins from *Zea mays*, *Phytochemistry*, 12, 2445, 1973.
46. **Klambt, D., Thies, G., and Skoog, F.,** Isolation of cytokinins from *Corynebacterium fascians*, *Proc. Natl. Acad. Sci. U.S.A.*, 56, 52, 1966.
47. **Helgeson, J. P. and Leonard, N. J.,** Cytokinins: identification of compounds isolated from *Corynebacterium fascians*, *Proc. Natl. Acad. Sci. U.S.A.*, 56, 60, 1966.

48. **Letham, D. S. and Wettenhall, R. E. H.,** Transfer RNA and cytokinins, in *The Ribonucleic Acids*, Stewart, P. R. and Letham, D. S., Eds., Springer, New York, 1977, 129.
49. **Iacobellis, N. S., Evidente, A., and Surico, G.,** Isolation of plant growth substances from *Pseudomonas amygdali, Experientia*, 44, 70, 1988.
50. **Miller, C. O.,** Zeatin and zeatin riboside from a *Mycorrhizal* fungus, *Science*, 157, 1055, 1967.
51. **Zhang, W., Yamane, H., Chapman, D. J., and Phinney, B. O.,** Identification of a cytokinin in the green alga, *Chara globularis, Phytochemistry*, 28, 337, 1989.
52. **Farooqi, A. H., Yogendra, N. S., Shukla, A., and Bhakuni, D. S.,** Cytokinins from marine organisms, *Phytochemsitry*, 29, 2061, 1990.
53. **Dyson, W. H. and Hall, R. H.,** N^6-(Δ^2-isopentenyl)adenosine: its occurrence as a free nucleoside in an autonomous strain of tobacco tissue, *Plant Physiol.*, 50, 616, 1972.
54. **Summons, R. E., Entsch, B., Letham, D. S., Gollnow, B. I., and MacLeod, J. K.,** Regulators of cell division in plant tissues. XXVIII. Metabolites of zeatin in sweet corn kernels: purifications and identifications using high-performance liquid chromatography and chemical ionization mass-spectrometry, *Planta*, 147, 422, 1980.
55. **Scott, I. M., Horgan, R., and McGaw, B. A.,** Zeatin-9-glucoside, a major endogenous cytokinin of *Vinca rosea* L. crown gall tissue, *Planta*, 149, 472, 1980.
56. **Grayling, A. and Hanke, D. E.,** Cytokinins in exudates from leaves and roots of red *Perilla, Phytochemistry*, 31, 1863, 1992.
57. **Wang, T. L. and Horgan, R.,** Dihydrozeatin riboside, a minor cytokinin from the leaves of *Phaseolus vulgaris, Planta*, 140, 151, 1978.
58. **Palni, L. M. S. and Horgan, R.,** Cytokinins from the culture medium of Vinca rosea crown gall tumour tissue, *Plant Sci. Lett.*, 24, 327, 1982.
59. **Letham, D. S. and Zhang, R.,** Cytokinin translocation and metabolism in *Lupinus* species. II. New nucleotide metabolites of cytokinins, *Plant Sci.*, 64, 161, 1989.
60. **Mornet, R. and Laloue, M.,** Synthesis of 6-(E)-4-acetoxy-3-methylbut-2-enylamino-9-β-D-ribofuranosylpurine, a new metabolite of zeatin riboside, *J. Chem. Res.*, (S), 382, 1989.
61. **Evidente, A., Iacobellis, N. S., Vellone, R., Sisto, A., and Surico, G.,** 2′-Deoxyzeatin riboside and other cytokinins in culture filtrates of *Pseudomonas amygdalis, Phytochemistry*, 28, 2603, 1989.
62. **Evidente, A., Surico, G., Iacobellis, N. S., and Randazzo, G.,** 1′-Methyl zeatin, an additional cytokinin from *Pseudomonas syringae, Phytochemistry*, 25, 525, 1986.
63. **Parker, C. W., Letham, D. S., Cowley, D. E., and MacLeod, J. K.,** Raphanatin, an unusual purine derivative and a metabolite of zeatin, *Biochem. Biophys. Res. Commun.*, 49, 460, 1972.
64. **Cowley, D. E., Duke, C. C., Liepa, A. J., MacLeod, J. K., and Letham, D. S.,** The structure and synthesis of cytokinin metabolites. I. The 7- and 9-β-D-glucofuranosides and pyranosides of zeatin and 6-benzylaminpurine, *Aust. J. Chem.*, 31, 1095, 1978.
65. **Entsch, B., Letham, D. S., Parker, C. W., Summons, R. E., and Gollnow, B. I.,** Metabolites of cytokinins, in *Plant Growth Substances 1979*, Skoog, F., Ed., Springer-Verlag, Berlin, 1979, 109.
66. **Babcock, D. F. and Morris, R. O.,** Quantitative measurements of isoprenoid nucleosides in tRNA, *Biochemistry*, 9, 3701, 1970.
67. **Burrows, W. J., Skoog, F., and Leonard, N. J.,** Isolation and identification of cytokinins located in the transfer ribonucleic acid of tobacco callus grown in the presence of 6-benzylaminopurine, *Biochemistry*, 10, 2189, 1971.
68. **Einset, J. W. and Skoog, F.,** Isolation and identification of ribosyl-*cis*-zeatin from transfer RNA of *Corynebacterium fascians, Biochem. Biophys. Res. Commun.*, 79, 1117, 1977.
69. **Watanabe, N., Yokota, T., and Takahashi, N.,** *cis*-Zeatin riboside: its occurrence as a free nucleoside in cones of the hop plant, *Agric. Biol. Chem.*, 42, 2415, 1978.
70. **McCalla, D. R., Moore, D. J., and Osborne, D. J.,** The metabolism of a kinin, benzyladenine, *Biochem. Biophys. Acta.*, 55, 522, 1962.
71. **Letham, D. S., Summons, R. E., Parker, C. W., and MacLeod, J. K.,** Identification of an amino-acid conjugate of 6-benzylaminopurine formed in *Phaseolus vulgaris* seedlings, *Planta*, 146, 71, 1979.
72. **Parker, C. W., Letham, D. S., Gollnow, B. I., Summons, R. E., Duke, C. C., and MacLeod, J. K.,** Regulators of cell division in plant tissues. XXV. Metabolism of zeatin in lupin seedlings, *Planta*, 142, 239, 1978.
73. **Wilson, M. M., Gordon, M. E., Letham, D. S., and Parker, C. W.,** The metabolism of 6-benzylaminopurine in radish cotyledons and seedlings, *J. Exp. Bot.*, 25, 725, 1974.

74. **Letham, D. S., Wilson, M. M., Parker, C. W., Jenkins, I. D., MacLeod, J. K., and Summons, R. E.,** The identity of an unusual metabolite of 6-benzylaminopurine, *Biochem. Biophys. Acta,* 399, 61, 1975.
75. **Fox, J. E., Cornette, J., Deleuze, G., Dyson, W., Giersak, G., Niu, P., Zapata, J., and McChesney, J.,** The formation, isolation and biological activity of a cytokinin 7-glucoside, *Plant Physiol.*, 52, 627, 1973.
76. **Letham, D. S. and Gollnow, B. I.,** Regulators of cell division in plant tissues. XXX. Cytokinin metabolism in relation to radish cotyledons, *J. Plant Growth Regul.*, 4, 129, 1985.
77. **Blakesley, D., Lenton, J. R., and Horgan, R.,** Benzyladenine ribosylglucoside, a metabolite of benzyladenine in *Gerbera jamesonii, Phytochemistry,* 30, 387, 1991.
78. **Ernst, D., Schafer, W., and Oesterhelt, D.,** Isolation of a new naturally occurring cytokinin 6-benzylamino purine riboside from an anise cell culture, *Pimpinella anisum* L., *Planta,* 159, 222, 1983.
79. **Horgan, R., Hewett, E. W., Horgan, J. M., Purse, J. G., and Wareing, P. F.,** A new cytokinin from *Populus robusta, Phytochemistry,* 14, 1005, 1975.
80. **Chaves das Neves, H. J. and Pais, M. S.,** A new cytokinin from the fruits of *Zantedeschia aethiopica, Tet. Lett.*, 21, 4387, 1980.
81. **Horgan, R.,** A new cytokinin metabolite, *Biochem. Biophys. Res. Commun.*, 65, 358, 1975.
82. **Letham, D. S., Parker, C. W., Duke, O. C., Summons, R. E., and MacLeod, J. K.,** *O*-glucosylzeatin and related compounds — a new group of cytokinin metabolites, *Ann. Bot.*, 41, 261, 1977.
83. **Duke, C. C., MacLeod, J. K., Summons, R. E., Letham, D. S., and Parker, C. W.,** The structure and synthesis of cytokinin metabolites. II. Lupinic acid and *O*-β-D-glucopyranosylzeatin from *Lupinus angustifolius, Aust. J. Chem.*, 31, 1291, 1978.
84. **Wang, T. L., Thompson, A. G., and Horgan, R.,** A cytokinin glucoside from the leaves of *Phaseolus vulgaris* L., *Planta,* 135, 285, 1977.
85. **Turner, J. E., Mok, D. W. S., Mok, M. C., and Shaw, G.,** Isolation and partial purification of the enzyme catalysing the formation of *O*-xylosylzeatin in *Phaseolus vulgaris* embryos, *Proc. Natl. Acad. Sci. U.S.A.*, 84, 3714, 1987.
86. **Shaw, G., Mok, M. C., and Mok, D. W. S.,** Purines, pyrimidines and imidazoles. Part 63. Synthesis of *O*-β-D-xylopyranosyl-(E)-zeatin, a new cytokinin metabolite, *J. Chem. Res.*, (S), 110, 1987.
87. **Morris, R. O.,** Mass spectroscopic identification of cytokinins. Glucosyl zeatin and glucosyl ribosylzeatin from *Vinca rosea* crown gall, *Plant Physiol.*, 59, 1029, 1977.
88. **Summons, R., Entsch, B., Parker, C. W., and Letham, D. S.,** Mass spectrometric analysis of cytokinins in plant tissues. III. Quantitation of the cytokinin glycoside complex of lupin pods by stable isotope dilution, *FEBS Lett.*, 107, 21, 1979.
89. **Letham, D. S.,** Isolation and probable identity of a third cytokinin in sweet corn extracts, *Life Sci.*, 5, 1999, 1966.
90. **Shaw, G., Smallwood, B. M., and Wilson, D. V.,** Purines, pyrimidines and imidazoles. Part 28. Synthesis of 9-β-D-ribofuranosylzeatin-5′-phosphate, a naturally occurring adenylic acid derivative with plant cell-division promoting activity, and a new synthesis of 6-chloro-9-β-D-ribofuranosylpurine 5′-phosphate, *J. Chem. Soc.*, (C), 1516, 1968.
91. **Letham, D. S. and Palni, L. M. S.,** The biosynthesis and metabolism of cytokinins, *Ann. Rev. Plant Physiol.*, 34, 263, 1983.
92. **Skoog, F., Hamzi, H. Q., Szweykowska, A. M., Leonard, N. J., Carraway, K. L., Fujii, T., Helgeson, J. P., and Leoppky, R. N.,** Cytokinins: structure–activity relationships, *Phytochemistry,* 6, 1169, 1967.
93. **Matsubara, S.,** Structure–activity relationships of cytokinins, *Phytochemistry,* 19, 2239, 1980.
94. **Shaw, G., Smallwood, B. M., and Steward, F. C.,** The structure and physiological activity of some N^6-substituted adenines and related compounds, *Phytochemistry,* 10, 2329, 1971.
95. **Klee, H. and Estelle, M.,** Genetic approaches to plant hormone biology, *Ann. Rev. Plant Physiol.*, 42, 529, 1991.
96. **Van Staden, J. and Drewes, S. E.,** Identification of zeatin and zeatin riboside in coconut milk, *Physiol. Plant.*, 34, 106, 1975.
97. **Van Staden, J. and Menary, R. C.,** Identification of cytokinins in the xylem sap of tomato, *Z. Pflanzenphysiol.*, 78, 262, 1976.
98. **Purse, J. G., Horgan, R., Horgan, J. M., and Wareing, P. F.,** Cytokinins of sycamore spring sap, *Planta,* 132, 1, 1976.
99. **Krasnuk, M., Witham, F. H., and Tegley, J. R.,** Cytokinins extracted from Pinto bean fruit, *Plant Physiol.*, 48, 320, 1971.
100. **Young, H.,** Identification of cytokinins from natural sources by GC-MS, *Ann. Biochem.*, 79, 226, 1977.
101. **Letham, D. S.,** Regulators of cell division in plant tissues. A comparison of the activities of zeatin and other cytokinins in five bioassays, *Planta,* 74, 228, 1967.

102. **Horgan, R., Hewett, E. W., Purse, J. G., Horgan, J. M., and Wareing, P. F.**, Identification of a cytokinin in sycamore sap by gas chromatography-mass spectrometry, *Plant Sci. Lett.*, 1, 321, 1973.
103. **Lorenzi, R., Horgan, R., and Wareing, P. F.**, Cytokinins in *Picea sitchensis* Carriere, identification and relation to growth, *Biochem. Physiol. Pflanzen*, 168, 333, 1975.
104. **Peterson, J. B. and Miller, C. O.**, Glucosyl zeatin and glucosyl ribosylzeatin from *Vinca rosea* L. crown gall tissue, *Plant Physiol.*, 59, 1026, 1977.
105. **Dakiya, J. S. and Tewari, J. P.**, Plant growth factors produced by the fungus *Alternaria brassicae*, *Phytochemistry*, 30, 2825, 1991.
106. **Armstrong, D. J., Scarborough, E., Skoog, F., Cole, D. C., and Leonard, N. J.**, Cytokinins in *Corynebacterium fascians* cultures. Isolation and identification of 6-(4-hydroxy-3-methyl-cis-2-butenylamino)-2-methylthiopurine, *Plant Physiol.*, 58, 749, 1976.
107. **Cohen, J. D. and Bandurski, R. S.**, Chemistry and physiology of the bound auxins, *Ann. Rev. Plant Physiol.*, 33, 403, 1982.

Chapter 3

Chemistry of Phenylurea Cytokinins

Koichi Shudo

TABLE OF CONTENTS

I. INTRODUCTION

Early in the history of tissue and cell culture experiments, coconut milk was often used as a growth promoter in media formulation. One of the active components of coconut milk was identified as *N,N'*-diphenylurea (**1**, see Figure 1a) by Shantz and Steward.[1] Although several investigators obtained only sporadic activity with this compound, while Skoog et al. excluded compounds without a purine-based structure in their definition of cytokinins,[2] biological activity has been observed consistently with some phenylureas by Bruce and Zwar.[3]

With the development of two synthetic compounds, *N*-(2-chloro-4-pyridyl)-*N'*-phenylurea (**2**, forchlorofenuron, 4PU-30, Fulmet(c)*)[4] and *N*-phenyl-*N'*-(1,2,3-thiadiazol-5-yl)urea (**3**, thidiazuron, SN49537, Dropp(c)**),[5,6] as strong positive regulators of cell division and shoot formation, cytokinins can be structurally classified into at least two categories: the adenine cytokinins and the urea cytokinins. This chapter describes the structure–activity relationships and some biological aspects of the urea cytokinins.

*Registered trademark of Kyowa Hakko Kogyo, Ohtemachi, Tokyo, Japan.

**Registered trademark of Schering A.G., Berlin, Germany.

II. DIPHENYLUREAS

N,N'-diphenylurea was first reported as a cell division stimulator for carrot root tissue.[1] The activity of a number of phenylureas and diphenylureas was tested by Bruce and Zwar[3] and Bruce et al.[7] The bioassays adopted were senescence retardation, pea lateral-bud development, cell division activity, and lettuce-seed germination. Of 500 ureas examined, a large number showed some activity, though the bioassays seem to be semiquantitative. Several compounds that have consistently high activity (<1 μg/ml, approximately 10^{-5} *M* in the initiation of cell division of tobacco pith) are *N*-3-fluorophenyl-*N'*-4-fluorophenylurea (**4a**), *N*-4-nitrophenyl-*N'*-phenylurea (**4b**) and *N*-4-nitro-3-trifluoromethylphenyl-*N'*-phenylurea (**4c**). Moderately active ones are *N*-3,4-dichlorophenyl-*N'*-phenylurea (**4d**), *N*-3,5-dimethylphenylurea (**5a**), *N,N'*-diphenylurea (**1**), and *N*-phenyl-*N'*-4-pyridylurea (**6a**). Phenylurea (**5b**) itself is also weakly active. An NHCONH bridge conferred higher activity than an NHCSNH linkage, and any other arrangement, such as $N(CH_3)CONH$, CH_2CONH, or OCONH, gave inactive compounds.

Biological activities of these diphenylureas, evaluated in terms of growth promotion of second-generation tobacco callus derived from newly prepared callus from tobacco pith (the standard tobacco callus bioassay),[4] were not quantitatively identical with the results obtained by Bruce and Zwar.[3] Phenylurea was not active, while 4-chlorophenylurea (**5c**) was weakly active in the

0-8493-6252-0/94/$0.00+$.50

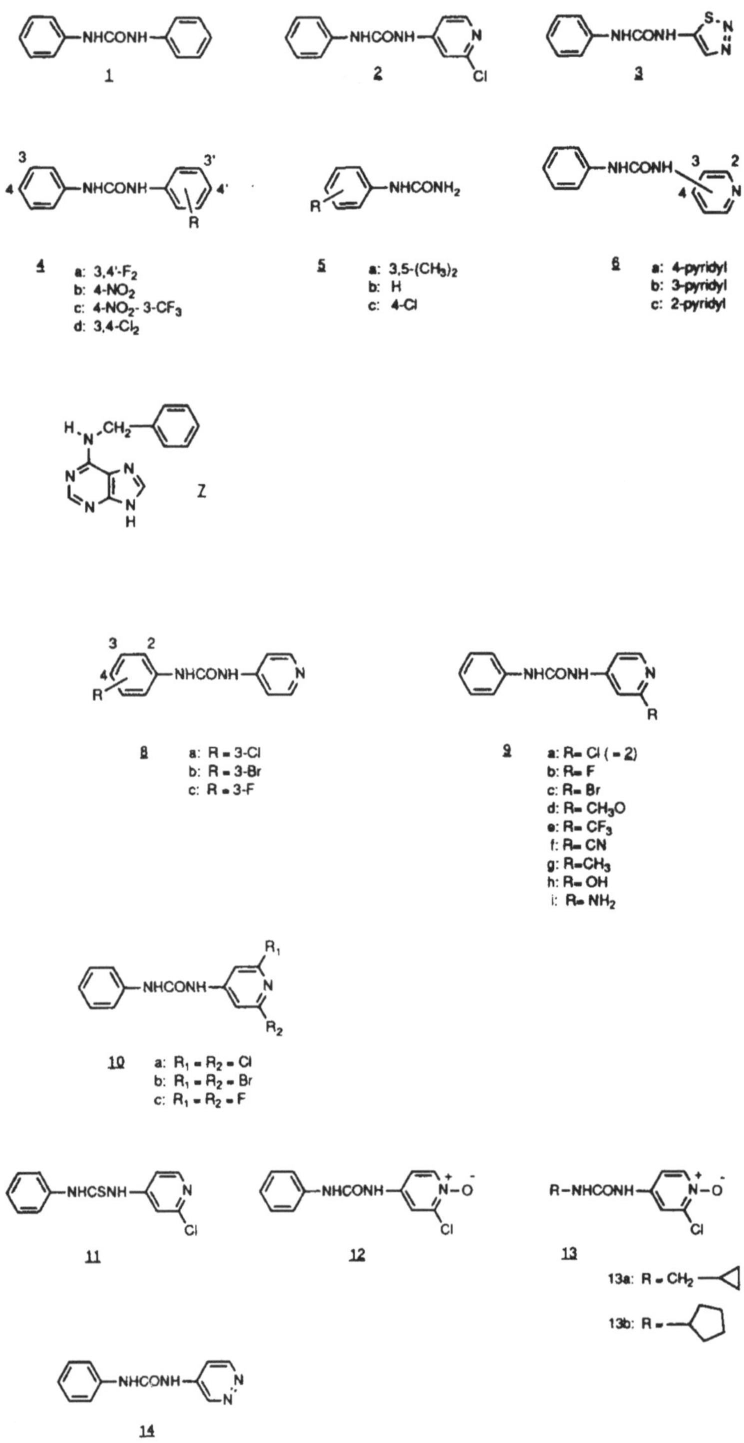

Figure 3-1a Structures of phenylureas (see text for details).

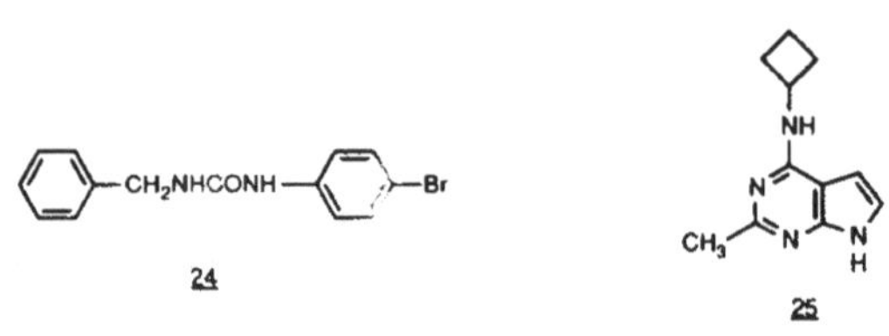

Figure 3-1b Structures of phenylureas (see text for details).

standard tobacco callus assay. Ring substitution at the *meta* position with electron-withdrawing substituents generally increased the activity. Compounds with two substituents on both phenyl rings of *N,N'*-diphenylurea generally had lower activity.[4]

The quantitative structure–activity relationship was analyzed by Iwamura et al.[8] The results suggest that position-specific steric and hydrophobic effects of aromatic substituents participate in the variation in activity, and the general trend of the activity is *meta* > *para* > *ortho* derivatives. The electronic effect is significant for the activity of diphenylureas, but not for that of phenylureas.

Though *N,N'*-diphenylurea and some substituted derivatives definitely show biological activity in the tobacco callus and other bioassays, the highest activity so far found is less than a hundredth of that for 6-benzylaminopurine **(7)**. The very low solubility of these compounds also makes the correct evaluation of activity difficult.

III. PHENYLPYRIDYLUREAS

Three isomers of *N*-phenyl-*N'*-pyridylureas **(6a–c)** were tested by Bruce and Zwar, and one, *N*-phenyl-*N'*-(4-pyridyl)urea **(6a)** (Figure 1b), was evaluated as being as active as *N,N'*-diphenylurea, by the tobacco pith disk assay.[3] Independently, the same compound was designed as a link to combine the features of *N,N'*-diphenylurea and isonicotinoylanilide (*vide infra*), and its cytokinin activity was tested by the standard tobacco callus bioassay.[9] In contrast to the result with the tobacco pith disk assay, *N*-phenyl-*N'*-(4-pyridyl)urea was a hundred times more active than *N,N'*-diphenylurea; the optimum concentration for cell division is 4.7×10^{-7} *M*, while the optimum concentration for 6-benzylaminopurine is 4.4×10^{-8} *M*. The callus yield is good, and the rate of proliferation is generally faster than with 6-benzylaminopurine.

Substituent effects on the 4-pyridylurea were examined.[4,10] Monosubstitution on the phenyl ring with CH_3, OCH_3, OH, or COOR caused a loss of activity. Substitution with Cl or Br at the *ortho* or *para* position also decreased the activity. The decrease caused by *meta* substitution with Cl or Br (i.e., compounds **8a,b**) was minimal, and only fluorine substitution at the *meta* position (compound **8c**) did not cause a decrease of the activity. Though introduction of a substituent on the phenyl ring of *N*-phenyl-*N'*-(4-pyridyl)urea did not yield any more potent compound, substitution of Cl, F, Br, CH_3O, CF_3, or CN at the 2-position of the pyridyl ring (i.e., **9a–f**) greatly enhanced the activity. Substitution with CH_3, OH, or NH_2 (i.e., compounds **9g–i**) did not alter or lower the activity. Thus, *N'*-(2-chloro-4-pyridyl)-*N'*-phenylurea **(2)** exhibits high cytokinin activity (optimum concentration 10^{-9} *M*), which is about 10, 100, and 10,000 times higher than that of 6-benzylaminopurine, *N*-phenyl *N'*-(4-pyridyl)urea **(6a)** or *N,N'*-diphenylurea **(1)**, respectively. Additional introduction of a substituent on the phenyl ring decreased the activity, except in the case of *meta*-fluorine.

It is clear that introduction of an electron-withdrawing and hydrophobic substituent is favorable for increasing the activity. This is well represented in the quantitative structure–activity relationship described in Equation 1.

$$\log(1/C) = 3.81(\pm 1.01)\sigma_m + 0.52(\pm 0.32)\pi + 6.50(\pm 0.28) \quad (1)$$

where C is the optimum concentration for the standard tobacco callus growth. The electronic effect (Hammett σ_m) is the major contributor, and the hydrophobicity (Hansch π) is less dominant. There is no significant correlation with the steric effect. The correlation to σ_m may reflect influence on the pK value of the ring nitrogen, and possibly on the acidity of the 4-amino group, too. Since the steric effect of the 2-substituent is not relevant to the activity, 2,6-disubstituted compounds **(10)** were tested: *N*-(2,6-dichloro-4-pyridyl)-*N*-phenylurea **(10a)** showed the highest callus yield in the range of 10^{-8} to 10^{-13} *M* (throughout the range, even at 10^{-13} *M*, the callus production was maximized, but any suppressive effect was not observed up to 10^{-7} *M*). The 2,6-dibromo compound **(10b)** was also very active, but the 2,6-difluoro compound **(10c)** was less active than expected. It seems interesting that further introduction of fluorine (2,3,4,5-tetrafluoro-, 2,3,4,5,2',3',4',5',6'-nanofluoro, and 2,6,2',3',4',5',6'-heptafluoro-pyridylphenylureas) resulted in diminished activity.[11] Activity optimization for a variety of *N'*-alkyl-substituted *N*-4-(2-chloro-4-pyridyl)ureas was performed for the chlorophyll retention activity.[12,13] Optimal moieties on the urea nitrogen are 2-chloroethyl and propyl. It is interesting that the phenyl group attached to one urea nitrogen can be replaced by an alkyl group instead of the phenyl group.

The modification of the urea moiety by *N*-methylation abolished the activity, because of the drastic stereochemical change of the urea conformation.[14] The *cis* stereochemistry of *N*-methylureas has been proved by X-ray crystallography and nuclear magnetic spectral analysis. In any attempt at molecular superposition of different types of cytokinins, the *trans* structure should be used for the active conformation of the urea cytokinins.[15]

The corresponding thiourea, *N*-(2-chloro-4-pyridyl)-*N*′-phenylthiourea **(11)** (Figure 1b), is as active as the urea **(2)**, although *N,N*-diphenylthiourea is inactive.[16] *N*-oxides of the pyridylphenylureas **(12)** are less active than corresponding pyridyl *N*′-arylureas, in terms of the shoot-forming activity using tobacco callus.[17] This suggested that the observed activity is probably due to metabolic reduction of the *N*-oxide to the parent urea compound. On the other hand, the chlorophyll retention induced by the *N*-oxides seems to be greater than that by the parent compounds, though the substituent effect is common in both series.[12,13] A chlorophyll retention assay of various *N*-oxides of *N*-alkyl-*N*′-pyridy-lureas **(13)** was reported: *N*-(2-chloro-4-pyridyl)-*N*′-cyclopropylmethylurea **(13a)** and *N*-(2-chloro-4-pyridyl)-*N*′-cyclopentylurea **(13b)** were active. The quantitative structure–activity relationship for *N*-oxides gives a different equation from that for pyridylphenylureas.

IV. THIADIAZOLYL UREAS AND RELATED COMPOUNDS

N-(1,2,3-thiadiazol-5-yl)-*N*′-phenylurea **(3)** was discovered as a growth retardant and developed as an efficient defoliant (Dropp®) for cotton.[5] The compound induced defoliation without apparent breakdown of chlorophyll. This suggested that it might have cytokinin activity. Mok et al. confirmed this activity in the standard *Phaseolus* callus bioassay.[6,18] In the assay, the compound exhibited activity as strong as that of *N*-(2-chloro-4-pyridyl)-*N*′-phenylurea **(2)**. The compound is also active in the tobacco callus assay at concentrations of 10^{-13} to 10^{-8} *M*.[19] Its various biological activities have been evaluated.[20-23] It seems to be one of the strongest cytokinins, being as potent as *N*-(2,6-dichloropyridyl)-*N*′-phenylurea. Replacement of the phenyl ring of **(3)** with the thiadiazol-5-yl, 2-pyridyl, benzyl, or furfuryl group resulted in a reduction in activity.[6] Since the sulfur atom in a five-membered aromatic structure may be an isoster of a C=C bond, the thiadiazol urea **(3)** is isosteric with *N*-(4-pyridazinyl)-*N*′-phenylurea **(14)** which exhibited strong activity in the tobacco callus assay, though it is sparingly soluble.[19] This may suggest that the 2-nitrogen atom of the thiadiazole urea and the 1-nitrogen atom of 4-pyridyl- and 4-pyridazinylureas can be regarded as equivalent from the viewpoint of molecular superposition.

V. STRUCTURAL RELATIONSHIPS BETWEEN UREA AND ADENINE CYTOKININS

Though, at first glance, adenine cytokinins represented by zeatin, benzylaminopurine and kinetin, and urea cytokinins, such as **(2)** (forchlorfenuron) and **(3)** (thidiazuron), seem to be structurally very different, there exists a variety of compounds that link them (Figure 2). These compounds can be classified into two groups. One consists of imidazole carbanilide **(15)** and isonicotinoylanilide **(16)**.[9] Both compounds exhibit a weak but definite activity in the tobacco callus bioassay. Imidazole carbanilide is easily superposed on *N*-phenyladenine **(17)**, which is significantly active.[24] Fusion of isonicotinoylanilide and diphenylurea yielded *N*-phenyl-*N*′-(4-pyridyl)urea **(6a)**.[9] The second group consists of 4-aminopyrimidine derivatives. N^4-substituted pyrimidines such as N^4-isopentenylamino

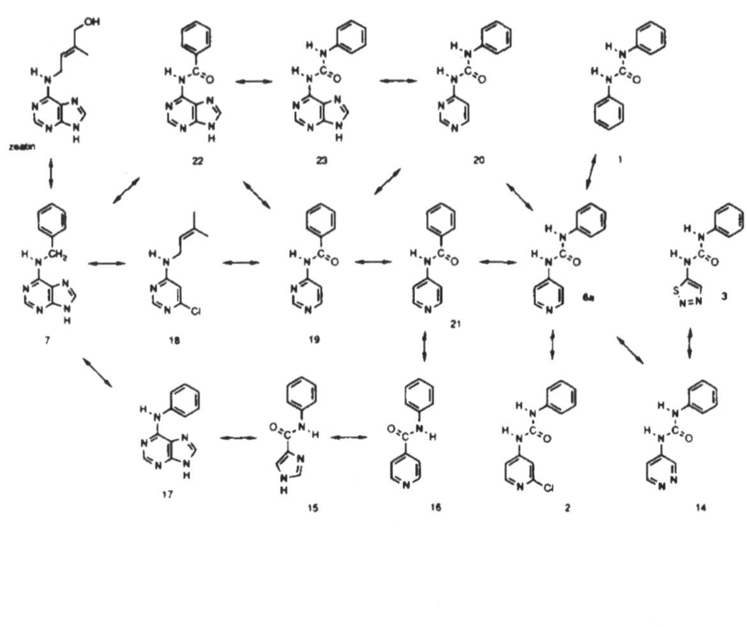

Figure 3-2 Structural correlation between purine cytokinins and urea cytokinins.

(**18**) or N^4-benzoylaminopyrimidine (**19**) are active, and correlate to *N*-pyriminyl-*N′*-phenylureas (**20**).[25] Obviously, the pyrimidine moiety shares a part of the purine ring. Aminopyrimidines also correlate with cytokinin-active phenyl carboxamide (**21**).[26] Compounds (**19**) and (**21**) are readily correlated to 6-benzylaminopurine (**7**) through 6-benzoylaminopurine (**22**)[27] and then zeatin. Combination of (**21**) and (**16**) also leads to the pyridylurea (**6a**). This is also clearly shown by the activity of phenylureidopurine (**23**), synthesized by Leonard's group.[28,29] This compound also structurally links the purine and urea cytokinins. As evidenced by Iwamura et al., quantitative structure–activity relationships of purine and urea cytokinins can be interpreted in terms of a hypothetical map for the mode of interaction with the site of action.[8] Such structural superposition strongly suggests that adenine cytokinins and urea cytokinins are agonists and that both cytokinins act through a common receptor.

VI. BIOLOGICAL CORRELATION BETWEEN ADENINE AND UREA CYTOKININS

Kefford et al. reported that the substitution of a benzyl group for the phenyl group of *N,N′*-diphenylureas converts them from active cytokinins to antagonists in the chlorophyll retention bioassay.[30] The inhibition by the antagonists is competitively reversed by purine cytokinins as well as by agonistic diphenylurea cytokinins. This was confirmed in the standard callus bioassay.[31] *N*-benzyl-*N′*-(4-bromophenyl)urea (**24**) inhibits the growth of tobacco callus induced by *N*-phenyl-*N′*-(4-pyridyl)urea and also by 6-benzylaminopurine. This inhibition is restored competitively by the addition of a higher concentration of the urea or the adenine cytokinin. Further, a strong antagonist, 4-cyclobutylamino-2-methylpyrrolo[2,3-d]pyrimidine (**25**), developed by Iwamura et al.,[32] also inhibits the callus growth promoted by *N*-phenyl-*N′*-(4-pyridyl)urea as well as that promoted by 6-benzylaminopurine. This inhibition is also competitively reversed. Lineweaver-Burk plots of each response curve also clearly suggested that the response site is common. These studies strongly suggest that purine cytokinins and urea cytokinins work through interaction with a common cellular target, the cytokinin receptor.

This conclusion will be very useful in the identification of the entity bearing the interaction site, the receptor. Any candidate receptor should bind both to adenine cytokinins and to urea cytokinins. Several binding proteins have been reported so far where the binding capacity has been tested only for adenine cytokinins. We have tried to identify the cytokinin receptor by using highly labeled *N*-(2-chloro-4-pyridyl)-*N′*-phenylurea, and a protein whose apparent molecular weight is about 30 K has been found in etiolated seedlings of *Phaseolus*.[33] It binds to the urea cytokinin with a high association constant of $K_a = 4 \times 10^{10}\ M^{-1}$, and the binding is competitively inhibited by the cold pyridylphenylurea as well as by 6-benzyl-aminopurine ($K_i = 3 \times 10^9\ M^{-1}$), zeatin ($K_i = 3 \times 10^8\ M^{-1}$), and the 4-cyclobutylamino inhibitor ($K_i = 1 \times 10^9\ M^{-1}$). The purification and identification of the protein are in progress.

VII. BIOLOGICAL ACTIVITY OF UREA CYTOKININS

N-Phenyl-*N′*-(4-pyridyl)urea, *N*-(2-chloro-4-pyridyl)-*N′*-phenylurea and *N*-phenyl-*N′*-(1,2,3-thiadiazol-5-yl)urea exhibit essentially the same activities as 6-benzylaminopurine or zeatin, i.e., growth promotion of callus of tobacco, pea, ginseng, and datura and stimulation of shoot formation of tobacco, azalea, petunia, mulberry, and torenia.[20,34] The shoot-forming activity in tobacco callus by urea cytokinins is pronounced at concentrations of 100 times the optimum concentrations for callus growth, and the lower the concentration of auxin, the better the shoot formation.[16,26,35,36] This is also observed in case of purine cytokinins.[36,37] They inhibit senescence of excised leaves or of intact plants. The inhibition of flower shedding is complete in grape vines and melons.[38,39] Fruit sizes of grapes, kiwi fruit, and apples are dramatically increased.[40,41] The effect of *N*-(2-chloro-4-pyridyl)-*N′*-phenylurea on wheat yield has been studied by Karanov's group.[42-44] Retardation of the aging of the flag leaf could increase crop yield. The average increase over three seasons ranged from 10.6 to 11.8%. The preservation of the normal function of the flag leaves apparently maintained nutritional supply to the developing grains during the last weeks of the growth season. This hypothesis was substantiated by the higher seed weight of the tested plants.

VIII. SUMMARY

The cytokinin activities of several representative urea cytokinins, in terms of the optimum concentrations for tobacco callus growth, are shown in Figure 3. The similar structure–relationships between the adenine cytokinins and the urea cytokinins, the similar biological activity characteristic to cytokinins, and the occurrence of a common, highly specific binding protein indicate that the phenylurea compounds described here possess intrinsic cytokinin properties.

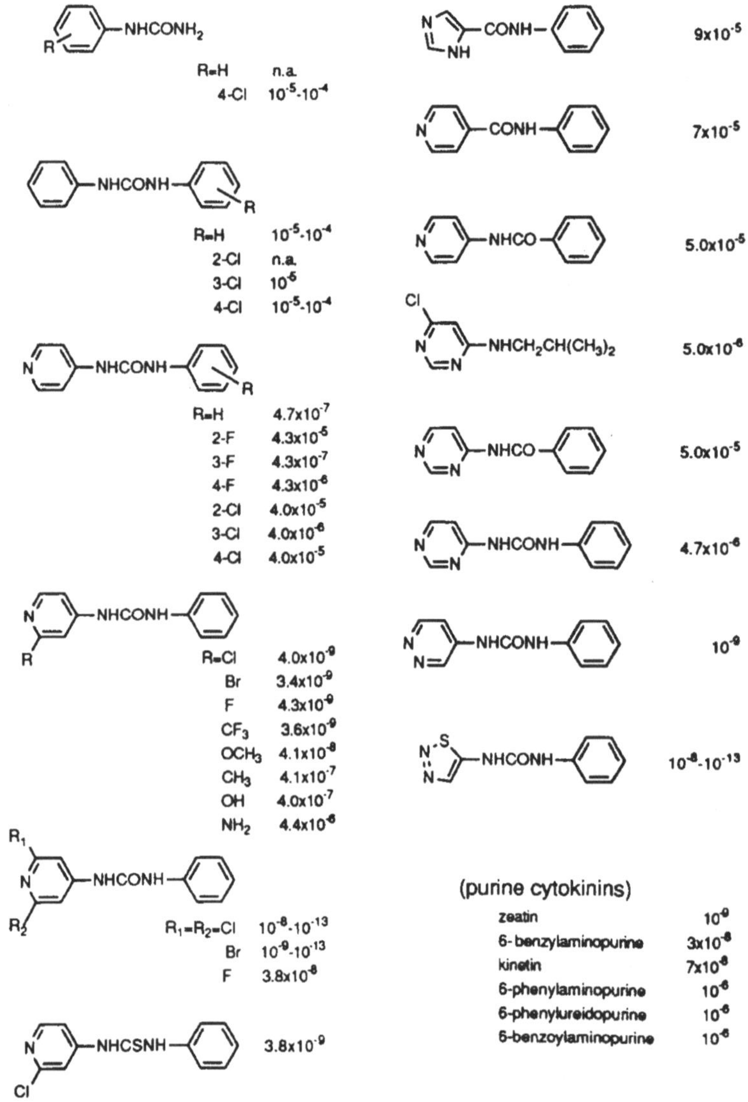

Figure 3-3 Biological activity of selected cytokinins evaluated by the tobacco callus bioassay (optimum concentration in *M*).

ACKNOWLEDGMENTS

The research on phenylurea cytokinins originated from the work of Dr. Torigoe supervised by Professor Okamoto and Professor Isogai.[45,46] Since the early study,[9] the most extensive work was performed by Dr. Takahashi. The very accurate callus bioassay was done by Ms. Yamada and Ms. Kawachi. The agricultural experiments were designed and organized by Mr. Oshima and Mr. Yamada. The author acknowledges and appreciates their contributions and assistance.

REFFERENCES

1. **Shantz, E. M. and Steward, F. C.,** The identification of compound A from coconut milk as 1,3-diphenylurea, *J. Am. Chem. Soc.*, 74 6133, 1955.
2. **Skoog, F., Strong, F. M., and Miller, C. O.,** Cytokinins, *Science*, 148, 532, 1965.
3. **Bruce, M. I. and Zwar, J. A.,** Cytokinin activity of some substituted ureas and thioureas, *Proc. Soc. London (Ser B.)*, 165, 245, 1966.

4. **Takahashi, S., Shudo, K., Okamoto, T., Yamada, K., and Isogai, Y.**, Cytokinin activity of N-phenyl-N′-(4-pyridyl)urea derivatives, *Phytochemistry*, 17, 1201, 1978.
5. **Arndt, F., Rusch, R., and Stilfried, H. V.**, SN49537, a new cotton defoliant, *Plant Physiol.*, 57, 99, 1976.
6. **Mok, M. C., Mok, D. W. S., Armstrong, D. J., Shudo, K., Isogai, Y., and Okamoto, T.**, Cytokinin activity of N-phenyl-N′-1,2,3-thiadiazol-5-ylurea (thidiazuron), *Phytochemistry*, 21, 1509, 1982.
7. **Bruce, M. I., Zwar, J. A., and Kefford, N. P.**, Chemical structure and plant kinin activity. The activity of urea and thiourea derivatives, *Life Sci.*, 4, 461, 1965.
8. **Iwamura, H., Fujita, T., Koyama, S., Koshimizu, K., and Kumazawa, Z.**, Quantitative structure–activity relationship of cytokinin-active adenine and urea derivatives, *Phytochemistry*, 19, 1309, 1980.
9. **Okamoto, T., Shudo, K., and Isogai, Y.**, Cytokinin activity of aromatic heterocyclic urea and amide derivatives, in *Plant Growth Substances 1973*, Tamura, S., Ed., Hirokawa Publishing, 1974, 447.
10. **Okamoto, T., Shudo, K., Takahashi, S., Kawachi, E., and Isogai, Y.**, 4-Pyridylureas are surprisingly potent cytokinins. The structure–activity relationship, *Chem. Pharm. Bull.*, 29, 3748, 1981.
11. **Shudo, K., Okamoto, T., Isogai, Y., and Ohshima, K.**, Structure-activity relationships and agricultural applications of synthetic cytokinins, in *Physiology and Biochemistry of Cytokinins in Plants*, Kamínek, M., Mok, D. W. S., and Zazímalová, E., Eds., SPB Academic Publishing, The Hague, 1992, 189.
12. **Henrie, R. N., Green, C. M., Yeager, W. H., and Ball, T. F.**, Activity optimization of pyridyl N-oxide urea cytokinin mimics, *J. Agric. Food Chem.*, 36, 626, 1988.
13. **Henrie, R. N., Chaguturu, R., Stenzel, J. I., Ball, T. F., and Yeager, W. H.**, QSAR of cytokinin-active N-pyridyl-N′-alkylureas and N-oxides, *J. Agric. Food Chem.*, 37, 513, 1989.
14. **Yamaguchi, K. and Shudo, K.**, Crystallographic studies on urea cytokinins, *J. Agric. Food Chem.*, 39, 793, 1991.
15. **Yamaguchi, K., Muramatsu,G., Kagechika, H., Azumaya, I., Ito, Y., Itai, A., and Shudo, K.**, Aromatic architecture. Use of the N-methylamide structure as a molecular splint, *J. Am. Chem. Soc.*, 113, 5474, 1991.
16. **Isogai, Y., Kawachi, E., Takahashi, S., Shudo, K., and Okamoto, T.**, Shoot formation effect of ureas, pyrimidines and related compounds on tobbaco pith disk cultured *in vitro*, *Sci. Pap. Coll. Gen. Educ.* (Univ. of Tokyo), 30, 151, 1980.
17. **Isogai,Y.**, Shoot formation effect of cytokinins on tobacco callus cultures *in vitro* and related problems, *Chem. Regul. Plants.*, 17, 27, 1982.
18. **Mok, D. W. S., Mok, M. C., and Armstrong, D.J.**, Cytokinin activity of N-phenyl-N′-1,2,3-thiadiazol-5-ylurea and its effects on cytokinin autonomy in callus cultures of *Phaseolus*, *Plant Physiol.*, 65, 24, 1980.
19. **Okamoto, T.**, unpublished results.
20. **Reynolds, J. F.**, Chemical regulation in tissue culture: an overview, *Hortscience*, 22, 1192, 1987.
21. **Saxena, P. K., Malik, K. A., and Gill, R.**, Induction by thidiazuron of somatic embryogenesis in intact seedlings of peanut, *Planta*, 87, 421, 1992.
22. **Suttle, J. C.**, Effect of the defoliant thidiazuron on ethylene evolution from mung bean hypocotyl segments, *Plant Physiol.*, 75, 902, 1984.
23. **Elstner, E. F., Keller, G., and Paradies, I.**, Contrasting effects of the cotton defoliant thidiazuron and aminoethyoxyvinylglycine, an inhibitor of ethylene formation in bean (*Phaseolus vulgaris*) leaves, *Ber. Deutsch. Bot. Ges.*, 96, 459, 1983.
24. **Skoog, F., Hamzi, H. O., Szweykowska, A. M., Leonard, N. J., Carraway, K. L., Fujii, T., Helgeson, J. P., and Loeppky, R. N.**, Cytokinins: structure–activity relationships, *Phytochemistry*, 6, 1169, 1967.
25. **Takahashi, S., Yatsunami, T., Shudo, K., Okamoto, T., Yamada, K., and Isogai, Y.**, Cytokinin activity of pyrimidine derivatives, *Chem. Pharm. Bull.*, 26, 2286, 1978.
26. **Isogai, Y., Yamada, K., Takahashi, S., Shudo, K., and Okamoto, T.**, Shoot formation effect of ureas, pyrimidines and related compounds on tobacco callus cultured *in vitro*, *Sci. Pap. Coll. Gen. Educ.* (Univ. of Tokyo), 28, 93, 1978.
27. **Letham, D. S., Parker, C. W., and Gordon, M. E.**, Regulators of cell division in plant tissues. XIV. The cytokinin activities and metabolism of 6-acylaminopurines, *Physiol. Plant.*, 27, 285, 1972.
28. **Dyson, W. H., Hall, R. H., Dutta, C. I., and Chheda, G. B.**, Cytokinin activity of ureidopurine derivatives related to the t-RNA component, N-(purin-6-yl-carbamoyl)-threonines, *Can. J. Biochem.*, 50, 237, 1972.

29. **Hong, C. I., Leonard, N. J., Schmitz, R. Y., and Skoog, F.,** Cytokinins: synthesis and biological activity of ureidopurines, *Phytochemistry*, 10, 1429, 1971.
30. **Kefford, N. P., Zwar, J. A., and Bruce, M. I.,** Antagonism of purine and urea cytokinin activities by derivatives of benzylurea, in *Biochemistry and Physiology of Plant Growth Substances*, Wightman, F. and Setterfield, G., Eds., Runge Press, Ottawa, 1968, 61.
31. **Kurosaki, F., Shudo, K., Okamoto, T., and Isogai, Y.,** Structural and biological links between urea- and purine-cytokinins, *Chem. Pharm. Bull.*, 29, 3751, 1981.
32. **Iwamura, H., Masuda, N., Koshimizu, K., and Matsubara, S.,** Cytokinin-agonistic and antagonistic activities of 4-substituted-2-methyl-pyrrolo[2,3-d]pyrimidines, 7-deaza analogs of cytokinin-active adenine derivatives, *Phytochemistry*, 18, 217, 1979.
33. **Nagata, R., Kawachi, E., Hashimoto, Y., and Shudo, K.,** Cytokinin-specific binding protein in etionated mung bean seedlings, *Biochem. Biophys. Res. Commun.*, 191, 543, 1993.
34. **Mok, M. C., Mok, D. W. S., Turner, J. E., and Mujer, C. V.,** Biological and biochemical effects of cytokinin active phenylurea derivatives in tissue culture systems, *Hortscience*, 22, 1194, 1987.
35. **Okamoto, T., Shudo, K., Takahashi, S., Yatsunami, T., Isogai, Y., and Yamada, K.,** Shoot formation of tobacco callus by various cytokinin-active ureas and pyrimidines, *Chem. Pharm. Bull.*, 26, 3250, 1978.
36. **Isogai, Y., Shudo, K., and Okamoto, T.,** Effect of N-phenyl-N′-(4-pyridyl)urea on shoot formation in tobacco pith disk and callus cultured *in vitro*, *Plant Cell Physiol.*, 17, 591, 1976.
37. **Isogai, Y.,** Effect of 6-(3-methyl-2-butenylamino)purine, zeatin, and 6-benzylamino-purine on tobacco pith callus and disk cultured *in vitro*, *Sci. Pap. Coll. Educ.* (Univ. of Tokyo), 27, 9, 1977.
38. **Ogata, R., Saito, T., and Ohshima, K.,** Effect of N-phenyl-N′-phenylurea (4PU) on fruit size: apples, Japanese pear, grape-vine, and kiwi fruit, *Acta Hort.*, 239, 395, 1989.
39. **Nickell, L. G.,** Effects of N-(2-chloro-4-pyridyl)-N′-phenylurea on grapes and other crops, in *Proc. 13th Ann. Meet. Plant Growth Soc. of America*, 1987, 236.
40. **Nasegawa, K. and Nakajima, Y.,** The effect of KT-30 and GA paste treatment on the growth of peach fruits, *Res. Rep. Kochi Univ. Agric. Sci.*, 37, 141, 1988.
41. **Iwahori, S., Tominaga, S., and Yamasaki, T.,** Stimulation of fruit growth of kiwi fruit, *Actinidia sinensis* Planch, by *N*-(2-chloro-4-pyridyl)-*N′*-phenylurea, a diphenylurea derivative cytokinin, *Sci. Hort.*, 35, 109, 1988.
42. **Alexieva, V. S. and Karanov, E. N.,** Effect of N-(2-chloro-4-pyridyl)-N′-phenylurea (4PU-30) on wheat yield, in *Physiology and Biochemistry of Cytokinins in Plants*, Kamínek, M., Mok, D. W. S., and Zazímalová, E., Eds., SPB Academic Publishing, The Hague, 1992, 249.
43. **Karanov, E., Iliev, L., Georgiev, G. T. S., Tsolova, M., Alexieva, V., and Puneva, I.,** Physiology and application of phenylurea cytokinins, in *Progress in Plant Growth Regulation*, Karssen, C. M., van Loon, L. C., and Vreugdenhil, D., Eds., Kluwer Academic Publishers, The Netherlands, 1992, 842.
44. **Karanov, E. N., Iliev, L. K., and Alexieva, V. S.,** Phenylurea cytokinins, chemistry, physiology and application, in *Physiology and Biochemistry of Cytokinins in Plants*, Kamínek, M., Mok, D. W. S., and Zazímalová, E., Eds., SPB Academic Publishings, The Hague, 1992, 199.
45. **Okamoto, T., Torigoe, Y., Sato, M., and Isogai, Y.,** Synthesis of s-triazolo(1,5-a)-pyrazine derivatives and cytokinin activity, *Chem. Pharm. Bull.*, 16, 1154, 1968.
46. **Torigoe, Y., Akiyama, M., Hirobe, M., Okamoto, T., and Isogai, Y.,** Cytokinin activity of azaindene, azanaphthalene, naphthalene and indole derivatives, *Phytochemistry*, 11, 1623, 1972.

Chapter 4

Cytokinin Antagonists: Synthesis and Biological Activity

Hajime Iwamura

TABLE OF CONTENTS

I. INTRODUCTION

The development of antagonists, as well as agonists, is of potential value in studies of the mechanism of action of biologically active compounds. First, the development of antagonists contributes to an understanding of the structural basis associated with intrinsic activity, since both the agonists and antagonists are structurally related entities and, in theory, should bind to common sites. Second, antagonists extend the study of agonist action when used to block the function of the agonist in biological systems. This approach is of particular importance in plant systems, since antagonists can block the action of a single hormone rendering the plant more responsive to the effects of agonists. Third, selective suppression of multiple effects of cytokinins with antagonists may isolate the individual functions of cytokinins. Such chemicals would also represent novel growth regulators.

In general, searches for cytokinin antagonists were conducted among purine analogs. The simplest N^6-substituted adenine cytokinin, 6-methylaminopurine,[1] antagonizes the action of the agonist only when applied at high concentrations.[2,3] Compounds known as metabolic inhibitors have also been the subject of interest. The inhibition of soybean callus growth by 8-azaguanine and 8-azaadenine[4] and the inhibition of mitosis of the phytoflagellate *Polytoma uvella* by 8-azaguanine[5] were lessened by kinetin. However, the counteraction was slight, and the basis for the competitive antagonism was not known. The synthetic growth inhibitors chloramphenicol,[6] (2-chloroethyl) trimethylammonium chloride (CCC),[7] and 1-(2-methylcyclohexyl)-3-phenylurea,[8] a herbicide known as siduron, have been examined as well, but evidence of antagonism was not found. A search among analogs of substituted ureas, some of which

0-8493-6252-0/94/$0.00+$.50

have cytokinin properties, was made by Kefford et al.[9] Their pioneering work, somewhat obscured at the time by controversies regarding whether urea derivatives were cytokinins (see Section III.A), was not pursued further. About a decade after the discovery of kinetin, a systematic search relying on chemical synthesis began, and the antagonistic nature of the resulting compounds has been examined extensively.

II. DEVELOPMENT OF ADENYLATE ANTICYTOKININS

A. PYRAZOLO[*4,3-d*]PYRIMIDINES

Many compounds with structures related to kinetin have been prepared since its discovery. Most were N^6-side-chain analogs. Few analogs had alterations in the purine ring, probably because modification of the ring system is difficult and, thus far, resulted in low activity. The heterocyclic analogs have a structure similar in size to that of the parent cytokinins. Cytokinin antagonists may be found among analogs, since they might satisfy requirements as antagonists; i.e., similarity in structure to cytokinins, which allows for competition for the same receptor, but renders them ineffective as cytokinins. Based on this principle, 7-substituted 3-methylpyrazolo[*4,3-d*]pyrimidines were developed by Hecht et al.[10,11] Activity was assayed with tobacco callus (unless otherwise noted, cytokinin-dependent *Nicotiana tabacum* L. Wisconsin No. 38 callus), and the strongest inhibitor was the *n*-amyl derivative (**6**) (Table 1). At suboptimal levels of cytokinin, inhibition by compound (**6**) was alleviated by further addition of cytokinins, allowing regrowth. On the other hand, at supraoptimal concentrations (inhibitory to growth), the addition of (**6**) relieved the inhibition and restored growth. These observations were taken as evidence supporting competing action with cytokinins.

The anticytokinin activity was affected by the structure of the N^7 side chain, with linear substituents conferring more activity than the corresponding branched or cyclized isomers. Compound (**2**), with an isopentenyl side chain, had cytokinin activity, but was not antagonistic. When the 3-methyl group of (**2**) was removed, the compound (**1**) became an even more effective cytokinin. These findings suggest that the receptor has critical structural requirements.

B. PYRROLO[*2,3-d*]PYRIMIDINES

1. 4-Substituted 7(β-D-ribofuranosyl) pyrrolo[*2,3-d*]pyrimidines

In the early 1970s, biochemical studies of cytokinins showed that a variety of plants convert cytokinins to their 7-glucosides.[12,13] These metabolites act as cytokinins and resist enzymatic degradation, and it was suggested that the 7-glucosides might be the storage form of active cytokinins.[12,13] If metabolic conversion is indispensable for cytokinin activity, then the 7-deaza analogs in which the 7-nitrogen atom of adenylate cytokinins is replaced by a carbon atom may be cytokinin inhibitors because the structure is similarenough to allow participation in the same kind of enzyme–substrate complex with cytokinins. The absence of the nitrogen atom at the 7-position might prevent glucosylation from occurring. With this hypothesis in mind, 4-substituted 7-(β-D-ribofuranosyl) pyrrolo[*2,3-d*]pyrimidines (**7–15**) (Figure 1) were synthesized.[14,15] These compounds had the expected growth-retarding effects, but soon afterwards the "active form" hypothesis was disproven, as described in Section IV.B. Thus, the compounds are thought to have acted as competitive antagonists at the site of cytokinin action, rather than as glycosylation inhibitors.

This set of compounds did not promote growth of tobacco callus, but at concentrations of 3 µ*M* or

Table 1. Structures of 7-substituted pyrazolo[4,3-*d*]pyrimidines and their effects on growth of tobacco callus.

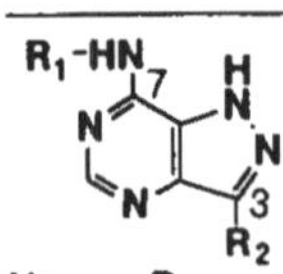

No	R1	R2	Cytokinin activity: min. conc.(µM) for max. growth	No	R1	R2	Inhibition[a] min. conc. (µM) for: detection	lethal dosage
1	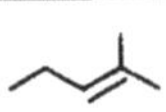	H	1.0	4	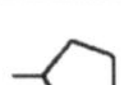	Me	0.2	0.73
2	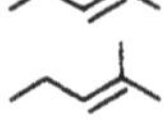	Me	6.6	5	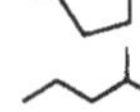	Me	0.03	0.5
3	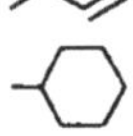	Me	-	6	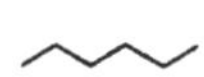	Me	0.03	0.2

[a] Tested at 0.003 µM i⁶Ade.

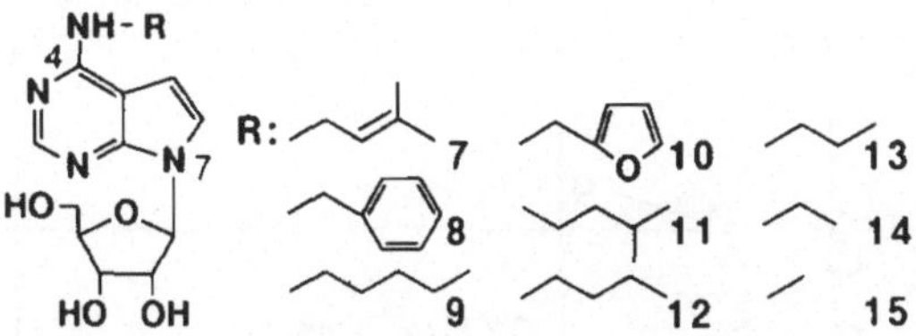

Figure 4-1 Structures of 4-substituted 7-(β-D-ribofuranosyl)pyrrolo[*2,3-d*]pyrimidines.

more, each inhibited the growth induced by kinetin. At concentrations lower than 3 μ*M*, the inhibition was counteracted by kinetin.[14] The compounds with isopentenyl, benzyl, and *n*-pentyl chains (**7–9**, Figure 1) were the strongest inhibitors, and the activity decreased in the following order: furfuryl (**10**), isopentyl (**11**), *n*-butyl (**12**), *n*-propyl (**13**), and ethyl (**14**) compounds. The methyl derivative (**15**) was not inhibitory. Thus, the order of strength of inhibition appeared to coincide with the order of cytokinin activity of the corresponding N^6-substituted adenines. This finding suggested that the compounds may share the site of action with active cytokinins.[15]

2. 4-Substituted Pyrrolo[*2,3-d*]-pyrimidines

Soon after the development of pyrrolo[*2,3-d*]-pyrimidine riboside anticytokinins,[14,15] Skoog et al. prepared 4-substituted 2-methylthiopyrrolo[*2,3-d*]-pyridines, 7-deaza analogs of 2-methylthioadenine cytokinins.[16] Among these derivatives, the cyclopentylamino (**19**) and cyclohexylamino analogs (**21**) strongly inhibited callus growth caused by N^6-isopentenyladenine (Table 2). The analogs with the 4-substituents have the same order of cytokinin activity when attached to the N^6 position of the purine base. The inhibition was reversed by the addition of cytokinin (Figure 2). These results were taken to demonstrate that the derivatives were specific anticytokinins.[16]

Compounds (**23**) and (**24**) (Figure 3), in which the 2-methylthio substituent of (**16**) and (**20**) is deleted, were found to have cytokinin activity by the same group.[17] Similarly, compounds (**25–29**), in which the 7-ribosyl moiety of the anticytokinins (**7–15**) was deleted, were prepared and also found to act as cytokinins.[18] Consequently, the conclusion was reached that the introduction of a substituent into the 2 or 7 position was probably important for the conversion of 4-substituted pyrrolo[*2,3-d*]-pyrimidines into anticytokinins. Hence, compounds with the simplest substituent, methyl at the 2 position, 4-substituted 2-methylpyrrolo[*2,3-d*]pyrimidines, were prepared.[19] As summarized in Table 3, the compounds with the highest activity were cyclobutyl and cyclopentyl derivatives (**36**) and (**35**). The compounds with a side chain of strong adenine cytokinins, isopentenyl (**30**) and benzyl (**31**), were not anticytokinins, but cytokinins. To study the antagonistic nature of the compounds, the method of Lineweaver and Burk,[20] which has been used previously to establish the specificity of antiauxins, was used.[21] The results of the treatment with the cyclobutyl derivative (**36**) are shown in Figure 4, with the reciprocal of the growth response plotted against the reciprocal of the concentration of kinetin added. The common intercept of linear lines obtained indicated the competitive inhibition of kinetin by compound (**36**).

4-*n*-Alkoxy analogs of 2-methylthiopyrrolo[*2,3-d*]pyrimidine anticytokinins have been prepared by Wilcox and Wain.[22] *n*-Pentyloxy (**37**) and *n*-hexyloxy (**38**) (Figure 5) compounds exhibited strong inhibition of the effects of kinetin at 1 μ*M*. 4-Cyclopentyloxy- and 4-cyclohexyloxy-2-methylpyrrolo[*2,3-d*]pyrimidines, (**39**) and (**40**), prepared to

Table 2. Structures of 4-substituted 2-methylthiopyrrolo[2,3-*d*]pyrimidines and their effects on growth of tobacco callus.

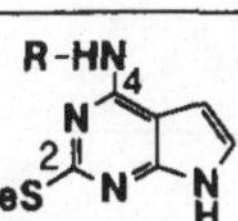

No.	R	Growth inhibition[a] min. conc. (μM) for: detection	complete inhibition	No.	R	Growth inhibition[a] min. conc. (μM) for: detection	complete inhibition
16	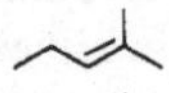	0.24	2.2	**20**		6.6	>20
17	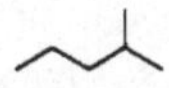	0.40	2.0	**21**	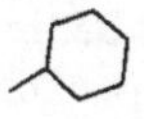	0.05	0.60
18	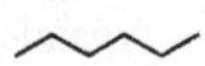	0.1	1	**22**	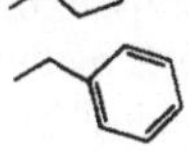	10	>20
19	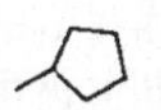	0.009	0.05				

[a]Tested at 0.003 μM i^6Ade.

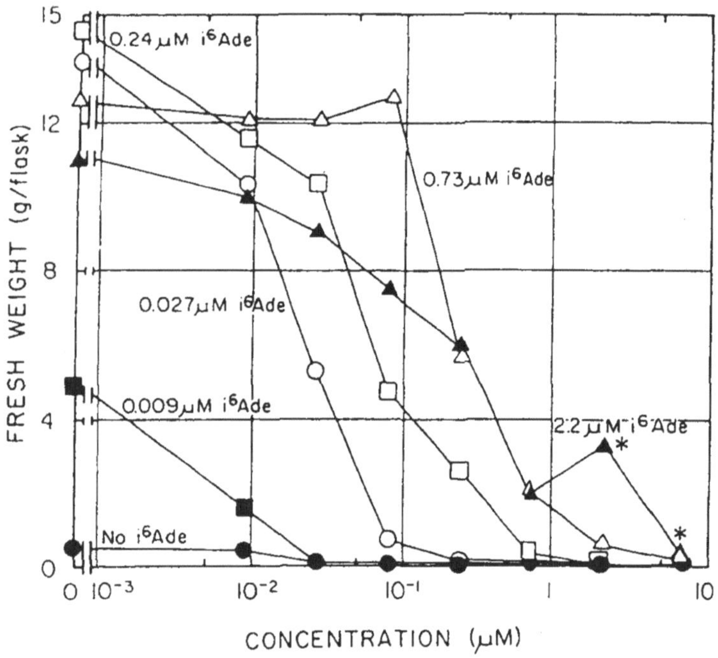

Figure 4-2 Effects of serial combination of compound **(17)** and N^6-iso-pentenyl-adenine on the growth of tobacco callus. (From Skoog, F., Schmitz, R. Y., Hecht, S. M., and Frye, R. B., *Proc. Natl. Acad. Sci. U.S.A.*, 72, 3508, 1975. With permission.)

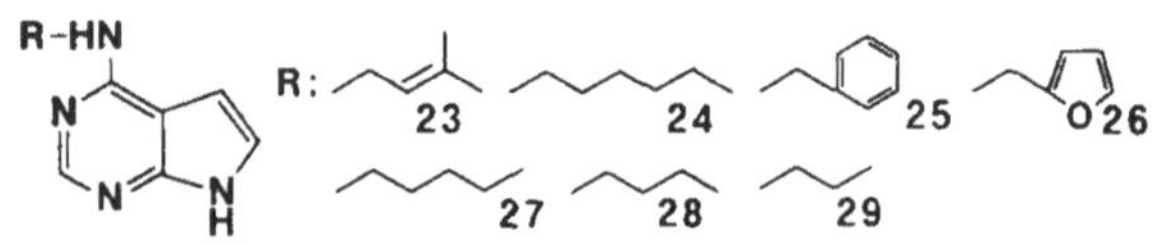

Figure 4-3 Structures of 4-substituted pyrrolo[*2,3-d*]pyrimidines.

Table 3. Structures of 4-substituted 2-methylpyrrolo[2,3-*d*]pyrimidines and their effects on growth of tobacco callus.

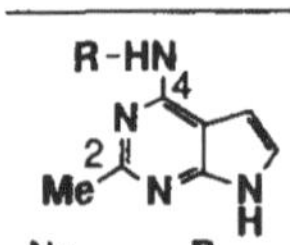

No.	R	Cytokinin activity, E_{50} (μM)	No.	R	Anticytokinin activity,[a] I_{50} (μM)
30	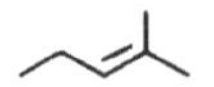	4.0	**33**	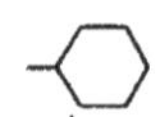	0.9
31	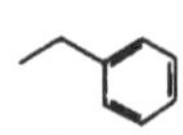	10.0	**34**	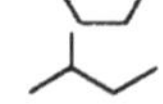	0.3
32	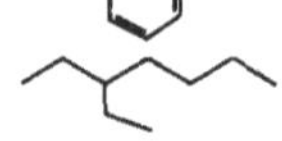	10.0	**35**	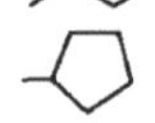	0.07
			36	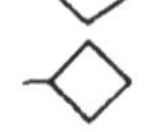	0.06

[a]With 0.05 μM kinetin

examine the role of the N^{4}H function, were found to be 10 to 100 times less active than the corresponding 4-imino compounds (**35** and **33**, respectively, in Table 3) when tested against 0.05 μ*M* kinetin.[23] The activity of 4-*n*-alkoxy-2-methylthio and cycloalkoxy-2-methyl analogs, against the same concentration of cytokinins, has not been determined; therefore, it is not known if the difference in potency resides with stereochemical or physicochemical features of the substituents.

C. PYRIDO[*2,3-d*]PYRIMIDINES

The system of fused rings, one five-membered and one six-membered, of the 4-substituted 2-methylthiopyrrolo[*2,3-d*]pyrimidine anticytokinins listed in Table 2 was expanded to a system of fused rings that were both six-membered to give 4-substituted 2-methylthiopyrido[*2,3-d*]pyrimidines (Table 4).[24] These compounds retarded tobacco callus growth, and their specific anticytokinin property was again confirmed kinetically by

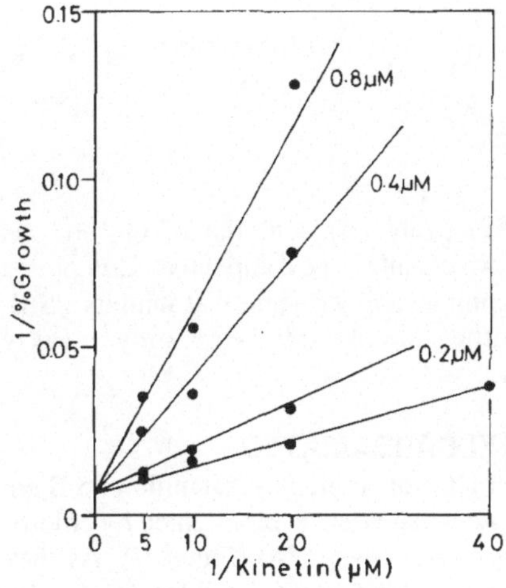

Figure 4-4 Reciprocal of the growth rate of tobacco callus plotted as a function of the reciprocal of the concentration of kinetin alone (bottom line) and in the presence of compound **(35)**. (From Iwamura, H., Masuda, N., Koshimizu, K., and Matsubara, S., *J. Med. Chem.*, 26, 838–844, 1983. With permission.)

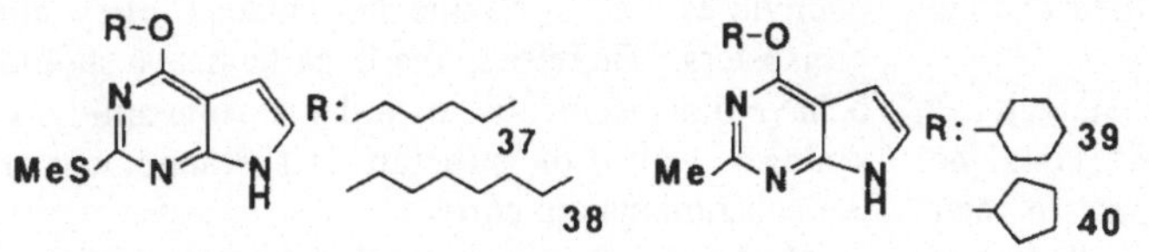

Figure 4-5 Structures of 4-alkoxy-2-methylthio- and 4-alkoxy-2-methylpyrrolo[*2,3-d*]pyrimidines.

Lineweaver-Burk plots. Analogs with two six-membered rings of the 2-methylpyrrolo[*2,3-d*]pyrimidine anticyto-kinins listed in Table 3 were not made, due to difficulties in their synthesis.

III. DEVELOPMENT OF NONADENYLATE ANTICYTOKININS

A. *N*-BENZYL-*N'*-PHENYLUREAS

In the 1960s, at the start of the study of anticytokinins, Kefford et al.[9] sought "antagonists" among urea derivatives that had been tested in a number of bioassays (e.g., chlorophyll retention, radish leaf expansion, and tobacco callus, etc).[25,26] They chose analogs that were either inactive or inhibitory. In addition, compounds that reduced cytokinin action, but bear little structural similarity, were also identified as antagonists. The *N*-benzyl-*N'*-(3,4- and 4-substituted phenyl)ureas **(47a–b)** and the related compound **(48)** (Figure 6) were found to retard the retention of chlorophyll mediated by cytokinin-active ureas and adenines, applied at 10^{-5} *M*.

Direct experimental evidence of antagonism was not examined, except for the reduction of the effects of agonists. This was due, in part, to the prevailing thought at the time that such urea derivatives were not cytokinins, but were molecules either acting on separate pathways from those of purine cytokinins or indirectly affecting the purine cytokinin level in tissues.[9]

B. *s*-TRIAZINES

Except for *N*-benzyl-*N'*-phenylureas,[9] all of the anticytokinins described above have structures similar to those of adenine cytokinins. Compounds with a nonadenine structure, but with clearly anticytokinin properties, have been sought based

Table 4. Structures of 4-substituted 2-methylthiopyrido[2,3-*d*]pyrimidines and their effects on tobacco callus growth.

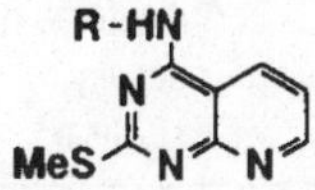

No.	R	Anticytokinin activity, I_{50} (μM)[a]	No.	R	Anticytokinin activity, I_{50} (μM)[a]
41	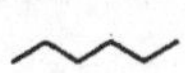	0.22	**44**	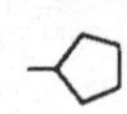	0.46
42	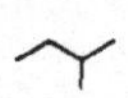	0.49	**45**	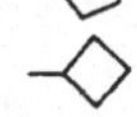	0.65
43		0.22	**46**	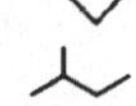	0.81

[a]With 0.05 μM kinetin.

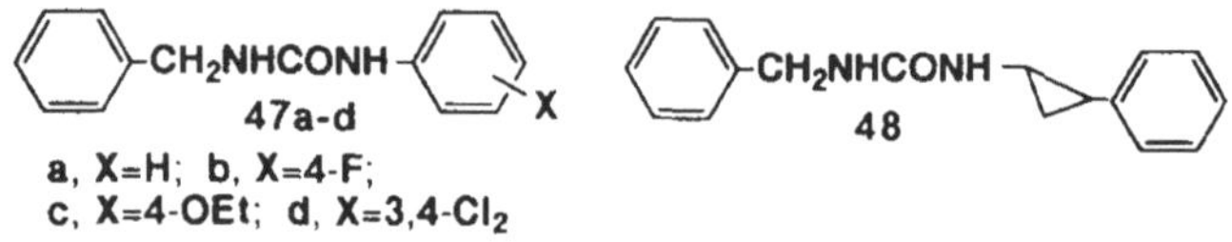

Figure 4-6 Structures of *N*-substituted *N'*-benzylureas.

on considerations of the bioisosterism between N^6-substituted adenine and phenylurea cytokinins and also between phenylurea and *s*-triazine herbicides.[27] Appropriately substituted *s*-triazines such as shimazine **(49)** and atrazine **(50)** (Figure 7) inhibit the photosystem (PS) II electron flow in chloroplasts and are herbicidal.[28] Phenylurea herbicides such as monuron **(51)** and diuron **(52)** also act at the site where *s*-triazine herbicides bind.[29] This means that appropriately substituted *s*-triazines and phenylureas are bioisosters in their binding to the site of action in chloroplasts, despite their distinct structures.

As phenylureas have cytokinin activity,[26] *s*-triazines, if appropriately substituted, could be bioisosteric to phenylureas, with respect to their binding to the cytokinin receptor as well. It was anticipated that the *s*-triazines may bind antagonistically rather than agonistically. As expected, these compounds inhibited kinetin-induced growth of tobacco callus.[27] Most potent were *n*-butyl **(55)**, cyclobutyl **(57)**, cyclohexyl **(58)**, and cyclopropylmethyl **(59)** (Table 5), and the inhibition was confirmed kinetically to be competitive. Compound **(50)** is a commercial herbicide that inhibits PS II, but the other compounds were only weakly herbicidal.

C. *N*-ARYLCARBAMATES

Another class of herbicides that inhibit PS II are the *N*-phenylcarbamates, as represented by chloropropham **(53)** and Swep **(54)** (Figure 7). As they share a common binding site in chloroplasts with phenylurea and *s*-triazine herbicides,[29] they are bioisosters. Therefore, these carbamates should behave bioisosterically toward *s*-triazine anticytokinins as well if the structure and arrangement of the substituents are correct.

N-phenylcarbamates retarded the growth of cultured tobacco cells caused by kinetin, and the mechanism was verified using the method of Lineweaver and Burk.[30] Both *O*-alkyl and *O*-phenyl derivatives were active (Table 6), but only moderately so, the I_{50} being 1 to 5 μM. The *O*-alkyl structure resembled

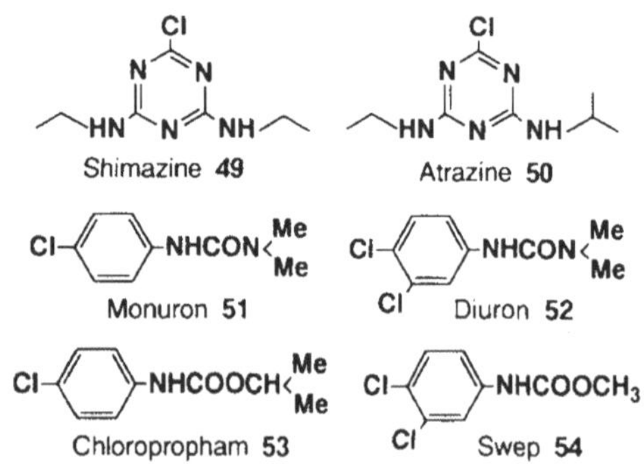

Figure 4-7 Structures of photosystem II inhibitory herbicides.

Table 5. Structures of 2-substituted 4-chloro-6-ethylamino-*s*-triazines and their effects on tobacco callus growth.

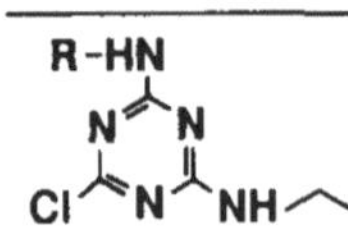

No.	R	Anticytokinin activity, I_{50} (μM)[a]	No.	R	Anticytokinin activity, I_{50} (μM)[a]
50	 (atrazine)	1.10	57	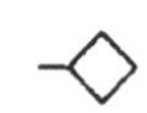	0.29
55	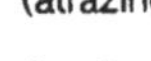	0.44	58	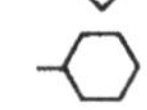	0.47
56	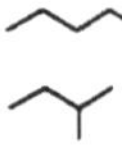	1.05	59	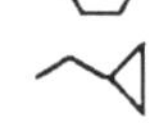	0.47

[a]With 0.05 μM kinetin.

that of the herbicidal carbamates and caused some PS II inhibition. The *O*-phenyl carbamates have not been reported to be herbicidal, and **(64)** and **(65)** had only weak herbicidal activity.

A search of structures with higher anticytokinin activity was made, keeping in mind the much higher cytokinin activity of *N*-(4-pyridyl)-*N'*-phenylureas compared with that of *N,N'*-diphenylureas.[31,32] *N*-(4-pyridyl)carbamates were prepared. The potency of the most active compounds had an I_{50} value of 0.3 to 0.5 μ*M*,[30] suggesting that ureas and carbamates have a similar mode of interaction with cytokinin receptor(s), with ureas being agonistic and the carbamates being antagonistic.

IV. STRUCTURE–ACTIVITY RELATIONSHIPS

A. CHEMICAL MODE OF ACTION OF CYTOKININ AGONISTS AND ANTAGONISTS

Comprehensive studies of the structure–activity relationships of N^6-adenines and ureas have been reported by Skoog et al.[1] and Bruce and Zwar,[26] respectively. Based on those studies, the effects of substituents of both classes of cytokinins were analyzed quantitatively using physicochemical parameters and regression analysis.[33] The results were then combined with those for anticytokinin compounds **(23)**, **(27)**, and **(34)** to give the mode-of-action map, as shown in Figure 8.[30] It is proposed that there is a stereochemical framework, indicated by fringed lines and hatched ovals, that regulates binding. The importance of H-bond interactions of the central NH proton with a basic group (:B) of the receptor is shown. Hydrophobic interaction at the region is indicated by the shaded circle. The fitting into this model of adenine and urea cytokinins (**A**) and of anticytokinins (**B, C**), based on the quantitative analyses, showed different modes of action of agonists and antagonists with the shared receptor; anticytokinins are close to the upper-right boundary of the framework, while cytokinins are close to the upper-left.

B. IN RELATION TO THE "ACTIVE FORM" OF CYTOKININS

In the early 1970s, scientists in Australia discovered that exogenous cytokinins were converted to stable 7-glucosyl derivatives in tissues of various plants.[12,13,35-37] They suggested that the derivatives might be an active (protected) form of cytokinins (see Section II.B.1). This interesting hypothesis was, however, soon disproved by the synthesis and assay of heterocyclic analogs of cytokinins.

Hecht et al. reported on the cytokinin activity of pyrazolo[*3,4-d*]pyrimidine **(68)** (Figure 9) and pyrrolo[*2,3-d*]pyrimidines **(23)** and **(24)** (Figure 3), all of which have a carbon, rather than a nitrogen, at the position corresponding to 7 in purines, and thus would be metabolically stable.[17,38] They suggested that 7-glucosylation is not a prerequisite for the expression of cytokinin activity.[38] Similarly, **(23)** and a series of 4-substituted pyrrolo[*2,3-d*]pyrimidines,[25-29] which were deribosylated analogs of anticytokinins **(7–15)** (Figure 1), were found to have cytokinin activity.[18]

Other candidates for the active form that has long been postulated are 9-ribosyl adenines.[39] It is clear that the pyrrolo[*2,3-d*]pyrimidines **(23–29)** mentioned above have cytokinin activity without ribosylation, although they have a nitrogen atom at the position corresponding to 9 in purines. If ribosylation had occurred, the resulting compounds should have behaved like anticytokinins **(7–15)**

Table 6. Structures of *N*-phenyl- and *N*-pyridylcarbamates and their effects on tobacco callus growth.

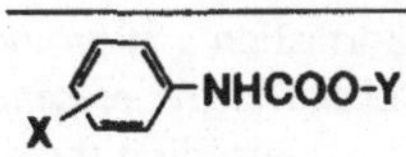			Anticytokinin activity,				Anticytokinin activity,
No.	X	Y	I_{50} (μM)[a]	No.	X	Y	I_{50} (μM)[a]
60	3,4-Cl_2		14.8	**64**	3-Cl	4-Cl	1.0
61	3-Cl		2.9	**65**	3-F	4-F	1.8
62	3,4,5-Cl_3		1.8				
63	2,5-Cl_2		2.1				
				66		4-F	0.45
				67		4-Cl	0.28

[a]With 0.05 μM kinetin.

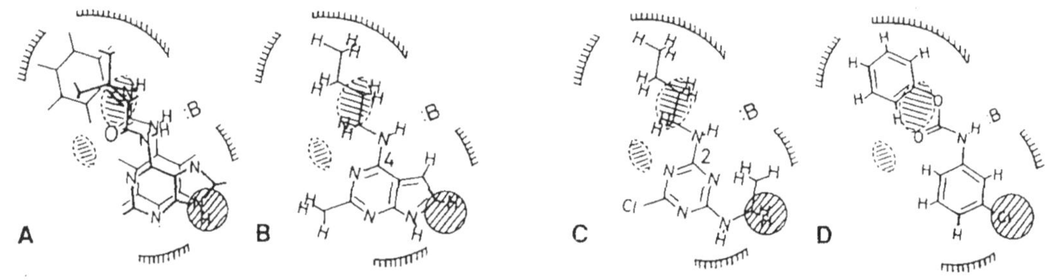

Figure 4-8 Fitting of (**A**) *N*⁶-substituted adenine cytokinins (bold lines), *N,N'*-diphenylurea cytokinins (light lines); (**B**) pyrrolo[*2,3-d*]pyrimidine anticytokinin; (**C**) *s*-triazine anticytokinin; and (**D**) carbamate anticytokinin into the mode-of-action map of cytokinins. The solid lines with a fringe represent the steric interaction site or receptor walls on the page plane, and the hatched ovals are those above or below it. The shaded circle shows the hydrophobic region, and :B shows the electron interaction site or the basic group of the receptor. In (**A**), the ends of the bars represent hydrogen atoms, and the double bonds of the aromatic rings are omitted. (From Shimizu, R., Iwamura, H., and Fujita, T., *Z. Naturforsch.*, 45c, 89, 1990. With permission.)

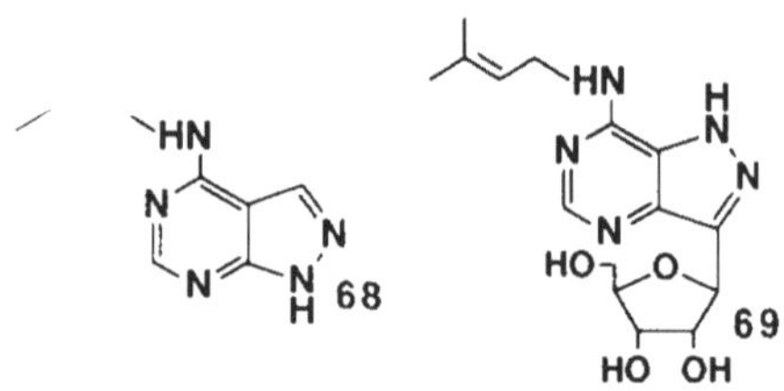

Figure 4-9 Structures of 4-isopentenylamino-pyrazolo-[*3,4-d*]pyrimidine and 7-isopentenylamino-3-β-D-ribofuranosylpyrazolo[*4,3-d*]pyrimidine.

(Figure 1).[15] Cytokinin activity was found in compounds without a nitrogen atom, but with a carbon atom, at the position corresponding to 9 of adenines; these compounds are 7-substituted pyrazolo[*4,3-d*]pyrimidines (**1** in Table 1 and its analogs).[17] Therefore, ribosylation at the 7 carbon is not required for cytokinin activity. However, the possibility that 9-ribosides or ribotides act *in situ* as active species is still open. 7-Isopentenylamino-3-*β*-D-ribofuranosylpyrazolo[*4,3-d*]pyrimidine **(69)** (Figure 9), connected to its ribosyl moiety via a C–C bond, making it metabolically stable, has been prepared and found to have cytokinin activity.[17] Thus, endogenously 9-ribosylated adenine species may also fit into the cytokinin receptor.

V. BIOLOGICAL EFFECTS

A. EFFECTS ON CYTOKININ-INDEPENDENT TOBACCO CALLUS

The activity of anticytokinin (**5**) of Table 1 was tested in cytokinin-autonomous tobacco callus[11] that produces zeatin and i⁶Ade from [¹⁴C]adenine.[40] Compound (**5**) inhibited the growth of the callus, and the inhibition was reversed when cytokinin was added, suggesting that cytokinin-autonomous tissue uses endogenous cytokinins.[11] The results indicate that anticytokinins can be used to block the action of endogenous cytokinins in autonomous cultures.

B. EFFECTS ON BUDDING OF TOBACCO CALLUS

In the course of assaying anticytokinins, using cultured tobacco cells, 4-cyclopentylamino- and 4-cyclohexylamino-2-methylthiopyrrolo[*2,3-d*] pyrimidines (**19** and **21** in Table 2) were found to cause bud formation in the presence of 1 to 10 μ*M* cytokinins and 11.4 μ*M* of IAA, the hormonal regime usually promoting callus growth.[16] Selected results of these experiments are presented in Figure 2. 4-(2-Ethyl-*n*-hexylamino)-2-methylpyrrolo[*2,3-d*]-pyrimidine (**32**) (Table 3), which has cytokinin activity, also induced budding at 40 μ*M*, but the more active cytokinin (**30**) did not.[19] These observations suggest that bud formation and promotion of callus growth may be related to the presence of different cytokinin receptors controlling the two events.[19,41]

C. EFFECTS ON *AMARANTHUS* BETACYANIN SYNTHESIS

Cytokinins induced the formation of betacyanins in the cotyledons and hypocotyls of derooted seedlings of *Amaranthus caudatus* when the cuttings were incubated with tyrosine in the dark.[42,43] The effects of the 2-methylpyrrolo[*2,3-d*]pyrimidine compounds listed in Table 3, on betacyanin formation, were examined in the presence and the absence of added cytokinin.[19] Compounds (**30–32**), with cytokinin activity, and (**35**), with anticytokinin activity, induced the synthesis of pigments, but the anticytokinins (**33**) and (**34**) were inactive. Thus, compounds that act as anticytokinins in one system, tobacco callus, did not necessarily do so in other systems, indicating species and developmental differences in the receptors.[19]

D. EFFECTS ON GERMINATION

The seeds of *Lactuca sativa* cvs. Grand Rapid and New York[44,45] normally germinate only in the light. However, when incubated with cytokinins, germination can occur in the dark. Some of the 2-methylpyrrolo[*2,3-d*]pyrimidines listed in Table 3 were tested for this kind of cytokinin activity.[19] Compounds (**30–32**), which had cytokinin activity in the tobacco assay, as well as compound (**33**), which had anticytokinin activity in the same assay, had agonist activity here. Thus, these results again indicate a complexity of the receptor molecules, depending on the bioassays used. Pyrazolo[*4,3-d*]pyrimidine (**5**) (Table 1), an anticytokinin, had no effect on germination, but reduced root initiation and development in wheat and radish seedlings and *Coleus* cuttings. Similar effects were observed with the cytokinins N^6-isopentenyladenine and benzyladenine.[11]

E. EFFECTS ON ADVENTITIOUS BUD FORMATION

Treating *Torenia* stem segments with cytokinins resulted in active meristematic divisions of the epidermis, and these meristematic zones (MZs) developed into adventitious buds, with no further application of growth regulators.[46] To elucidate the roles of cytokinins in such *in vitro* organogenesis, Tanimoto and Harada examined the effects of 7-substituted 2-methylpyrrolo[*2,3-d*]pyrimidine anticytokinins.[47] They found that 4-cyclobutyl (**36**) (Table 3) at 1.5 to 50 μ*M* caused 50 to 80% inhibition of MZ formation in the presence of 0.5 μ*M* benzyladenine (Table 7), but did not inhibit adventitious bud formation. The *Torenia* segments also formed the MZs when treated with cAMP and traumatic acid, but the number of MZs was not reduced by the application of (**36**).[48] It appears that mechanisms stimulating MZ formation or adventitious bud initiation by cAMP and traumatic acid were separate from that of cytokinins.

Calcium intensified several cytokinin-regulated responses, such as the retardation of senescence and ethylene production.[49-51] The calcium ionophore A23187 was found to induce meristem division in the absence of cytokinins.[52] This induction was not inhibited by anticytokinins, but the elimination of calcium from the medium caused 50% inhibition of MZ initiation stimulated by either A23187 or cytokinins. Treatment with cAMP, traumatic acid, or A23187 did not result in formation of adventitious buds. These observations let Tanimoto and Harada to propose three steps in the process of organogenesis. The first step involves the elicitation of cell division, by cutting the plants, and is followed by the formation of MZs. These two steps are stimulated by cAMP, traumatic acid, and A23187, but in a different way than with cytokinins, with calcium ion being the final mediator for the effects of all four compounds. The third step is the development of adventitious buds, a process promoted by cytokinins only.

F. RETARDATION OF PLANT GROWTH

The development of anticytokinins led to their practical application as plant growth regulators or herbicides. 7-Isopentylamino-3-methylpyrazolo[*4,3-d*]pyrimidine (**5**) was reported to cause severe wilting and death of tomato seedlings, but older tomato and tobacco exhibited no damage upon spraying.[11]

Table 7. Effects of anticytokinins on meristematic zone (MZ) formation in *Torenia* stem segments[a]

		Anticytokinin 36	
Compd.	Conc. (μM)	Conc. (μM)	MZ formation (%)[b]
BA	0.5	5	34
cAMP	3.0	10	119
Traumatic acid	1.0	10	134
A23187	10.0	5	99

[a]Formulated from the data in the literature (46 - 48).

[b]Calculated on the basis of MZ numbers obtained with the application of respective chemicals without **36**.

7-(β-D-Ribofuranosyl)pyrrolo[*2,3-d*]pyrimidines in Figure 1 were tested for herbicidal activity.[53] When applied to germinating rice, most of the compounds caused inhibition at 100 μ*M*, while isopentenyl (**7**), benzyl (**8**), and *n*-butyl (**12**) were effective at 25 μ*M*. In pre-emergence tests, 90 to 100% inhibition was observed for rice and sawa millet (*Panicum crusgalli* L.), and somewhat lower inhibition was observed for cucumber. These three compounds were less inhibitory in the submerged tests and had little effect when applied as foliar treatment on mature plants. Their deribosylated analogs, the 4-substituted pyrrolo[*2,3-d*]pyrimidines (**23**), (**25**), and (**28**) (Figure 3), which are not anticytokinins, but exhibited cytokinin activity in the tobacco callus assay (Section II.B.2),[18] had as much herbicidal activity as the parent nucleosides in pre-emergence tests.[54] The series of 2-methylthio- and 2-methylpyrrolo[*2,3-d*]pyrimidines, which are potent anticytokinins that are not nucleosidal (Tables 2 and 3), were only weakly active in these three kinds of tests.

Some of the anticytokinins were herbicidal in some tests, but their effects were not greater than those of existing commercial herbicides. Nevertheless, compounds of greater potency could be obtained with further modifications and with appropriate evaluation instead of relying only on *in vitro* assays.

G. FLOWER-INDUCING ACTIVITY

Abe and Kameya observed flower induction in seedlings of *Asparagus officinalis,* by a triazine herbicide (atrazine **50**) and urea herbicide (diuron **52**).[55] Asparagus is a perennial plant, and normally, flowering does not occur until 2 to 3 years after seeding. However, soaking seeds for 12 days in solutions of these chemicals resulted in flower formation at the top of the plants 10 to 15 days after planting. The flowering rate was 20 to 40%. However, the treated plants died soon after, due to the herbicidal effects of the compounds. Atrazine and diuron are PS II herbicides and are structural analogs of the *s*-triazine anticytokinins and urea cytokinins, respectively. Based on the bioisosteric nature of the PS II herbicides with cytokinins and anticytokinins (mentioned in Sections III.B and C), a series of *s*-triazine and carbamate anticytokinins were tested (Table 8).[56,57] 4-Cyclohexylamino-*s*-triazine (**58**), a strong anticytokinin in the tobacco assay, was found to induce the highest rate of flowering (45%) at 200 μ*M* without killing the plants.[56] Carbamate compounds produced even higher flowering rates, but they were not necessarily strong anticytokinins (Table 8).[57]

The flowering of asparagus occurs only once at the top of the plants after treatment, suggesting that the effect is triggered at early stages of germination.[58] To examine the correlation between anticytokinin and the flower-inducing property, the effect of benzyladenine was tested. However, the evaluation of antagonism, if any, was hampered by the severe toxicity of the cytokinins, which caused malformation of plants and early seedling death. Subsequently, changes in the levels of various endogenous compounds were studied. An increase in flowering correlated with a decrease in the concentrations of chlorogenic acid and other secondary metabolites. Therefore, antagonism between benzyladenine and carbamate was measured in terms of chlorogenic acid levels. Benzyladenine at 10^{-5} to 10^{-3} μ*M* reversed the decrease of chlorogenic acid caused by carbamate (**60**, Figure 10). Antagonism at higher concentrations of benzyl-adenine could not be determined, due to the toxicity of the cytokinins (Figures 10A and B).[58] Thus, the interaction between flower-inducing compounds and cytokinins was identified, but whether the effect is antagonistic is not known.

VI. CONCLUDING REMARKS

The development of antagonists is a challenge to chemists, but the products are of interest to physiologists and biochemists. In some cases testing of such compounds provided unique insights into problems that were difficult to solve by physiology or biochemistry. Research regarding the active form of cytokinins (Section IV.B) and cytokinin receptors (Sections V.B, C, and D) are some examples.

Research concerning the biological properties of antagonists (Section V) was performed primarily in the chemists' laboratories where the compounds were developed. Few biologists have utilized antagonists in their specialized fields of research. This is due, in part, to the unavailability of the compouds commercially. In addition, the effects of antagonists have been defined primarily in tobacco callus bioassays, and their anticytokinin property in other plant systems is uncertain. Neverthless, anticytokinins should be uniquely suited for studies involving cytokinin receptors or the separate identification of multiple functions of cytokinins. I hope that this chapter will stimulate the interest of biologists in employing anticytokinins for basic and applied research. The utilization of such chemicals will, in turn, encourage chemists to develop additional compounds.

Table 8. Flower-inducing activity of *s*-triazine and carbamate compounds in *Asparagus* seedlings.[a]

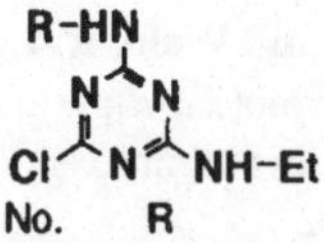 No.	R	Flowering, %[b]	I_{50} (μM)[c]	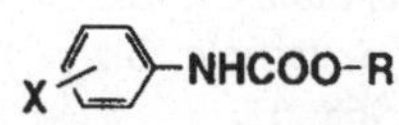 No.	R	X	Flowering, % μM 200	400	I_{50} (μM)[c]
50	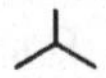	20	1.10	**60**		3,4-Cl_2	84	67	14.8
57	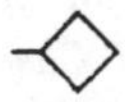	32	0.29	**70**	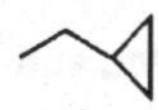	3,4-Cl_2	69	92	11.5
58	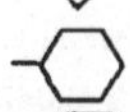	45	0.47	**71**	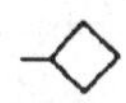	3,4-Cl_2	62	90	14.8
68	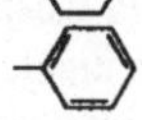	20	1.91	**72**		4-Cl	95	64	20.0

[a]Seeds were soaked in a solution of a compound for 12 days at 25 °C, and seeds that germinated were planted in soil or Vermiculite and grown for 13 days more at 25 °C.

[b]At 200 μM.

[c]Tobacco callus growth induced by 0.05 μM kinetin.

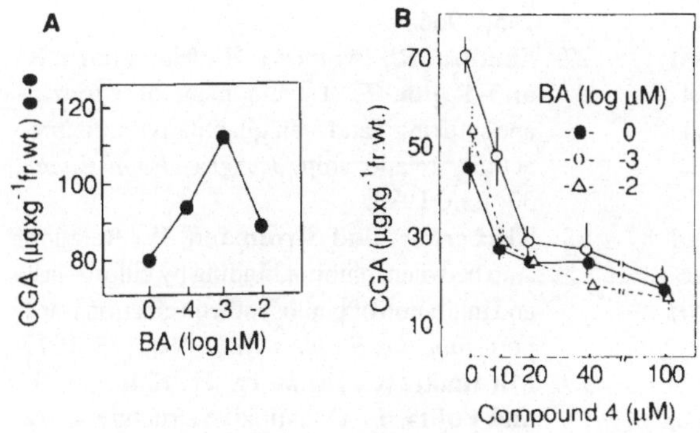

Figure 4-10 Effects of N^6-benzyl-adenine on the chlorogenic acid (CGA) content of *Asparagus officinalis* seedlings. (A) Effects of N^6-benzyladenine alone, and (B) effects of N^6-benzyladenine and compound **(60)**. The assay was done 8 days after seeding and chemical treatment. (From Yanosaka, K., Iwamura, H., Shinozaki, M., and Yoshida, K., *Plant Cell Physiol.*, 32, 447, 1991. With permission.)

REFERENCES

1. **Skoog, F., Hamzi, H. Q., Szweykowska, A. M., Leonard, N. J., Carraway, K. L., Fujii, T., Helgeson, J. P., and Loeppky, R. N.,** Cytokinins: structure/activity relationships, *Phytochemistry*, 6, 1169, 1967.
2. **Klambt, D., Thies, G., and Skoog, F.,** Isolation of cytokinins from *Corynebacterium fascians*, *Proc. Natl. Acad. Sci. U.S.A.*, 56, 52, 1966.
3. **Helgeson, J. P. and Leonard, N. J.,** Cytokinins: identification of compounds isolated from *Corynebacterium fascians*, *Proc. Natl. Acad. Sci. U.S.A.*, 56, 60, 1966.
4. **Blaydes, D. F.,** Interaction of kinetin and various inhibitors in the growth of soybean tissue, *Physiol. Plant.*, 19, 748, 1966.
5. **Moewus, F.,** Competitive antagonism between kinetin and 8-azaguanine in *Polytoma uvella*, *Science*, 130, 921, 1959.
6. **Laloraya, M. M. and Banerji, D.,** Effect of chloramphenicol on the expansion of isolated cotyledons of *Cucurbita pepo* in light and in presence of kinetin, *Experientia*, 21, 516, 1965.
7. **Teltscherova, L. and Havlickova, H.,** Reversal of the inhibitory effect of (2-chloroethyl)trimethylammonium chloride on the flowering of *Chenopodium rubrum* L. by kinetin, *Experientia*, 23, 861, 1967.
8. **Splittstoesser, W. E. and Hopen, H. J.,** Root growth inhibition by siduron and its relief by kinetin, *Physiol. Plant.*, 23, 964, 1970.

9. **Kefford, N. P., Zwar, J. A., and Bruce, M. I.,** Antagonism of purine and urea cytokinin activities by derivatives of benzylurea, in *Biochemistry and Physiology of Plant Growth Substances*, Whitman, F. and Setterfield, G., Eds., Runge Press, Ottawa, 1968, 61.
10. **Hecht, S. M., Bock, R. M., Schmitz, R. Y., Skoog, F., and Leonard, N. J.,** Cytokinins: development of a potent antagonist, *Proc. Natl. Acad. Sci.U.S.A.*, 68, 2608, 1971.
11. **Skoog, F., Schmitz, R. Y., Bock, R. M., and Hecht, S. M.,** Cytokinin antagonists: synthesis and physiological effects of 7-substituted 3-methylpyrazolo[*4,3-d*]pyrimidines, *Phytochemistry*, 12, 25, 1973.
12. **Fox, J. E., Cornette, J., Deleuze, G., Dyson, W., Giersak, C., Niu, P., Zapata, J., and McChesney, J.,** The formation, isolation, and biological activity of a cytokinin 7-glucoside, *Plant Physiol.*, 52, 627, 1973.
13. **Parker, C. W. and Letham, D. S.,** Regulators of cell division in plant tissues. XVI. Metabolism of zeatin by radish cotyledons and hypocotyls, *Planta*, 114, 199, 1973.
14. **Iwamura, H., Ito, T., Kumazawa, Z., and Ogawa, Y.,** Anticytokinin activity of 4-furfurylamino-7-(β-D-ribofuranosyl) pyrrolo[*2,3-d*]pyrimidine, *Biochem. Biophys. Res. Commun.*, 57, 412, 1974.
15. **Iwamura, H., Ito, T., Kumazawa, Z., and Ogawa, Y.,** Synthesis and anticytokinin activity of 4-substituted-7-(β-D-ribofuranosyl) pyrrolo[*2,3-d*]pyrimidines, *Phytochemistry*, 14, 2317, 1975.
16. **Skoog, F., Schmitz, R. Y., Hecht, S. M., and Frye, R. B.,** Anticytokinin activity of substituted pyrrolo[*2,3-d*]pyrimidines, *Proc. Natl. Acad. Sci. U.S.A.*, 72, 3508, 1975.
17. **Hecht, S. M., Frye, R. B., Werner, D., Hawrelak, S. D., Skoog, F., and Schmitz, R. Y.,** On the "activation" of cytokinins, *J. Biol. Chem.*, 250, 7343, 1975.
18. **Iwamura, H., Kurimoto, M., Matsubara, S., Ogawa, Y., Koshimizu, K., and Kumazawa, Z.,** Effect of 4-substituted pyrrolo[*2,3-d*]pyrimidines on the tobacco callus growth, *Agric. Biol. Chem.*, 40, 1885, 1976.
19. **Iwamura, H., Masuda, N., Koshimizu, K., and Matsubara, S.,** Cytokinin-agonistic and antagonistic activities of 4-substituted-2-methylpyrrlo[*2,3-d*]pyrimidines, 7-deaza analogs of cytokinin-active adenine derivatives, *Phytochemistry*, 18, 217, 1979.
20. **Lineweaver, H. and Burk, D.,** The determination of enzyme dissociation constants, *J. Am. Chem. Soc.*, 56, 658, 1934.
21. **McRae, D. H. and Bonner, J.,** Chemical structure and antiauxin activity, *Physiol Plant.*, 6, 485, 1953.
22. **Wilcox, E. J. and Wain, R. L.,** 4-Alkoxy-2-methylthiopyrrolo[*2,3-d*]pyrimidines. A new series of anti-cytokinins, *Ann. Appl. Biol.*, 100, 399, 1982.
23. **Iwamura, H., Masuda, N., Koshimizu, K., and Matsubara, S.,** Quantitative aspects of the receptor binding of cytokinin agonists and antagonists, *J. Med. Chem.*, 26, 838, 1983.
24. **Iwamura, H., Murakami, S., Koga, J., Matsubara, S., and Koshimizu, K.,** Quantitative analysis of anticytokinin activity of 4-substituted-2-methylthiopyrido[*2,3-d*]pyrimidines, *Phytochemistry*, 18, 1265, 1979.
25. **Bruce, M. I., Zwar, J. A., and Kefford, N. P.,** Chemical structure and plant kinin activity: the activity of urea and thiourea derivatives, *Life Sci.*, 4, 461, 1965.
26. **Bruce, M. I. and Zwar, J. A.,** Cytokinin activity of some substituted ureas and thioureas, *Proc. R. Soc. Lond. [Biol.]*, 165, 245, 1966.
27. **Shimizu, R., Iwamura, H., Matsubara, S., and Fujita, T.,** Development of *s*-triazine anticytokinins and their quantitative structure–activity relationship, *J. Agric. Food Chem.*, 37, 236, 1989.
28. **Tischer, W. and Strotmann, H.,** Relationship between inhibitor binding by chloroplasts and inhibition of photosynthetic electron transport, *Biochim. Biophys. Acta*, 460, 113, 1977.
29. **Mitsutake, K., Iwamura, H., Shimizu, R., and Fujita, T.,** Quantitative structure–activity relationship of photosystem II inhibitors in chloroplasts and its link to herbicidal action, *J. Agric. Food Chem.*, 34, 725, 1986.
30. **Shimizu, R., Iwamura, H., and Fujita, T.,** Anticytokinin activity of *N*-phenyl- and *N*-pyridylcarbamates, *Z. Naturforsch.*, 45c, 89, 1990.
31. **Takahashi, S., Shudo, K., Okamoto, T., Yamada, K., and Isogai, Y.,** Cytokinin activity of *N*-phenyl-*N'*-(4-pyridyl)urea derivatives, *Phytochemistry*, 17, 1201, 1978.
32. **Okamoto, T., Shudo, K., Matsubara, S., and Kawachi, E.,** 4-Pyridylureas are surprisingly potent cytokinins. The structure–activity relationship, *Chem. Pharm. Bull.*, 29, 3748, 1981.
33. **Iwamura, H., Fujita, T., Koyama, S., Koshimizu, K., and Kumazawa, Z.,** Quantitative structure–activity relationship of cytokinin-active adenine and urea derivatives, *Phytochemistry*, 19, 1309, 1980.

34. **Iwamura, H., Murakami, S., Koshimizu, K., and Matsubara, S.,** Quantitative structure–activity relationships in cytokinin agonistic and antagonistic pyrido[*2,3-d*]pyrimidine derivatives: insights into receptor topology, *J. Med. Chem.*, 28, 577, 1985.
35. **Parker, C. W., Letham, D. S., Cowley, D. E., and MacLeod, J. K.,** Raphanatin, an unusual purine derivative and a metabolite of zeatin, *Biochem. Biophys. Res. Commun.*, 49, 460, 1972.
36. **Parker, C. W. and Letham, D. S.,** Regulators of cell division in plant tissues. XVIII. Metabolism of zeatin in *Zea mays* seedlings, *Planta*, 115, 337, 1974.
37. **Deleuze, G. G., McChesney, J. D., and Fox, J. E.,** Identification of a stable cytokinin metabolite, *Biochem. Biophys. Res. Commun.*, 48, 1426, 1972.
38. **Hecht, S. M., Bock, R. M., Schmitz, R. Y., Skoog, F., Leonard, N. J., and Occolowitz, J. L.,** Question of the ribosyl moiety in the promotion of callus growth by exogenously added cytokinins, *Biochemistry*, 10, 4224, 1971.
39. **Letham, D. S. and Palni, L. M. S.,** The biosynthesis and metabolism of cytokinins, *Ann. Rev. Plant Physiol.*, 34, 163, 1983.
40. **Einset, J. W. and Skoog, F.,** Biosynthesis of cytokinins in cytokinin-autotrophic tobacco callus, *Proc. Natl. Acad. Sci. U.S.A.*, 70, 658, 1973.
41. **Hecht, S. M.,** Cytokinin antagonists: regulation of the growth of plant and animal cells, in *Bioorganic Chemistry*, Vol. IV, van Tamelen, E. E., Ed., Academic Press, New York, 1978, chap. 6.
42. **Biddington, N. L. and Thomas, T. H.,** A modified *Amaranthus betacyanin* bioassay for the rapid determination of cytokinins in plant extracts, *Planta*, 111, 183, 1973.
43. **Conrad, K.,** Ein sensibilisierter Amaranthus-cytokinintest (AT 74), *Biochem. Physiol. Pflanzen*, 165, 531, 1974.
44. **Miller, C. O.,** Similarity of some kinetin and red light effects, *Plant Physiol.*, 31, 318, 1956.
45. **Matsubara, S., Koshimizu, K., and Nakahira, R.,** Cytokinin activities of dihydrozeatin in several bioassays, *Sci. Rep. Kyoto Pref. Univ.*, 19, 19, 1968.
46. **Tanimoto, S. and Harada, H.,** Studies on the initial process of adventitious bud differentiation in *Torenia* stem segments cultured *in vitro*. I. Effects of cytokinin, *Biochem. Physiol. Pflanzen*, 177, 222, 1982.
47. **Tanimoto, S. and Harada, H.,** Effects of cytokinin and anticytokinin on the initial stage of adventitious bud differentiation in the epidermis of *Torenia* stem segments, *Plant Cell Physiol.*, 23, 1371, 1982.
48. **Tanimoto, S. and Harada, H.,** Stimulation of adventitious bud initiation by cyclic AMP and traumatic acid in *Torenia* stem segments, *Biol. Plant.*, 26, 337, 1984.
49. **Poovaiah, B. W. and Leopold, A. C.,** Deferral of leaf senescence with calcium, *Plant Physiol.*, 52, 236, 1973.
50. **Lau, O. I. and Yang, S. F.,** Interaction of kinetin and calcium in relation to their effect on stimulation of ethylene production, *Plant Physiol.*, 55, 738, 1975.
51. **Green, J.,** The effect of potassium and calcium on cotyledon expansion and ethylene production induced by cytokinins, *Physiol. Plant.*, 57, 57, 1983.
52. **Tanimoto, S. and Harada, H.,** Involvement of calcium in adventitious bud initiation in *Torenia* stem segments, *Plant Cell Physiol.*, 27, 1, 1986.
53. **Iwamura, H., Kumazawa, Z., Nagato, S., and Okuda, S.,** Plant growth retarding activity of the 4-substituted-7-(β-D-ribofuranosyl) pyrrolo[*2,3-d*]pyrimidine anticytokinins, *Agric. Biol. Chem.*, 40, 1653, 1976.
54. **Iwamura, H., Ito, T., Masuda, N., Mizuno, A., Koshimizu, K., Kumazawa, Z., Nagato, S., and Okuda, S.,** Effects of substituted pyrrolo[*2,3-d*]pyrimidine derivatives on plant growth, *J. Pest. Sci.*, 6, 9, 1981.
55. **Abe, T. and Kameya, T.,** Promotion of flower formation by atrazine and diuron in seedlings of *Asparagus*, *Planta*, 169, 289, 1986.
56. **Abe, T., Shimizu, R., Iwamura, H., and Kameya, T.,** Flower induction by trazine analogs in seedlings of *Asparagus officinalis*, *Physiol. Plant.*, 70, 228, 1987.
57. **Yanosaka, K., Iwamura, H., and Fujita, T.,** Flower induction in seedlings of *Asparagus officinalis* L. by *N*-phenylcarbamates, *Z. Naturforsch.*, 44c, 226, 1989.
58. **Yanosaka, K., Iwamura, H., Shinozaki, M., and Yoshida, K.,** Correlation between biosynthesis of chlorogenic acid and flowering in *Asparagus* seedlings induced by carbamate compounds, *Plant Cell Physiol.*, 32, 447, 1991.

Chapter 5

Cytokinins as Phytohormones — Sites of Biosynthesis, Translocation, and Function of Translocated Cytokinin

David S. Letham

TABLE OF CONTENTS

ABBREVIATIONS

Abbreviations not defined in the text include:
BAP — 6-benzylaminopurine
DZ — dihydrozeatin
DZ7G — dihydrozeatin 7-glucoside
DZR — dihydrozeatin riboside
DZNT — dihydrozeatin nucleotides
iP — N^6-(Δ^2-isopentenyl)adenine
iPA — N^6-(Δ^2-isopentenyl)adenosine
OGZ — *O*-glucosylzeatin
OGZR — *O*-glucosylzeatin riboside
OGDZ — *O*-glucosyldihydrozeatin
OGDZR — *O*-glucosyldihydrozeatin ribosid
Z — *trans*-zeatin
ZR — zeatin riboside
Z7G — zeatin 7-glucoside
ZNT — zeatin nucleotides

I. INTRODUCTION

The widely held view that plant hormones move from a site of production to one of action to control plant development has been questioned in recent years. Thus, Trewavas considers that biosynthesis of plant hormones is not localized,[1,2] and hence, movement to sites of action cannot occur, and that plant development is not controlled by change in hormone concentration, but by alteration in sensitivity to these regulators. Hence, it is now important and timely to reevaluate our current knowledge of cytokinin biosynthetic sites and translocation in higher plants. The validity of the views of Trewavas can then be reassessed with respect to cytokinins.

In this chapter sites of cytokinin biosynthesis, cytokinin translocation (especially from identified

0-8493-6252-0/94/$0.00+$.50

sites of synthesis), and the function of this translocated cytokinin are assessed. These considerations form a basis for elucidation of the coordinating role of cytokinin in plant development.

II. SITES OF CYTOKININ BIOSYNTHESIS

Cytokinins have been found in roots, stems, leaves, flowers, fruits, and seeds, and are probably present in all living cells of intact higher plants. However, establishing unequivocally which particular organs or tissues are sites of cytokinin biosynthesis has proved to be a perplexing problem. High levels of cytokinin in a particular tissue do not necessarily indicate a site of biosynthesis, and the tissue may only constitute a sink or storage site. For example, the reported high levels of cytokinin in root tips could originate (wholly or partially) in the shoot and accumulate at the root tip, the basal terminus of the phloem system. Since cytokinins and other hormones circulate in the vascular system, the end points of this system, the meristems, might well exhibit considerable accumulation of growth substances, just as they do of certain nutrients.

Because of the ability of cytokinins to induce cell division, tissues with high meristematic activity have been proposed as likely sites of cytokinin biosynthesis. Thus, based on an early study with isolated flax roots, Goldacre proposed that cytokinins are formed in lateral root primordia and induce cell division in neighboring pericycle tissues.[3] It is an open question as to whether the high cytokinin level in meristematic tissues is the cause of continuing cell division or the product of such division. While cells require cytokinin for the induction of cell division, dividing cells probably produce cytokinins that could activate adjacent cells to divide. This may be the mechanism of self-perpetuation of meristems.

Recent studies with immobilized cells show that exogenous cytokinins induce secretion of endogenous cytokinins, and this occurs when cell growth is suppressed as a consequence of immobilization.[4] The concentration of exogenous cytokinin (BAP) supplied and the levels of Z-type cytokinins secreted were positively correlated. Exogenous cytokinins also elevate endogenous cytokinin levels in tissue culture systems[5,6] and in excised cotyledons.[7] Based on studies of cytokinin habituation, Meins and Hansen concluded that cytokinins induce their own synthesis.[8] Exogenous applications of any cytokinin to cytokinin-requiring tissues could evoke an autoinductive mechanism that induces synthesis of cytokinins.

Because cytokinins occur in tRNA and may be released during tRNA degradation, Sheldrake proposed that free cytokinins are formed in the dying, autolyzing cells that occur in some tissues often regarded as sites of cytokinin biosynthesis.[9] However, there is little evidence to support this view. If cytokinins are released during cell death, the cytokinin level in leaves should rise during senescence. However, cytokinin levels usually decline during leaf senescence (see review by Van Staden et al.[10]). Throughout the senescence of detached and attached tobacco leaves, the levels of cytokinin bases and ribosides declined markedly.[11,12] During the senescence of these leaves, degradative metabolism of cytokinins is not promoted;[13] hence, release of cytokinin from tRNA would be expected to be reflected in an elevation of cytokinin level.

In concluding this general introduction to sites of biosynthesis, it should be pointed out that this is a difficult area with two major problems: the extremely low levels of the endogenous cytokinins in plant tissues, and the central role of the most likely precursors (adenine, its nucleoside, or its nucleotides) in cellular metabolism. Thus, a major portion of the supplied radioactive precursor is incorporated into the common purine metabolites, and only a very small fraction is incorporated into cytokinins. This poses considerable technical problems regarding the isolation and purification of minute quantities of putatively labeled cytokinins from the complex of other highly labeled purines and their numerous degradation products. When radiolabeled adenine is supplied to plant tissues, labeled compounds are produced with chromatographic properties that very closely mimic those of cytokinins. (For further discussion of these problems, the reader is referred to Palni et al.[14] and Horgan.[15]) There are numerous reports of incorporation of radioactive adenine into cytokinins in tissues of higher plants. Most of these reports must be questioned because of insufficient purification of the fractions containing cytokinins.

A. THE ROOT AS A BIOSYNTHETIC ORGAN

An unambiguous demonstration of conversion of isotopically labeled precursors into cytokinins in excised roots would provide convincing evidence for cytokinin biosynthesis in this organ. In several studies incorporation of [^{14}C]adenine into cytokinins in roots has been claimed as follows:

1. Excised tomato roots cultured with adenine for 12 days[16]
2. Decapitated root systems of 5-week-old tomato plants[17]

3. Pea roots cultured aseptically on a medium containing [^{14}C]adenine for 21 days[18]

In studies 1 and 2, purification of ^{14}C-labeled compounds is obviously insufficient to establish that labeled cytokinins were formed. The same comment appears to apply to study 3. The data presented for these studies suggest that several percent of the extracted radioactivity was attributed to cytokinin bases plus ribosides plus nucleotides. Comparison with biosynthesis studies in plant tumor tissue indicates that this is an unrealistic level of incorporation. The tumor tissues contain greatly elevated cytokinin levels, relative to those in typical normal tissues, and are characterized by abnormally high rates of cytokinin biosynthesis.[19] Yet when radiolabeled adenine is supplied to such tissue, about 1% of the label is found to be incorporated into the total cytokinins after rigorous purification.[20,21] In normal tissue the incorporation would be expected to be much less than in the tumor tissue. Finally, regarding study 2, the claim that [^{14}C]iP and [^{14}C]iP nucleotide were formed from [^{14}C]adenine in a cell-free system prepared from tomato roots is invalid. The claim rests on the observation that a nucleotide-like fraction was converted by acid hydrolysis (1 N HCl at 100°C) to a ^{14}C-labeled product that cochro-matographed with iP riboside during TLC, while another ^{14}C product considered to be iP was not affected. However, this merely proves that the putative iP and iP nucleotide were not these compounds because acid hydrolysis of both cytokinins would modify the iP moiety and would yield 6-(3-hydroxy-3-methylbutylamino)-purine.[22]

While attempts to demonstrate precursor incorporation into cytokinins in excised roots have not established that this organ is a site of cytokinin biosynthesis, there is strong evidence of other types that indicate that this is so. First, during aseptic culture of rice root tips[23] and maize and tomato roots,[24] cytokinins were released into the culture medium. The excised maize roots accumulated cytokinin, and their endogenous level greatly exceeded that of equivalent intact roots. After growth of excised tomato root tips for 7 days, the cytokinin level in the medium was eight times that in the root tissue.[25] When these root tips were subcultured eight times, the cytokinin level in the medium after subculture was approximately constant.

Second, marked cytokinin production follows adventitious root formation. Root development on petioles of excised bean leaves results in marked increases in the cytokinin level in the laminae. In a classical study, Engelbrecht[26] could not detect cytokinin activity in bean leaf blades prior to root formation on the petioles. However, immediately after roots developed, the cytokinin activity in the petioles increased very markedly, and soon afterward, activity was detected in the blades. Cytokinin activity accumulated there until it exceeded that in the roots plus the petioles. Sap from the rooted petioles contained considerable cytokinin activity. When the blades began to accumulate cytokinin activity, they exhibited the characteristics of cytokinin-treated leaves. In further studies with excised bean leaves, cytokinin production by newly formed adventitious roots appeared to commence before the roots ruptured the epidermis of the petiole. After the roots penetrated the epidermis, cytokinin synthesis increased very markedly.[27] Formation of root primordia on hypocotyls of tomato seedling cuttings was followed by a marked (about 100-fold) increase in the hypocotyl cytokinin level, determined by ELISA after high-pressure liquid chromatography (HPLC).[28] However, cytokinin levels in the other parts of the seedlings did not change. Immunocytochemical studies showed that the primordia and the apices of the emerging roots were localizations with high levels of Z- and iP-type cytokinins.[29]

Third, formation of lateral root primordia in response to decapitation of pea roots resulted in a pronounced increase in root cytokinin level.[30] Thus, both lateral roots and adventitious roots on petioles and stems initiate cytokinin production at a very early stage of development, and this synthesis is very marked by the time the roots emerge from the parent organ.

Fourth, cytokinins have been detected in xylem exudates of the roots of plants of diverse species, and environmental and soil factors unfavorable for root growth decrease cytokinin levels in xylem sap and, presumably, cytokinin production by roots. The stresses of heat, low temperature, flooding, water deficiency, and salinity all affect root growth and all appear to result in reduction of cytokinin production by roots (see reviews by Skene[31] and Letham[32]). During the flooding of sunflower plants, Burrows and Carr observed a correlation between the metabolic activity of root apices and cytokinin content of xylem sap.[33] A similar correlation was found in a recent study of plants with hypoxic roots.[34]

Fifth, studies of the maintenance of cytokinin production by root systems of detopped plants, and studies of this production relative to the cytokinin content of the roots, indicate that plant roots are a cytokinin-producing organ. Thus, the level of cytokinins in xylem exudate of sunflowers did not decrease appreciably for 4 days after excision of the plant shoot.[35] Preparation of root exudate from decapitated sunflower plants by suction, instead of by normal bleeding, increased the amount of exudate obtained fivefold, while the total amount of exuded cytokinin activity was increased to an even greater extent.[36] The amount of cytokinin in xylem exudate of the detopped root system of cocklebur plus the amount in the roots after exudation was determined by Henson and Wareing.[37] This summed amount appeared to be in excess of that in the intact root at the time of top excision, indicating production of cytokinin by the root during the period (24 h) of sap collection.

Sixth, in seedlings and cuttings, exogenous cytokinins will often, completely or partially, substitute for roots in inducing or maintaining growth and/or physiological responses in the shoot.[38] In recent studies using soybean stem explants, it has been shown that exogenous cytokinin will replace roots in maintaining foliar function[39] (see also Section IV.C).

B. THE ROOT TIP AS A BIOSYNTHETIC SITE

When considered collectively, the above six lines of evidence provide convincing support for the view that roots are a site of cytokinin biosynthesis. One question that remains is, where within the root does this biosynthesis occur? The view is widely held that cytokinin biosynthesis in a root is localized at the tip. The evidence most frequently cited in support of this is the two reports, based on bioassays, that root tips contain very high cytokinin levels relative to those of more proximal regions of the root. Thus, Short and Torrey reported that the terminal (tip) 1-mm segment of pea roots contained 44 times more free cytokinin, on a fresh weight basis, than the next 1- to 5-mm root segment, while no cytokinin was detected in extracts from other segments.[40a] However, this conclusion is based on purification procedures that are no longer acceptable as quantification methods, namely, precipitation from a crude fraction with $AgNO_3$, and Dowex 50 chromatography. Repetition of this method with pea root extract, using [^{3}H]DZ as a recovery marker, gave very low and variable recoveries;[40b] the $AgNO_3$ step precipitated purine bases only while ribosides remained in solution. In the second relevant study,[41a] cytokinins were detected in extracts of sunflower root apices (0–1 mm from tip), the level being 100 times that in segments 1 to 3 mm from the tip. However, these extracts were insufficiently purified to give reliable bioassay data.

Determination of the cytokinin level, by reliable methods, along primary roots is important in relation to the site of cytokinin biosynthesis and the role of cytokinin in control of the site of lateral root initiation. Hence, in unpublished studies,[41b] the gradients of cytokinins along 6-cm pea roots were determined. The levels of ZR and DZR, the major cytokinins, were maximal at the root tip, then declined to a minimum about 2 cm from the tip, and eventually rose again over the region of the lateral root primordia. Although the very large differences between tips and more proximal segments, reported by the early workers, were not observed, the results indicated a basipetally decreasing gradient of cytokinin over the upper region of the root. Such a gradient was also observed in a recent immunocytochemical study of cytokinins in the tip region of tomato roots.[29]

Studies of cytokinin levels in roots following root cap removal also provide evidence that cytokinin biosynthesis is concentrated at the tip. After removal of the root cap from *Zea mays* roots, the total cytokinin level in the terminal millimeter of the decapped root decreased by a factor of 6.4 over a 24-h period, while the content of the principal cytokinins (probably zeatin and zeatin riboside) was reduced to 4.9% of the initial value.[42] In further studies excision of the root cap plus quiescent center was found to result in a marked reduction in the cytokinin level of the tissue of the proximal meristem.[43] If cytokinins extracted from the root apex came from the more proximal regions of the root, cytokinin levels in the terminal millimeter, following decapping, would be expected to remain constant. The decline in cytokinin level from the root tip toward the root base, discussed earlier, coupled with the marked effect of cap removal on cytokinin production, leads to the reasonable conclusion that cytokinin synthesis within the root is predominantly localized at the tip when lateral root primordia are absent.

The exact location of cytokinin biosynthesis in the root tip remains unclear. Torrey suggested that the quiescent center is the main site of this synthesis and, as a result, has supraoptimal levels of cytokinin that induce the quiescent state of the cells.[44] However, immunocytochemical studies indicate that the quiescent center does not contain cytokinin levels that exceed those of adjacent tissue.[29,45] In studies with tomato root tips, using both ZR and iPA antibodies, no distinction could be made between the cap initials, the quiescent center, and the meristem; however, the root cap was less immunoreactive.[29] Hence, it is improbable that cytokinin biosynthesis is confined to the quiescent center. Studies based on culture of excised maize root tip tissues[46] and studies involving the determination of cytokinin levels in maize root tips following excision of the quiescent center or reduction in the size of this center[42,43] indicate that cytokinin biosynthesis in the root apex is the result of an interaction between the quiescent center and the adjoining proximal meristem. Thus, if the proximal meristem and the quiescent center are cultured in the association existing in the root tip, but with root cap removed, a new root regenerates in the absence of exogenous cytokinin. However, when cultured separately, cytokinin must be added to the medium for root regeneration to occur.[46] After excision of the quiescent center or reduction in its size induced by root cap removal, cytokinin levels decline markedly in the remaining root tip, but regeneration of the quiescent center to its normal size is associated with the restoration of normal cytokinin levels.[42,43]

Hence, in summary, plant roots produce cytokinins; the synthesis is localized in the tip and is the result of an interaction between the quiescent center and the proximal meristem.

C. BIOSYNTHESIS IN THE SHOOT

Excision of roots results in a reduction of cytokinin levels in the shoot[47-49] and markedly suppresses several aspects of shoot development[38,48,49] that can often be normalized by exogenous cytokinin applied to the derooted plants (see Section IV.C). Hence, shoot tissues, apart from developing seed, appear to be largely dependent on root-produced cytokinin, for the development and retardation of leaf senescence. However, the vegetative organs of the shoot probably have a limited capacity for cytokinin biosynthesis. The reported incorporation of [^{14}C]adenine into cytokinins of apparently rootless tobacco plantlets derived from callus tissue[50] is frequently cited as evidence for such biosynthesis. However, it was noted that root primordia may have been formed during the 17 to 28 days of labeling,[50] and such primordia are now considered to be sites of cytokinin production. In these studies about 4% of the extracted radioactivity was attributed to the four cytokinins iP, iPA, Z, and ZR. If other cytokinins, including glucosides, were considered, the incorporation could be considerably higher. Because of the high level of incorporation (see Section II.A), the ^{14}C assigned to cytokinins should be examined by HPLC and after derivatization, before definite conclusions are made regarding biosynthesis. Evidence for cytokinin biosynthesis in specific regions and organs of the shoot is summarized below.

1. Cambium, Shoot Apex, and Lateral Buds

Because these regions contain actively dividing tissues, they are likely sites of cytokinin biosynthesis. However, the evidence to support this view is far from conclusive. Purified extracts of the cambial zone of the poplar tree exhibit strong cytokinin activity,[51] and this and other cambial tissues can be cultured *in vitro* without exogenous cytokinin,[52] indicating that they synthesize ample amounts of this hormone. When the bark of *Salix babylonica*, with the cambium exposed, was cultured *in vitro*, rapid cell division occurred, and the level of cytokinins increased considerably.[53]

Immunocytochemical studies show that the meristematic cells of the apex and young leaf primordia of tomato contain high cytokinin levels relative to those in adjacent tissues. Below the apex there was a basipetally decreasing gradient of immunoreactivity located mainly in the medullary meristem and procambial strands and further below in vascular bundles.[29] These observations indicate that the apex is a likely site of cytokinin biosynthesis. In 7-week-old maize stems with eight internodes, the Z plus ZR level (determined by ELISA) was highest in the topmost internodes and then declined to a minimum at node 4, a result consistent with the view that the apex and/or very young leaves produce cytokinins.[54]

In two studies the amount of cytokinin that diffuses from excised shoot apices into agar has been measured in attempts to demonstrate cytokinin biosynthesis. Thus, Kannangara and Booth measured cytokinin diffusion from *Dahlia* shoot apices, and cytokinin levels in root xylem exudate and in apices before and after diffusion.[55] Because their calculation suggested that the supply from the roots would be insufficient to account for the observed cytokinin content of the apices, it was suggested that the apices were probable sites of cytokinin biosynthesis. However, the above calculation involves uncertain assumptions. Furthermore, the amount of cytokinin diffused plus that extracted from the apices after diffusion was less than that in the original apices. Hence, viewed in another way, the data could indicate that the apices were not sites of cytokinin biosynthesis.

However, the second study, using asparagus shoot apices, provided strong evidence that these apices were active in cytokinin biosynthesis.[56] The cultured apices, which do not develop adventitious roots, diffused constant amounts of cytokinins into the agar medium during five subcultures, while the cytokinin content of the apices remained constant throughout the culture period. The growth rate of apices decreased during the culture period, but growth was partially restored by exogenous cytokinin. Hence, the cytokinin produced by the shoot apices was insufficient to maintain growth, which may depend on additional cytokinin from the root. The cytokinin produced per asparagus root tip was six times that produced per shoot apex, indicating that the latter was a relatively minor source of cytokinin for the developing asparagus shoot.[56]

Cytokinin levels increase in buds of excised poplar shoots that lack roots about the time of bud burst.[57] While this increase could be due to synthesis within the bud, there are obviously other possibilities. Some evidence that cytokinins are synthesized within buds is provided by studies in relation to lateral bud outgrowth and apical dominance. Hadacidin, an inhibitor of adenylosuccinate synthetase, inhibits lateral bud growth on pea[58] and tomato plants.[59] The inhibition could be reversed by cytokinin, but not by adenine, applied directly to the buds.[58] These studies indicated that lateral bud growth is dependent on the synthesis of cytokinin in the bud itself. However, as discussed by Van Staden and Davey,[60] in a number of situations bud development is apparently largely dependent on root-produced cytokinin or on cytokinin stored in the shoot. In an interesting study with rootless almond shoots maintained at low temperature,[61] buds appeared to utilize cytokinins derived from within the shoot (probably the bark including the cambium), and disbudding elevated the cytokinin level considerably in both xylem and bark tissue. In this study buds were clearly not sites of cytokinin synthesis.

2. Leaves

Probably the first evidence of cytokinin production by leaves came from the studies of Salama and Wareing.[62] They detected, by bioassay, an increase in the level of active cytokinins in detached leaves of *Helianthus annuus* when supplied with a nitrate-containing nutrient solution. Similarly, supply of NH_4NO_3 to excised tobacco leaves caused a rapid increase in the cytokinin level.[63] In a significant study by Vonk,[64] excision of leaves from *Yucca* plants caused a rapid decline in the level of cytokinin in exudates of detopped inflorescences; however, excision of all the roots and part of the rhizome caused only a transient decrease in cytokinin exudation and then an increase to normal levels, which were maintained. This increase must be attributed to cytokinins derived from the leaves. While it has been proposed that *de novo* synthesis of cytokinins occurs in *Yucca* leaves,[65] and the maintenance of cytokinin levels in the exudate from the inflorescence in the absence of roots is suggestive of this, the leaf-derived cytokinin could also be released from *O*-glucosides known to occur in *Yucca* leaves.[66] More certain evidence that leaves synthesize cytokinins comes from studies of root xylem exudates and leaf exudates (phloem sap) from *Perilla*.[65] In the former, Z-type cytokinins predominated, and iP-type cytokinins were relatively negligible; in the latter exudate, iP-type cytokinins often predominated over Z-type cytokinins. Since Z-type cytokinins are derived from iP-type compounds, and not vice versa, *de novo* cytokinin synthesis must have occurred in the *Perilla* leaves. Although Van Staden and Forsyth[67] and Horgan[15] could not detect ^{14}C incorporation into cytokinins when bean leaf explants were supplied with ^{14}C-adenine or -adenosine, Chen et al. reported that labeled cytokinins were formed during incubation of pea leaves with [^{14}C]adenine under aseptic conditions for 21 days.[18] Although this study is often regarded as unequivocal evidence that leaves can synthesize cytokinin, it is better regarded with caution. First, culture of the leaves for a period of 21 days on an artificial medium is unphysiological and may have influenced cytokinin metabolism. Second, root primordia may have formed on the leaves. In a recent study Singh et al. demonstrated incorporation of [^{14}C]adenine into Z in small, developing tobacco leaves after excision, but no incorporation was found in mature or early senescent leaves.[13] Although the incorporation was established rigorously, it was very low (0.006% of extracted ^{14}C was due to Z), and it is not known where within the leaf the synthesis occurred. It could represent incorporation in the vascular and cambial tissue of the mid rib or main veins, rather than synthesis in tissues that are distinctive to leaves.

3. Fruits and Developing Seeds

The appreciable degree of fruit and seed development in plants derooted at flowering, and with adventitious root primordia formation prevented, indicates fruit development is not entirely dependent on continuing cytokinin production by the roots.[68] The cytokinin requirement might be satisfied by biosynthesis within the fruit or elsewhere in the shoot, but since stem tissues appear to accumulate root-produced cytokinin,[48] the requirement could be met by this cytokinin reserve. The development of many fruits is characterized by a short initial period of cell division, and subsequent growth is largely due to cell enlargement. The cytokinin levels in some fruits (e.g., apple[69]) are high and maximal during the period of cell division. Hence, fruitlets at this stage of development have either accumulated cytokinin from elsewhere in the plant or have become active in cytokinin biosynthesis. Evidence presented below indicates that the developing seed is a site of active cytokinin biosynthesis, although synthesis in other parts of the fruit may also occur, especially in tissues undergoing active cell division. The cytokinin activities of developing apple[70] and avocado seeds[71,72] are markedly greater than those of the receptacles and mesocarp, respectively. In cytokinin bioassays, extracts of developing apple seeds are also much more active than extracts of all other parts of the developing apple fruit bud.[70] Avocado cotyledon tissue derived from the seed grows *in vitro* without exogenous cytokinin and actually produces cytokinins.[73] The reported lower endogenous cytokinin levels in genetically parthenocarpic fruits, compared to seeded fruits,[74] also suggest that seeds produce cytokinins. Excision of the very small developing seeds from apple fruitlet explants[75] renders the growth of the explants by active cell division much more dependent on exogenous cytokinin,[76] indicating production of cytokinin by the seed.

In a more direct approach, an attempt was made to quantify cytokinin production by pea pods excised 10 days after anthesis and grown in sterile culture on medium lacking cytokinin.[77] The pea seeds developed normally, and their cytokinin content rose over 30-fold, during 18 days in culture. However, in similar experiments with cultured pea pods, the increase in the cytokinin level during seed growth was not confirmed by Krechting et al.[78] or by Van Staden and Button.[79] Unfortunately, all three studies have undesirable features. In the first, the final cytokinin level is obviously very low relative to that found in pea seed developed *in vivo*; in the second, only one Sephadex LH 20 fraction was assayed for activity; in the third (pods excised 5 days after pollination for culture), the pea seed did not develop normally, and the pod walls increased in weight only during the first 5 days in culture. The pod walls in lupin contain high cytokinin activity,[80,81] and it is possible that in experiments by Hahn et al.,[77] the observed increase in cytokinin levels in developing pea seeds could have originated from the pod walls. Unfortunately, no clear results can be reached from this potentially useful approach, and it should be noted that pea pods cultured *in vitro* are now known to require exogenous gibberellin or cytokinin, for normal elongation.[82] The experiments concerning cytokinin biosynthesis in cultured pea pods merit repetition using culture media based on those of Barratt.[82] Furthermore, very immature soybean seeds and intact pods have now been cultured on hormone-free media to yield normal mature seeds and pods, respectively.[83,84] This provides additional useful systems to study cytokinin biosynthesis and suggests that these seeds are able to produce the cytokinin they require. Attempts to demonstrate incorporation of radioactive precursors into cytokinins in developing seeds and fruits have so far been unsuccessful.[15,85-87]

To summarize, in the intact plant shoot, no sites of cytokinin synthesis have been established unequivocally. However, developing seeds, cambial tissue, and the shoot apex are probable sites of synthesis, and certain leaves may have some biosynthetic capacity.

D. BIOSYNTHESIS IN GERMINATING SEEDS

Germinating seeds appear to possess the ability to synthesize cytokinins because exogenous cytokinins can replace the embryonic axis of several species of dicot seeds in inducing enzyme activities in the cotyledons.[88] Further evidence that the embryonic axis may act as a site of cytokinin biosynthesis is provided by studies of cytokinin levels in cotyledons cultured with the axis attached or detached. Removal of the axis has been reported to cause a considerable reduction in the cytokinin level in bean cotyledons[89,90] and chickpea cotyledons.[91a]

Axis excision also resulted in a reduction of the cytokinin level in yellow lupin (*Lupinus luteus*) cotyledons,[91b] and studies of [^{14}C]adenine incorporation into cytokinins established that the embryonic axis was a site of cytokinin biosynthesis in the germinating yellow lupin seed.[92] After incubation with [^{14}C]adenine, imbibed intact seeds and excised embryos and cotyledons were extracted. Incorporation of ^{14}C into DZR and its nucleotide was evident in extracts of intact cotyledons, intact embryos, and excised embryos. The incorporation was established rigorously by column chromatography (several steps), TLC, HPLC, and chemical derivatization. No incorporation into any cytokinin was found with extracts of excised cotyledons. The results also indicated that the synthesized cytokinin was translocated from embryo to cotyledon.

In a recent study the excised embryos of germinating *Zea mays* seeds were shown to incorporate [^{14}C]adenine into zeatin nucleotide, indicating that the embryo of monocots is also capable of cytokinin biosynthesis.[93]

III. TRANSLOCATION

In early studies it was noted that when kinetin was applied to the surface of leaves and stems, its effects were very localized, and this suggested that cytokinins were immobile. However, numerous reports soon appeared (1964 onward) concerning movement of ^{14}C-labeled BAP and kinetin along segments of petioles, immature stems, epicotyls, and coleoptiles, from a donor to a receiver agar block (see references in the review by Letham[32]). These studies were modeled on the experimental systems used so successfully to investigate auxin transport. Depending on the species of the plants used (and sometimes on their age) and on whether or not IAA was also applied, cytokinin movement was either apolar or basipetal, in the above studies.

There was sometimes a lack of agreement between workers, concerning polarity of movement. Two factors may have contributed to this. In some experiments when basipetal polarity was observed, the concentration of BAP was so high (about 200 μM) that it was unphysiologic. The other factor was probably the auxin status of the tissue. Exogenous auxin was frequently found to enhance movement and sometimes to affect polarity. The rate of cytokinin movement (about 1 mm h^{-1}) in the above studies was usually much less than that of auxin in such systems (about 10 mm h^{-1}), and the physiological relevance of the observed cytokinin translocation may be questioned. The nature of the effect of IAA is also uncertain; the auxin may be just enhancing cytokinin uptake from the agar block or affecting cytokinin metabolism rather than promoting actual transport. Auxin is known to have marked effects on cytokinin metabolism and can inhibit alanine conjugation,[94] a form of metabolism of BAP and kinetin, that would be expected to predominate in the *Phaseolus* tissue used in many of the above translocation experiments. In a more recent study, movement of Z was examined in *Coleus* petiole segments in which kinetin moved at 1.1 mm h^{-1}, but without polarity. Z also exhibited apolar movement.[95]

Fortunately, most workers have moved away from the above type of "transport system" that yields meaningful results with auxins, but may be physiologically irrelevant as far as cytokinins are concerned. Cytokinins now appear to move, principally, in the xylem and phloem transport streams, and this was first indicated by the work of Lagerstedt and Langston,[96] who studied translocation of [8-^{14}C]kinetin in tobacco and other intact plants. Cytokinins in xylem sap are largely derived from the root[38] and therefore represent translocation from this site of biosynthesis. This translocation, translocation in the phloem, and translocation from the embryonic axis of seed, another established site of synthesis, are discussed below.

A. TRANSLOCATION IN THE XYLEM

1. Endogenous Xylem Cytokinins

Cytokinin activity was first detected in xylem sap, by Kulaeva,[97] and since then cytokinins have been detected and sometimes characterized in root xylem exudates of diverse species (see Letham[32] and Van Staden and Davey[60] for early reports). In recent years xylem cytokinins have been rigorously identified by mass spectrometry, while other reliable identifications have been based on RIA or ELISA, after extensive purification. Recent studies of these types include those of root exudates from the following species (the principal cytokinins present are given in parentheses): rice (Z nucleotide);[98] lupin (ZR and DZR);[99] *Phaseolus vulgaris* whole plants (Z and DZ nucleotides, probably);[100] disbudded *Phaseolus vulgaris* rooted cuttings (DZR);[101] *Urtica dioica*, stinging nettle (Z, Z nucleotide, iP, iPA, and iP nucleotide);[102] wheat (Z, ZR, DZ, DZR, OGZR, OGDZR);[103,104] *Eucalyptus cinerea* (iP, iPA);[105] *Gymnocladus dioica*, Kentucky coffee tree (ZR, DZR, iPA);[106] *Perilla frutescens* (ZR);[65] and tobacco (ZR, DZR).[13]

In early studies of xylem exudates, riboside cytokinins appeared to predominate, and when cytokinin bases were supplied to roots, the corresponding ribosides were the only form in the xylem sap.[107,108] The mobility of kinetin riboside in apple xylem is much greater than that of kinetin or BAP.[109] These results lead to the view that ribosides were of particular significance in root-to-shoot translocation. While this is probably so, it is now clear that a wide range of cytokinins, including nucleotides and *O*-glucosides, can also occur in some saps, and the nucleotides and glucosides can sometimes be dominant.

Since stem tissue is known to contain considerable cytokinin activity,[48] cytokinin interchange between the xylem and adjacent tissue could contribute to the diversity of cytokinins found in xylem sap. While various lines of evidence indicate that the xylem cytokinins are produced principally in the root,[38] root cytokinins appear to be released into the xylem with great selectivity. Thus, OGZ is the major cytokinin in roots of bean plants,[110] but is absent in the xylem sap of rooted bean cuttings where the dominant cytokinin is DZR.[101] While Z7G is the principal metabolite of exogenous [^{3}H]Z in radish roots, [^{3}H]ZR is the only labeled cytokinin present in the xylem sap.[107]

2. Distribution of Cytokinins Moving in the Xylem

Studies of the distribution and metabolic fate of radioactive cytokinins moving in the xylem are of considerable relevance to our understanding of plant development. Early (pre-1985) studies were of two types. In the first, radioactive cytokinins were applied to roots of intact plants: notably, BAP to *Citrus aurantium*[111] and bean roots,[108] kinetin and BAP to pea roots,[112-114] Z to *Alnus glutinosa* root nodules,[115] and Z to roots of radish seedlings before leaf development.[107] In the second type of study, the labeled cytokinin was applied in a single application at high concentration directly to the exposed xylem of the stem: notably, Z to *Lupinus albus*[116] and to bean,[117] and kinetin to willow.[118]

With the exception of the experiments with radish,[107] all these studies have one or more of the following limitations:

1. BAP and kinetin differ in their metabolism from that of Z-type cytokinins, and hence their behavior may not reflect that of the endogenous hormones.
2. The concentration applied was unphysiologically high and may even have been toxic in some cases when millimolar concentrations were applied to the stem.
3. The identities of the radioactive compounds in the shoot tissues were not established.

In studies of cytokinin transport from root to shoot, it is obviously important first to quantify the endogenous cytokinins in the xylem exudate and then to introduce these natural compounds in radioactive forms directly into the xylem, in concentrations similar to those occurring endogenously. The discussion below emphasizes such studies.

In the first study of this type, [^{3}H]ZR was supplied to derooted lupin plants (*Lupinus angustifolius*) via the transpiration stream and to intact plants via a cotton thread inserted into the xylem.[99] Noteworthy features of this study included

1. Over 50% of the extracted radioactivity that was recovered from the stems where DZR was a major metabolite
2. The direct lateral movement of ZR and/or related cytokinins from xylem to bark, which was established by bark-ringing experiments
3. The relatively high level of radioactivity per unit of tissue fresh weight in developing lateral shoots (major metabolites were cytokinin nucleotides)
4. The lack of appreciable movement to the seed, although the cytokinin reached the pod walls and was partially conserved therein
5. The radioactivity level in petioles, which exceeded that of laminae and this effect became more pronounced as leaf age increased

From a related study in which [^{3}H]ZR and [^{3}H]DZR of high specific activity were introduced into the xylem of primary laterals of the lupin plants, it was concluded that xylem cytokinins accounted for only a small portion of the high endogenous cytokinin content in the seed.[119] Lack of appreciable movement of xylem cytokinin to developing seeds was also observed in studies with soybean stem explants.[120,121] The developing seeds of legumes do not act as a sink for xylem cytokinin and, hence, do not cause a foliar cytokinin deficiency and, as a result, leaf senescence, as sometimes proposed by plant physiologists. Xylem-supplied cytokinins present in seeds occurred largely in the seed coats.[119-121] The seed coats of legume seeds appear to shield the embryos from the low proportion of xylem cytokinins that reach the seeds, while the common cytokinin breakdown product adenosine appears to move readily from the xylem to the embryo.[120]

While the above studies provide no evidence that xylem cytokinin moves preferentially into the seeds either directly or indirectly via the leaves, developing fruit and vegetative organs may compete for xylem cytokinin. Thus, removal of leaves from tomato plants increased cytokinin levels in the developing fruit,[122] while removal of young grape berries elevated cytokinin levels in the leaves.[123] Removal of the apex of poplar shoots elevated cytokinin levels in leaves.[124] While these findings have been interpreted in terms of competition for xylem cytokinin, this conclusion may not be valid. Disbudding can elevate the cytokinin level of xylem exudate,[125] and the excision of other organs might have similar effects.

In sequential leaf senescence, the lower leaves experience senescence first, and senescence then progresses towards the shoot tip. The cause of this phenomenon is unknown, but it might be explained by a differential supply of xylem cytokinin to the leaves along the stem or by differences in the rate of cytokinin inactivation in the leaves. However, when [^{3}H]ZR or [^{3}H]DZR were supplied through the xylem to tobacco plants, leaves of differing maturity did not differ appreciably in the level of radioactivity or in the metabolism of cytokinins.[13] However, relative to upper leaves, the older, lower leaves contained a much greater proportion of their total radioactivity in the petioles, a result in accord with the studies with lupin.[99]

In the studies mentioned above concerning xylem translocation of [^{3}H]ZR in lupin,[99] about 50% of the recovered ^{3}H was derived from the stem, and cytokinin movement from xylem to bark was established. A similar degree of stem retention of natural cytokinins moving in the xylem was also indicated in studies with radish seedlings[107] and flowering tobacco plants.[13] Following the application of [^{14}C]Z to root nodules of *Alnus glutinosa*[115] and of [^{3}H]ZR to pea root nodules,[126] radioactivity was present in stems and in leaves in similar amounts. A proportion of the stem radioactivity was due to cytokinins: Z and ZR in the case of *Alnus glutinosa,* and DZR, Z nucleotide, and Z glucosides, in the case of pea seedlings. Retention by stem tissues of radiolabeled BAP or its metabolites, moving in the xylem, has also been reported,[111,127] and in these studies more radioactivity was recovered from the stem than from the leaves. Radial translocation of [^{14}C]kinetin from the transpiration stream to the bark of willow stems was also suggested in early work.[118] Lateral movement of cytokinin from the xylem to adjacent stem tissue, and its sequestration therein, may explain the appreciable cytokinin levels reported for stems of some plants. For example, the cytokinin activity level in stems of *Phaseolus vulgaris* exceeds that in roots and leaves,[48] and these stems can rapidly metabolize cytokinin bases to nucleotides.[128] Stems may provide a cytokinin reserve for supply to the leaf[48] and appear to have the ability to remove, modify, and possibly contribute to cytokinins in transit to the shoot. Stems appear to have a significance in cytokinin translocation far beyond their role as a channel for root-produced metabolites.

When radiolabeled Z and DZ were supplied to cut carnation flowers via the transpiration stream, the cytokinins were rapidly translocated to, and accumulated in, the receptacle. This appeared to act as a cytokinin reservoir that slowly released cytokinins to the petals and back into the stems.[129] Similarly, [8-^{14}C]BAP, when supplied to these cut flowers, accumulated, initially, in the receptacles.[130]

In concluding this section, it merits mention that the auxin status of the shoot apex or the apical end of a stem cutting appears to markedly influence metabolism and translocation of exogenous cytokinin and thus controls cytokinin distribution in the shoot.[32,114,131] In the absence of an apical bud, the cytokinin applied to the stem base or root appears to accumulate in lateral buds, but this is suppressed by application of auxin to the apical region.

3. Metabolic Fate of Xylem Cytokinins

Metabolites derived from labeled natural cytokinins after translocation in the xylem of stems have been identified unambiguously, or with reasonable certainty, in only a few plant systems, which are listed in Table 1, with the metabolites formed. In another relevant study, Z was supplied via the transpiration stream to excised poplar leaves, and the principal metabolites were identified by mass spectrometry as *O*-glucosides.[134] In rooted bean cuttings, the DZR present in the xylem sap appears to accumulate in the leaves as OGDZ.[101,135]

While the initial metabolites of xylem cytokinins formed in the leaves and stems of some species include nucleotides (Table 1), these compounds are subsequently converted to glucosides and adenine-type compounds (Table 1).[121] *O*- and *N*-glucosides are clearly the major foliar metabolites in which an isoprenoid side chain is conserved. The *O*-glucosides are probably not active per se and are compartmentalized so that they do not release active cytokinin by glucosidase action. Glucosides appear to be remarkably stable during leaf maturation and senescence[39,135] and are, presumably, lost to the plant at abscission.

One important feature of foliar metabolism of xylem cytokinin is its rapidity. Thus, when [^{3}H]ZR was supplied to soybean explants for 1 h, 83% of the radioactivity in the leaves was due to metabolites.[121] After transfer of the explants to water only, for a further 24 h, ZR, DZR, Z, DZ, Z nucleotide, and DZ nucleotide could not be detected with certainty. Hence, xylem cytokinins in active forms do not accumulate in leaves, and change in xylem cytokinin flux has the potential to constitute a sensitive root-to-shoot signal.

B. TRANSLOCATION IN THE PHLOEM

Our understanding of cytokinin movement in the phloem is extremely limited. While cytokinin activity has been detected in phloem sap from both herbaceous and woody plants,[32,60] only one phloem cytokinin has been identified with some certainty. Thus, the cytokinin activity of *Yucca* phloem exudate is due mainly to a phosphate of ZR, presumably the 5′-phosphate.[136] When hydrolyzed with phosphatase, the cytokinin yields ZR (identified by GC-MS), and it appears to be derived from zeatin derivatives, including OGZ, present in *Yucca* leaves.[66] Bioassay and chromatographic studies indicate that a cytokinin *O*-glucoside, possibly *O*-glucosylzeatin, is a major cytokinin in aphid honey dew of willow trees.[137] Compounds that crossreact with ZR and iPA antibodies were detected in phloem exudates from 14 tree species of 8 genera.[138] iP-type cytokinins were found in concentrations much higher than those of Z-type cytokinins. However, these studies were done with unpurified phloem exudate, and the results are questionable. Phloem sap from excised *Sinapis alba* leaves has been obtained by exudation into a solution containing ethylenediaminetetraacetic acid (EDTA). Bioassay detected cytokinin activities corresponding to the HPLC retention times of Z, ZR, iP, and iPA,[139] and a single long day elevated the Z content of exudate markedly. Similarly, leaf phloem exudate of *Perilla* was shown by HPLC and ELISA to contain a number of cytokinins, including iPA and iP 9-glucoside.[65] The amounts of iPA and ZR exudated from leaves that had been exposed to short days exceeded those exuded from leaves kept in long days or night-interrupted short days.[65]

Phloem exudate can be readily obtained in small amounts from various positions on white lupin (*Lupinus alba*) plants, which provide an ideal system

Table 5-1 **Principal Metabolites of Cytokinins Supplied to the Xylem of the Stem of Various Plants.**
The Nucleotides of *O*-acetyl-Z and *O*-acetyl-DZ are Largely Confined to the Bark Tissue of Lupin Stems.

		Initial Metabolites[a,c]		**Final Metabolites[c]**	
Plants	Labeled cytokinin supplied in xylem	Stem	Laminae	Stem	Laminae
Derooted radish seedlings[132]	Z	ZNT	Z7G	ZNT,AMP	Z7G
Intact radish seedlings[107]	ZR[b]	ZNT	Z7G	AMP	Z7G
Derooted lupin plants[99,133]	ZR	DZR,DZ,ZNT,DZNT,ACNT	ND	DZR,*O*-glucosides,Ado	Ado,OGDZ
Soybean stem explants[121]	ZR	ZNT,DZNT	ZNT	Ado	Ado,OGDZ
	DZR	ND	DZ	ND	OGDZ
Tobacco plants[13]	ZR	Ade,Ado	Ade,Ado	Ade,Ado	Ade,Ado,Z7G
	DZR	ND	DZ7G	ND	DZ7G

[a] Metabolites present after a few hours; [b] The [^{3}H]ZR in the xylem was derived from [^{3}H]Z supplied to the roots and was the only labeled cytokinin in the root xylem exudate; [c] Abbreviations: ACNT, nucleotides of *O*-acetyl Z and DZ; Ade, adenine; Ado, adenosine; ND, not determined.

to study phloem translocation of cytokinins and xylem-phloem interchange. Phloem sap was collected from several locations on this plant, and cytokinin activity was detected in all locations; however, cytokinin activity level was highest in phloem sap from the base of the inflorescence, and this activity increased during fruit development.[140] These studies suggested movement of cytokinins from leaves into the developing inflorescence and fruit, rather than direct translocation to the fruit via the xylem sap.

In another study using white lupin, DZ was found to be the main labeled cytokinin in phloem sap exuded by pods of plants that had been supplied with [^{3}H]ZR through the roots.[141] This indicates that phloem cytokinin in this species is, to some extent at least, xylem derived, and it is relevant to note that DZ is a major endogenous cytokinin (identified by GC-MS) in white lupin mixed xylem plus phloem sap.[140] The observation that labeled cytokinins can move from the transpiration stream in the stem of *Lupinus angustifolius* into bark tissue also suggests that phloem can derive cytokinins in this way.[99] In early work, Bowen and Wareing[118] observed lateral movement of ^{14}C from xylem to bark in willow after application of [^{14}C]kinetin to the xylem, and movement in the reverse direction was also found. However, these results must be interpreted with caution for several reasons. Kinetin, an unnatural cytokinin, was applied as a methanol solution to bark or to exposed xylem, at a concentration of 3 m*M*, which is certainly unphysiological and may even be toxic. Furthermore, a complex of kinetin metabolites was detected in the bark, but none were identified, and hence there is no evidence that kinetin itself actually moved radially to the bark, as claimed.

While radioactive cytokinins applied to leaves are invariably strongly retained within the treated leaf, a small proportion of the label is translocated to other plant parts. This translocation can be either acropetal or basipetal and is presumably via the phloem. In the examples cited below, translocated radioactivity appears to be due to the cytokinin supplied to the leaf, or to a metabolite in which the N^6 moiety is conserved; the cytokinin applied, and the organ to which it moved, is listed: BAP to apical bud, other leaves, stem, and root of cocoa plants (relative movement is dependent on the stage of the flush cycle);[142] Z7G to roots of radish;[143] Z to axillary buds of rose shoots;[144] Z to side shoots of lupin, while negligible movement to the fruits occurred;[145] ZR to stems and nodules of pea plants;[126] and Z to the inflorescence stalk of *Yucca* before flowering.[66] In addition, Z appears to be translocated from the receptacle of cut carnation flowers to the stem.[129] These findings indicate movement of exogenous foliar cytokinin to several vegetative organs and also to developing inflorescences. If endogenous cytokinins exhibit the same types of phloem translocation, then long-distance movement of cytokinins presents a complex picture, with acropetal movement in the xylem, bidirectional movement in the phloem, and an interchange between xylem and phloem.

C. TRANSLOCATION WITHIN GERMINATING SEEDS

In Section II.D the embryonic axis of a germinating lupin seed was shown to have the capacity for cytokinin biosynthesis and [^{14}C]adenine was found to be incorporated into DZR and its nucleotide.[92] The translocation of this riboside (^{3}H-labeled of high specific activity) at physiological levels in the lupin seed has been studied during the early stages of germination (within 16 h from the start of imbibition) before radicle emergence.[146] There was considerable translocation of [^{3}H]DZR from the embryo into the cotyledons, but there was no detectable movement from cotyledons to the embryo. Hence, there was a highly polarized movement of cytokinin during germination of the lupin seed, and the cotyledons of the germinating seed probably derive their cytokinin from the axis where its synthesis is apparently localized.[92]

A number of reports substantiate the view that cotyledons are supplied with cytokinin from the axis. Excision of the embryonic axis resulted in decreased levels of cytokinins in cotyledons of seeds of several species (see Section II.D). Rapid translocation of [^{14}C]Z, from the embryonic radicle to the cotyledons of germinating bean seeds, was observed by Hutton and Van Staden.[147] Transport in the reverse direction did not occur during the early stages of germination, but after radicle elongation, such movement was observed, probably as part of a general nutrient mobilization to the rapidly growing embryos. Movement of [^{3}H]ZR from the embryo to the endosperm was also found during maize seed germination, but movement in the other direction was negligible.[148]

In a study of the translocation of [^{14}C]Z within seeds of the gymnosperm *Podocarpus henkelii*, movement from the epimatium to the female gametophyte (site of main seed reserves), and vice versa, was observed.[149] Unfortunately, it was not feasible to apply cytokinin to the embryo sporophyte to determine if cytokinin moved from it to the storage site of reserves, as occurs in monocot and dicot seed.

IV. FUNCTIONS OF TRANSLOCATED CYTOKININS

A. CYTOKININS FROM THE EMBRYONIC AXIS

The embryonic axis of a germinating yellow lupin seed is now known to be a site of cytokinin biosynthesis,[92] and the cytokinin produced is translocated to the cotyledons (see Section III.C).[146] The function of this translocated cytokinin merits consideration. Exogenous BAP or DZ completely replaced the axis in inducing cotyledon expansion and chlorophyll formation in the cotyledons. GA_3 was without effect.[150] It is reasonable to propose that the endogenous cytokinins, which are produced by the axis and which are translocated to the cotyledons, also promote these effects, which are well-established responses to exogenous cytokinin. However, this is the first occasion when they have been linked to embryo-derived cytokinin.

In germinating dicotyledonous seeds, the embryonic axis induces the development of enzyme activities (notably amylolytic and proteolytic) in the cotyledons. Exogenous cytokinin will frequently, and exogenous gibberellin occasionally, replace the axis in this function. Exogenous IAA does not replace the axis in any of the systems tested. The above results stemmed from the work of Wiley and Ashton[151] and Penner and Ashton.[152] The former showed that if excised squash cotyledons were incubated in culture medium with dissected embryonic axes, their proteolytic activity increased up to the level in attached cotyledons in the intact seedling. This work indicated that a factor that induced proteolytic activity was released from the axes into the medium and taken up by the cotyledons. The work of Penner and Ashton indicated that this factor was probably cytokinin.[152] These results and numerous others[88] involving replacement of the axis by cytokinin have led to the view that cytokinins released by the embryonic axis promote the development of enzyme activities in the cotyledons of diverse species. In the case of amylase activity in bean cotyledons, this view is supported by the close similarity in the effects of the embryonic axis and of exogenous cytokinins, on the various amylolytic fractions in the cotyledons.[153] Exogenous cytokinin also replaces the embryonic axis in promoting protease activity in yellow lupin cotyledons,[154] and endogenous embryo-derived cytokinins could have the same effect. When the axis is excised from a germinating chickpea seed, the cytokinin level in the cotyledons declines markedly, and the levels of amylase and protease also decrease considerably. However, when cytokinins are supplied exogenously to reproduce the normal levels found in intact cotyledons, enzyme activity levels are reported to be restored to near normal.[91]

While the above studies collectively constitute strong evidence for the control of enzyme activity in cotyledons by cytokinin derived from the embryonic axis, this has been a controversial area, and some have considered that the axis merely acts as a sink that withdraws from the cotyledons otherwise inhibitory products formed by hydrolysis of protein or carbohydrate reserves.[88,155] In the absence of the "axial sink", the products are proposed to accumulate, and by a feedback mechanism, further enzyme action is suppressed. While there is evidence that suggests that such a mechanism can operate, to some extent, in certain cases, it does not negate the strong evidence now accumulated for a hormonal control mechanism involving cytokinins. The "sink" theory is clearly not supported by a recent study with *Agrostemma* seeds[156] or by earlier evidence reviewed by Ilan and Gepstein.[88]

B. PHLOEM CYTOKININS

The role of cytokinins moving in the phloem is unclear, but there are indications that they are involved in the induction of flowering. The LD plant *Sinapis alba* can be induced to flower with a single LD that causes shoot apical changes in the meristem. One of the earliest changes is a mitotic wave induced by a factor derived from the leaves.[157] This mitotic activation can also be induced by a single application of cytokinin to the apical bud. One LD markedly elevated the cytokinin level in xylem sap and also in phloem sap exuded from leaves. This phloem cytokinin is likely to be the mitotic stimulant.[139] Exposure of *Perilla* to short days to induce flowering also elevates the cytokinin level in leaf phloem exudate.[65]

Phloem cytokinin is probably involved in the formation of vascular tissue associated with root thickening. Thickening begins near the root base of many higher plants by initiation of cell division in the procambial and pericycle tissues. The shoot has a definite influence on these changes, and this involves much more than the supply of carbohydrates. In addition to auxin, other growth substances from the shoot appear to be involved.[158] When supplied to the basal end of excised radish roots in sterile culture, auxin plus cytokinin rapidly induced vascular cambium activity and then secondary vascular tissue formation. The requirement for these hormones was absolute, but the response was enhanced by the supply of *myo*-inositol.[158-160] It should be emphasized that the hormones must be supplied to the basal end of the excised root. Studies with turnip roots, using basal feeding methods,[161] have confirmed all the results mentioned above. Excision of the root tip from dark-grown seedlings had no effect on the formation of cambial derivatives, but removal of the shoot, or one or both cotyledons, reduced the formation of vascular tissue markedly.[161]

The cytokinin produced by the root tip is not effective in the induction of radial growth at the base of radish and turnip roots, and the cytokinin required must be derived from the shoot. This suggestion is supported by studies of the endogenous cytokinins in developing radish roots.[162] Three cytokinins (I, II, and III) were detected by thin-layer chromatography (TLC) and bioassay. Cytokinins I (? Z nucleotide) and II (? Z or ZR) increased in amount at the start of cambial activity and induced cambial activity when supplied to the basal end of excised roots in culture. Cytokinin III did not induce cambial activity, was located in the xylem tissue, and increased in level well after onset of cambial activity. Radin and Loomis proposed that I and II are derived from the shoot, while III is formed in the root tip and translocated in the xylem.[162]

C. XYLEM CYTOKININS

Because the transpiration rate and the flow of xylem sap vary greatly due to frequent changes in environmental factors, Trewavas[163] proposed that xylem transported cytokinin could not constitute a coordinating signal between root and shoot. However, in making this proposal, the firmly established principles of ion transport in the xylem were ignored. Thus, although the linear rate of ion transport in the xylem is greatly affected by the rate of transpiration, the amount of solute transported in unit time is largely independent of the rate of xylem sap flow. The concentration of xylem sap decreases with an increase in transpiration rate, and vice versa. For example, there was little difference in the amounts of ions accumulated in shoots of barley when the nutrients were provided only at night, or only during the day, despite very large differences in the transpiration rate induced by light.[164] By analogy, while the rate of xylem flow may vary, the cytokinin delivered to the shoot organs over a period of time should remain constant unless the synthesis rate in the root or xylem loading changes. Calculation of cytokinin flux is usually based on flow rate of, and cytokinin level in, xylem sap exuded from roots in a period of over 12 h.[165] This flux is unlikely to equal that in the intact plant. To assess the latter, the volume of exudate should be small enough to be sap in transit at the time of shoot excision. Cytokinin level can then be multiplied by transpiration flow of the intact plant.[39]

1. Shoot Development

In early studies (see Goodwin et al.[38]), because of the ability of exogenous cytokinins to substitute for roots in inducing growth or biochemical events in the shoot, it was proposed that root-produced cytokinin promoted DNA synthesis and growth in the apical region of soybean hypocotyls, growth and auxin production in coleoptiles, conversion of a stolon into a leafy shoot, inflorescence development on grape cuttings, inflorescence branching of *Carex flacca*, and CO_2 fixation and the activity of photosynthetic enzymes. A similar experimental approach indicated that root-produced cytokinin promoted microbody enzyme activities,[166] chlorophyll synthesis,[167] and photosynthetic rate.[168] In intact bean plants, BAP application to the primary leaves caused a reduction in root development, but enhanced the net photosynthetic rate of the primary leaves.[169] Root-produced cytokinins appear to play an important role in the differentiation of fibers, vessels, and sieve-tubes in vascular tissues.[170] Using soybean explants, in which defined solutions can be substituted for the roots, Noodén et al.[171] showed that cytokinins and nutrients translocated in the xylem control seed coat thickness and pod development. When cytokinin mixtures were supplied via the transpiration stream to oat leaves at the concentration of occurrence in xylem sap, transpiration was promoted and senescence delayed.[172]

Bean seedlings, when grown in very small pots, developed as dwarf plants with retarded root development. This dwarfing was not a consequence of nutrient deficiency or of water stress. Exogenous cytokinin applied to the shoots markedly promoted growth of the primary leaves, and a combination of GA_3 and BAP completely restored the growth of stem and leaves.[173] A limited root system restricts shoot growth, through a hormonal mechanism involving cytokinins.

The above observations indicate that root-produced cytokinin influences many aspects of shoot development. The effects on photosynthesis are probably particularly important. It should be noted that in other systems, exogenous cytokinins (although not xylem supplied) stimulate chloroplast biogenesis, promoting synthesis of RuBPCase, chloroplast differentiation, chlorophyll synthesis, and the expression of genes encoding the small subunit of RuBPCase and the light-harvesting chlorophyll a/b binding proteins. Increases in free endogenous cytokinin levels have been associated with changes in leaf photosynthesis.[174] This formidable list of foliar functions attributable to cytokinins indicates a key role for these root-derived hormones in the regulation of the development of photosynthetic potential.

The marked increases in cytokinin activity in xylem sap of deciduous trees, which precede bud burst (for a recent example see Tromp and Ovaa[175]), coupled with the ability of exogenous cytokinins to break bud dormancy, have led to the idea that xylem cytokinins transported into the buds are involved in the induction of bud burst. However, some observations are contrary to this view,[38] and it is likely that the elevated cytokinin level in xylem sap may serve to support or induce the rapid growth that follows bud burst. Developing apple buds seem to utilize xylem cytokinins. Xylem cytokinin levels decreased markedly in chilled trees as bud break occurred, but not in unchilled trees that did not show bud break.[176]

Cytokinins play a key role in the release of lateral buds from apical dominance. However, the role of xylem cytokinin and roots in the development of lateral buds is unclear. In early studies the presence of roots and the application of cytokinins were reported to have similar effects on lateral shoot development in cuttings of *Solanum andigena*, suggesting that root-produced cytokinin controls the development of these laterals.[177-179] However, in later studies with similar cuttings, this view was questioned, and the shoots appeared to be capable of maintaining their level of active cytokinins in the absence of roots.[180] Studies with pigeonpea plants inoculated with wild-type and mutant strains of *Rhizobium* are relevant to the question of xylem cytokinins and apical dominance. When inoculated with a *Rhizobium* strain that was a cytokinin overproducer in culture, the plants had an elevated cytokinin level in xylem sap and exhibited release from apical dominance, and proliferation of lateral-buds.[181] This work confirmed the importance of xylem cytokinins in lateral bud development and provides a new system to study their role in the release from apical dominance. Control of the distribution and metabolism of root-produced cytokinin, by auxin derived from the shoot apex, could contribute to the maintenance of apical dominance (see Section III.A.2).

Cessation of vegetative growth, and the subsequent onset of flowering and then senescence, are precisely timed phenomena in annual plants. There appears to be a close correlation between these developmental phases in the shoot and cytokinin translocated from the roots. Thus, the cytokinin level in xylem sap decreases considerably during flower-bud development or floral induction in several species, namely, sunflower, white lupin, cocklebur (a SD species; decrease induced by one SD), and tomato (see references in Van Staden and Davey,[60] also Taylor et al.[140]). Low cytokinin level appears to be a prerequisite for onset of flowering, but during flowering and fruit development, the levels tend to rise again. Xylem cytokinins also appear to play a key role in sex determination in some species. Formation of adventitious roots on derooted hemp (a SD dioecious species), spinach (a LD dioecious species), and cucumber (monoecious species with unisex flowers) plants markedly promotes female sex expression, while continual removal of the developing roots produces mainly male sex expression.[49] Supply of cytokinin through the transpiration stream to these plants without adventitious roots induces formation of female plants or flowers. Hence, xylem cytokinin is a feminizing hormone in these species and interacts with gibberellin (a hormone inducing male sex expression, apparently derived from the leaves) to determine the balance of sex expression. This conclusion was confirmed by application of BAP through the roots of intact hemp, spinach, and cucumber plants and by determination of endogenous cytokinin levels in shoots and roots of the dioecious plants.[49] Application of BAP to intact corn plants dramatically promoted female sex expression, and 78% of the plants developed ears within the tassels.[49] The natural location of ears on the lower part of the corn stem could be related to the preferential accumulation of root-produced cytokinin at this location.

2. Leaf Senescence

Application of cytokinin to laminae of detached and attached leaves delays senescence markedly, and endogenous cytokinin levels in leaves usually decline before or during senescence (see the review by Van Staden et al.[10]). Hence, changes in cytokinin flux to leaves via the xylem could influence the onset of leaf senescence. Evidence supporting this view is summarized below. First, rooting of excised tobacco[182] and bean leaves[26] delays leaf senescence, and the cytokinin level in bean petioles and laminae is increased markedly. Second, partial excision of roots from soybean, pea, and wheat plants induces senescence, which can be prevented by application of cytokinin to the leaves.[183-185] Third, a variety of treatments applied to roots reduce cytokinin activity in xylem sap and usually induce leaf senescence (see below). Fourth, disbudding of *Perilla*,[186] tomato, and tobacco plants,[125] which delayed leaf senescence, resulted in an increase in both the concentration and the flux of cytokinin in the xylem sap. When vegetative *Perilla* plants were disbudded, sequential leaf senescence was delayed, and regreening of yellow leaves at the base of the plants occurred.[186] These changes were also observed when *Perilla* plants were exposed to short days (floral induction), which also elevated the cytokinin level in the xylem sap.[186] Fifth, the studies with soybean explants, outlined below, provide convincing evidence that the levels of xylem cytokinins play a key role in the control of monocarpic leaf senescence in soybeans.

The leaf senescence of soybean explants (one leaf, associated pod/pods, and the subtending internode) closely resembles that of intact plants. The senescence in the explant is dependent on the presence of the pod and is markedly delayed by cytokinin supplied via the xylem.[39] Early in soybean pod development, the cytokinin flux through the xylem declines markedly, and this decrease is reflected in a reduction in the foliar cytokinin activity level. The decrease in xylem flux appears to sensitize the leaf to a senescence-inducing factor derived from the seed.[39] However, if the supply of exogenous cytokinin to the explants is maintained, it will largely override the senescence signal from the seed. Excision of pods from soybean plants at early pod fill prevents leaf senescence and induces a large increase in the cytokinin level in the xylem sap, while depodding at late pod fill does not.[187] By supplying cytokinins via the xylem to the explants, at a range of concentrations, it can be shown that the changes in the endogenous cytokinin level in xylem sap, during pod development and after depodding, are sufficient to affect onset of leaf senescence.[39,187] Participation of xylem cytokinins in the control of leaf senescence in monocarpic plants is now a firmly established concept in plant physiology.

3. Response to Stress Conditions

When plant roots are subjected to stress conditions, the cytokinin level in the xylem sap is often reduced. The stress conditions include (see reviews by Letham,[32] Goodwin et al.,[38] Van Staden and Davey[60]) abnormal temperature, drought, flooding, excess salinity, low pH, low nitrogen status of soil, phosphorus deficiency, boron deficiency, and application of the growth regulator ethrel. The leaf senescence associated with these root stresses is partially attributable to a reduction in the cytokinin supply. Water stress applied to the shoots of tobacco plants, by increased evaporation, also reduced cytokinin activity in xylem sap.[32] In plants with hypoxic roots, xylem sap fluxes of cytokinins were reduced markedly relative to controls,[34] and this was associated with a reduced growth rate and stomatal conductance of the leaves. Water deficiency is the root stress that is best understood in relation to cytokinins and is of great interest; it is discussed further (below) in terms of root-produced cytokinin.

Water deficiency could inhibit cytokinin production in roots and inhibit cytokinin transport out of roots via the xylem. Studies with sunflower plants grown in an aeroponic culture indicated that water deficiency did not reduce the cytokinin level in roots and suggest that the latter mechanism contributes to reduction in the shoot cytokinin level.[188] There is good evidence that a chemical signal is transmitted from roots to leaves to induce stomatal closure when roots sense drying of the soil.[189] Exogenous cytokinins increase the stomatal aperture and transpiration rate in numerous types of plants and particularly in older leaves[190] and can overcome the effect of partial root drying, on stomata. It has been suggested that the above root-to-leaf signal could be a reduction in cytokinin flux through the xylem. However, it is now clear that a "negative" signal of this type cannot be the message that induces initial closure of the stomata.[189] However, reduction in cytokinin supply could amplify the effect of abscisic acid or other inhibitors that appear to be the "positive" signal that initiates stomatal closure.[189,191]

When the nitrogenous nutrients supplied to roots are reduced to suboptimal levels, the cytokinin level in xylem sap also declines, and this change would be expected to alter N metabolism in the shoot. Thus, in studies of wheat plants with a split root system, kinetin was applied to one root that received no N, while the other root received a suboptimal N level.[192] The retranslocation of N from shoot to root was reduced by the flow of kinetin to the shoot. Hence, the distribution of N between root and shoot is determined in the shoot under the control of xylem-translocated cytokinin.[192]

V. CONCLUDING REMARKS

The studies reviewed show that cytokinin biosynthesis is localized in developed plants and also in germinating seeds. Although some cytokinin may be produced in the shoot, various types of evidence indicate that this is probably minor relative to the cytokinin flux from the root system. This matter has been considered recently by Incoll and Jewer,[193] who calculated that the number of active meristems (sites of cytokinin synthesis) in the root system of a young wheat plant is at least 50 times that in the shoot. The cytokinin produced by the root and by the embryonic axis moves to other tissues to control diverse aspects of development, by concentration-dependent mechanisms. Hence, contrary to the views of Trewavas, cytokinins are growth regulators that are coordinators of growth and that are phytohormones in the classical sense originally proposed in 1937 by Went and Thimann.[194] Changes in sensitivity to cytokinins could occur, but these may well just reflect changes in the concentration of another hormone that interacts with cytokinin, a concept not considered by Trewavas.[1,2] Plant development is an expression of the interaction between plant hormones, including translocated cytokinin.

An appropriate hormonal balance between root and shoot, as well as nutrient interdependence, is responsible for the close correlation between root and shoot growth. Hormonal signals that move from root to shoot probably modulate those translocated from shoot to root, and vice versa. A positive feedback relation between shoot and root development and between auxin production by the shoot and cytokinin production by the root has been proposed by Sachs[195] and further discussed.[38,196] This feedback relation depends on correlations between auxin production and shoot apical growth and between cytokinin production and development of root primordia and apices. Increased auxin flux from the shoot would promote development of root apices that would then elevate cytokinin flux to the shoot. Such effects on xylem cytokinin, although important, would be indirect and delayed. However, some signals from the shoot, probably hormonal, appear to rapidly suppress cytokinin translocation from the root.[38] For example, when *Xanthium* plants were subjected to one short day, the cytokinin level in the xylem sap decreased considerably, and bark ringing prevented this decrease.[197] Depodding before appreciable seed development[39] and disbudding[38,125] also alter cytokinin supply from the root, presumably through a shoot-to-root signal. Elucidation of the nature of such shoot-to-root signals presents a challenging project for the future.

REFERENCES

1. **Trewavas, A.,** How do plant growth substances work? *Plant Cell Environ.*, 4, 203, 1981.
2. **Trewavas, A. J.,** Growth substance sensitivity: the limiting factor in plant development, *Physiol. Plant.*, 55, 60, 1982.
3. **Goldacre, P. L.,** Potentiation of lateral root induction by root initials in isolated flax roots, *Aust. J. Biol. Sci.*, 12, 388, 1959.
4. **Vankova, R., Kamínek, M., Eder, J., and Vanek, T.,** Dynamics of production of *trans*-zeatin and *trans*-zeatin riboside by immobilized cytokinin-autonomous and cytokinin-dependent tobacco cells, *J. Plant Growth Regul.*, 6, 147, 1987.
5. **Hansen, C. E., Meins, F., and Milani, A.,** Clonal and physiological variation in the cytokinin content of tobacco cell lines differing in cytokinin requirement and capacity for neoplastic growth, *Differentiation*, 29, 1, 1985.
6. **Hansen, C. E., Meins, F., and Aebi, R.,** Hormonal regulation of zeatin riboside accumulation by cultured tobacco cells, *Planta*, 172, 520, 1987.
7. **Marziani Longo, G. P., Rossi, G., and Longo, C. P.,** Endogenous levels of three cytokinins in attached and excised watermelon cotyledons, in *Physiology and Biochemistry of Cytokinins in Plants*, Kamínek, M., Mok, D. W. S., and Zazímalová, E., Eds., SPB Academic Publishing, The Hague, 1992, 357; see also *Physiol. Plant.*, 83, 647, 1991.
8. **Meins, F. and Hansen, C. E.,** Epigenetic and genetic factors regulating the cytokinin requirement of cultured cells, in *Plant Growth Substances 1985*, Bopp, M., Ed., Springer-Verlag, Berlin, 1986, 333.
9. **Sheldrake, A. R.,** The production of hormones in higher plants, *Biol. Rev.*, 48, 509, 1973.
10. **Van Staden, J., Cook, E. L., and Noodén, L. D.,** Cytokinins and senescence, in *Senescence and Aging in Plants*, Noodén, L. D. and Leopold, A. C., Eds., Academic Press, San Diego, 1988, 281.
11. **Even-Chen, Z., Atsmon, D., and Itai, C.,** Hormonal aspects of senescence in detached tobacco leaves, *Physiol. Plant.*, 44, 377, 1978.
12. **Singh, S., Letham, D. S., and Palni, L. M. S.,** Cytokinin biochemistry in relation to leaf senescence. VII. Endogenous cytokinin levels and exogenous applications of cytokinins in relation to sequential leaf senescence of tobacco, *Physiol. Plant.*, 86, 388, 1992.
13. **Singh, S., Letham, D. S., and Palni, L. M. S.,** Cytokinin biochemistry in relation to leaf senescence. VIII. Translocation, metabolism and biosynthesis of cytokinins in relation to sequential leaf senescence of tobacco, *Physiol. Plant.*, 86, 398, 1992.
14. **Palni, L. M. S., Nandi, S. K., Singh, S., and Letham, D. S.,** An overview of cytokinin biosynthesis, in *Plant Growth Substances 1988*, Pharis, R. P. and Rood, S. B., Eds., Springer-Verlag, Berlin, 1990, 258.
15. **Horgan, R.,** Cytokinin biosynthesis, in *Cytokinins: Plant Hormones in Search of a Role*, Horgan, R. and Jeffcoat, B., Eds., British Plant Growth Regulator Group, Bristol, 1987, 1.
16. **Van Staden, J. and Forsyth, C.,** Adenine incorporation into cytokinins in aseptically cultured tomato roots, *J. Plant Physiol.*, 117, 249, 1984.
17. **Dickinson, J. R., Forsyth, C., and Van Staden, J.,** The role of adenine in the synthesis of cytokinins in tomato plants and in cell-free root extracts, *Plant Growth Regul.*, 4, 325, 1986.

18. **Chen, C. M., Ertl, J. R., Leisner, S. M., and Chang, C. C.,** Localization of cytokinin biosynthetic sites in pea plants and carrot roots, *Plant Physiol.*, 78, 510, 1985.
19. **Morris, R. O.,** Genes specifying auxin and cytokinin biosynthesis in phytopathogens, *Ann. Rev. Plant Physiol.*, 37, 509, 1986.
20. **Palni, L. M. S., Horgan, R., Darrall, N. M., Stuchbury, T., and Wareing, P. F.,** Cytokinin biosynthesis in crown-gall tissue of *Vinca rosea*: the significance of nucleotides, *Planta*, 159, 50, 1983.
21. **Palni, L. M. S., Tay, S. A. B., Nandi, S. K., Pianca, D. J., de Klerk, G. J. M., Wong, O. C., Letham, D. S., and Macleod, J. K.,** Cytokinin biosynthesis in plant tumour tissues, *Biol. Plant.*, 27, 195, 1985.
22. **Robins, M. J., Hall, R. H., and Thedford, R.,** N^6-(Δ^2-isopentenyl)adenosine. A component of the transfer ribonucleic acid of yeast and of mammalian tissue, methods of isolation, and characterization, *Biochemistry*, 6, 1837, 1967.
23. **Yoshida, R. and Oritani, T.,** Studies on nitrogen metabolism in crop plants. XIII. Effects of nitrogen top-dressing on cytokinin content in the root exudate of rice plants, *Proc. Crop Sci. Soc. (Jpn)*, 43, 47, 1974.
24. **Van Staden, J. and Smith, A. R.,** The synthesis of cytokinins in excised roots of maize and tomato under aseptic conditions, *Ann. Bot.*, 42, 751, 1978.
25. **Koda, Y. and Okazawa, Y.,** Cytokinin production by tomato root: occurrence of cytokinins in staled medium of root culture, *Physiol. Plant.*, 44, 412, 1978.
26. **Engelbrecht, L.,** Cytokinins in leaf-cuttings of *Phaseolus vulgaris* L. during their development, *Biochem. Physiol. Pflanz.*, 163, 335, 1972.
27. **Featonby-Smith, B. C. and Van Staden, J.,** Endogenous cytokinins and rooting of leaf cuttings of *Phaseolus vulgaris*, *Z. Pflanzenphysiol.*, 102, 329, 1981.
28. **Maldiney, R., Pelese, F., Pilate, G., Sotta, B., Sossountzov, L., and Miginiac, E.,** Endogenous levels of abscisic acid, indole-3-acetic acid, zeatin and zeatin riboside during the course of adventitious root formation in cuttings of Craigella and Craigella lateral suppressor tomatoes, *Physiol. Plant.*, 68, 426, 1986.
29. **Sossountzov, L., Maldiney, R., Sotta, B., Sabbagh, I., Habricot, Y., Bonnet, M., and Miginiac, E.,** Immunocytochemical localization of cytokinins in Craigella tomato and a sideshootless mutant, *Planta*, 175, 291, 1988.
30. **Forsyth, C. and Van Staden, J.,** The effects of root decapitation on lateral root formation and cytokinin production in *Pisum sativum*, *Physiol. Plant.*, 51, 375, 1981.
31. **Skene, K. G. M.,** Cytokinin production by roots as a factor in the control of plant growth, in *The Development and Function of Roots*, Torrey, J. G. and Clarkson, D. T., Eds., Academic Press, London, 1975, chap. 17.
32. **Letham, D. S.,** Cytokinins, in *Phytohormones and Related Compounds: A Comprehensive Treatise*, Vol. 1, Letham, D. S., Goodwin, P. B., and Higgins, T. J. V., Eds., Elsevier/North Holland, Amsterdam, 1978, 205.
33. **Burrows, W. J. and Carr, D. J.,** Effects of flooding the root system of sunflower plants on the cytokinin content in the xylem sap, *Physiol. Plant.*, 22, 1105, 1969.
34. **Neuman, D. S., Rood, S. B., and Smit, B. A.,** Does cytokinin transport from root-to-shoot in the xylem sap regulate leaf responses to root hypoxia? *J. Exp. Bot.*, 41, 1325, 1990.
35. **Kende, H. and Sitton, D.,** The physiological significance of kinetin- and gibberellin-like root hormones, *Ann. N. Y. Acad. Sci.*, 144, 235, 1967.
36. **Wagner, H. and Michael, G.,** Einfluss unterschiedlicher Stickstoffversorgung auf die Cytokinin-bildung in Wurzeln von Sonnenblumenpflanzen, *Biochem. Physiol. Pflanzen*, 162, 147, 1971.
37. **Henson, I. E. and Wareing, P. F.,** Cytokinins in *Xanthium strumarium* L., distribution in the plant and production in the root system, *J. Exp. Bot.*, 27, 1268, 1976.
38. **Goodwin, P. B., Gollnow, B. I., and Letham, D. S.,** Phytohormones and growth correlations, in *Phytohormones and Related Compounds — A Comprehensive Treatise*, Vol. 2, Letham, D. S., Goodwin, P. B., and Higgins, T. J. V., Eds., Elsevier/North Holland, Amsterdam, 1978, chap. 4.
39. **Noodén, L. D. and Letham, D. S.,** Cytokinin metabolism and signalling in the soybean plant, *Aust. J. Plant Physiol.*, 20, 639, 1993.

40a. **Short, K. C. and Torrey, J. G.,** Cytokinins in seedling roots of pea, *Plant Physiol.*, 49, 155, 1972.

40b. **Letham, D. S.,** Unpublished data.

41a. **Weiss, C. and Vaadia, Y.,** Kinetin-like activity in root apices of sunflower plants, *Life Sci.*, 4, 1323, 1965.

41b. **Wightman, F. and Letham, D. S.,** Unpublished study.

42. **Feldman, L. J.,** Cytokinins and quiescent center activity in roots of *Zea*, in *The Development and Function of Roots*, Torrey, J. G. and Clarkson, D. T., Eds., Academic Press, London, 1975, 55.
43. **Feldman, L. J.,** Cytokinin biosynthesis in roots of corn, *Planta*, 145, 315, 1979.
44. **Torrey, J. G.,** On the initiation of organization in the root apex, in *Dynamics of Meristem Cell Populations*, Miller, M. W. and Kuehnert, C. C., Eds., Plenum Press, New York, 1972, 1.
45. **Zavala, M. E. and Brandon, D. L.,** Localization of a phytohormone using immunocytochemistry, *J. Cell Biol.*, 97, 1235, 1983.
46. **Feldman, L. J.,** The proximal meristem in the root apex of *Zea mays* L., *Ann. Bot.*, 43, 1, 1979.
47. **Henson, I. E. and Wareing, P. F.,** Cytokinins in *Xanthium strumarium* L.: the metabolism of cytokinins in detached leaves and buds in relation to photoperiod, *New Phytol.*, 78, 27, 1977.
48. **Carmi, A. and Van Staden, J.,** Role of roots in regulating the growth rate and cytokinin content of leaves, *Plant Physiol.*, 73, 76, 1983.
49. **Chailakhyan, M. K. and Khrianin, V. N.,** *Sexuality in Plants and its Hormonal Regulation*, Springer-Verlag, New York, 1987, chaps. 3 and 4.
50. **Chen, C. M. and Petschow, B.,** Cytokinin biosynthesis in cultured rootless tobacco plants, *Plant Physiol.*, 62, 861, 1978.
51. **Nitsch, J. P. and Nitsch, C.,** Presence d'une phytokinine dans le cambium, *Bull. Soc. Bot. Fr.*, 112, 1, 1965.
52. **Gautheret, R. J.,** *La Culture des Tissue Vegetaux*, Masson, Paris, 1959, 863.
53. **Van Staden, J. and Choveaux, N. A.,** Cytokinins in internodal stem segments of *Salix babylonica*, *Z. Pflanzenphysiol.*, 96, 153, 1980.
54. **Hansen, C. E., Wenzler, H., and Meins, F.,** Concentration gradients of *trans*-zeatin riboside and *trans*-zeatin in the maize stem, *Plant Physiol.*, 75, 959, 1984.
55. **Kannangara, T. and Booth, A.,** Diffusible cytokinins in shoot apices of *Dahlia variabilis*, *J. Exp. Bot.*, 25, 459, 1974.
56. **Koda, Y. and Okazawa, Y.,** Cytokinin production by *Asparagus* shoot apex cultured *in vitro*, *Physiol. Plant.*, 49, 193, 1980.
57. **Hewett, E. W. and Wareing, P. F.,** Cytokinins in *Populus* x *robusta*: changes during chilling and bud burst, *Physiol. Plant.*, 28, 393, 1973.
58. **Lee, P. K., Kessler, B., and Thimann, K. V.,** The effect of hadacidin on bud development and its implications for apical dominance, *Physiol. Plant.*, 31, 11, 1974.
59. **Tucker, D. J.,** Hormonal regulation of lateral bud outgrowth in the tomato, *Plant Sci. Lett.*, 8, 105, 1977.
60. **Van Staden, J. and Davey, J. E.,** The synthesis, transport and metabolism of endogenous cytokinins, *Plant Cell Environ.*, 2, 93, 1979.
61. **Van Staden, J. and Dimalla, G. G.,** The production and utilisation of cytokinins in rootless, dormant almond shoots maintained at low temperature, *Z. Pflanzenphysiol.*, 103, 121, 1981.
62. **Salama, A. M. and Wareing, P. F.,** Effect of mineral nutrition on endogenous cytokinins in plants of sunflower (*Helianthus annuus* L.), *J. Exp. Bot.*, 30, 971, 1979.
63. **Singh, S., Letham, D. S., Zhang, X., and Palni, L. M. S.,** Cytokinin biochemistry in relation to leaf senescence. VI. Effect of nitrogenous nutrients on cytokinin levels and senescence of tobacco leaves, *Physiol. Plant.*, 84, 262, 1992.
64. **Vonk, C. R.,** Origin of cytokinins transported in the phloem, *Physiol. Plant.*, 46, 235, 1979.
65. **Grayling, A. and Hanke, D. E.,** Cytokinins in exudates from leaves and roots of red *Perilla*, *Phytochemistry*, 31, 1863, 1992.
66. **Vonk, C. R. and Davelaar, E.,** 8-^{14}C-Zeatin metabolites and their transport from leaf to phloem exudate of *Yucca*, *Physiol. Plant.*, 52, 101, 1981.
67. **Van Staden, J. and Forsyth, C.,** Cytokinin biosynthesis in *Phaseolus vulgaris* leaf explants, *J. Plant Physiol.*, 119, 159, 1985.
68. **Peterson, C. A. and Fletcher, R. A.,** Formation of fruits on rootless plants, *Can. J. Bot.*, 51, 1899, 1973.
69. **Letham, D. S.,** Regulators of cell division in plant tissues. I. Inhibitors and stimulants of cell division in developing fruits: their properties and activity in relation to the cell division period, *N. Z. J. Bot.*, 1, 336, 1963.
70. **Letham, D. S. and Williams, M. W.,** Regulators of cell division in plant tissues. VIII. The cytokinins of the apple fruit, *Physiol. Plant.*, 22, 925, 1969.
71. **Blumenfeld, A. and Gazit, S.,** Cytokinin activity in avocado seeds during fruit development, *Plant Physiol.*, 46, 331, 1970.
72. **Gazit, S. and Blumenfeld, A.,** Cytokinin and inhibitor activities in the avocado fruit mesocarp, *Plant Physiol.*, 46, 334, 1970.

73. **Blumenfeld, A. and Gazit, S.,** Growth of avocado fruit callus and its relation to exogenous and endogenous cytokinins, *Physiol. Plant.*, 25, 369, 1971.
74. **Mapeli, S.,** Changes in cytokinin in the fruits of parthenocarpic and normal tomatoes, *Plant Sci. Lett.*, 22, 227, 1981.
75. **Letham, D. S.,** Cultivation of apple-fruit tissue *in vitro*, *Nature*, 182, 473, 1958.
76. **Letham, D. S. and Bollard, E. G.,** Stimulants of cell division in developing fruits, *Nature*, 191, 1119, 1961.
77. **Hahn, H., De Zacks, R., and Kende, H.,** Cytokinins formation in pea seeds, *Naturwissenschaften*, 61, 170, 1974.
78. **Krechting, H. C. J. M., Varga, A., and Bruinsma, J.,** Absence of cytokinin biosynthesis in pea seeds developing *in vitro*, *Z. Pflanzenphysiol.*, 87, 91, 1978.
79. **Van Staden, J. and Button, J.,** The cytokinin content of aseptically cultured pea fruits, *Z. Pflanzenphysiol.*, 87, 129, 1978.
80. **Davey, J. E. and Van Staden, J.,** Cytokinin activity in *Lupinus albus*. III. Distribution in fruits, *Physiol. Plant.*, 43, 87, 1978.
81. **Summons, R. E., Entsch, B., Parker, C. W., and Letham, D. S.,** Mass spectrometric analysis of cytokinins in plant tissues. III. Quantitation of the cytokinin glycoside complex of lupin pods by stable isotope dilution, *FEBS Lett.*, 107, 21, 1979.
82. **Barratt, D. H. P.,** *In vitro* pod culture of *Pisum sativum*, *Plant Sci.*, 43, 223, 1986.
83. **Obendorf, R. L., Rytko, G. T., and Byrne, M. C.,** Soya bean seed growth and maturation by in vitro pod culture, *Ann. Bot.*, 51, 217, 1983.
84. **Hsu, F. C. and Obendorf, R. L.,** Compositional analysis of in vitro matured soybean seeds, *Plant Sci. Lett.*, 27, 129, 1982.
85. **Dickens, C. W. S. and Van Staden, J.,** The metabolism of ^{14}C-adenine and ^{14}C-isopentenyl pyrophosphate by soybean explants flowering and fruiting *in vitro*, *J. Plant Physiol.*, 129, 73, 1987.
86. **Van Staden, J. and Forsyth, C.,** The metabolism of adenine and zeatin in immature caryopses of maize, *J. Plant Physiol.*, 124, 299, 1986.
87. **Van Staden, J. and Vos, J. E.,** [8-^{14}C] Adenine feeding and cytokinin production in pea fruits, *J. Plant Physiol.*, 135, 114, 1989.
88. **Ilan, I. and Gepstein, S.,** Hormonal regulation of food reserve breakdown in germinating dicotyledonous seeds, *Israel J. Bot.*, 29, 193, 1981.
89. **Van Onckelen, H. A., Caubergs, R., and DeGreef, J.A.,** Effect of light treatment and endogenous growth regulators on α- and β-amylase activities in cotyledons of *Phaseolus vulgaris* L., *Plant Cell Physiol.*, 18, 1029, 1977.
90. **Hutton, M. J., Van Staden, J., and Davey, J. E.,** Cytokinins in germinating seeds of *Phaseolus vulgaris* L. I. Changes in endogenous levels within the cotyledons, *Ann. Bot.*, 49, 685, 1982.
91a. **Munoz, J. L., Martin, L., Nicolas, G., and Villalobos, N.,** Influence of endogenous cytokinins on reserve mobilization in cotyledons of *Cicer arietinum* L., *Plant Physiol.*, 93, 1011, 1990.
91b. **Letham, D. S.,** Unpublished data.
92. **Nandi, S. K., Palni, L. M. S., Letham, D. S., and Knypl, J.S.,** The biosynthesis of cytokinins in germinating lupin seed, *J. Exp. Bot.*, 39, 1649, 1988.
93. **Hocart, C. H. and Letham, D. S.,** Biosynthesis of cytokinin in germinating seeds of *Zea mays*, *J. Exp. Bot.*, 41, 1525, 1990.
94. **Zhang, R., Letham, D. S., Wong, O. C., Noodén, L. D., and Parker, C.W.,** Cytokinin biochemistry in relation to leaf senescence. II. The metabolism of 6-benzylaminopurine in soybean leaves and the inhibition of its conjugation, *Plant Physiol.*, 83, 334, 1987.
95. **Jacobs, W. P.,** Apolar movement of zeatin through *Coleus* petioles and *Pisum* roots as estimated by bioassay and radioactive labeling, *Am. J. Bot.*, 63, 571, 1976.
96. **Lagerstedt, H. B. and Langston, R. G.,** Translocation of radioactive kinetin, *Plant Physiol.*, 42, 611, 1967.
97. **Kulaeva, O. N.,** The effect of roots on leaf metabolism in relation to the action of kinetin on leaves, *Sov. Plant Physiol.*, 9, 182, 1962.
98. **Murofushi, N., Inoue, A., Watanabe, N., Ota, Y., and Takahashi, N.,** Identification of cytokinins in root exudate of the rice plant, *Plant Cell Physiol.*, 24, 87, 1983.
99. **Jameson, P. E., Letham, D. S., Zhang, R., Parker, C. W., and Badenoch-Jones, J.,** Cytokinin translocation and metabolism in lupin species. I. Zeatin riboside introduced into the xylem at the base of *Lupinus angustifolius* stems, *Aust. J. Plant Physiol.*, 14, 695, 1987.
100. **Palmer, M. V. and Wong, O. C.,** Identification of cytokinins from xylem exudate of *Phaseolus vulgaris* L., *Plant Physiol.*, 79, 296, 1985.

101. **Griggs, P., Stuchbury, T., and Wang, T. L.,** Dihydrozeatin riboside from *Phaseolus* sap: quantification by radioimmunoassay and GC-MS, *Phytochemistry*, 27, 1583, 1988.
102. **Fusseder, A., Wagner, B., and Beck, E.,** Quantification by ELISA of cytokinins in root-pressure exudates of *Urtica dioica* plants grown under different nitrogen levels, *Botanica Acta*, 101, 214, 1988.
103. **Parker, C. W., Badenoch-Jones, J., and Letham, D. S.,** Radioimmunoassay for quantifying the cytokinins *cis*-zeatin and *cis*-zeatin riboside and its application to xylem sap samples, *J. Plant Growth Regul.*, 8, 93, 1989.
104. **Badenoch-Jones, J., Parker, C. W., and Letham, D. S.,** Unpublished data, 1993.
105. **Hall, P. J., Badenoch-Jones, J., Parker, C. W., Letham, D. S., and Barlow, B.A.,** Identification and quantification of cytokinins in the xylem sap of mistletoes and their hosts in relation to leaf mimicry, *Aust. J. Plant Physiol.*, 14, 429, 1987.
106. **Hautala, E., Stafford, A., Corse, J., and Barker, P. A.,** Cytokinin variation in the sap of male and female *Gymnocladus dioica*, *J. Chromatogr.*, 351, 560, 1986.
107. **Gordon, M. E., Letham, D. S., and Parker, C. W.,** Regulators of cell division in plant tissues. XVII. The metabolism and translocation of zeatin in intact radish seedlings, *Ann. Bot.*, 38, 809, 1974.
108. **Ramina, A.,** Aspects of [8-^{14}C]benzylaminopurine metabolism in *Phaseolus vulgaris*, *Plant Physiol.*, 63, 298, 1979.
109. **Hill-Cottingham, D. G. and Lloyd-Jones, C. P.,** Relative mobility of some organic nitrogenous compounds in the xylem of apple shoots, *Nature*, 220, 389, 1968.
110. **Scott, I. M. and Horgan, R.,** Root cytokinins of *Phaseolus vulgaris* L., *Plant Sci. Lett.*, 34, 81, 1984.
111. **Mozes, R. and Altman, A.,** Characteristics of root-to-shoot transport of cytokinin 6-benzylaminopurine in intact seedlings of *Citrus aurantium*, *Physiol. Plant.*, 39, 225, 1977.
112. **Morris, D. A. and Winfield, P. J.,** Kinetin transport to axillary buds of dwarf pea, *J. Exp. Bot.*, 23, 346, 1972.
113. **Morris, D. A.,** Distribution and metabolism of root-applied cytokinins in *Pisum sativum*, *Physiol. Plant.*, 52, 251, 1981.
114. **Prochazka, S., Navratilova, D., and Blazkova, J.,** Transport of benzyl-8-^{14}C-adenine from roots into shoots of pea seedlings with regard to outgrowth of lateral buds, *Beitr. Biol. Pflanzen.*, 58, 421, 1977.
115. **Henson, I. E. and Wheeler, C. T.,** Hormones in plants bearing nitrogen-fixing root nodules: cytokinin transport from the root nodules of *Alnus glutinosa* (L.) Gaertn, *J. Exp. Bot.*, 28, 1099, 1977.
116. **Davey, J. E. and Van Staden, J.,** Cytokinin activity in *Lupinus albus*. V. Translocation and metabolism of [8-^{14}C]zeatin applied to the xylem of fruiting plants, *Physiol. Plant.*, 51, 45, 1981.
117. **Hutton, M. J. and Van Staden, J.,** Transport and metabolism of labelled zeatin applied to the stems of *Phaseolus vulgaris* at different stages of development, *Z. Pflanzenphysiol.*, 114, 331, 1984.
118. **Bowen, M. R. and Wareing, P. F.,** The interchange of ^{14}C-kinetin and ^{14}C-gibberellic acid between the bark and xylem of willow, *Planta*, 89, 108, 1969.
119. **Zhang, R. and Letham, D. S.,** Cytokinin translocation and metabolism in lupin species. III. Translocation of xylem cytokinin into the seeds of lateral shoots of *Lupinus angustifolius*, *Plant Sci.*, 70, 65, 1990.
120. **Noodén, L. D. and Letham, D. S.,** Translocation of zeatin riboside and zeatin in soybean explants, *J. Plant Growth Regul.*, 2, 265, 1984.
121. **Singh, S., Letham, D. S., Jameson, P. E., Zhang, R., Parker, C. W., Badenoch-Jones, J., and Noodén, L. D.,** Cytokinin biochemistry in relation to leaf senescence. IV. Cytokinin metabolism in soybean explants, *Plant Physiol.*, 88, 788, 1988.
122. **Varga, A. and Bruinsma, J.,** The growth and ripening of tomato fruits at different levels of endogenous cytokinins, *J. Hort. Sci.*, 49, 135, 1974.
123. **Hoad, G. V., Loveys, B. R., and Skene, K. G. M.,** The effect of fruit removal on cytokinins and gibberellin-like substances in grape leaves, *Planta*, 136, 25, 1977.
124. **Hewett, E. W. and Wareing, P. F.,** Cytokinins in *Populus robusta*: qualitative changes during development, *Physiol. Plant.*, 29, 386, 1973.
125. **Colbert, K. A. and Beever, J. E.,** Effect of disbudding on root cytokinin export and leaf senescence in tomato and tobacco, *J. Exp. Bot.*, 32, 121, 1981.
126. **Badenoch-Jones, J., Rolfe, B. G., and Letham, D. S.,** Phytohormones, *Rhizobium* mutants, and nodulation in legumes. V. Cytokinin metabolism in effective and ineffective pea root nodules, *Plant Physiol.*, 74, 239, 1984.

127. **Van Staden, J. and Mallett, J. A.**, Metabolism and tranport of [8-^{14}C]benzyladenine applied to rootless shoots and the roots of intact tomato plants, *Plant Physiol. Biochem.*, 26, 47, 1988.
128. **Palmer, M. V., Horgan, R., and Wareing, P. F.**, Cytokinin metabolism in *Phaseolus vulgaris* L. Identification of endogenous cytokinins and metabolism of [8-^{14}C] dihydrozeatin in stems of decapitated plants, *Planta*, 153, 297, 1981.
129. **Van Staden, J. and Bosse, C. A.**, Cytokinins in cut carnation flowers. VI. The transport and metabolism of zeatin and dihydrozeatin, *J. Plant Physiol.*, 135, 160, 1989.
130. **Van Staden, J., Bayley, A. D., Upfold, S. J., and Drewes, F. E.**, Cytokinins in cut carnation flowers. VIII. Uptake, transport and metabolism of benzyladenine and the effect of benzyladenine derivatives on flower longevity, *J. Plant Physiol.*, 135, 703, 1990.
131. **Harrison, M. A. and Kaufman, P. B.**, The role of hormone transport and metabolism in apical dominance in oats, *Bot. Gaz.*, 145, 293, 1984.
132. **Parker, C. W. and Letham, D. S.**, Regulators of cell division in plant tissues. XVI. Metabolism of zeatin by radish cotyledons and hypocotyls, *Planta*, 114, 199, 1973.
133. **Letham, D. S. and Zhang, R.**, Cytokinin translocation and metabolism in lupin species. II. New nucleotide metabolites of cytokinins, *Plant Sci.*, 64, 161, 1989.
134. **Duke, C. C., Letham, D. S., Macleod, J. K., Parker, C. W., and Summons, R.E.**, The complex of *O*-glucosylzeatin derivatives formed in *Populus* species, *Phytochemistry*, 18, 819, 1979.
135. **Palmer, M. V., Horgan, R., and Wareing, P. F.**, Cytokinin metabolism in *Phaseolus vulgaris* L. I. Variations in cytokinin levels in leaves of decapitated plants in relation to lateral bud outgrowth, *J. Exp. Bot.*, 32, 1231, 1981.
136. **Vonk, C. R.**, Formation of cytokinin nucleotides in a detached inflorescence stalk and the occurrence of nucleotides in phloem exudate from attached *Yucca* plants, *Physiol. Plant.*, 44, 161, 1978.
137. **Van Staden, J.**, Occurrence of a cytokinin glucoside in the leaves and in honeydew of *Salix babylonica*, *Physiol. Plant.*, 36, 225, 1976.
138. **Weiler, E. M. and Ziegler, H.**, Determination of phytohormones in phloem exudate from tree species by radioimmunoassay, *Planta*, 152, 168, 1981.
139. **Lejeune, P., Kinet, J. M., and Bernier, G.**, Cytokinin fluxes during floral induction in the long day plant *Sinapis alba* L., *Plant Physiol.*, 86, 1095, 1988.
140. **Taylor, J. S., Thompson, B., Pate, J. S., Atkins, C. A., and Pharis, R. P.**, Cytokinins in the phloem sap of white lupin (*Lupinus albus* L.), *Plant Physiol.*, 94, 1714, 1990.
141. **Summons, R. E., Letham, D. S., Gollnow, B. I., Parker, C. W., Entsch, B., Johnson, L. P., Macleod, J. K., and Rolfe, B. G.**, Cytokinin translocation and metabolism in species of the *Leguminosae* — studies in relation to shoot and nodule development, in *Metabolism and Molecular Activities of Cytokinins*, Guern, J. and Peaud-Lenoel, C., Eds., Springer-Verlag, Berlin, 1981, 69.
142. **Abo-hamed, S., Collin, H. A., and Hardwick, K.**, Biochemical and physiological aspects of leaf development in cocoa. VIII. Export and distribution of ^{14}C auxin and kinin from the young and mature leaves, *New Phytol.*, 97, 219, 1984.
143. **Letham, D. S., Tao, G. Q., and Parker, C. W.**, An overview of cytokinin metabolism, in *Plant Growth Substances 1982*, Wareing, P. F., Ed., Academic Press, London, 1982, 143.
144. **Van Staden, J.**, Transport of [8-^{14}C]-zeatin from mature rose leaves after shoot decapitation, *Bot. Gaz.*, 143, 201, 1982.
145. **Van Staden, J. and Davey, J. E.**, Cytokinin activity in *Lupinus alba*. VI. Translocation and metabolism of [8-^{14}C]zeatin applied to the leaves and fruits of fruiting plants, *Physiol. Plant.*, 51, 49, 1981.
146. **Nandi, S. K. and Palni, L. M. S.**, Transport and metabolism of dihydrozeatin riboside in germinating lupin seeds, *J. Exp. Bot.*, 40, 615, 1989.
147. **Hutton, M. J. and Van Staden, J.**, Cytokinins in germinating seeds of *Phaseolus vulgaris* L. II. Transport and metabolism of 8[^{14}C]t-zeatin applied to the radicles, *Ann. Bot.*, 49, 693, 1982.
148. **Hocart, C. H., Letham, D. S., and Parker, C. W.**, Metabolism and translocation of exogenous zeatin riboside in germinating seeds and seedlings of *Zea mays*, *J. Exp. Bot.*, 41, 1517, 1990.
149. **Dodd, M. C. and Van Staden, J.**, Cytokinins in *Podocarpus henkelii* seeds. II. Transport and metabolism of 8[^{14}C]zeatin during seed germination, *Z. Pflanzenphysiol.*, 108, 401, 1982.

150. **Nandi, S. K., Palni, L. M. S., and Letham, D. S.,** Axial control of cotyledon expansion and chlorophyll formation in germinating lupin seed, *Plant Sci.*, 60, 181, 1989.
151. **Wiley, L. and Ashton, F. M.,** Influence of the embryonic axis on protein hydrolysis in cotyledons of *Cucurbita maxima*, *Physiol. Plant.*, 20, 688, 1967.
152. **Penner, D. and Ashton, F. M.,** Hormonal control of proteinase activity in squash cotyledons, *Plant Physiol.*, 42, 791, 1967.
153. **Gepstein, S. and Ilan, I.,** Cytokinin-induced amylolytic activity in bean cotyledons: identification of the regulated enzyme, *Plant Cell Physiol.*, 20, 1603, 1979.
154. **Nandi, S. K.,** Studies of cytokinin biosynthesis, metabolism and function, Ph.D. thesis, Australian National University, Canberra, 1988.
155. **Chapman, J. M. and Davies, H. V.,** Control of the breakdown of food reserves in germinating dicotyledonous seeds — a reassessment, *Ann. Bot.*, 52, 594, 1983.
156. **de Klerk, G. J.,** Axial control of amylase synthesis and secretion in cotyledons of *Agrostemma githago*, *Acta Bot. Neerl.*, 35, 143, 1986.
157. **Bernier, G., Kinet, J. M., Bodson, M., Rouma, Y., and Jacqmard, A.,** Experimental studies on the mitotic activity of the shoot apical meristem and its relation to floral evocation and morphogenesis in *Sinapis alba*, *Bot. Gaz.*, 135, 345, 1974.
158. **Loomis, R. S. and Torrey, J. G.,** Chemical control of vascular cambium initiation in isolated radish roots, *Proc. Natl. Acad. Sci. U.S.A.*, 52, 3, 1964.
159. **Torrey, J. G. and Loomis, R. S.,** Auxin-cytokinin control of secondary vascular tissue formation in isolated roots of *Raphanus*, *Am. J. Bot.*, 54, 1098, 1967.
160. **Webster, B. D. and Radin, J. W.,** Growth and development of cultured radish roots, *Am. J. Bot.*, 59, 744, 1972.
161. **Peterson, R. L.,** Control of cambial activity in roots of turnip (*Brassica rapa*), *Can. J. Bot.*, 51, 475, 1973.
162. **Radin, J. W. and Loomis, R. S.,** Changes in the cytokinins of radish roots during maturation, *Physiol. Plant.*, 25, 240, 1971.
163. **Trewavas, A. J.,** Is plant development regulated by changes in the concentration of growth substances or by changes in the sensitivity to growth substances? *Trends Biochem. Sci.*, 8, 354, 1983.
164. **Sutcliffe, J. F.,** Regulation in the whole plant, in *Transport in Plants. II. Tissues and Organs, Encyclopedia of Plant Physiology*, Vol. 2B, Luttge, U. and Pitman, M. G., Eds., Springer-Verlag, Berlin, 1967, 394.
165. **Dixon, R. K., Garrett, H. E., and Cox, G. S.,** Cytokinins in the root pressure exudate of *Citrus jambhiri* Lush. colonized by vesicular-arbuscular mycorrhizae, *Tree Physiol.*, 4, 9, 1988.
166. **Theimer, R. R., Anding, G., and Matzner, P.,** Kinetin action on the development of microbody enzymes in sunflower cotyledons in darkness, *Planta*, 128, 41, 1976.
167. **Carmi, A. and Koller, D.,** Effects of the roots on the rate of photosynthesis in primary leaves of bean (*Phaseolus vulgaris* L.), *Photosynthetica*, 12, 178, 1978.
168. **Wareing, P. F., Khalifa, M. M., and Treharne, K. J.,** Rate-limiting processes in photosynthesis at saturating light intensities, *Nature*, 220, 453, 1968.
169. **Carmi, A.,** Effects of cytokinins and root pruning on photosynthesis and growth, *Photosynthetica*, 20, 1, 1986.
170. **Aloni, R.,** The role of cytokinin in organized differentiation of vascular tissues, *Aust. J. Plant Physiol.*, in press, 1993.
171. **Noodén, L. D., Blakley, K. A., and Grzybowski, J. M.,** Control of seed coat thickness and permeability in soybean, *Plant Physiol.*, 79, 543, 1985.
172. **Parker, C. W., Badenoch-Jones, J., and Letham, D. S,** Unpublished results, 1993.
173. **Carmi, A. and Heuer, B.,** The role of roots in control of bean shoot growth, *Ann. Bot.*, 48, 519, 1981.
174. **Jesko, T. and Vizarova, G.,** Changes of free endogenous cytokinins during transitorily increased photosynthetic rate initiated by formation of the first two whorls of nodal roots in *Zea mays*, *Photosynthetica*, 14, 83, 1980.
175. **Tromp, J. and Ovaa, J. C.,** Seasonal changes in the cytokinin composition of xylem sap of apple, *J. Plant Physiol.*, 136, 606, 1990.
176. **Young, E.,** Cytokinin and soluble carbohydrate concentrations in xylem sap of apple during dormancy and budbreak, *J. Am. Soc. Hort. Sci.*, 114, 297, 1989.
177. **Woolley, D. J. and Wareing, P. F.,** The role of roots, cytokinins and apical dominance in the control of lateral shoot form in *Solanum andigena*, *Planta*, 105, 33, 1972.

178. **Woolley, D. J. and Wareing, P. F.,** The interaction between growth promoters in apical dominance. I. Hormonal interaction, movement and metabolism of a cytokinin in rootless cuttings, *New Phytol.*, 71, 781, 1972.

179. **Woolley, D. J. and Wareing, P. F.,** The interaction between growth promoters in apical dominance. II. Environmental effects on endogenous cytokinin and gibberellin levels in *Solanum andigena*, *New Phytol.*, 71, 1015, 1972.

180. **Wang, T. L. and Wareing, P. F.,** Cytokinins and apical dominance in *Solanum andigena*: lateral shoot growth and endogenous cytokinin levels in the absence of roots, *New Phytol.*, 82, 19, 1979.

181. **Upadhyaya, N. M., Parker, C. W., Letham, D. S., Scott, K. F., and Dart, P. J.,** Evidence for cytokinin involvement in *Rhizobium* (IC 3342)-induced leaf curl syndrome of pigeonpea, *Plant Physiol.*, 95, 1019, 1990.

182. **Mothes, K. and Engelbrecht, L.,** On the activity of a kinetin-like root factor, *Life Sci.*, 2, 852, 1963.

183. **McDavid, C. R., Sagar, G. R., and Marshall, C.,** The effect of root pruning and 6-benzylaminopurine on the chlorophyll content, CO_2 fixation and the shoot/root ratio in seedlings of *Pisum sativum* L., *New Phytol.*, 72, 465, 1973.

184. **Hsia, C. P. and Kao, C. H.,** The importance of roots in regulating the senescence of soybean primary leaves, *Physiol. Plant.*, 43, 385, 1978.

185. **Herzog, H.,** *Source and Sink during the Reproductive Period of Wheat. Development and its Regulation with Special Reference to Cytokinins,* Adv. Agron. Crop. Sci., Paul Parey Publishers Berlin, 1986, chap. 2.

186. **Beever, J. E. and Woolhouse, H. W.,** Increased cytokinin export from the roots of *Perilla frutescens* following disbudding or floral induction, in *Mechanisms of Regulation of Plant Growth,* Bieleski, R. L., Ferguson, A. R., and Cresswell, M. M., Eds., Royal Society of New Zealand, Wellington, 1974, 681.

187. **Noodén, L. D., Singh, S., and Letham, D. S.,** Correlation of xylem sap cytokinin levels with monocarpic senescence in soybean, *Plant Physiol.*, 93, 33, 1990.

188. **Hubick, K. T., Taylor, J. S., and Reid, D. M.,** The effect of drought on levels of abscisic acid, cytokinins, gibberellins and ethylene in aeroponically-grown sunflower plants, *Plant Growth Regul.*, 4, 139, 1986.

189. **Davies, W. J. and Zhang, J.,** Root signals and the regulation of growth and development of plants in drying soil, *Ann. Rev. Plant Physiol. Plant. Mol. Biol.*, 42, 55, 1991.

190. **Incoll, L. D. and Jewer, P. C.,** Cytokinins and stomata, in *Stomatal Function*, Zeiger, E., Farquhar, G. D., and Cowan, I. R. Eds., Stanford University Press, Palo Alto, CA, 1987, chap. 12.

191. **Incoll, L. D., Ray, J. P., and Jewer, P. C.,** Do cytokinins act as root to shoot signals? in *Importance of Root to Shoot Communication in the Responses to Environmental Stress*, Davies, W. J. and Jeffcoat, B., Eds., British Society for Plant Growth Regulation, Bristol, 1990, 185.

192. **Simpson, R. J., Lambers, H., and Dalling, M. J.,** Kinetin application to roots and its effect on uptake, translocation and distribution of nitrogen in wheat grown with a split root system, *Physiol. Plant.*, 56, 430, 1982.

193. **Incoll, L. D. and Jewer, P. C.,** Cytokinins and the water relations of whole plants, in *Cytokinins: Plant Hormones in Search of a Role*, Horgan, R. and Jeffcoat, B., Eds., British Plant Growth Regulator Group, Bristol, 1986, 85.

194. **Went, F. and Thimann, K. V.,** Phytohormones, Macmillan, New York, 1937.

195. **Sachs, T.,** A possible basis for apical organization in plants, *J. Theor. Biol.*, 37, 353, 1972.

196. **Sachs, T.,** Cellular interactions in tissue and organ development, in *Plasticity in Plants*, Symp. Soc. Exp. Biol. no. 40, Jennings, D. H. and Trewavas, A. J., Eds., Company of Biologists, Cambridge, 1986, 181.

197. **Henson, I. E. and Wareing, P. F.,** Cytokinins in *Xanthium strumarium* L., some aspects of the photoperiodic control of endogenous levels, *New Phytol.*, 78, 35, 1977.

Chapter 6

Cytokinin Biosynthetic Enzymes in Plants and Slime Mold

Chong-maw Chen and John R. Ertl

TABLE OF CONTENTS

I. INTRODUCTION

An understanding of the biosynthetic pathway(s) of cytokinins would provide basic knowledge for regulating the hormone in plant cell division and differentiation. Two biosynthetic pathways have been reported for cytokinins: the *de novo* pathway[1-5] and the tRNA pathway.[6-7] While the turnover of cytokinin containing tRNAs may be one of the sources of cytokinins, there is no conclusive evidence that this alone could represent an adequate source to supply the need for cytokinin in plant cells, especially considering the rapid metabolic turnover rate of free cytokinins in cells. It is very likely that the *de novo* biosynthetic pathway reported in plants,[2,8-11] *Dictyostelium*,[1,12] and some microorganisms[3-5,13-17] plays an important role in cytokinin biosynthesis.

In the *de novo* biosynthetic pathway, the key enzyme, Δ^2-ipp:5′-AMP-Δ^2-isopentenyltransferase (isopentenyltransferase or iptase) (EC 2.5.1.8), catalyzes the formation of i^6AMP from 5′AMP and Δ^2ipp. The enzyme is extremely unstable,[2,5,8,12] but has been partially purified from plants[2,5,11] and slime mold.[12] This paper reviews the current status of the purification of iptase and provides hypothetical approaches to the isolation of the iptase gene from plants.

II. ISOLATION OF ENZYMES

A. CELLS AND CELL CULTURES

The cytokinin-autotrophic tobacco pith tissue culture used by Chen and Melitz[2] and Chen[8] was originally isolated by Fox[18] and has been grown on a culture medium devoid of cytokinin and auxin.[18] Nishinari and Syono used tobacco strain XD6S cells,[9] isolated from the stem of *Nicotiana tabacum* cv. Xanthi, for cell-free cytokinin biosynthesis. Developing corn kernels were used for the partial purification of iptase, by Reinecke et al.[10,11] The slime mold *Dictyostelium discoideum*, employed by Ihara et al.[12] for iptase purification, was cultured on nutrient agar plates containing Murashige Skoog (MS) medium.[19]

B. ENZYME EXTRACTION

Plant tissues or slime mold were frozen with liquid N_2 and ground to powder and homogenized with a buffer containing appropriate reagents, as shown in Table 1. Plant tissues contain very low levels of cytokinins;[20] thus, it is expected that the enzymes involved in cytokinin biosynthesis are present at extremely low levels. The biosynthetic enzyme, iptase, is very unstable and the substrates ipp and 5′AMP, as well as the reaction product

0-8493-6252-0/94/$0.00+$.50

Table 6-1 **Composition of Enzyme Extraction Buffers**

Source of Cells	Extraction Buffer	Reference
Cytokinin-autotrophic tobacco tissue	30 m*M* Tris.HCl, pH 7.0 15 m*M* magnesium acetate 6 m*M* 2-mercaptoethanol 1 m*M* EDTA, 25% glycerol	2
Tobacco XD6S strain cells	10 m*M* Tris.HCl, pH 8.0 10 m*M* magnesium acetate 60 m*M* KCl, 1 m*M* EDTA 6 m*M* 2-mercaptoethanol 10% glycerol	9
Immature corn kernels	50 m*M* HEPES buffer, pH 7.5 15–30 m*M* $MgCl_2$, 10 m*M* 2-mercaptoethanol, 5% PVP, protease inhibitors	11
Slime mold	50 m*M* Tris.HCl, pH 7.5 1 m*M* EDTA, 0.1 m*M* DTT 10% glycerol, 100 m*M* KCl 10 m*M* magnesium acetate	12

i^6AMP, are targets for phosphatase attack. This is one of the reasons for erratic experimental results.

C. ENZYME ASSAY

The activity of iptase is assayed by measuring either the amount of radioactive i^6AMP formed from Δ^2ipp and radiolabeled 5′AMP or the amount of radioactive i^6Ado present after the dephosphorylation of the i^6AMP.[2,3,5,8,9,11,12]

The following reaction mixture, with slight variation, has been used by various laboratories for iptase assays: 50 m*M* Tris.HCl buffer (pH 7.0 or 7.5) (or 50 m*M* HEPES buffer, pH 7.0); 5 to 30 m*M* $MgCl_2$ (or 30 m*M* magnesium acetate); 0.15 m*M* Δ^2iPP; ^{3}H- or ^{14}C-labeled 5′AMP; 5 to 10 m*M* KF; 5 m*M* 2-mercaptoethanol; and the enzyme preparation. After 30 min incubation at 25 or 37°C, the reaction is stopped either by (1) adding one volume of 95% ethanol or by (2) boiling in a water bath for 2 min. If procedure (1) is used, the reaction product is re-extracted with 95% ethanol and then analyzed by paper electrophoresis; in procedure (2) the reaction mixture is cooled, and alkaline phosphatase is added to convert radioactive i^6AMP to i^6Ado and extracted with 2 volumes of water-saturated ethylacetate. Unlabeled i^6Ado can be added to the reaction mixture, as a carrier before ethylacetate extraction, to improve the recovery of radioactive i^6Ado. AMP, Ado, and i^6AMP are present in the aqueous phase, while i^6Ado is extracted into the organic phase. The amount of radioactive i^6Ado can be determined by liquid scintillation counting. Extraction of i^6Ado with water-saturated ethyl-acetate is less time consuming and can be used for a quick enzyme assay, but it is not an accurate quantification method.

One unit of enzyme is defined as the amount of enzyme that produces 1 pmol of i^6AMP per minute under described assay conditions.

III. IDENTIFICATION OF BIOSYNTHESIZED CYTOKININS

Several techniques or combinations of techniques have been developed to identify biosynthesized cytokinins.

A. PAPER ELECTROPHORESIS

Paper electrophoretic analysis of the extracted products, with unlabeled markers, will tentatively identify the biosynthesized radioactive cytokinins.[2] Further analysis is needed to fully characterize the biosynthesized cytokinins.

B. COLUMN, PAPER, AND HIGH-PERFORMANCE LIQUID CHROMATOGRAPHY

The reaction product extracted with water-saturated ethylacetate is reduced to a small volume, fractionated over a column of Sephadex LH-20, and further characterized by paper chromatography, in various solvent systems,[2] and/or by high-performance liquid chromatography (HPLC).[21]

C. HPLC/IMMUNOAFFINITY CHROMATOGRAPHY

A short ODS column (Altex; 3 μm, 4.6 × 75 mm) is used to remove proteins from the reaction product, and biosynthesized cytokinins are separated by HPLC.[5] Alternatively, biosynthesized cytokinin nucleotide is dephosphorylated and isolated by immunoaffinity (anti-i^{6}Ado antibody-cellulose) chromatography.[5]

D. UNIQUE CHEMICAL REACTION OF CYTOKININ NUCLEOSIDES

i^{6}Ado or io^{6}Ado undergoes a unique reaction in the presence of HCl.[22] Biosynthesized radioactive cytokinin nucleotide is dephosphorylated and mixed with unlabeled i^{6}Ado, and the mixture is hydrolyzed, with HCl, in a sealed tube in boiling water for 15 min. The reaction product is identified by paper chromatography.[2]

IV. PARTIAL PURIFICATION OF IPTASE

Due to the instability of iptase, very limited information on the purification of this enzyme is available. Thus, only the procedures used in the partial purification of iptase from slime mold and plant tissues are described here.

A. PARTIAL PURIFICATION OF IPTASE FROM SLIME MOLD

The enzyme was purified 6687-fold from the fruiting body of the cellular slime mold,[12] using various separation procedures, including ammonium sulfate fractionation, ion-exchange columns, gel filtration, and 5′AMP$^{ox\text{-}red}$-Sepharose 4B affinity column chromatography. The final step of purification, DEAE-cellulose column chromatography, resulted in a specific activity of 10,700 units per milligram of protein. This partially purified enzyme was very unstable and lost most of its activity in a day. However, when the enzyme was less purified (about 1680 units per milligram of protein; or 1000-fold purification), the enzyme preparation was stable for about a month at 4°C.

Starting with 11,200 mg of protein in a crude extract, the total protein recovered after DEAE-cellulose chromatography was 0.006 mg.

B. PARTIAL PURIFICATION OF IPTASE FROM PLANTS

There is no report of the total purification of iptase from plants. This enzyme has been found in cytokinin-autotrophic tissue cultures,[2,8] tobacco strain XD6S cells,[9] and developing corn kernels.[10,11] Attempts to purify this enzyme from plant sources have been plagued by the problem of loss of activity. This biosynthetic enzyme was partially purified from cytokinin-autotrophic tobacco tissue cultures (the degree of purification was about 50-fold), using the procedure of ammonium sulfate fractionation, ion-exchange column chromatography, and gel filtration column chromatography.[2,8] Continual efforts have been made to further purify this enzyme, using various techniques, including DEAE-Sephacel, Bio-Gel HPT, Affi-gel Blue, AMP-affinity column chromatography and hydroxyapatite HPLC, as well as using various enzyme stabilizing agents; but purification of this enzyme to near homogeneity has thus far not been achieved.[23a] If this enzyme could be purified about 3000-fold from cytokinin-autotrophic tobacco tissue, total purification might be achieved by immunoaffinity chromatography prepared from cytokinin-heterotrophic tobacco tissue, using a subtractive immunoabsorption technique.[23b]

C. KINETICS AND PROPERTIES OF IPTASE

A partially purified enzyme has been used for studying the kinetics and properties of the enzyme (Table 2).

V. ALTERNATIVE APPROACHES TO THE STUDY OF IPTASE

Since the purification of iptase has, thus far, proven difficult, alternative approaches to the isolation of this enzyme have been considered. If the plant gene for iptase could be isolated and expressed, the enzyme could then be further characterized. Attempts to identify the iptase gene in plants, by heterologous probing with bacterial iptase sequences, have been unsuccessful.[24] However, since both enzymes catalyze the same reaction, they may share amino acid motifs involved in substrate binding or catalytic activity. Comparisons of the bacterial iptase genes have been made,[14,24] and additional sequences have since been deter-mined.[25,26a] An ipp binding motif observed in polyprenyl synthetases is not well conserved among the known iptases;[27] however, multiple sequence alignments reveal patterns that are.[26b] Degenerate oligomers designed against these regions may allow identification of an iptase clone in a library screening, or amplification of a portion of the iptase gene by the polymerase chain reaction.

There are several other approaches that might be useful to investigate plant iptase, for example:

1. Use hybrid-substraction to enrich a cDNA library for the iptase message and screen for expression in *Escherichia coli* of iptase activity.
2. Rescue gene function in interspecies hybrid mutants with genomic clones.

Table 6-2 **Kinetics and Properties of iptase**

	iptase of Plant[a]	Slime Mold[b]
Molecular mass	52 kDa	40 kDa
Substrate	5′-AMP and Δ^2iPP	5′-AMP or ADP and Δ^2iPP
Optimum pH	7.0, 50 m*M* Tris.HCl buffer, 30 m*M* Mg^+	7.0, 50 m*M* PIPES-KOH buffer, 1 m*M* $ZnSO_4$, 6.5–8.5, (MES, pH 6.5; PIPES, 7.0; Tris.HCl, 7.5–8.5), 10 m*M* $MgCl_2$
Metal ions	Mg^{2+}, can be partially replaced by Mn^{2+}	Mg^{2+} or Zn^{2+}
Inhibitors	Hg^{2+}, iodoacetamide	
Km	4.7 μM for 5′-AMP	0.1 μM for 5′-AMP 2.2 μM for Δ^2iPP

3. Use likely similarities with the bacterial iptase by employing cross-reactive antibodies against the bacterial iptase to aid in purifying the plant form.

ACKNOWLEDGMENTS

Part of the authors' research referred to in this review and the preparation of the review have been supported by National Science Foundation grant No. DCB-9018899 and by the University of Wisconsin System Research Fellowship (to C. M. C.).

REFERENCES

1. **Taya, T., Tanaka, Y., and Nishimura, S.,** 5′AMP is a direct precursor of cytokinin in *Dictyostelium discoideum, Nature*, 271, 545, 1978.
2. **Chen, C. M. and Melitz, D. K.,** Cytokinin biosynthesis in a cell-free system from cytokinin-autotrophic tobacco tissue cultures, *FEBS Lett.*, 107, 15, 1979.
3. **Murai, N.,** Cytokinin biosynthesis and its relationship to the presence of plasmids in strains of *Corynebacterium fascians*, in *Metabolism and Molecular Activities of Cytokinins*, Guern, J. and Péaud-Lenoël, C., Eds., Springer-Verlag, New York, 1981, 17.
4. **Barry, G. F., Rogers, S. G., Fraley, R. T., and Brand, L.,** Identification of a cloned cytokinin biosynthetic gene, *Proc. Natl. Acad. Sci.* U.S.A., 81, 4776, 1984.
5. **Hommes, N. G., Akiyoshi, D. E., and Morris, R. O.,** Assay and partial purification of the cytokinin biosynthetic enzyme dimethylallyl pyrophosphate: 5′AMP transferase, *Methods Enzymol.*, 110, 340, 1985.
6. **Hall, R. H.,** N^6-(Δ^2-isopentenyl) adenosine: chemical reactions, biosynthesis, metabolism and significance to the structure and function of tRNA, in *Progress in Nucleic Acid Research and Molecular Biology*, Vol. 10, Davidson, J. N. and Cohn, W. E., Eds., Academic Press, New York, 1970, 57.
7. **Skoog, F. and Armstrong, D. J.,** Cytokinins, *Ann. Rev. Plant Physiol.*, 21, 359, 1970.
8. **Chen, C. M.,** Cytokinin biosynthesis in cell-free systems, in *Plant Growth Substance 1982*, Wareing, P. F., Ed., Academic Press, London, 1982, 155.
9. **Nishinari, N. and Syono, K.,** Biosynthesis of cytokinins by tobacco cell cultures, *Plant Cell Physiol.*, 21, 1143, 1980.
10. **Reinecke, D. M., Brenner, M. L., and Rubenstein, I.,** Cytokinin biosynthesis in developing corn kernels, *Plant Physiol.*, 96(suppl.), 77, 1991.
11. **Reinecke, D. M., Brenner, M. L., and Rubenstein, I.,** Cytokinin biosynthesis in developing *Zea Mays* kernels, *Plant Physiol*, 99(Suppl.), 66, 1992.
12. **Ihara, M., Taya, Y., Nishimura, S., and Tanaka, Y.,** Purification and some properties of Δ^2-isopentenyl pyrophosphate: 5′AMP Δ^2-isopentenyltransferase from the cellular slime mold *Dictyostelium discoideum*, *Arch. Biochem. Biophys.* 230, 652, 1984.
13. **Buchmann, I., Marner, F. J., Schroder, G., Waffenschmidt, S., and Schroder, J.,** Tumor genes in plants: T-DNA encoded cytokinin biosynthesis, *EMBO J.*, 4, 853, 1985.
14. **Powell, G. K. and Morris, R. O.,** Nucleotide sequence and expression of a *Pseudomonas savastanoi* cytokinin biosynthetic gene: homology with *Agrobacterium tumefaciens* tmr and tzs loci, *Nucleic Acids Res.*, 14, 2555, 1986.

15. **Akiyoshi, D. E., Klee, H., Amasino, R. M., Nester, E. W., and Gordon, M. P.,** T-DNA of *Agrobacterium tumefaciens* encodes an enzyme of cytokinin biosynthesis, *Proc. Natl. Acad. Sci.* U.S.A., 81, 5994, 1984.
16. **Beaty, J. S., Powell, G. K., Lica, L., Regier, D. A., MacDonald, E. M. S., Hommes, N. G., and Morris, R. O.,** Tzs, a nopaline Ti plasmid gene, from *Agrobacterium tume-faciens* associated with trans-zeatin biosynthesis, *Mol. Gen. Genet.*, 203, 274, 1986.
17. **Regier, D. A. and Morris, R. O.,** Secretion of *trans*-zeatin by *Agrobacterium tumefaciens:* a function determined by the nopaline Ti plasmid, *Biochem. Biophys. Res. Commun.*, 104, 1560, 1982.
18. **Fox, J. E.,** Growth factor requirements and chromosome number in tobacco tissue cultures, *Physiol. Plant.*, 16, 793, 1963.
19. **Sussman, M.,** Cultivation and serial transfer of the slime mold, *Dictyostelium discoideum* in liquid nutrient medium, *J. Gen. Microbiol.*, 25, 375, 1961.
20. **Horgan, R.,** Cytokinins, in *Advanced Plant Physiology,* Wilkins, M. B., Ed., Pitman Press, London, 1984, 53.
21. **Chen, C. M.,** Characterization of cytokinin and related compounds by HPLC, in *High Performance, Liquid Chromatography in Plant Sciences,* Linskens, H. F. and Jackson, J. F., Eds., Springer-Verlag, New York, 1987, 23.
22. **Robins, M. J., Hall, R. H., and Thedford, R.,** N^2-(Δ^2-isopentenyl)adenosine. A component of the transfer ribonucleic acid of yeast and of mammalian tissue, methods of isolation, and characterization, Biochemistry, 6, 1837, 1967.
23a. **Chen, C. M., et al.,** Unpublished data.
23b. **Sato, T. and Theologis, A.,** Cloning the mRNA encoding 1-aminocyclopropane-1-carboxylate synthase, the key enzyme for ethylene biosynthesis in plants, *Proc. Natl. Acad. Sci. U.S.A.*, 86, 6621, 1989.
24. **Morris, R. O.,** Genes specifying auxin and cytokinin biosynthesis in prokaryotes, in *Plant Hormones and Their Role in Plant Growth and Development,* Davies, P. J., Ed., Martinus Nijhoff, Boston, 1987, 636.
25. **Crespi, M. D., Messens, E., Caplan, A. B., Van Montagu, M. C., and Desner, J. G.,** Fasciation induction by the phytopathogen *Rhodococcus fascians* depends upon a linear plasmid encoding a cytokinin synthase gene, *EMBO J.*, 11, 795, 1992.
26a. **Henkle, K. J., Liebau, E., Meuller, S., Bergmann, B., and Wlater, R. D.,** Characterization and molecular cloning of a copper-zinc superoxide dismutase from the human parasite *Onchocercha-volvulus, Infect. Immun.*, 59, 2063, 1991.
26b. **Ertl, J. R. and Chen, C. M.,** Unpublished observation.
27. **Ashby, M. N., Spear, D. H., and Edwards, P. A.,** Prenyltransferase: from yeast to man, in *Molecular Biology of Atherosclerosis,* Attie, A. D., Ed., Elsevier, New York, 1990, 27.

Chapter 7

Cytokinin Biosynthesis in tRNA and Cytokinin Incorporation into Plant RNA

Norimoto Murai

TABLE OF CONTENTS

I. INTRODUCTION: CYTOKININ STIMULATES PLANT RNA SYNTHESIS

The chemistry, occurrence, biosynthesis, and function of cytokinin in tRNA have been summarized previously.[1-10] This chapter will review articles that appeared since the paper by Greene,[2] with special emphasis on the biosynthesis of cytokinin in tRNA and cytokinin incorporation into plant RNA.

The discovery of cytokinin in tRNA in 1966 initiated extensive and productive research efforts in this area, culminating in comprehensive understanding of this knowledge.[11,12] During the last 27 years, a number of models have been presented concerning the relationship between cytokinin in tRNA, the mode of actions of cytokinin, and the biosynthesis of cytokinin. One of the most widely tested models proposed that diverse actions of cytokinin on plant growth and development may be mediated through the function of cytokinin in tRNA on protein synthesis. This model predicted that free cytokinins might serve as direct precursors to the structurally identical ribonucleosides in tRNA. Alternatively, it was argued that cytokinins might regulate the biosynthesis of cytokinin-active ribonucleoside in tRNA, modulating the concentration of a particular subset of tRNA species and thus affecting the rate of protein synthesis. Another influential model suggested that cytokinin residues in tRNA might serve as direct precursors of free cytokinin after degradation of tRNA molecules. These and other models have been tested in the past 27 years and modified to accommodate new experimental findings.

Since the initial work by Osborne in 1962, it has been known that cytokinins stimulate RNA synthesis.[13] However, it is still unclear whether the stimulation is due to a specific effect of cytokinin on this process or due to a result of increased general anabolic activity in response to cytokinin. Stimulation of RNA synthesis by cytokinins has been observed for all RNA fractions and with relatively slight difference in the magnitude of the effect on individual fractions. Srivastava examined the effect of kinetin on the incorporation of inorganic ^{32}P into tobacco pith tissue RNAs fractionated on Methylated Albumin Kieselguhr (MAK) columns.[14] Kinetin stimulated the incorporation into heavy rRNA, slightly more than into other RNA fractions. Carpenter and Cherry demonstrated that low concentrations of N^6-benzyladenine (b^6Ade) promoted the incorporation of inorganic ^{32}P into all RNA fractions of excised peanut cotyledons, to approximately the same extent.[15] High concentrations of b^6Ade inhibited the incorporation into all RNAs fractionated on MAK columns. However, the lowest concentration of b^6Ade (90 μM) specifically increased one of the soluble RNA fractions. Trewavas demonstrated that b^6Ade maintained the

0-8493-6252-0/94/$0.00+$.50

rRNA content, but slightly increased the salt-soluble nucleic acid content, including DNA and tRNA, in *Lemma minor* grown in distilled water.[16] The rates of synthesis and degradation of rRNA were increased to the same extent. Cytokinin increased the rate of synthesis of soluble nucleic acid, twice as much as the rate of degradation.

An effect of cytokinin on the relative abundance of particular tRNA species was reported for soybean seedlings.[17,18] Anderson and Cherry fractionated Leu-$tRNA^{Leu}$ from cotyledons and from hypocotyls into six species, on Freon columns.[17] These tRNA species were referred to as $tRNA_1^{Leu}$, $tRNA_2^{Leu}$, etc. in order of their elution from the column. A cytokinin, b^{6}Ade, increased the relative abundance of Leu-$tRNA_5^{Leu}$ and Leu-$tRNA_6^{Leu}$ and decreased the level of Leu-$tRNA_1^{Leu}$, in the hypocotyl. Bick et al. found that the relative abundance of Leu-$tRNA_5^{Leu}$ and Leu-$tRNA_6^{Leu}$ from the cotyledon increased with age.[18] Treatment of soybean seeds with b^{6}Ade increased the relative abundance of the two Leu-$tRNA^{Leu}$ species. However, the b^{6}Ade had no effect on changes with age in the level of Tyr-$tRNA^{Try}$ species. The effect of cytokinin on the relative abundance of Leu-$tRNA_5^{Leu}$ and Leu-$tRNA_6^{Leu}$ is of particular interest, as *cis*-ribosylzeatin (c-io^{6}A) was identified as a constituent of $tRNA_6^{Leu}$ in peas.[19-21] This selective effect of cytokinin implies that the hormone may influence the synthesis or degradation of certain tRNA species containing cytokinin-active ribonucleoside.

II. CYTOKININ AS A CONSTITUENT OF tRNA MOLECULE

A. CYTOKININ IS LOCATED ADJACENT TO THE 3′-END OF THE ANTICODON OF U_1-CODON tRNA

The elucidation of the nucleotide sequence of two $tRNA^{Ser}$ species from brewer's yeast in 1966 led Zachau et al. to the discovery that a cytokinin-active ribonucleoside N^6-(Δ^2-isopentenyl)adenosine (i^{6}A) was located adjacent to the 3′-end of the anticodon.[11,12] Subsequently, Skoog et al. and Hall et al. reported cytokinin activity in hydrolysates of tRNA from *Escherichia coli*, yeast, animal, and plant sources.[22,23] Cytokinin-active ribonucleosides have been reported in tRNA from virtually all organisms tested.[2,9]

Sequence determinations of other cytokinin-containing tRNA species demonstrated that i^{6}A or its derivative always occurs at the same position, generally the nucleoside residue 38 adjacent to the third A residue of the anticodon of tRNA. This has been shown in *E. coli* for a $tRNA^{Phe}$, a $tRNA^{Ser}$, a $tRNA^{Tyr}$, and a $tRNA^{Trp}$ species, in yeast for two $tRNA^{Ser}$, a $tRNA^{Cys}$, and a $tRNA^{Tyr}$ species, and in pea for a $tRNA^{Leu}$ species.[2,8,10]

The presence of a cytokinin-active ribonucleoside is restricted to tRNA species corresponding to each codon with the initial letter U or U_1 codon, as initially demonstrated in *E. coli* tRNA species, by Armstrong et al.[24] Approximately 1 mole of methylthio N^6-(Δ^2-isopentenyl)adenosine (ms^{2}i^{6}A) or i^{6}A was identified in *E. coli* for $tRNA^{Phe}$ (UUY), $tRNA^{Leu}$ (UUG), $tRNA_4^{Ser}$ (UCR), $tRNA^{Tyr}$ (UAY), $tRNA^{Cys}$ (UGY), and $tRNA_2^{Trp}$ (UGG) species.[24-26] Cytokinin activity was detected in *Saccharomyces cerevisiae* and *S. lactis* for two $tRNA^{Ser}$ (UCY) species and a $tRNA^{Tyr}$ (UAY) and a $tRNA^{Cys}$ (UGY) species, in *Staphylococcus epidermidis* for one (UCU) of four $tRNA^{Ser}$ species, in *Drosophila melanogaster* for three (UCU, UCG) of five $tRNA^{Ser}$ species, and in wheat germ for $tRNA^{Ser}$.[27-30] Radioactivity from [2-^{14}C] mevalonic acid was incorporated into U_1 codon tRNA species of *Lactobacillus acidophilus*.[31]

The chemical nature of cytokinin in tRNA from various organisms, ranging from prokaryotes to higher eukaryotes, has been rigorously characterized based on mass spectrometry (MS) and chromatographic mobilities. The prototype molecule i^{6}A is formed after the transfer of the isopentenyl side chain to the N^6-position of the adenosine residue 37 in the preformed tRNA.[32,33] i^{6}A has been found in the tRNA of virtually all organisms examined and was identified as the sole cytokinin detected in the tRNA from yeast, *L. acidophilus*, and animal sources, including *D. melanogaster*, calf and human liver, and chick embryo.[2,9,29,32,33] The methylthiolation of i^{6}A at the 2 position of the purine ring yields ms^{2}i^{6}A in tRNA of microorganisms and higher plants. *E. coli* and *S. epidermidis* tRNA contained both i^{6}A and methylthio N^6-(Δ^2-isopentenyl)adenosine (ms^{2}i^{6}A).[28,34] The hydroxylation of the methyl group at the 4 position of the isopentenyl side chain of i^{6}A creates either a *cis* or a *trans* configuration of ribosylzeatin (io^{6}A). The *cis*-isomer of io^{6}A generally predominates in the tRNA of higher plants and microorganisms that form association with higher plants.[32,33,35-40] However, the *trans* isomer of io^{6}A was also detected in the tRNA from *Pisum sativum* shoots and *Spinacea oleracea* leaves,[38,39] but the ratio of *cis*- to *trans*-isomers was 40:1. *trans*-Ribosylzeatin was also characterized in the tRNA of two plant pathogens: *Agrobacterium tumefaciens,* which incites crown gall disease, and *Corynebacterium fascians,* which causes witches' broom.[41,42] The 2-methylthio derivative of, presumably, the *cis*-isomer of io^{6}A has been reported in the tRNA of higher plants and plant-associating microorganisms. Both *trans*- and *cis*-isomers of ms^2io^6A occur in the tRNA of *P. sativum* shoots and *S. oleracea* leaves.[39,43] The occurrence of methylthio derivatives of io^6A is restricted to the tRNA of chloroplasts in *Euglena gracilis*.[44] Thus, plant tRNA has the greatest diversity of cytokinin-active ribonucleosides, that is, i^6A, ms^2i^6A, *c*-io^6A, *c*-ms^2io^6A, *t*-io^6A, and *t*-ms^2io^6A, among which *c*-io^6A and i^6A are the most abundant.[35-39]

B. CYTOKININ IS SYNTHESIZED IN THE PREFORMED tRNA AFTER THE SIDE CHAIN TRANSFER

The *in vivo* incorporation of exogenous mevalonic acid into i^6A of tRNA has been demonstrated in *Lactobacillus* species and tobacco callus tissue.[32,45-47] *L. acidophilus*, which requires acetate or mevalonate for growth, was incubated in the presence of [2-^{14}C]mevalonate.[32,45] The radioactivity in the tRNA hydrolysate was found to be mainly associated with the position of i^6A in a paper chromatogram developed with several solvent systems. This radioactive compound underwent degradation reactions characteristic of i^6A and thus was reported as i^6A.[48] The level of incorporation corresponded to the occurrence of one i^6A residue in 12 tRNA molecules. The quantity of labeled i^6A in the tRNA of *L. platarum* was found to be approximately one fifth of that in tRNA of *L. acidophilus*.[45] A stimulative effect of mevalonic acid on the growth of cytokinin-dependent tobacco callus tissue in the absence of cytokinin has been reported by some, but could not be confirmed by other investigations.[46,47,49] Chen and Hall isolated radioactive tRNA from tobacco callus tissue grown on media containing [2-^{14}C]mevalonate.[46] The radioactivity in tRNA was found to be exclusively localized in i^6A and present to the extent of one labeled i^6A in every 100 molecules of tRNA. The simultaneous addition of i^6A and [2-^{14}C]mevalonate doubled the fresh weight yield, but reduced by 50% the amount of radioactivity in tRNA. This might be due to an effect of i^6A on the incorporation of mevalonic acid into tRNA. It should be noted that io^6A, i^6A, b^6A, and ms^2io^6A were identified in the tRNA hydrolysate of tobacco callus cultured in the presence of b^6Ade.[36] In fact, Murai et al. demonstrated that *cis*-io^6A was the predominant radioactive ribonucleoside in tobacco callus tRNA, containing approximately 40% of the label from [2-^{14}C]mevalonate incorporated into tRNA.[47] They found that only a small peak of radioactivity was associated with the chromatographic position corresponding to i^6A and ms^2io^6A. Fractionation of tobacco callus tRNA on a benzoylated DEAE (BD)-cellulose column demonstrated that both cytokinin activity and radioactivity from [2-^{14}C]mevalonate were associated with fractions containing at least two major and some minor tRNA species. The failure to label i^6A with [2-^{14}C]mevalonate in the tRNA of *E. coli* and yeast can be explained by the poor incorporation of mevalonate into these cells.[32,45] Fittler et al. reported that [2-^{14}C]acetate, a precursor of mevalonate, labels the isopentenyl side chain of i^6A in yeast tRNA.[45] The incorporation of a small quantity of radioactivity from [2-^{14}C]mevalonate into rRNA has been reported and found to be not associated with cytokinin-active ribonucleoside.[32,46,47]

Studies on the *in vitro* biosynthesis of i^6A in tRNA demonstrated that the isopentenyl group of Δ^2-isopentenyl-pyrophosphate attaches to the adenine residue adjacent to the 3′ end of the anticodon in the preformed tRNA molecule. The enzyme catalyzing this reaction has been named Δ^2-isopentenyl pyrophosphate: tRNA Δ^2-isopentenyl transferase. Enzyme activity was reported in crude cell free extracts from yeast, rat liver, and tobacco callus tissue.[46,50,51] The purified enzyme preparations from yeast and *E. coli* have been shown to require tRNA and Δ^3-isopentenyl-pyrophosphate as substrates and to use pig liver isomerase, Mg^{2+}, and mercaptoethanol as cofactors.[52-55] The molecular weight of the enzyme from *E. coli* was estimated to be 55,000 Da.[53] Among various tRNA preparations serving as substrates for the *in vitro* enzyme reaction, permanganate-treated tRNA has been the preferred substrate because of its convenience. This treatment stripped up to 30% of the isopentenyl side chain of i^6A to yield adenosine and also reduced approximately 25% of the radioactivity in labeled i^6A of *L. acidophilus* grown in the presence of [2-^{14}C]mevalonic acid.[48,52] However, it is possible that this treatment causes some damage to the conformation of the anticodon loop of the tRNA. The naturally occurring tRNA substrates have the advantages of being homologous and without nonspecific chemical modification. These tRNA substrates include *Mycoplasma* sp. (Kid) tRNA deficient in cytokinin-active ribonucleoside, *E. coli* suppressor $tRNA^{Tyr}$ containing unmodified adenosine adjacent to the 3′ end of the anticodon, and *S. cerevisiae* $tRNA^{Ser}$ and $tRNA^{Tyr}$ from an i^6A-deficient antisuppressor mutant.[53,54,56] Using permanganate-treated tRNA as a substrate, Kline et al. reported that the isopentenyl transferase from yeast was most effective with the homologous tRNA substrate, but it also catalyzed the reaction of *E. coli* tRNA or rat liver tRNA.[52] $tRNA^{Tyr}$, in which some of the i^6A has been stripped of the isopentenyl side chain, was shown to be five times more efficient as a substrate than unfractionated tRNA. Rosenbaum and Gefter, utilizing modification-deficient suppressor $tRNA^{Tyr}$ as a substrate, reported that the saturation level of the *E. coli* enzyme would modify only 42% of this tRNA molecule.[54] This enzyme seemed to require additional secondary and tertiary structures to the oligonucleotides of the anticodon loop, for recognition. Nucleotide sequence analysis of the modified tRNA localized the modification to either adenosine residue 38 or 39, of which the former is known to occur modified *in vivo*. Zubay et al. developed a DNA-directed cell-free system from *E. coli*, which synthesizes suppressor $tRNA^{Tyr}$.[55] The biological competence of this tRNA was demonstrated by its ability to support *in vitro* synthesis of β-galactosidase, the activity of which increased in proportion to the amount of tRNA added. The addition of Δ^2-isopentenyl pyrophosphate seems to be essential for the synthesis of biologically active tRNA.

The biosynthesis of the methylthio group of ms^2i^6A in *E. coli* $tRNA^{Tyr}$ seems to occur after the attachment of the isopentenyl side chain to adenosine residues in tRNA.[57] The *in vivo* incorporation study demonstrated that the sulfur moiety of the methylthio group can be labeled with either $^{35}SO_4^{2-}$ or cysteine-^{35}S, and that the methyl group can be labeled with methionine-methyl-^{14}C.[44,58,59] The *in vitro* synthesis of ms^2i^6A in methyl-deficient $tRNA^{Tyr}$ demonstrated that S-adenosylmethionine serves as a precursor of the methyl group of ms^2i^6A.[59] The iron appears to catalyze the initial reaction (oxygenation) in the formation of the methylthio group.[60]

C. CYTOKININ RESIDUE PROMOTES THE CODON–ANTICODON INTERACTION

The presence of a cytokinin-active ribonucleoside at a strategic location in U_1-codon tRNA species suggests an important function of this ribonucleoside in the translational process. Gefter and Russell clearly demonstrated that the cytokinin-active ribonucleoside functions to support efficient binding of tRNA to the mRNA-ribosome complex.[57] After infection of *E. coli* with bacteriophage, three forms of suppressor $tRNA^{Tyr}$ were produced. The three forms of $tRNA^{Tyr}$ had the same primary nucleotide sequence, but differed in the extent of modification of the adenosine residue adjacent to the 3′ end of the anticodon. The three tRNA subspecies contained either unmodified adenosine, i^6A, or ms^2i^6A, next to the anticodon. All three had a similar ability to accept tyrosine, but showed marked differences in the ability to support *in vitro* protein synthesis and in the ability to bind to the mRNA-ribosome complex. Their relative ribosome-binding efficiencies were in the order stated 14/54/100 ($A/i^6A/ms^2i^6A$). Furthermore, these values corresponded, roughly, to their efficiencies in supporting *in vitro* protein synthesis. Therefore, the presence of the cytokinin-active ribonucleoside adjacent to the 3′ end of the anticodon is essential for efficient binding of tRNA to the mRNA-ribosome complex and is thus important for protein synthesis.

The same conclusion was reached in studies of chemical modifications specific to a cytokinin-active ribonucleoside. The presence of an allylic double bond in the isopentenyl side chain of i^6A and its derivatives seemed to make this molecule the most reactive among the seven ribonucleotides in the anticodon loop, which is the most exposed part of the tRNA molecule.[61] The aqueous iodine treatment of yeast $tRNA^{Ser}$ containing i^6A left its serine acceptor activity intact, but reduced by 75% its ability to bind to the mRNA-ribosome complex.[62] Yeast $tRNA^{Phe}$ containing base Y was not affected by the aqueous iodine treatment. The treatment of yeast $tRNA^{Tyr}$ with radioactive bisulfite followed by weak alkali was reported to incorporate radioactivity only in the i^6A adjacent to the 3′ end of the anticodon.[63] This did not affect the capacity of the tRNA to accept tyrosine, but reduced by 65% its ability to bind to the mRNA-ribosome complex.

On the other hand, it has been reported that *L. acidophilus* tRNA that was 50% deficient in i^6A functioned normally to support *in vitro* protein synthesis.[64] *Mycoplasma* sp. *(Kid)* tRNA apparently deficient in cytokinin-active ribonucleoside was as efficient as *E. coli* tRNA in amino acid incorporation.[65] Although no data were available for estimating the efficiency of i^6A-deficient tRNA in the ribosome-binding test, the above results were used as an argument against the functioning of cytokinin-active ribonucleoside in promoting efficient binding of tRNA to the mRNA-ribosome complex.

In Fuller and Hodgson's structural model of the anticodon loop, the cytokinin residue functions not only to prevent the wobbling interaction between the third anticodon base and the first base in the codon, but also to increase hydrophobic stabilization of this stacked conformation.[66] Crystallographic studies on i^6Ade and ms^2i^6Ade, in concurrence with the model, demonstrated that the isopentenyl side chain of the cytokinins is pointing away from the adenine imidazole ring.[67,68] This prevents the cytokinin-active ribonucleoside from complementary base pairing.

N-[9-(β-D-Ribofuranosyl)purin-6-ylcarbamoyl]-threonine (t^6A), another hypermodified ribonucleoside, is located adjacent to the 3′ end of the anticodon of tRNA species responding to codons with the initial letter A.[69,70] Radioactive threonine was shown to be incorporated into t^6A of *E. coli* tRNA.[71] *E. coli* $tRNA^{Ile}$ that was 50% deficient in t^6A had only 50% of the binding response to the mRNA-ribosome complex, compared with normal tRNA. Other functions of the tRNA were not affected by the t^6A deficiency.[72] It should be noted that the ureido group (R_1-N-CO-N-R_2) ($R_{1,2}$:ring) present in t^6A represents a structural requirement for biological activity of *N, N′*-diphenylurea and its derivatives.[37] One such derivative, 6-phenylureidopurine, is a moderately active cytokinin in the tobacco bioassay.[73] Base Y, a third hypermodified ribonucleoside, is also located adjacent to the 3′ end of the anticodon of $tRNA^{Phe}$ from yeast and wheat germ.[74,75] Partial modification of base Y in wheat germ tRNA resulted in a 70% reduction of tRNA-catalyzed polyphenylalanine formation.[75] Complete removal of base Y from yeast tRNA altered the configuration of the anticodon and, consequently, changed its coding properties.[76] Therefore, the presence of these hypermodified ribonucleosides adjacent to the 3′ end of the anticodon also apparently functions to increase the efficient and proper binding of tRNA to the mRNA-ribosome complex.

Recent results from analysis of the tRNA and aminoacyl-tRNA synthase interaction indicated that the anticodon loop and modified bases of some tRNA species may play a crucial role in recognizing the identity of tRNA species by aminoacyl-tRNA synthases. For example, substitution of the Arg anticodon CCC for the Met anticodon CAU converted the identity of $tRNA^{Met}_m$ from Met to Arg and resulted in the formation of arg-$tRNA^{Met}_m$ by *E. coli* arginyl-tRNA synthase.[77] Similarly, the substitution of the Thr anticodon GGU for the Met anticodon CAU allowed the threonylation of $tRNA^{Met}_m$ by threonyl-tRNA synthase.[78] The anticodon of yeast $tRNA^{Asp}$ was also found to be an important determinant of the tRNA identity by aspartyl-tRNA synthase.[79] Furthermore, the refined crystal structure of *E. coli* glutamyl-tRNA synthase complexed with $tRNA^{Gln}$ and ATP suggests that the entire anticodon loop provides essential sites for glutamyl-tRNA synthase to discriminate among tRNA species.[80] Posttranscriptional conversion of the first anticodon position from an unmodified C to a modified nucleoside lysidine changed the identity of *E. coli* $tRNA^{Ile}$, resulting in the acceptance of isoleucine instead of methionine.[81] Removal of modified ribonucleosides from $tRNA^{Asp}$ led to a relaxation of the specificity of tRNA recognition by aminoacyl-tRNA synthases.[82] These observations suggest that cytokinin residue in the anticodon loop of U1 tRNA species may play a critical structural role in recognizing the identity of tRNA species, by aminoacyl-tRNA synthases, in addition to promoting the accuracy and efficiency of tRNA binding to the mRNA/ribosome complex.

The results discussed above suggest that the cytokinin residue in tRNA may function in the regulation of protein synthesis. The possibility that structural modification of particular tRNA species plays a critical regulatory role in protein synthesis has been proposed, but conclusive evidence has not yet been provided.

III. FREE CYTOKININ INCORPORATION INTO PLANT RNA

A. FREE CYTOKININ IS CONVERTED TO THE SUBSTRATE OF RNA SYNTHESIS

Free cytokinin N^6-benzyladenine (b^6Ade) appears to be extensively converted to the corresponding ribonucleoside (b^6A), ribonucleotide (b^6A-5'-P), and glucoside, in the tissues of the several plants that have been studied. The time course of b^6Ade metabolism has been studied in both soybean and tobacco callus tissues.[83,84] Martin reported that a barium precipitable fraction, presumably b^6A-5′-P, was a major metabolite in tobacco callus tissue during the first 25 h, but after 48 h b^6A became dominant and no b^6Ade remained.[83] Dyson et al. found that b^6A, b^6A-5′-P, and an unidentified compound Y were major metabolites of [methyl-^{14}C]b^6Ade in soybean callus tissue after a 12 h incubation.[84] The compound Y was later characterized as 6-benzylamino-7-glucofuranosylpurine.[85] The formation of this glucoside was reported to be metabolically related to the disappearance of b^6A-5′-P.[84] It remained as a stable metabolite over a long period of time, long after b^6Ade, b^6A, and b^6A-5′-P had disappeared. These facts led Dyson et al. to speculate that this glucoside is an active or protected form of b^6Ade and that b^6A and b^6A-5′-P are intermediate in its formation.[84]

Evidence for conversion of cytokinins to the ribonucleoside 5′-triphosphate was obtained by Laloue et al., using rapidly growing tobacco and *Acer pseudoplatanus* suspension cultures.[86] After 3-h incubation in cytokinin-requiring and -nonrequiring tobacco cell lines [8-^{14}C]i^6A was converted to the cytokinin 5′-triphosphate, diphosphate, and monophosphate, representing 77% of the total extracted radioactivity. In cytokinin-independent *Acer* cells i^{6}ATP/i^{6}ADP and i^{6}AMP represented 9.6 and 43.2%, respectively, of the total extracted radioactivity. Another cytokinin, b^{6}A, supplied either as [8-^{14}C]b^{6}Ade or [benzyl-^{14}C]b^{6}Ade, was converted to the cytokinin ribonucleotides. b^{6}ATP/b^{6}ADP and b^{6}AMP represented 6 and 28%, respectively, of the recovered radioactivity in tobacco cells. These results indicated that free cytokinins are converted to the substrate of RNA synthesis in plant cells.

B. b^{6}Ade IS INCORPORATED INTO PLANT tRNA AS AN INTACT MOLECULE

The incorporation of b^{6}Ade into tRNA has been studied most extensively in tobacco and soybean callus tissues, as well as in detached tobacco leaves and in *Lemna* fronds. Evidence for the incorporation of b^{6}Ade into tRNA preparations was obtained only when tobacco or soybean callus tissues were grown on solid media for more than 20 days in the presence of b^{6}Ade or grown in liquid suspension culture media for 11 to 55 h.[87-89] Fox initially reported the incorporation of b^{6}Ade into the tRNA fraction of soybean and tobacco callus, and Fox and Chen confirmed this report.[87,88] Total RNA was extracted from the tissue grown in the presence of b^{6}Ade labeled with ^{14}C, either in the side chain or in the purine moiety. Total RNA hydrolysate gave one radioactive peak with Rf values corresponding to b^{6}Ade, regardless of the position in which ^{14}C was inserted. This is not conclusive, but suggestive of the incorporation of the intact b^{6}Ade molecule into RNA. Total RNA extracted from the tissue grown in the presence of [methylene-^{14}C]b^{6}Ade was fractionated on an MAK column. The radioactivity appeared preferentially in the tRNA fraction, but low counts were associated with the tail of the rRNA fraction. Further separation of the tRNA fraction on an *O*-benzoyl-*O*-DEAE cellulose column demonstrated that the radioactivity was confined to only one of six subfractions. This suggests that b^{6}Ade is incorporated into a specific fraction of the tRNA preparation. However, there was no demonstration that the radioactivity was incorporated into the tRNA rather than associated contaminants. The work of Martin and Dyson indicates that Fox's tRNA fraction may have contained radioactive UV-absorbing material other than tRNA.[83,84] Attempts to repeat Fox's observation yielded results contradictory to his, namely, a higher proportion of the radioactivity in the total RNA was found in the rRNA fraction than in the tRNA fraction.[83,84]

Conclusive evidence for the incorporation of b^{6}Ade into b^{6}A of the tRNA preparation was obtained in tobacco callus tissue grown for 21 days in the presence of b^{6}Ade.[36] In addition, Burrows et al. isolated three cytokinin-active ribonucleosides from the tRNA hydrolysate of the tobacco callus tissue and rigorously characterized them as io^{6}A, i^{6}A, and ms^{2}io^{6}A.[36] They argued that the biosynthesis of b^{6}A may involve the transfer of the benzyl side chain of b^{6}Ade to adenosine residues in preformed tRNA, in conformity with the formation of i^{6}Ade.[46,47] However, the results with a double-labeled b^{6}Ade, ^{3}H in the benzene ring and ^{14}C at the 8 position of the purine ring, provided unequivocal evidence that the incorporation of the b^{6}Ade into tRNA involves the intact molecule.[90] The ^{3}H/^{14}C ratio of radioactive b^{6}A recovered from the tobacco callus tRNA preparation was essentially equivalent to the ratio for b^{6}Ade originally supplied to the medium. When a sample of tRNA was slowly heated to a denaturating temperature in the presence of unlabeled b^{6}A and b^{6}A-5′-MP and cooled to room temperature, no detectable radioactivity was found to be released to the cytokinin ribonucleoside and ribonucleotide. This suggests that the b^{6}Ade is covalently incorporated into the tRNA molecule and is not present as a nonpolynucleotide contaminant such as b^{6}A or b^{6}A-5′-MP. The labeled tRNA was fractionated by the BD-cellulose column chromatography to compare the distribution of radioactive b^{6}A and cytokinin-active ribonucleoside in tRNA fractions.[91] While the occurrence of endogenous *c*-io^{6}A was restricted to two tRNA fractions, radioactive b^{6}A was recovered from essentially all tRNA fractions, in quantities roughly proportional to their 260-nm absorbance. This observation is contrary to the reported restricted distribution of b^{6}A in one of six tRNA fractions by Fox and Chen.[88] The above results suggest the nonspecific nature of b^{6}Ade incorporation into tRNA species, which is contrary to the limited distribution of endogenous cytokinin ribonucleoside in a few tRNA species.

It should be noted that the amount of b^{6}Ade incorporated into the tRNA preparation of tobacco callus is very small, on the order of 1 molecule of b^{6}Ade per 10,000 molecules of tRNA, and this could have prevented the detection of the incorporation of b^{6}Ade in other reports.[92-95] Kende and Tavares grew soybean callus tissue for 2 weeks in the presence of [methylene-^{14}C]b^{6}Ade-9-methyl and found little radioactivity in any RNA fractionated on an MAK column, but the specific activity of the cytokinin used was stated to detect its incorporation only at a level of one molecule per 20 or fewer molecules of tRNA.[92] Richmond et al. incubated detached tobacco leaves for 9 days in the presence of [methylene-^{14}C]b^{6}Ade and found no appreciable incorporation of radioactivity into any RNA fractions.[93] Their experimental conditions allowed detection of b^6Ade incorporation down to the level of one b^6Ade in 200 tRNA molecules. Bezemer-Sybrandy and Valdstra reported that no radioactivity was associated with b^6Ade in alkaline and acid hydrolysates of tRNA and rRNA preparations from *Lemna* grown in the presence of ^{14}C-labeled b^6Ade.[94] However, their analytic procedures could only maximally detect the incorporation of one b^6Ade per 100 tRNA molecules. Elliott and Murray incubated soybean callus tissue for 8 days in the presence of [U-^{3}H]b^6Ade and found no radioactivity associated with b^6A in enzymic hydrolysate of the tRNA preparation.[95] However, radioactivity was found with an Rf value corresponding to b^6A in the tRNA hydrolysate from the tissue grown in the presence of b^6Ade and [2,8-^{3}H]Ade. This low radioactivity could be due to the randomization of the ^{3}H label.

Attempts to detect incorporation of [8-^{14}C: methylene-^{3}H]i^6Ade, i^6A-8-^{14}C, or *t*-io^6Ade-8-^{14}C into tRNA were unsuccessful.[3,96]

C. FREE CYTOKININS ARE PREFERENTIALLY INCORPORATED INTO PLANT rRNA

The initial observation by Dyson in 1969 suggested that cytokinin might also be incorporated into the higher-molecular-weight RNA as well as into the tRNA fraction.[84] Armstrong et al. presented, in 1976, the first evidence for preferential incorporation of the intact b^6Ade molecule into rRNA.[91] To this end they isolated a large amount of tRNA and rRNA preparations from tobacco callus grown for 26 days in the presence of [8-C^{14}]b^6Ade or [8-C^{14}: benzene-^{3}H]b^6Ade. As previously shown for tRNA, the double-labeled cytokinin was used to demonstrate that cytokinin incorporation into rRNA involves the intact molecule. Over 90% of the radioactive b^6A recovered from the total RNA preparation was associated with the rRNA fraction. On a specific activity basis (decompositions per minute, or dpm, of b^6A per A_{260} unit RNA), the b^6A content of the rRNA preparation was approximately 3.5-fold greater than that of the tRNA preparation. Radioactive b^6A was released from the rRNA preparation after treatment with venom phosphodiesterase and alkaline phosphatase, ribonuclease T_2 and phosphatase, or ribonuclease T_2 and 3′-nucleotidase. However, b^6A was not released from the rRNA preparation by treatment with either ribonuclease T_2 or phosphatase alone or by successive treatment with ribonuclease T_2 and 5′-nucleotidase. These enzyme digestions and other earlier experiments demonstrated that b^6A is present as a constituent of the rRNA molecule and not as a nonpolynucleotide contaminant.

Similarly, another synthetic cytokinin, kinetin or fr^6Ade, was also found to be incorporated into tobacco callus rRNA. Murai et al. showed that approximately 0.7% of the radioactivity in the [8-^{14}C]fr^6Ade added to the medium was recovered as fr^6A in the rRNA and tRNA preparations.[97] The rRNA contained over 90% of fr^6A moieties. The extent of kinetin incorporation was found to be four times greater than that observed for b^6Ade. The chemical nature of incorporated cytokinin in rRNA was rigorously characterized as fr^6A, by GC-MS. When the labeled rRNA preparation was fractionated into 4-5S, 18S, and 25S rRNA components by preparative polyacrylamide gel electrophoresis, [8-^{14}C]fr^6A was associated with both the 18S and 25S components, in quantities roughly proportional to their 260-nm absorbance.[98] This pattern of apparently nonspecific association was not affected by prior denaturation of the rRNA with formamide. To determine whether the [8-^{14}C]fr^6A moiety is located in any particular sequence of rRNA, the labeled rRNA was digested with ribonuclease T_1 and pancreatic ribonuclease A, and the digest was fractionated by a DEAE-cellulose column in the presence of 7 *M* urea. The [8-^{14}C]fr^6A was distributed in fractions from mono- to oligonucleotides, and its content varied from 46 to 210 μmol/mol of adenosine. These observations indicated that incorporation of cytokinin into rRNA lacks the specificity that would be expected from the limited localization of cytokinin to a certain rRNA component and a particular nucleotide sequence, as is the case with cytokinin in tRNA.

The above evidence for the preferential incorporation of cytokinin into rRNA raised the obvious question of whether endogenous cytokinins may occur in the rRNA molecule. Taller et al. demonstrated the presence of endogenous cytokinin-active ribonucleosides in the rRNA preparation from pea epicotyl and wheat germ.[99] Cytokinin activity in rRNA represented up to 10% of cytokinin activity in tRNA, even though the specific cytokinin activity (activity per A_{260} unit) of the tRNA was over 40-fold higher than that of rRNA. Cytokinin-active ribonucleosides in pea rRNA were rigorously identified as ms^2io^6A and tentatively characterized as i^6A and io^6A. The relative cytokinin activity of these three ribonucleosides differs between rRNA and tRNA, 58/24/14 (ms^2io^6A/i^6A/io^6A) for rRNA and 12/52/37 for tRNA. After enzymic digestion of rRNA and oligonucleotide fractionation, most of the cytokinin activity appeared in fractions corresponding to oligonucleotides larger than five nucleotides in length. However, similarly treated tRNA yielded cytokinin-active fractions corresponding to four and five nucleotides. This is expected, since cytokinin residues in tRNA invariably occur in the sequence Ai^{6}AA within the anticodon loop of U_1-codon tRNA species. These observations were taken to suggest that the pea rRNA preparation is not likely to be contaminated by cytokinin-containing tRNA.

To further understand the extent and kinetics of cytokinin incorporation into RNA, Jouanneau et al. used rapidly growing tobacco cells in a liquid suspension culture.[89,100] Incorporation of [*p*-^{3}H]b^{6}Ade into RNA increased with the increasing culture duration and reached a plateau after a 50-h incubation at either 0.4 or 10 μM.[101] The extent of cytokinin incorporation into RNA fractions increased with the increasing concentration of cytokinin applied to the medium, from 0.04 to 10.0 μM; 0.4 μM is the optimal concentration for cell division and RNA synthesis, and 10 μM is the cytostatic overdose concentration with essentially no cell division and reduced RNA synthesis. At the optimal cytokinin concentrations, b^{6}Ade and i^{6}Ade incorporation into 18S and 25S rRNA components was two- to threefold greater than incorporation into 4S and 5S RNA.[101,102] Furthermore, cytokinin was found incorporated into both the transcribed region and poly (A) tails of mRNA, to a similar extent.[103] The extent of cytokinin incorporation at the overdose concentration of cytokinin (one b^{6}A per 1300 nucleotides) was tenfold higher than that at the growth-stimulating cytokinin concentration (one b^{6}A per 15,000 to 20,000 nucleotides) at 0.4 μM. Thus, at the overdose concentration of cytokinin, b^{6}Ade appeared to act as a base analog inhibiting the processing of rRNA precursors to the mature 18S/25S components and reducing the translational capacity of higher-molecular-weight mRNA.

IV. CONCLUDING REMARKS

A number of models have been presented concerning the relationship between cytokinins in RNAs, and the mode of action and the biosynthesis of cytokinins. One of the most tested models was that diverse actions of cytokinin on plant growth development may be mediated through the function of cytokinin in tRNA on protein synthesis. This model predicted that free cytokinins might serve as direct precursors to structurally identical ribonucleosides in tRNA. However, there is no clear evidence to suggest a direct precursor–product relationship between free cytokinin and cytokinin in tRNAs. Even though free cytokinins are incorporated into plant RNA, the incorporation involves the intact molecule and lacks the specificity that would be expected from the limited localization of cytokinin to a specific rRNA component, or a particular nucleotide sequence, as in the case with cytokinin in tRNA. Alternatively, the model argues that cytokinins might regulate the biosynthesis of cytokinin-active ribonucleosides in tRNA. Thus far there is no direct evidence to suggest that cytokinin stimulates the biosynthesis of cytokinin-active ribonucleotides in tRNA.

Another model suggests that cytokinin residues in tRNA might serve as direct precursors of free cytokinin after degradation of tRNA molecules. In plants, the majority of the cytokinins in tRNAs are *cis*-zeatin riboside and i^{6}A, and the former is biologically inactive. However, the occurrence of *cis-trans*-zeatin isomerase (see Chapter 10) may allow further testing of this model.

A recent model of cytokinin action, by Romanov, predicts that free cytokinin and cytokinin residue in tRNA compete for a binding site in the specific receptor protein, thus modulating the availability of cytokinin-containing tRNA species in the translation of mRNA to protein.[104] However, the evidence against this model has been presented based on the analysis of the frequency of Leu codon UUG and UUA in plant genes.[105]

Cytokinins occur in tRNAs of virtually all organisms examined, including animals, plants, and microorganisms. These observations suggest the importance of cytokinin in the translational process, presumably increasing the accuracy and efficiency of tRNA binding to the mRNA/ribosome complex. Cytokinin residues may also mediate the structural recognition of tRNA molecules by aminoacyl-tRNA synthases.

Thus far, free cytokinins are limited to higher plants and microorganisms that form associations with higher plants. Unequivocal demonstration of hormonal effects of cytokinin is also restricted to higher plants. Therefore, the occurrence of free cytokinin in higher plants may have an origin independent from their occurrence in tRNAs. Nevertheless, essentially the same species of cytokinins occur in the tRNA of higher plants and plant-associated microorganisms,[42] which may suggest an exchange of genetic information during evolution, as demonstrated by the cytokinin biosynthetic genes in T-DNA of *A. tumefaciens* Ti plasmid. Comparing the phylogenic relationship between IPP-tRNA isopentenyl transferase of higher plants and the IPP-AMP isopentenyltransferase of higher plants and plant-associated microorganisms may provide additional information on the evolutionary relationship between the two cytokinin biosynthetic pathways.

REFERENCES

1. **Burrows, W. J.,** Mechanisms of action of cytokinins, *Curr. Adv. Plant Sci.*, 7, 837, 1975.
2. **Greene, E. M.,** Cytokinin production by microorganisms, *Bot. Rev.*, 46, 25, 1980.
3. **Hall, R. H.,** Cytokinins in the transfer RNA: their significance to the structure of tRNA, *Biochemistry and Physiology of Plant Growth Substances*, Runge Press, Ottawa, 1968, 47.
4. **Hall, R. H.,** Cytokinins as a probe of developmental process, *Ann. Rev. Plant Physiol.*, 24, 415, 1973.
5. **Kende, H.,** The cytokinins, *Int. Rev. Cytol.*, 31, 301, 1971.
6. **Letham, D. S.,** Chemistry and physiology of kinetin-like compounds, *Ann. Rev. Plant Physiol.*, 18, 349, 1967.
7. **Letham, D. S. and Wettenhall, R. E. H.,** Transfer RNA and cytokinins, in *The Ribonucleic Acids*, Springer-Verlag, Berlin, 1975, 129.
8. **Nishimura, S.,** Minor components in transfer RNA: their characterization, location and function, *Prog. Nucl. Acid Res. Mol. Biol.*, 10, 49, 1972.
9. **Skoog, F. and Armstrong, D. J.,** Cytokinins, *Ann. Rev. Plant Physiol.*, 21, 359, 1970.
10. **Soll, D.,** Enzymatic modification of transfer RNA, *Science*, 173, 293, 1971.
11. **Zachau, H.G., Dutting, D., and Feldmann, H.,** Nucleotide sequence of two serine-specific transfer ribonucleic acids, *Angew. Chem.*, 78, 392, 1966.
12. **Bieman, K., Tsunakawa, S., Sonnenbickler, J., Feldmann, H., Dutting, D., and Zachau, H. G.,** Structure of an odd nucleoside from serine-specific transfer ribonucleic acid, *Angew. Chem.*, 78, 600, 1966.
13. **Osborne, D. J.,** Effect of kinetin on protein and nucleic acid metabolism in *Xanthium* leaves during senescence, *Plant Physiol.*, 37, 595, 1962.
14. **Srivastava, B. I. S.,** Effect of kinetin on biochemical changes in excised barley leaves and in tobacco pith tissue culture, *Ann. N. Y. Acad. Sci.*, 144, 260, 1967.
15. **Carpenter, W. J. G. and Cherry, J. H.,** Effects of benzyladenine on accumulation of ^{32}P into nucleic acids of peanut cotyledons, *Biochem. Biophys. Acta.*, 114, 640, 1966.
16. **Trewavas, A. J.,** The turnover of nucleic acids in *Lemna minor*, *Plant Physiol.*, 45, 742, 1970.
17. **Anderson, M. and Cherry, J. H.,** Differences in leucyl-transfer RNA's and synthetase in soybean seedlings, *Proc. Nat. Acad. Sci. U.S.A.*, 62, 202, 1969.
18. **Bick, M. D., Liebke, H., Cherry, J. H., and Strehler, B. L.,** Changes in leucyl- and tyrosyl-tRNA of soybean cotyledons during plant growth, *Biochem. Biophys. Acta.*, 204, 175, 1970.
19. **Babcock, D. F. and Morris, R. D.,** Specific degradation of plant leucyl transfer ribonucleic acid by a factor in the homologous synthetase preparation, *Plant Physiol.*, 52, 292, 1973.
20. **Einset, J. W., Swaminathan, S., and Skoog, F.,** Ribosyl-*cis*-zeatin in a leucyl transfer RNA species from peas, *Plant Physiol.*, 58, 140, 1976.
21. **Lester, B. R., Morris, R. O., and Cherry, J. H.,** Purification of leucine tRNA isoaccepting species from soybean cotyledons. II. RPC-2 purification, ribosome binding, and cytokinin content, *Plant Physiol.*, 63, 87, 1979.
22. **Skoog, F., Armstrong, D. J., Cherayil, J. D., Hampel, A. E., and Bock, R. M.,** Cytokinin: localization in transfer RNA preparations, *Science*, 154, 1354, 1966.
23. **Hall, R. H., Csonka, L., David, H., and McLennan, B.,** Cytokinins in the soluble RNA of plant tissues, *Science*, 156, 69, 1967.
24. **Armstrong, D. J., Burrows, W. J., Skoog, F., Roy, K. L., and Soll, D.,** Cytokinins: distribution in transfer RNA species of *Escherichia coli*, *Proc. Natl. Acad. Sci. U.S.A.*, 63, 834, 1969.
25. **Bartz, J., Soll, D., Burrows, W. J., and Skoog, F.,** Identification of the cytokinin-active ribonucleoside in pure *Escherichia coli* tRNA species, *Proc. Nat. Acad. Sci. U.S.A.*, 67, 1448, 1970.
26. **Yamada, Y., Nishimura, S., and Ishikura, H.,** The presence of 2-methylthio-N^6-(Δ^2-isopentenyl)adenosine in leucine, tryptophan and cysteine tRNA from *Escherichia coli*, *Biochem. Biophys. Acta*, 247, 170, 1971.
27. **Armstrong, D. A., Skoog, F., Kirkegaard, L. H., Hampel, A. E., Bock, R. M., Gillan, I., and Tener, G. M.,** Cytokinins: distribution in species of yeast transfer RNA, *Proc. Nat. Acad. Sci. U.S.A.*, 63, 504, 1969.
28. **Armstrong, D. A., Evans, P. K., Burrows, W. J., Skoog, F., Petit, J. F., Dahl, J. L., Steward, T., Strominger, J. L., Leonard, N. J., Hecht, S. M., and Occolowitz, J.,** Cytokinins. Activity and identification in *Staphylococcus epidermidis* transfer ribonucleic acid, *J. Biol. Chem.*, 245, 2922, 1970.
29. **White, B. N., Dunn, R., Gillam, I., Tener, G. M., Armstrong, D. J., and Skoog, F.,** An analysis of five serine transfer ribonucleic acids from *Drosophila*, *J. Biol. Chem.*, 250, 515, 1975.

30. **Struxness, L. A., Armstrong, D. J., Gillam, I., Tener, G. M., Burrows, W. J., and Skoog, F.,** Distribution of cytokinin-active ribonucleosides in wheat germ tRNA species, *Plant Physiol.*, 63, 35, 1979.
31. **Peterkofsky, A. and Jesensky, C.,** The localization of N^6-(Δ^2-isopentenyl)adenosine among acceptor species of transfer ribonucleic acid of *Lactobacillus acidophilus*, *Biochemistry*, 8, 3798, 1969.
32. **Peterkofsky, A.,** The incorporation of mevalonic acid into the N^6-(Δ^2-isopentenyl)adenosine of transfer ribonucleic acid in *Lactobacillus acidophilus*, *Biochemistry*, 7, 472, 1968.
33. **Hall, R. H.,** N^6-(Δ^2-Isopentenyl)adenosine: chemical reactions, biosynthesis, metabolism and significance to the structure and function of tRNA, *Prog. Nucl. Acid Res. Mol. Biol.*, 10, 57, 1970.
34. **Burrows, W. J., Armstrong, D. J., Skoog, F., Hecht, R. M., Boyle, J. T. A., Leonard, N. J., and Occolowitz, J.,** The isolation and identification of two cytokinins from *Escherichia coli* transfer ribonucleic acid, *Biochemistry*, 8, 3071, 1969.
35. **Burrows, W. J., Armstrong, D. J., Kamínek, M., Skoog, F., Bock, R. M., Hecht, S. M., Dammann, L. G., Leonard, N. J., and Occolowitz, J.,** Isolation and identification of four cytokinins from wheat germ transfer ribonucleic acid, *Biochemistry*, 9, 1867, 1970.
36. **Burrows, W. J., Skoog, F., and Leonard, N. J.,** Isolation and identification of cytokinins located in the transfer ribonucleic acid of tobacco callus grown in the presence of 6-benzylaminopurine, *Biochemistry*, 10, 2189, 1971.
37. **Burrows, W. J.,** Mode of action of N′,N′-diphenylurea: the isolation and identification of the cytokinins in the transfer RNA from tobacco callus grown in the presence of N′, N′-diphenylurea, *Planta*, 130, 313, 1976.
38. **Vreman, H. J., Skoog, F., Frihart, C. R., and Leonard, N. J.,** Cytokinins in *Pisum* transfer ribonucleic acid, *Plant Physiol.*, 49, 848, 1972.
39. **Vreman, H. J., Thomas, R., Corse, J., Swaminathan, S., and Murai, N.,** Cytokinins in tRNA obtained from *Spinacia oleracea* L. leaves and isolated chloroplasts, *Plant Physiol.*, 61, 296, 1978.
40. **Swaminathan, S. and Bock, R. M.,** Isolation and identification of cytokinins from *Euglena gracilis* transfer ribonucleic acid, *Biochemistry*, 16, 1355, 1977.
41. **Chapman, R. W., Morris, R. O., and Zaerr, J. B.,** Occurrence of *trans*-ribosylzeatin in *Agrobacterium tumefaciens* tRNA, *Nature*, 262, 153, 1976.
42. **Murai, N., Skoog, F., Doyle, M. E., and Hanson, R. S.,** Relationships between cytokinin production, presence of plasmids, and fasciation caused by strains of *Corynebacterium fascians*, *Proc. Natl. Acad. Sci. U.S.A.*, 77, 619, 1980.
43. **Vreman, H. J., Schmitz, R. Y., Skoog, F., Playtis, A. J., Frihart, C. R., and Leonard, N. J.,** Synthesis of 2-methylthio-*cis*- and *trans*-zeatin and their isolation from *Pisum* tRNA, *Phytochemistry*, 13, 31, 1974.
44. **Swaminathan, S., Bock, R. M., and Skoog, F.,** Subcellular localization of cytokinins in transfer ribonucleic acid, *Plant Physiol.*, 59, 558, 1977.
45. **Fittler, F., Kline, L. K., and Hall, R. H.,** Biosynthesis of N^6-(Δ^2-isopentenyl)adenosine. The precursor relationship of acetate and mevalonate to the Δ^2-isopentenyl group of the transfer ribonucleic acid of microorganism, *Biochemistry*, 7, 940, 1968.
46. **Chen, C. M. and Hall, R. H.,** Biosynthesis of N^6-(Δ^2-isopentenyl)adenosine in the transfer of ribonucleic acid of cultured tobacco pith tissue, *Phytochemistry*, 8, 1687, 1969.
47. **Murai, M., Armstrong, D. J., and Skoog, F.,** Incorporation of mevalonic acid into ribosylzeatin in tobacco callus ribonucleic acid preparations, *Plant Physiol.*, 55, 853, 1975.
48. **Robins, M. J., Hall, R. H., and Thedford, R.,** N^6-(Δ^2-isopentenyl)adenosine. A component of the transfer ribonucleic acid of yeast and mammalian tissue, method of isolation, and characterization, *Biochemistry*, 6, 1837, 1967.
49. **McChesney, J. D.,** Stimulation of *in vitro* growth of plant tissues by *dl*-mevalonic acid, *Can. J. Bot.*, 48, 2357, 1970.
50. **Fittler, F., Kline, L. K., and Hall, R. H.,** N^6-(Δ^2-isopentenyl) adenosine: biosynthesis *in vitro* by an enzyme extract from yeast and rat liver, *Biochem. Biophys. Res. Commun.*, 31, 571, 1968.
51. **Vinkers, J. D. and Logan, D. M.,** N^6-(Δ^2-isopentenyl)adenosine: a heterologous system for *in vitro* synthesis of *E. coli* B transfer RNA, *Biochem. Biophys. Res. Commun.*, 41, 741, 1970.
52. **Kline, L. K., Fittler, F., and Hall, R. H.,** N^6-(Δ^2-Isopentenyl) adenosine. Biosynthesis in transfer ribonucleic acid *in vitro*, *Biochemistry*, 8, 4361, 1969.

53. **Bartz, J., Kline, L., and Soll, D.,** N^6-(Δ^2-Isopentenyl)adenosine: biosynthesis *in vitro* in transfer RNA by an enzyme purified from *Escherichia coli, Biochem. Biophys. Res. Commun.*, 40, 1481, 1970.
54. **Rosenbaum, N. and Gefter, M. L.,** Δ^2-Isopentenyl-pyrophosphate: transfer ribonucleic acid Δ^2-isopentenyl transferase from *Escherichia coli.* Purification and properties of the enzyme, *J. Biol. Chem.*, 247, 5675, 1972.
55. **Zubay, G., Cheong, L., and Gefter, M. L.,** DNA-directed cell-free synthesis of biologically active transfer RNA: SU^+_{III} tyrosyl tRNA, *Proc. Nat. Acad. Sci. U.S.A.*, 68, 2195, 1971.
56. **Laten, H., Gorman, J., and Bock, R. M.,** Isopentenyl adenosine deficient tRNA from an antisuppressor mutant of *Saccharomyces cerevisiae, Nucl. Acid Res.*, 5, 4329, 1978.
57. **Gefter, M. L. and Russell, R. L.,** Role of modifications in tyrosine transfer RNA: a modified base affecting ribosome binding, *J. Mol. Biol.*, 39, 145, 1969.
58. **Goodman, H. M., Abelson, J., Landy, A., Brenner, S., and Smith, J. D.,** Amber suppression: a nucleotide change in the anticodon of a tyrosine transfer RNA, *Nature*, 217, 1019, 1968.
59. **Gefter, M. L.,** The *in vitro* synthesis of 2-o-methylguanosine and 2-methylthio-N^6-(γ,γ-dimethylallyl) adenosine in transfer RNA of *Escherichia coli, Biochem. Biophys. Res. Commun.*, 36, 435, 1969.
60. **Rosenberg, A. H. and Gefter, M. L.,** An iron-dependent modification of several transfer RNA species in *Escherichia coli, J. Mol. Biol.*, 46, 581, 1969.
61. **Gupta, N. K. and Woodley, C. L.,** The structure and functions of transfer ribonucleic acid, *J. Sci. Ind. Res.*, 30, 513, 1971.
62. **Fittler, F. and Hall, R. H.,** Selective modification of yeast seryl-tRNA and its effect on the acceptance and binding function, *Biochem. Biophys. Res. Commun.*, 25, 441, 1966.
63. **Furuichi, Y., Wataya, Y., Hayatsu, H., and Ukita, T.,** Chemical modification of $tRNA^{Tyr}{}_{yeast}$ with bisulfite: a new method to modify isopentenyl-adenosine residue, *Biochem. Biophys. Res. Commun.*, 41, 1185, 1970.
64. **Litwack, M. D. and Peterkofsky, A.,** Transfer ribonucleic acid deficient in N^6-(Δ^2-isopentenyl)adenosine due to mevalonic acid limitation, *Biochemistry*, 10, 994, 1971.
65. **Hayashi, H., Fisher, H., and Soll, D.,** Transfer ribonucleic acid from *Mycoplasma, Biochemistry*, 8, 3680, 1969.
66. **Fuller, W. and Hodgson, A.,** Conformation of the anticodon loop in tRNA, *Nature*, 215, 817, 1967.
67. **Bugg, C. E. and Thewalt, U.,** Crystal structure of N^6-(Δ^2-isopentenyl)adenine, a base in the anticodon loop of some tRNAs, *Biochem. Biophys. Res. Commun.*, 46, 779, 1972.
68. **McMullan, R. K. and Sundaralingam, M.,** The crystal structure of N^6-(Δ^2-Isopentenyl)-2-methylthio-adenine, a modified base of tRNA with cytokinin activity, *Biochem. Biophys. Res. Commun.*, 43, 1158, 1971.
69. **Ishikura, H., Yamada, H., Murao, K., Saneyoshi, N., and Nishimura, S.,** The presence of N-9-(β-D-ribofuranosyl)purin-6-ylcarbamoyl threonine in serine, methionine and lysine transfer RNAs from *Escherichia coli, Biochem. Biophys. Res. Commun.*, 37, 990, 1969.
70. **Takemura, S., Murakami, M., and Miyazaki, M.,** The primary structure of isoleucine transfer ribonucleic acid from *Torulopsis utilis, J. Biochem. (Tokyo)*, 65, 553, 1969.
71. **Powers, D. M.,** Biosynthesis of N-(purin-6-ylcarbomoyl)threonine and its localization in tRNA's that recognize codon's beginning with adenine, *Fed. Proc.*, 31, 1319, 1972.
72. **Miller, J. P. and Schweizer, M. P.,** Requirement for the modified nucleoside adjacent to the 3′-end of the anti-codon for codon recognition by *E. coli* $tRNA^{Ile}$, *Fed. Proc.*, 31, 1318, 1972.
73. **McDonald, J. J., Leonard, N. J., Schmitz, R. Y., and Skoog, F.,** Cytokinins: synthesis and biological activity of ureidopurines, *Phytochemistry*, 10, 1429, 1971.
74. **Raj-Bhandary, U.L., Chang, S. H., Stewart, A., Faulkner, R. D., Hoskinson, R. M., and Khorana, H. G.,** Studies on polynucleotides. LXVIII. The primary structure of yeast phenylalanine transfer RNA, *Proc. Nat. Acad. Sci. U.S.A.*, 57, 751, 1967.
75. **Yoshikami, D. and Keller, E. B.,** Chemical modification of the fluorescent base in phenylalanine transfer ribonucleic acid, *Biochemistry*, 10, 2969, 1971.
76. **Gosh, K. and Gosh, H. P.,** Role of modified nucleoside adjacent to 3′-end of anticodon in codon-anticodon interaction, *Biochem. Biophys. Res. Commun.*, 40, 135, 1970.
77. **Schulman, L. H. and Pelka, H.,** The anticodon contains a major element of the identity of arginine transfer RNAs, *Science*, 246, 1595, 1989.

78. **Schulman, L. H. and Pelka, H.,** The anticodon change switches the identity of *E. coli* $tRNA^{Met}{}_{m}$ from methionine to threonine, *Nucl. Acids Res.*, 18, 285, 1989.
79. **Putz, J., Puglisi, J. D., Florentz, C., and Giege, R.,** Identity element for specific aminoacylation of yeast $tRNA^{Asp}$ by cognate aspartyl-tRNA synthase, *Science*, 252, 1696, 1991.
80. **Rould, M. A., Perona, J. J., and Steitz, T. A.,** Structural basis of anticodon loop recognition by glutaminyl-tRNA synthase, *Nature*, 352, 213, 1991.
81. **Muramatsu, T., Nishikawa, K., Nemoto, F., Kuchino, Y., Nishimura, S., Miyazawa, T., and Yokoyama, S.,** Codon and amino-acid specificities of a transfer RNA are both converted by a single post-transcriptional modification, *Nature*, 336, 179, 1988.
82. **Perret, V., Garcia, A., Grosjean, H., Ebel, J.-P., Florentz, C., and Glege, R.,** Relaxation of a transfer RNA specificity by removal of modified nucleotides, *Nature*, 344, 787, 1990.
83. **Martin, J. H.,** A study of cytokinin metabolism, Ph.D. thesis, University of Kansas, Lawrence, 1968.
84. **Dyson, W. H., Fox, J. E., and McChesney, J. D.,** Short-term metabolism of urea and purine cytokinins, *Plant Physiol.*, 49, 506, 1972.
85. **Deleuze, G. G., McChesney, J. D., and Fox, J. E.,** Identification of a stable cytokinin metabolite, *Biochem. Biophys. Res. Commun.*, 48, 1426, 1972.
86. **Laloue, M., Terrine, C., and Gawer, M.,** Cytokinin: formation of the nucleoside-5′-triphosphate in tobacco and *Acer* cells, *FEBS Lett.*, 46, 45, 1974.
87. **Fox, J. E.,** Incorporation of a kinin, N^6-benzyladenine into soluble RNA, *Plant Physiol.*, 41, 75, 1966.
88. **Fox, J. E. and Chen, C. M.,** Characterization of labeled ribonucleic acid from tissue grown on ^{14}C-containing cytokinin, *J. Biol. Chem.*, 242, 4490, 1967.
89. **Jouanneau, J. P., Gandar, J.-C., and Peaud-Lenoel, C.,** On the presence of benzyladenine nucleotide in the RNA of tobacco cells, *Plant Sci. Lett.*, 9, 77, 1977.
90. **Walker, G. C., Leonard, N. J., Armstrong, D. J., Murai, N., and Skoog, F.,** The mode of incorporation of 6-benzylaminopurine into tobacco callus transfer ribonucleic acid. A double labeling determination, *Plant Physiol.*, 54, 737, 1974.
91. **Armstrong, D. J., Murai, N., Taller, B. J., and Skoog, F.,** Incorporation of cytokinin N^6-benzyladenine into tobacco callus transfer ribonucleic acid and ribosomal ribonucleic acid preparations, *Plant Physiol.*, 57, 15, 1976.
92. **Kende, H. and Tavares, J. E.,** On the significance of cytokinin incorporation into RNA, *Plant Physiol.*, 43, 1244, 1968.
93. **Richmond, A., Back, A., and Sacks, B.,** A study of the hypothetical role of cytokinins in completion of tRNA, *Planta*, 90, 57, 1970.
94. **Bezemer-Sybrandy, S. M. and Valdstra, H.,** Investigation on cytokinins. IV. The metabolism of 6-benzylaminopurine in *Lemna minor*, *Plant Physiol.*, 25, 1, 1971.
95. **Elliott, D. C. and Murray, A. W.,** A quantitative limit for cytokinin incorporation into transfer ribonucleic acid by soya-bean callus tissue, *Biochem. J.*, 130, 1157, 1972.
96. **Murai, N.,** A study of cytokinin incorporation into ribonucleic acids in tobacco callus cultures, Ph.D. thesis, University of Wisconsin, Madison, 1973.
97. **Murai, N., Taller, B. J., Armstrong, D. J., and Skoog, F.,** Kinetin incorporated into tobacco callus ribosomal RNA and transfer RNA preparations, *Plant Physiol.*, 60, 197, 1977.
98. **Murai, N., Armstrong, D. J., Taller, B. J., and Skoog, F.,** Distribution of incorporated, synthetic cytokinins in ribosomal RNA preparations from tobacco callus, *Plant Physiol.*, 61, 318, 1978.
99. **Taller, B. J., Murai, N., and Skoog, F.,** Endogenous cytokinins in the ribosomal RNA of higher plants, *Plant Physiol.*, 83, 755, 1987.
100. **Teyssendier de la Serve, B. and Jouanneau, J. P.,** Preferential incorporation of an exogenous cytokinin, N^6-benzyladenine, into 18S and 25S ribosomal RNA of tobacco cells in suspension culture, *Biochimie*, 61, 913, 1979.
101. **Jouanneau, J. P., Teyssendier de la Serve, B., and Péud-Lenoël, C.,** Frequency of N^6-benzyladenine incorporation into major RNA fractions of tobacco cell suspension grown at stimulatory or cytostatic cytokinin concentration, *Plant Physiol.*, 74, 663, 1984.
102. **Teyssendier de la Serve, B., Jouanneau, J. P., and Péud-Lenoël, C.,** Covalent link of the N^6-(Δ^2-isopentenyl)adenosine nucleotide to the rRNA of cytokinin-requiring and of cytokinin-autonomous tobacco cells, *J. Plant Growth Regul.*, 1, 25, 1982.

103. **Teyssendier de la Serve, B., Jouanneau, J. P., and Péud-Lenoël, C.,** Incorporation of N^6-benzyladenine into messenger Poly(A)-RNA of tobacco cells incubated at stimulatory or cytostatic cytokinin concentration, *Plant Physiol.*, 74, 669, 1984.

104. **Romanov, G. A.,** Cytokinins and tRNAs: a hypothesis on their competitive interaction via specific receptor proteins, *Plant Cell Environ.*, 13, 751, 1990.

105. **Jansson, S.,** Cytokinins and tRNAs: arguments against a hypothesis of cytokinin action, *Plant Cell Environ.*, 15, 503, 1992.

Chapter 8

Distribution, Biosynthesis, and Function of Cytokinins in tRNA

Barbara J. Taller

TABLE OF CONTENTS

I. INTRODUCTION

Transfer RNA molecules are characterized by being highly decorated, with over 50 different modified nucleosides known to occur in tRNA.[1] Most tRNA modification mutants are viable, indicating that they are not essential for cell growth.[1] A tRNA lacking modified nucleosides was completely functional during *in vitro* translation, suggesting that modified nucleosides also are not required for protein synthesis.[2] Possible functions of modified nucleosides in tRNA include influencing tRNA structure, providing recognition sites, affecting the efficiency and accuracy of translation, and having a regulatory role. Position 34, the wobble position of the anticodon, and position 37, adjacent to the 3′ end of the anticodon, are modification "hot spots", in that they are often modified and contain the greatest diversity of modified nucleosides.[1] Among the modifications found in position 37 are the hypermodified, hydrophobic isopentenyladenosine (i^6A) derivatives known as cytokinins. Cytokinins occur only in tRNAs recognizing codons beginning with uridine (U-group tRNAs: Cys, Leu, Phe, Ser, Trp, Tyr). When present, cytokinins occur only once per tRNA molecule and only at position 37. Although cytokinins were discovered in tRNA almost 30 years ago,[3] relatively little is known about their distribution and function in tRNA, particularly in eukaryotic organisms. A number of reviews have been published since 1980 concerning tRNA[4,5] and tRNA modifications.[1,6-12] This chapter concerns the distribution, biosynthesis, and function of cytokinin-active nucleosides in tRNA. Special emphasis is given to the distribution of cytokinins among organisms, organelles, and tRNA species.

II. PHYLOGENETIC DISTRIBUTION

Cytokinins have been found in tRNA preparations from virtually all organisms.[13] However, the distribution of cytokinin-active modified bases among organisms seems to show inter-kingdom differences. For example, isopentenyladeno-sine (i^6A) is present in virtually all tRNA preparations and is thought to be the only cytokinin found in animal and yeast tRNA. The occurrence of hydroxylated derivatives, ribosylzeatin (io^6A) and methylthioribosylzeatin (ms^2io^6A), is characteristic of tRNA from plants[13] and plant-associated

0-8493-6252-0/94/$0.00+$.50

microbes.[6] The predominant cytokinins in prokaryotic tRNA are i^6A and methylthioisopentenyladenosine (ms^2i^6A),[6,13] although the occurrence of cytokinins has not been reported in archaebacterial tRNA.[14,15] Thus, except for bacteria that form symbiotic or pathogenic relationships with plants, hydroxylated cytokinins are thought not to occur, generally, in prokaryotic tRNA.[6]

As tRNA preparations are examined from additional species and as more tRNA molecules are sequenced, exceptions to the above distribution are becoming known. For example, ms^2io^6A has been reported to occur in the tRNA of *Pseudomonas aeruginosa*,[16] *Azotobacter vinelandii*,[17] and five of seven species of Enterobacteriaceae.[18-20] *Cis*-io^6A and ms^2io^6A have been found in cyanobacterial tRNA.[21] Thus, the concept that hydroxylated cytokinins typically do not occur in prokaryotes should be reevaluated. A possible exception to the exclusive occurrence of i^6A in animal tRNA is $tRNA^{Trp}$(UCA) from mitochondria of rat hepatoma cells and bovine liver, which reportedly contains ms^2i^6A.[14] Also, bovine liver mitochondrial $tRNA^{Ser}$(UCN) was sequenced and found to contain i^6A or ms^2i^6A.[22] Yeast tRNA has been reported to contain small amounts of *t*-io^6A in addition to i^6A.[23]

The greatest diversity of cytokinins occurs in plant tRNA preparations, from which as many as six different compounds have been isolated. These include *cis*- and *trans*-io^6A, i^6A, ms^2i^6A, and *c*- and *t*-ms^2io^6A. *cis*-io^6A is typically the most abundant cytokinin in plant tRNA, while ms^2i^6A, if present, occurs in trace amounts. A comparison of the cytokinin content of tRNA from green vs. nongreen tissue showed i^6A to be almost 100 times more abundant in green tissues.[24] Green tissue also contained a higher level (2.5–5.7%) of *t*-io^6A than nongreen tissue (0.4–0.8%).[24] The significance of the apparent differences in cytokinin distribution among organisms is unknown. Presumably they reflect differences in tRNA-modifying enzymes in various organisms, tissues, and organelles.[25] The diversity of cytokinin nucleosides in plant tRNA may reflect organelle-specific tRNAs in mitochondria and plastids.[13]

So far there is no good correlation between information concerning cytokinin content obtained from tRNA sequence analysis and that obtained by analysis of total or fractionated tRNA preparations. For example, *c*-io^6A is often the most abundant cytokinin in plant tRNA preparations, while ms^2i^6A is the least abundant. However, in those plant tRNAs that have been sequenced, io^6A has been identified only in *Lupinus luteus*[26] and perhaps in *Phaseolus vulgaris* chloroplast tRNA,[27] while ms^2i^6A has been reported in a number of plant chloroplast and mitochondrial tRNAs.[14,15] There are several reasons for such inconsistencies. Except for a few bacterial species and yeast, relatively few tRNAs from any one organism have been sequenced.[15] Differences in the relative abundance of tRNA species can influence the cytokinin content of tRNA preparations. Growth conditions or stage of development of the tRNA source have been shown to influence the relative abundance of tRNA species and the nature of the cytokinin-active bases present in a given tRNA species. Identification of cytokinins in tRNA sequences is commonly equivocal, as some methods may not distinguish between the different cytokinin moieties. Also, contamination of a tRNA preparation with rRNA may contribute to its cytokinin content.[28]

III. SUBCELLULAR DISTRIBUTION

A. CHLOROPLASTS

The variety of cytokinins in plant tRNA has led to studies concerning the possible subcellular localization of cytokinin-containing tRNA. The subcellular distribution of cytokinins was examined in *Euglena*, where tRNA was isolated from light-grown and dark-grown wild-type cultures and from two chloroplast-bleached mutants.[29,30] Both *c*-io^6A and i^6A were present in all the tRNA preparations, but the i^6A content was significantly decreased in the tRNA from both the dark-grown and mutant cells. Methylthiolated compounds, ms^2io^6A and ms^2i^6A, were only found in tRNA from green wild-type cells. Consistent with tRNA from green plant tissues, about 30% of the ms^2io^6A occurred as the *trans*-isomer. It was proposed that (1) i^6A synthesis was active in, but not unique to, chloroplasts; (2) methylthiolation probably occurred in the chloroplast; and (3) the chloroplast was probably not the site of hydroxylation.

Whether hydroxylation occurs in chloroplasts needs to be reconsidered in light of more recent evidence. It was assumed by analogy that hydroxylation might not occur in chloroplasts, since it was thought not to occur, generally, in prokaryotes. It was also assumed that hydroxylated cytokinins were imported from the cytosol and methylthiolated in the plastid to produce ms^2io^6A.[29] As discussed above, recent results have demonstrated the occurrence of hydroxylated cytokinins in tRNA from prokaryotes other than plant-associated microbes. Moreover, in *Salmonella typhimurium*, ms^2i^6A, not i^6A, is the preferred substrate for the hydroxylation reaction,[20] suggesting that hydroxylation may follow methylthiolation in the chloroplasts. Finally, it is now generally accepted that most chloroplasts produce all required tRNAs and do not import tRNAs from the cytosol.[5]

The cytokinin content of spinach leaf tRNA was compared with tRNA extracted from isolated chloroplasts.[31] io^{6}A was found in the leaf tRNA, but chloroplast tRNA did not contain a significant amount. Both i^{6}A and ms^{2}io^{6}A were found in chloroplast tRNA, but their possible presence in cytoplasmic tRNA could not be excluded. More recently, cytoplasmic tRNA^{Ser} from spinach has been shown to contain i^{6}A.[32] Surprisingly, the *trans*-isomer of ms^{2}io^{6}A was not detected in spinach chloroplast tRNA.[31] In tRNA from etiolated bean (*Phaseolus vulgaris*) seedlings, the most abundant cytokinin was *c*-io^{6}A, with i^{6}A and ms^{2}io^{6}A present in about sevenfold-smaller amounts.[33] These results are compatible with the proposed chloroplast localization of the latter two cytokinins, in this case presumably contributed by proplastids.

The first nucleotide sequence of organellar tRNA, chloroplast tRNA^{Phe} from *Euglena*, showed a modified base, "probably ms^{2}i^{6}A", adjacent to the 3′ end of the anticodon.[34,35] Since then, a number of chloroplast tRNAs have been sequenced. Chloroplast tRNA^{Phe} contains ms^{2}i^{6}A, while i^{6}A or ms^{2}i^{6}A occurs in tRNA^{Trp}.[14,15] *Phaseolus* tRNA^{Leu} species have been reported to contain i^{6}A or ms^{2}i^{6}A or their hydroxylated derivatives.[27] Only two nonorganellar plant tRNAs have so far been found to contain cytokinins: tRNA^{Ser} from spinach[32] and lupine seeds,[26] which contain i^{6}A and io^{6}A, respectively. These findings are in agreement with the chloroplast being the site of methylthiolation and with the above data on cytokinin localization. Confirmation of the subcellular localization of cytokinin-containing tRNAs is dependent on localization of the pertinent enzymes. Genes for tRNA-modifying enzymes have not been found in any of the sequenced chloroplast genomes,[5] although the function of a number of open reading frames is unknown. Most enzymes modifying chloroplast tRNA are probably encoded by nuclear genes, as is the case for enzymes modifying mitochondrial tRNA in yeast,[5] including the isopentenyl transferase (see below).

Data overwhelmingly favor the endosymbiotic, cyanobacterial origin of chloroplasts.[36] Thus, the cytokinin content of cyanobacterial tRNA was of interest. Transfer RNA preparations from the cyanobacterium *Agmenellum quadruplicatum* PR-6 contained three cytokinins, tentatively identified as i^{6}A, ms^{2}io^{6}A, and *c*-io^{6}A.[21] ms^{2}io^{6}A was the most abundant, while only a small amount of *c*-io^{6}A was found. *Agmenellum* tRNA^{Phe} has been sequenced and found to contain ms^{2}i^{6}A.[37] The presence of i^{6}A, ms^{2}i^{6}A, and ms^{2}io^{6}A in *Agmenellum* tRNA is consistent with the localization of these cytokinins in chloroplast tRNA, while the zeatin derivatives demonstrate cytokinin hydroxylation in prokaryotes. *Agmenellum* tRNA^{Phe} showed extensive sequence homology with prokaryotic and chloroplast tRNA^{Phe}, including that of *Euglena*, with lesser homology to eukaryotic cytoplasmic tRNAs.[37]

Comparison of cyanobacteria and chloroplasts is complicated by the fact that the ancestry of chloroplasts is not well understood. For example, there is disagreement as to whether the origin of chloroplasts was mono- or polyphyletic. It has been suggested that at least three separate endosymbiotic events gave rise to the chloroplasts of photosynthetic eukaryotes.[38] According to this proposal, phycobilin-containing cyanobacteria gave rise to Rhodophyta and Cryptophyta, while the Plantae, Chlorophyta, and Euglenophyta were derived from prokaryotes containing chlorophyll b. Alternatively, it has been suggested that the ancestors of cyanobacteria, prochlorophytes, and chloroplasts had two different light-harvesting systems: one utilizing phycobilins and the other chlorophyll a and b.[39] The loss of either pigment system could contribute to chloroplast diversity, consistent with a monophyletic chloroplast origin. Further complicating the issue of chloroplast evolution is the demonstrated occurrence of secondary endosymbiotic events involving endosymbiotic eukaryotic algae,[40,41] which also contributes to plastid diversity.

The fact that few algal tRNA preparations have been examined for cytokinin content does little to clarify the proposed relationships between algal chloroplasts and cyanobacteria. Only *Euglena* (Euglenophyta) and *Porphyra perforata* (Rhodophyta) tRNA preparations have been examined. In *Euglena* tRNA, *c*-io^{6}A was the most abundant, with approximately tenfold-smaller amounts of i^{6}A and ms^{2}io^{6}A.[30] Transfer RNA from *Porphyra* contained i^{6}A and *c*-io^{6}A, but *t*-io^{6}A and ms^{2}io^{6}A were not detected.[42] These data are in contrast to tRNA from the cyanobacterium *A. quadruplicatum*, where ms^{2}io^{6}A was the most abundant cytokinin.[21] Only two sequenced algal tRNAs, both from chloroplasts, have been found to contain cytokinin. *Euglena* tRNA^{Phe} and *Scenedesmus* tRNA^{Tyr} contain ms^{2}i^{6}A and i^{6}A, respectively.[14,15]

B. MITOCHONDRIA

The subcellular localization of cytokinin-containing tRNA in mitochondria has been little studied. When tRNA from aerobic, glucose-repressed, or neutral petite mutant strains of *Saccharomyces cerevisiae* was examined, only i^{6}A was found in all of the cultures.[43] In those cases where mitochondrial tRNAs have been sequenced, the cytokinins, if present, were i^6A or ms^2i^6A.[14,15,22] Thus, the cytokinin content of mitochondrial tRNA is consistent with the presumption that methylthiolation occurs in prokaryotes and semiautonomous organelles. The possible localization of cytokinin-containing tRNA in plant mitochondria may be difficult to assess, since the tRNAs are of three different genetic origins:[5] the mitochondrial genome, chloroplast DNA inserted into the mitochondrial genome during evolution, and the nuclear genome. In the latter case, tRNAs must be imported from the cytosol.

IV. DISTRIBUTION AMONG tRNA SPECIES

In those cases where tRNA has been sequenced, cytokinin-active nucleosides have only been found in U-group tRNAs, adjacent to the 3′ end of the anticodon in position 37. In *Escherichia coli* each codon within the U group was thought to be represented by a cytokinin-containing tRNA.[44-46] However, two serine tRNAs with a GGA anticodon were shown to lack modification of A37.[47] The nature and distribution of cytokinin-active ribonucleosides in *Agrobacterium tumefaciens* tRNA has been thoroughly characterized.[48] The distribution is similar to that in *E. coli*, in that cytokinins occurred in at least one isoaccepting tRNA species for each of the U-group amino acids.[48] Furthermore, the occurrence of cytokinins in the different U-group tRNA species appeared to be nonrandom; for example, *c*-io^{6}A was almost exclusively in one $tRNA^{Ser}$ species, while $tRNA^{Trp}$ contained only i^6A. Insufficient data exist to determine whether such a specific distribution of cytokinins occurs in the tRNAs of other organisms.

The distribution of modified bases in eukaryotic tRNA appears to differ from that of microbial tRNA, although tRNA has been examined from comparatively few sources. In eukaryotes relatively few U-group tRNAs contain cytokinin,[13,14] and the distribution of cytokinins in tRNA species varies with the organism. The bulk of the cytokinin activity in *Drosophila* tRNA occurs in three $tRNA^{Ser}$ species.[49] As discussed below, cytokinins in fractionated plant tRNA have been found in serine and leucine species. Cytokinin-containing tRNAs in yeast correspond to cysteine, serine, tryptophan, and tyrosine.[14,15]

Cytokinins in fractionated plant tRNAs have only been found in serine and leucine tRNA species. However, based on retention of tRNAs on an anti-i^6A immunoaffinity column, $tRNA^{Phe}$ and $tRNA^{Tyr}$ from etiolated cucumber seedlings were reported to contain cytokinin.[50] The distribution of cytokinins in tRNA species has been examined for wheat germ[51] and etiolated *Phaseolus vulgaris* seedlings.[52] In these studies cytokinins were found only in serine tRNAs and in minor $tRNA^{Leu}$ species. Leucine tRNA preparations from soybean cotyledons,[53,54] pea roots,[55] and pea seeds[56] were found to contain cytokinin, with $tRNA^{Leu}_6$ from peas containing *c*-io^{6}A.[55,56] All of these studies used source material where chloroplast development was minimal. It is possible that if the distribution of cytokinin-containing tRNAs from chlorophyllous tissue were examined, other cytokinin-containing tRNA species might be found.

Where leucine tRNAs have been sequenced, cytokinins have been found only in prokaryotic and chloroplast tRNAs, whereas most sequenced serine tRNAs found to contain cytokinin are eukaryotic and cytoplasmic.[14,15] In fact, except in yeast, $tRNA^{Ser}$ is the only eukaryotic cytosolic tRNA known to contain cytokinin.[14,15] Cytokinin-active bases in sequenced $tRNA^{Phe}$ and $tRNA^{Trp}$ have so far been found only in prokaryotes and semiautonomous organelles.[14,15] Transfer RNA^{Tyr} from prokaryotes, semiautonomous organelles, and yeast, but not other eukaryotes, has been found to contain cytokinins.[14,15] Cytokinins have been reported in sequenced $tRNA^{Trp}$ from chloroplasts of *Phaseolus*[57] and *Spinacia*,[58] and in $tRNA^{Tyr}$ from *Scenendesmus* chloroplasts.[59] In summary, cytokinins have been found in sequenced chloroplast $tRNA^{Leu}$, $tRNA^{Phe}$, $tRNA^{Trp}$, and $tRNA^{Tyr}$. Sequences of chloroplast cysteine or serine tRNAs have not yet been reported.[15] As previously proposed,[52] it appears that the distribution of cytokinin-containing tRNAs in chloroplasts may be similar to that in prokaryotes where most U-group tRNAs contain cytokinins. Cytokinins have been found in mitochondrial $tRNA^{Phe}$, $tRNA^{Ser}$, $tRNA^{Trp}$, and $tRNA^{Tyr}$, but not in $tRNA^{Leu}$. No mitochondrial tRNAs were among the few $tRNA^{Cys}$ that have been sequenced. Thus, it appears that in plant tRNAs found to contain cytokinins, $tRNA^{Leu}$ may be of plastid origin, while $tRNA^{Ser}$ is likely to be of cytoplasmic origin.

V. BIOSYNTHESIS

A. ISOPENTENYLATION

The first step in the modification of adenine at position 37 (A37) to a cytokinin-active nucleoside is catalyzed by Δ^2-isopentenyl pyrophosphate (IPP):tRNA isopentenyl transferase, which transfers the isopentenyl group to A37 in preformed tRNA, producing i^6A (see Chapter 7). The enzyme has been partially purified from *E. coli*,[60] *Lactobacillus*,[61] yeast,[62] and corn[63] and has been demonstrated to occur in animal tissues.[64] Isopentenyl transferase has been shown to be the product of the *miaA* gene (formerly *trpX*) in prokaryotes and *mod5* in yeast. *MiaA* has been located on the *E. coli/S. typhimurium* chromosomal map,[65,66] cloned,[67] and sequenced.[68] *MiaA* in *E. coli* has been found to be part of a complex operon, consistent with the arrangement of genes for several other tRNA- and rRNA-modifying enzymes.[69] Yeast *mod5* also has been cloned and sequenced.[70,71] Generally, it is assumed that proteins that perform similar functions in different cellular compartments are different. However, in yeast one nuclear gene affects the isopentenylation of both cytoplasmic and mitochondrial tRNA.[70,72] *Mod5* has at least two translation initiation sites and codes for at least two proteins.[71,73] Only the protein with an amino terminal extension is imported into the mitochondrion, although this protein is also found in the cytoplasm.[73] Consistent with most other enzymes that modify the anticodon loop,[1,74] isopentenyl transferase appears to be localized in the cytoplasm rather than in the nucleus.[73] The fact that isopentenylation takes place very late in tRNA maturation[75] is compatible with its occurrence in the cytoplasm.

Sequences of tRNA:isopentenyl transferase genes and proteins have been compared.[68,76] Significant homology exists between the prokaryotic *miaA* genes and also with yeast *mod5*. There is 52% identity at the nucleotide level between *miaA* genes from *Agrobacterium tumefaciens* and *E. coli*. Extensive homology exists among the three proteins, with 37% amino acid identity between the prokaryotic proteins, and 30% identity between the *Agrobacterium* and yeast proteins. There are four highly conserved regions in these proteins. One such conserved region is a putative ATP/GTP binding site close to the amino acid terminus.[68,77] A second well-conserved region is the proposed isopentenyl pyrophosphate binding site.[68] The significance of the other two highly conserved regions is not known.

There is also some homology with *Agrobacterium tumefaciens* isopentenyl transferase genes, *tzs* and *ipt*,[76] which are found on the Ti plasmid and are involved in the biosynthesis of free cytokinins. However, smaller regions of homology were found than among the tRNA:isopentenyl transferase proteins. Comparison of *Agrobacterium miaA* protein with that of *tzs* showed 33% identity over 57 amino acids, and 23% identity occurred over 163 amino acids in the *ipt* protein. This homology occurred near the amino terminus and included the putative ATP/GTP binding site.

B. MODIFICATION OF i^6A

i^6A can be further modified by hydroxylation and/or thiolation and methylation. Thiolation requires cysteine as the thio donor and is an iron-requiring reaction,[78,79] while the methyl donor is S-adenosyl methionine.[78] Evidence suggests that thiolation precedes methylation.[78,80] Thiolation activity is designated MiaB, and methylation, MiaC, but the nature of the enzyme(s) is not known. In most cases one gene governs the synthesis of one modified nucleoside. However, the complexity of nucleosides such as ms^2io^6A probably requires more than one enzyme. Hydroxylation of the isopentenyl side chain in *S. typhimurium* tRNA requires the presence of molecular oxygen, and ms^2i^6A is the preferred substrate.[20] Low levels of *c*-io^6A occurred in the tRNA of aerobically grown cells only when methylthiolation was restricted.[20] These data, as well as the phylogenetic distribution of cytokinin-containing tRNAs, suggest that isopentenylation and methylthiolation arose early in evolution, and hydroxylation developed later.

C. EFFECT OF GROWTH CONDITIONS

Growth conditions affect the cytokinin content of prokaryotic tRNA, with stress often causing undermodification. In *E. coli*, growth under iron restriction,[79] arginine,[10] limited aeration,[81] methionine deprivation,[82] phosphate limitation and low growth rate,[83] or late log phase[45] caused the accumulation of tRNA containing i^6A rather than ms^2i^6A. In contrast, in *Bacillus subtilis*, thiomethy-lation of i^6A occurred during the slowing of growth and under low phosphate concentrations, with ms^2i^6A being the predominant cytokinin in the $tRNA^{Tyr}$ of stationary cells and spores.[84,85] Oxygen restriction inhibited hydroxylation of ms^2i^6A to ms^2io^6A in *S. typhimurium*.[20] In *Azotobacter vinelandii* a nitrogen-fixing bacterium, nitrogen-free medium resulted in a reduced amount of methylthiolated cytokinin in the tRNA and increased the proportion of ms^2io^6A relative to ms^2i^6A.[17] The fact that growth conditions influence cytokinin synthesis has implications for both the analysis and function of modified nucleosides in tRNA.

D. STRUCTURAL REQUIREMENTS FOR tRNA MODIFICATION

There is evidence that tRNA-modifying enzymes recognize certain sequences surrounding the target nucleoside. The sequence A36–A37–A38 is involved in the recognition of A37 by the isopentenyl transferase, as cytokinins occur only in this sequence.[13,25] Gly-7 tRNA, a mutant of wild-type $tRNA^{Gly}$, contains A36 rather than the normal C36, giving it an A36–A37–A38 sequence and the ms^2i^6A modification.[86] Also consistent with the proposed structural requirements is the maturation of $tRNA^{Ser}$ (UCG), where isopentenylation does not occur until after removal of the intervening sequence adjacent to A37,[75] even though the extent of nucleotide modification in the precursor tRNA is high. In this tRNA species, splicing creates the requisite triple-A sequence.

However, triple-A sequences are not sufficient for isopentenylation, as they occur elsewhere in the tRNA molecule, without modification.[13] Sequence analysis of cytokinin-containing tRNAs has shown a requirement for a five-base-pair anticodon stem in addition to the triple-A sequence.[25,47,87,88] In *Tetrahymena*, where anticodon mutations have altered UAA and UAG termination codons to glutamine codons,[89] these unusual tRNAs have five-base-pair anticodon stems and the sequence A36–i^6A37–A38. In *E. coli*, two $tRNA^{Ser}$, anticodon GGA and recognizing codons starting with U, have an unmodified A37.[47] The triple-A sequence was present, but the five-base-pair anticodon stem was modified by an unusual G30-pseudouridine(Ψ)40 pair. Another exception to the triple-A rule occurs in T4 phage arginine-4 tRNA, where A37 is not modified. Again, the anticodon stem contains two unusual base pairs, U-Ψ and U-G.[25] Insertion of an extra A into the anticodon loop of $tRNA^{Trp}$ resulted in an unmodified A37, even though other structural requirements were met.[90] However, bovine mitochondrial $tRNA^{Ser}$ (UCN), which contains i^6A or ms^2i^6A, was described as having an unusual cloverleaf structure, including a proposed six-base-pair anticodon stem.[22] The above data indicate that both the primary sequence and the three-dimensional structure of the tRNA are important for recognition by tRNA-modifying enzymes.[1] Presumably, these same features, in addition to i^6A37, are recognized by the i^6A methylthiolating enzyme(s).[25]

VI. FUNCTION

A. TRANSLATION

The function of cytokinins in tRNA has been primarily studied in eubacteria where the abundance of cytokinin-containing tRNA and the availability of mutants has been effectively utilized. *MiaA* was discovered in *E. coli*, as a result of a derepressed *trp* operon,[91] in *S. typhimurium*, as an antisuppressor mutation,[66] and in *A. tumefaciens*, as a result of reduced virulence (*vir*) gene expresssion.[76] In most cases a deficiency in isopentenylation and/or methylthiolation results in effects on translation, which include tRNA–ribosome interaction,[92-94] codon–anticodon interaction,[95-97] codon context sensitivity,[98,99] and ribosome proofreading.[100,101] Cytokinins have not been shown to affect tRNA charging.[91,96] It appears that cytokinins are not required for translation or other functions, as they are lacking in a number of U-group tRNAs, particularly in eukaryotes. Transfer RNA^{Phe} from *Mycoplasma*, lacking i^6A, was fully active in poly(U)-directed protein synthesis.[102] Based on studies of the efficiency of suppressor tRNAs *in vivo*, it has been suggested that the importance of the cytokinin modification is tRNA dependent.[98]

Cytokinin nucleosides in tRNA have been shown to affect translation by modulating codon–anticodon interactions and/or tRNA–ribosome interactions. Studies of anticodon–anticodon interactions showed that ms^2i^6A had a strong stabilizing effect as compared with unmodified A37.[97] This is due to stacking of the hypermodified base on the anticodon–anticodon complex, with most of the stabilization attributable to the ms^2 group.[97] Consistent with this, yeast tRNAs, which contain i^6A, often form less-stable anticodon–anticodon complexes than the corresponding *E. coli* tRNAs, where ms^2i^6A predominates.[97] From data with *E. coli*, it was estimated that the average rate of translation *in vivo* of codons capable of utilizing cytokinin-modified tRNAs was reduced almost threefold in *miaA*⁻ strains.[100] The rate constants for peptide bond synthesis and GTP hydrolysis during poly(U) translation *in vitro* were insensitive to ms^2i^6A deficiency in $tRNA^{Phe}$.[101] ms^2i^6A deficiency reduced the efficiency of poly(U) translation, as a result of more aggressive proofreading of the undermodified tRNA.[103] It was suggested that ms^2i^6A may play a dual role, stabilizing the codon–anticodon interaction when the match in the first codon position is correct, but leading to destabilization if the match is incorrect.[101] ms^2i^6A in *E. coli* $tRNA^{Phe}$ was found to prevent misreading of the genetic code, by restricting first-position wobble and enhancing cognate codon–anticodon interactions.[96]

In yeast, antisuppressor mutations *mod5-1*[104] and *sin1*[105] resulted in a lack of the i^6A modification. The *mod5* locus was shown to code for the tRNA isopentenyl transferase.[70] *mod5-1* mutants grew normally, but failed to sporulate and showed reduced efficiency of nonsense suppression by a tyrosine-inserting suppressor tRNA.[104] Thus, as in prokaryotes, undermodification of A37 affects translation. It has been suggested that reduced nonsense suppression by undermodified A37 may be due to enhanced proofreading.[101]

Mutations in *miaA*, resulting in an unmodified A37, are highly pleiotropic. In *S. typhimurium* they result in up to a 50% decrease in growth rate, a 30% decrease in protein elongation, and up to a 50% decrease in cellular yield under limiting carbon sources.[66] Other effects of unmodified or undermodified A37 in eubacteria include decreased translational error and increased sensitivity to codon context,[98] enhanced proofreading,[100,101] decreased stability of codon–anticodon interactions,[95] and decreased termination at *trp* and *phe* operons.[91,106] Other reported phenotypes are increased aromatic amino acid transport[107] and synthesis,[108] increased enterochelin synthesis,[20] decreased leucine operon expression,[109] and decreased peptidyl-tRNA accumulation.[94] Many of these phenotypes are related to the role of cytokinin-active bases in influencing translational efficiency and accuracy, and codon context sensitivity.

B. REGULATION

Alteration in the degree of tRNA modification may function as a global regulatory mechanism, allowing cells to respond to environmental stress.[11,20] Stresses cause accumulation of undermodified tRNA molecules, which may affect the translation of regulatory proteins and thus influence gene expresssion. ms^2i^6A has been proposed to act as a physiological switch that modulates spontaneous mutation frequency and other metabolic functions.[68] In *E. coli* both insertion mutations in *miaA*, producing unmodified A37, or iron-limitation leading to undermodified A37 (i^6A), resulted in a mutator phenotype.[69] Thus, undermodification of A37 appears to increase spontaneous mutation frequency in response to environmental stresses.[69]

Some physiological conditions affect both tRNA modification and intermediary metabolism. In *E. coli*, iron limitation due to iron-deficient media[79] or the presence of iron-binding proteins[110] leads to tRNAs containing i^6A instead of the usual ms^2i^6A. Such undermodification increased aromatic amino acid synthesis,[108] enterochelin synthesis,[20] and aromatic amino acid transport.[107] It also has been proposed that hydroxylation of ms^2i^6A to ms^2io^6A, which requires molecular oxygen, may regulate aerobiosis in facultative aerobes such as *Salmonella typhimurium*.[20] Mechanisms coupling gene expression to translation, such as attenuation, probably mediate many changes in gene expression caused by A37 undermodification. It has been shown that attenuation is sensitive to A37 modification,[91,108,111] with undermodified $tRNA^{Trp}$ reducing the translation efficiency of the tandemly repeated control codons in the leader peptide coding region of the *trp* operon.[111] Such tandem codons appear to be a special case of codon context, since undermodified tRNAs essentially function normally when the same codons occur singly.[108]

C. RECOGNITION

There is evidence that cytokinins may also influence the subsequent modification of the anticodon loop, by functioning as recognition sites for some tRNA-modifying enzymes. In a study of the sequence specificity of such enzymes,[25] A36–ms^2i^6A37–A38 was suggested as a recognition sequence for two modification enzymes in *E. coli*, *S. typhimurium*, and T4 phage. Occurrence of 2′ -*O*-methylcytidine (Cm) and pseudouridine (Ψ), both at C32, which occurs in the anticodon loop, showed a strong correlation with the presence of the modified triple-A sequence. In fact, A36–ms^2i^6A37–A38 occurred in all tRNAs from the above organisms that contained Ψ32. However, based on sequence analysis, most tRNAs from other organisms did not require the modified triple-A sequence. Most eukaryotic and prokaryotic tRNAs containing Cm had the sequence A36–purine or modified purine at 37–A38. Consistent with the requirement for the triple-A sequence for C32 modification is the fact that nucleoside 32 in yeast $tRNA^{Ser}$ was modified after splicing.[75] No modified base has been shown to be required for tRNA splicing.[1]

VII. CONCLUSION

The distribution of cytokinins among organisms, organelles, and tRNA species has been studied to gain insight regarding the function of cytokinins in tRNA. However, as more tRNAs are examined, generalizations regarding the phylogenetic distribution of cytokinin-active nucleosides may be found to be inappropriate. Reports of hydroxylated cytokinins in prokaryotes and yeast, as well as methylthiolated cytokinins in yeast and animals, are inconsistent with long-held beliefs. There is now good evidence that the diversity of cytokinins in plant tRNA, and probably in eukaryotic tRNA generally, is due to the cytokinin content of organelle-specific tRNAs in plastids and mitochondria. The distribution of cytokinin-containing tRNAs in semiautonomous organelles resembles that of prokaryotes, where most U-group tRNAs contain cytokinins. In contrast, except in yeast, $tRNA^{Ser}$ is the only eukaryotic, cytosolic tRNA known to contain cytokinin.

Although cytokinins in tRNA have been found to influence various aspects of translation in prokaryotes, the function of cytokinins is not well understood. Little information is available concerning their role in eukaryotes. Cytokinin nucleosides in prokaryotic tRNA are not required for cell growth or protein synthesis, but instead appear to have a regulatory role. Cytokinin nucleosides in tRNA have been shown to affect translation by modulating codon–anticodon interactions and/or tRNA–ribosome interactions. Alteration in the degree of tRNA modification has been proposed to function as a global regulatory mechanism allowing eubacteria to adapt to stress. Such responses may involve mechanisms coupling gene expression to translation, such as attenuation. There is also evidence that cytokinins act as recognition sites influencing modification of the anticodon arm. Little is known about the role of cytokinin-active nucleosides in plant tRNA. Presumably, their function is analogous to that in prokaryotes and yeast.

ACKNOWLEDGMENTS

The assistance of Mr. W. A. Mulroy and Mr. W. J. Lamoreaux in the preparation of this manuscript is gratefully acknowledged.

REFERENCES

1. **Bjork, G. R. and Kohli, J.,** Synthesis and function of modified nucleosides in tRNA, in *Chromatography and Modification of Nucleosides*, Gehrke, C. N. and Kuo, K. C., Eds., Elsevier, Amsterdam, 1990, B13.
2. **Claesson, C., Samuelsson, T., Lustig, F., and Boren, T.,** Codon reading properties of an unmodified transfer RNA, *FEBS Lett.*, 273, 173, 1990.
3. **Zachau, H. G., Dutting, D., and Feldmann, H.,** Nucleotide sequence of two serine-specific transfer ribonucleic acids, *Angew. Chem.*, 78, 392, 1966.
4. **Dietrich, A., Weil, J. H., and Marechal-Drouard, L.,** Nuclear-encoded transfer RNAs in plant mitochondria, *Annu. Rev. Cell Biol.*, 8, 115, 1992.
5. **Marechal-Drouard, L., Weil, J. H., and Dietrich, A.,** Transfer RNAs and transfer RNA genes in plants, *Annu. Rev. Plant Physiol. Plant Mol. Biol.*, 44, 13, 1993.
6. **Greene, E. M.,** Cytokinin production by microorganisms, *Bot. Rev.*, 46, 25, 1980.
7. **Dirheimer, G.,** Chemical nature, properties, location, and physiological and pathological variations of modified nucleosides, *Recent Results Cancer Res.*, 84, 15, 1983.
8. **Letham, D. S. and Palni, L. M. S.,** The biosynthesis and metabolism of cytokinins, *Annu. Rev. Plant Physiol.*, 34, 163, 1983.
9. **Nass, G.,** Ed., *Modified Nucleosides and Cancer*, Springer-Verlag, Berlin, 1983.
10. **Kersten, H.,** On the biological significance of modified nucleosides in tRNA, *Prog. Nucl. Acid Res. Mol. Biol.*, 31, 59, 1984.
11. **Bjork, G. R., Ericson, J. U., Gustafsson, C. E. D., Hagervall, T. G., Jonsson, Y. H., and Wikstrom, P. M.,** Transfer RNA modification, *Annu. Rev. Biochem.*, 56, 263, 1987.
12. **Bjork, G. R.,** Modification of stable RNA, in *Escherichia coli and Salmonella typhimurium: Cellular and Molecular Biology*, Neidhardt, F. C., Ingraham, J. L., Low, K. B., Magasanik, B., Schaechter, M., and Umbarger, H. E., Eds., American Society for Microbiology, Washington, D.C., 1987, 719.
13. **Skoog, F. and Armstrong, D. J.,** Cytokinins, *Annu. Rev. Plant Physiol.*, 21, 359, 1970.
14. **Sprinzl, M., Hartmann, T., Weber, J., Blank, J., and Zeidler, R.,** Compilation of tRNA sequences and sequences of tRNA genes, *Nucleic Acid Res.*, 17(Suppl.), r1, 1989.
15. **Sprinzl, M., Dank, N., Nock, S., and Schon, A.,** Compilation of tRNA sequences and sequences of tRNA genes, *Nucleic Acid Res.*, 19, 2127, 1991.
16. **Thimmappaya, B. and Cherayil, J. D.,** Unique presence of 2-methylthio-ribosylzeatin in the transfer ribonucleic acid of the bacterium *Pseudomonas aeruginosa*, *Biochem. Biophys. Res. Commun.*, 60, 665, 1974.
17. **Ajitkumar, P. and Cherayil, J. D.,** Presence of 2-methylthioribosyl-*trans*-zeatin in *Azotobacter vinelandii* tRNA, *J. Bacteriol.*, 162, 752, 1985.
18. **Janzer, J. J., Raney, J. P., and McLennan, B. D.,** The transfer RNA of certain Enterobacteriaceae contain 2-methylthiozeatin riboside (ms^2io^6A) an isopentenyl adenosine derivative, *Nucleic Acid Res.*, 10, 5663, 1982.
19. **Buck, M., McCloskey, J. A., Basile, B., and Ames, B. N.,** *cis* 2-Methylthioribosylzeatin (ms^2io^6A) is present in the transfer RNA of *Salmonella typhimurium*, but not *Escherichia coli*, *Nucleic Acid Res.*, 10, 5649, 1983.
20. **Buck, M. and Ames, B. N.,** A modified nucleotide in tRNA as a possible regulator of aerobiosis: synthesis of *cis*-2-methylthioribosylzeatin in the tRNA of *Salmonella*, *Cell*, 36, 523, 1984.
21. **Taller, B. J., Mulroy, W. A., and Stevens, S. E., Jr.,** unpublished results, 1993.
22. **Yokogawa, T., Watanabe, Y.-I., Kumazawa, Y., Ueda, T., Hirao, I., Miura, K.-I., and Watanabe, K.,** A novel cloverleaf structure found in mammalian mitochondrial $tRNA^{Ser}$ (UCN), *Nucleic Acid Res.*, 19, 6101, 1991.
23. **Jameson, P. E. and Morris, R. O.,** Zeatin-like cytokinins in yeast: detection by immunological methods, *J. Plant Physiol.*, 135, 385, 1989.
24. **Vreman, H. J.,** Isolation of *cis* and *trans* ribosylzeatin and 2-methylthioribosylzeatin from plant tRNA preparations, Ph.D. thesis, University of Wisconsin, Madison, 1973.

25. **Tsang, T. H., Buck, M., and Ames, B. N.,** Sequence specificity of tRNA-modifying enzymes: an analysis of 258 tRNA sequences, *Biochim. Biophys. Acta*, 741, 180, 1983.
26. **Barciszewska, M., Keith, G., Dirheimer, G., and Barciszewska, J.,** The nucleotide sequence of $tRNA^{Ser}$ and $tRNA^{Ile}$ of yellow lupin seeds, *Nucleic Acid Res.*, 16, 8175, 1988.
27. **Osorio-Almeida, M. L., Guillemaut, P., Keith, G., Canaday, J., and Weil, J. H.,** Primary structure of three leucine transfer RNAs from bean chloroplast, *Biochem. Biophys. Res. Commun.*, 92, 102, 1980.
28. **Taller, B. J., Murai, N., and Skoog, F.,** Endogenous cytokinins in the ribosomal RNA of higher plants, *Plant Physiol.*, 83, 755, 1987.
29. **Swaminathan, S. and Bock, R. M.,** Subcellular localization of cytokinins in transfer ribonucleic acid, *Plant Physiol.*, 59, 558, 1977.
30. **Swaminathan, S. and Bock, R. M.,** Isolation and identification of cytokinins from *Euglena gracilis* transfer ribonucleic acid, *Biochemistry*, 16, 1355, 1977.
31. **Vreman, H. J., Thomas, R., Corse, J., Swaminathan, S., and Murai, N.,** Cytokinins in tRNA obtained from *Spinacia oleracea* L. leaves and isolated chloroplasts, *Plant Physiol.*, 61, 296, 1978.
32. **Francis, M. A. and Dudock, B. S.,** Nucleotide sequence of spinach cytoplasmic serine (IGA) tRNA, *Nucleic Acid Res.*, 17, 7996, 1989.
33. **Edwards, C. A., Armstrong, D. J., Kaiss-Chapman, R. W., and Morris, R. O.,** Cytokinin-active ribonucleosides in *Phaseolus* RNA. I. Identification in tRNA from etiolated *Phaseolus vulgaris* L. seedlings, *Plant Physiol.*, 67, 1181, 1981.
34. **Chang, S. H., Brum, C. K., Silberkland, M., RajBhandary, U. L. R., Hecker, L. I., and Barnett, W. E.,** The first nucleotide sequence of an organelle transfer RNA: chloroplast $tRNA^{Phe}$, *Cell*, 9, 717, 1976.
35. **Hecker, L. I., Uziel, M., and Barnett, W. E.,** Comparative base compositions of chloroplast and cytoplasmic $tRNA^{Phe}$ from *Euglena gracilis*, *Nucleic Acid Res.*, 3, 371, 1976.
36. **Gray, M. W.,** The evolutionary origins of organelles, *Trends Genet.*, 5, 294, 1989.
37. **Hecker, L. I., Barnett, W. E., Lin, F. K., Furr, T. D., Heckman, J. E., RajBhandary, U. L., and Chang, S. H.,** The nucleotide sequence of blue-green algae phenylalanine-tRNA and the evolutionary origin of chloroplasts, *Nucleic Acid Res.*, 10, 6433, 1982.
38. **Raven, P.,** A multiple origin for plastids and mitochondria, *Science*, 169, 641, 1970.
39. **Bryant, D. A.,** Puzzles of chloroplast ancestry, *Curr. Biol.*, 2, 240, 1992.
40. **Gibbs, S. P.,** The chloroplast of some algal groups may be evolved from endosymbiotic eukaryotic algae, *Ann. N. Y. Acad. Sci.*, 361, 193, 1981.
41. **Douglas, S., Murphy, C., Spencer, D., and Gray, M.,** Cryptomonad algae are evolutionary chimaeras of two phylogenetically distinct unicellular eukaryotes, *Nature*, 350, 148, 1991.
42. **Zhang, W., Yamane, H., Spray, C. R., Chapman, D. J., and Phinney, B. O.,** Cytokinins from the tRNA of the red alga *Porphyra perforata* J. Ag., *Plant Cell Physiol.*, 33, 197, 1992.
43. **Swaminathan, S.,** Studies on the structure, subcellular localization and biochemical effects of cytokinins in transfer ribonucleic acids, Ph.D. thesis, University of Wisconsin, Madison, 1975.
44. **Armstrong, D. J., Burrows, W. J., Skoog, F., Roy, K. L., and Soll, D.,** Cytokinins: distribution in transfer RNA species of *Escherichia coli*, *Proc. Natl. Acad. Sci., U.S.A.*, 63, 834, 1969.
45. **Bartz, J., Soll, D., Burrows, W. J., and Skoog, F.,** Identification of the cytokinin-active ribonucleosides in pure *Escherichia coli* tRNA species, *Proc. Natl. Acad. Sci. U.S.A.*, 67, 1448, 1970.
46. **Nishimura, S.,** Modified nucleosides in tRNA, in *Transfer RNA: Structure, Properties and Recognition,* Schimmel, P. R., Soll, D., and Abelson, J. N., Eds., Cold Spring Harbor Laboratory, Cold Spring Harbor, NY, 1979.
47. **Grosjean, H., Nicoghosian, K., Haumont, E., Soll, D., and Cedergren, R.,** Nucleotide sequences of two serine tRNAs with a GCA anticodon: the structure-function relationships in the serine family of *E. coli* tRNAs, *Nucleic Acid Res.*, 13, 5697, 1985.
48. **Morris, R. O., Regier, D. A., Olson, R. M., Struxness, L. A., and Armstrong, D. J.,** Distribution of cytokinin-active nucleosides in isoaccepting transfer ribonucleic acids from *Agrobacterium tumefaciens*, *Biochemistry*, 20, 6012, 1981.
49. **White, B. N., Dunn, R., Gillam, I., Tener, G. M., Armstrong, D. J., Skoog, F., Frihart, C. R., and Leonard, N. J.,** An analysis of five serine transfer ribonucleic acids from *Drosophila*, *J. Biol. Chem.*, 250, 515, 1975.

50. **Jayabaskaran, C. and Jacob, T. M.,** Isolation and estimation of cytokinins and cytokinin-containing transfer RNAs from *Cucumis sativus* L. var. Guntur seedlings, *Plant Physiol.*, 70, 1396, 1982.
51. **Struxness, L. A., Armstrong, D. J., Gillam, I., Tener, G. M., Burrows, W. J., and Skoog, F.,** Distribution of cytokinin-active ribonucleosides in wheat germ tRNA species, *Plant Physiol.*, 63, 35, 1979.
52. **Edwards, C. A. and Armstrong, D. J.,** Cytokinin-active ribonucleosides in *Phaseolus* RNA. II. Distribution in tRNA species from etiolated *P. vulgaris* seedlings, *Plant Physiol.*, 67, 1185, 1981.
53. **Lester, B. R. and Cherry, J. H.,** Purification of leucine tRNA isoaccepting species from soybean cotyledons. I. Benzoylated diethylamino cellulose fractionation, N-hydroxysuccinamide modification and characterization of product, *Plant Physiol.*, 63, 79, 1979.
54. **Lester, B. R., Morris, R. O., and Cherry, J. H.,** Purification of leucine tRNA isoaccepting species from soybean cotyledons. II. RPC-2 purification, ribosome binding and cytokinin content, *Plant Physiol.*, 63, 87, 1979.
55. **Babcock, D. F. and Morris, R. O.,** Specific degradation of a plant leucyl transfer ribonucleic acid by a factor in the homologous synthetase preparation, *Plant Physiol.*, 52, 31, 1973.
56. **Einset, J., Swaminathan, S., and Skoog, F.,** Ribosyl-*cis*-zeatin in a leucyl transfer RNA species from peas, *Plant Physiol.*, 58, 140, 1976.
57. **Guillemaut, P. and Keith, G.,** Primary structure of bean chloroplastic $tRNA^{Phe}$, *FEBS Lett.*, 84, 351, 1977.
58. **Canaday, J., Guillemaut, P., Gloeckler, R., and Weil, J.-H.,** The nucleotide sequence of spinach chloroplast tryptophan transfer RNA, *Nucleic Acid Res.*, 9, 47, 1981.
59. **Green, G. A. and Jones, D. S.,** The nucleotide sequences of a cytoplasmic and a chloroplast $tRNA^{Tyr}$ from *Scenedesmus obliquus*, *Nucleic Acid Res.*, 13, 1659, 1985.
60. **Bartz, L. and Soll, D.,** N^6-(Δ^2-isopentenyl)adenosine: biosynthesis *in vitro* in transfer RNA by an enzyme purified from *Escherichia coli*, *Biochimie*, 54, 31, 1972.
61. **Holtz, J. and Klambt, D.,** tRNA isopentenyltransferase from *Lactobacillus acidophilus* ATCC, *Hoppe Seyler's Z. Physiol. Chem.*, 356, 1459, 1975.
62. **Kline, L., Fittler, F., and Hall, R.**. N^6-(Δ^2-isopentenyl)adenosine biosynthesis in transfer RNA *in vitro*, *Biochemistry*, 8, 4361, 1969.
63. **Holtz, J. and Klambt, D.,** Characterization of the isopentenylation reaction of tRNA, oligo (A) and other nucleic acids, *Hoppe Seyler's Z. Physiol. Chem.*, 359, 89, 1978.
64. **Fittler, F., Kline, L. K., and Hall, R. H.,** N^6-(Δ^2-isopentenyl)adenosine: biosynthesis *in vitro* by an enzyme extract from yeast and rat liver, *Biochem. Biophys. Res. Commun.*, 31, 571, 1968.
65. **Gallagher, P. J., Schwartz, I., and Elseviers, D.,** Genetic mapping of *pheU*, an *Escherichia coli* gene for phenylalanine tRNA, *J. Bacteriol.*, 158, 762, 1984.
66. **Ericson, J. U. and Bjork, G. R.,** Pleiotropic effects induced by modification deficiency next to the anticodon of tRNA from *Salmonella typhimurium* LT2, *J. Bacteriol.*, 166, 1013, 1986.
67. **Caillet, J. and Droogmans, L.,** Molecular cloning of the *Escherichia coli mia*A gene involved in the formation of Δ2-isopentenyl adenosine in tRNA, *J. Bacteriol.*, 170, 4147, 1988.
68. **Connolly, D. M. and Winkler, M. E.,** Structure of *Escherichia coli* K-12 *mia*A and characterization of the mutator phenotype caused by *mia*A insertion mutations, *J. Bacteriol.*, 173, 1711, 1991.
69. **Connolly, D. M. and Winkler, M. E.,** Genetic and physiological relationships among the *mia*A gene, 2-methylthio-N^6-(Δ^2-isopentenyl)-adenosine tRNA modification, and spontaneous mutagenesis in *Escherichia coli* K-12, *J. Bacteriol.*, 171, 3233, 1989.
70. **Dihanich, M. E., Najarian, D., Clark, R., Gillman, E. C., Martin, N. C., and Hopper, A. K.,** Isolation and characterization of *MOD* 5, a gene required for isopentenylation of cytoplasmic and mitochondrial tRNAs of *Saccharomyces cerevisiae*, *Mol. Cell. Biol.*, 7, 177, 1987.
71. **Najarian, D., Dihanich, M. E., Martin, N. C., and Hopper, A. K.,** DNA sequence and transcript mapping of *MOD* 5: features of the 5′ region which suggest two translational starts, *Mol. Cell. Biol.*, 7, 185, 1987.
72. **Martin, N. C. and Hopper, A. K.,** Isopentenylation of both cytoplasmic and mitochondrial tRNA is affected by a single nuclear mutation, *J. Biol. Chem.*, 257, 10562, 1982.
73. **Gillman, E. C., Slusher, L. B., Martin, N. C., and Hopper, A. K.,** *MOD5* translation initiation sites determine N^6-isopentenyladenosine modification of mitochondrial and cytoplasmic tRNA, *Mol. Cell. Biol.*, 11, 2382, 1991.

74. **Carbon, P., Haumont, E., de Henau, S., Keith, G., and Grosjean, H.,** Enzymatic replacement *in vitro* of the first anticodon base of yeast $tRNA^{Asp}$: application to the study of tRNA maturation *in vitro*, after microinjection into frog oocytes, *Nucleic Acid Res.*, 10, 3715, 1982.
75. **Etcheverry, T., Colby, D., and Guthrie, C.,** A precursor to a minor species of yeast $tRNA^{Ser}$ contains an intervening sequence, *Cell*, 18, 11, 1979.
76. **Gray, J., Wang, J., and Gelvin, S. B.,** Mutation of the *mia*A gene of *Agrobacterium tumefaciens* results in reduced *vir* gene expression, *J. Bacteriol.*, 174, 1086, 1992.
77. **Walker, J. E., Saraste, M., Runswick, M. J., and Gay, N. J.,** Distantly related sequences in the *a*- and *b*-subunits of ATP synthase, myosin, kinases and other ATP-requiring enzymes and a common nucleotide binding fold, *EMBO J.*, 8, 945, 1982.
78. **Gefter, M. L.,** The *in vitro* synthesis of 2′-methylguanosine and 2-methylthio N^6-(γ,γ, dimethylallyl)adenosine in transfer RNA of *Escherichia coli*, *Biochem. Biophys. Res. Commun.*, 36, 435, 1969.
79. **Rosenberg, A. H. and Gefter, M. L.,** An iron-dependent modification of several transfer RNA species in *Escherichia coli*, *J. Mol. Biol.*, 46, 581, 1969.
80. **Agris, P. F., Armstrong, D. J., Schafer, K. P., and Soll, D.,** Maturation of a hypermodified nucleoside in transfer RNA, *Nucleic Acid Res.*, 2, 691, 1975.
81. **Wettstein, F. O. and Stent, G. S.,** Physiologically induced changes in the property of phenylalanine tRNA in *Escherichia coli*, *J. Mol. Biol.*, 38, 25, 1968.
82. **Isham, K. R. and Stulberg, M. P.,** Modified nucleosides in undermethylated phenylalanine transfer RNA from *Escherichia coli*, *Biochim. Biophys. Acta*, 340, 177, 1974.
83. **Mann, M. B. and Huang, P. C.,** New chromatographic form of phenylalanine transfer ribonucleic acid from *Escherichia coli* growing exponentially in a low phosphate medium, *J. Bacteriol.*, 118, 209, 1974.
84. **Keith, G., Rogg, H., Dirheimer, G., Menichi, B., and Heyman, T.,** Post-transcriptional modification of tyrosine tRNA as a function of growth in *Bacillus subtilis*, *FEBS Lett.*, 61, 120, 1976.
85. **Buu, A., Menichi, B., and Heyman, T.,** Thiomethylation of tyrosine transfer ribonucleic acid is associated with initiation of sporulation in *Bacillus subtilis*: effect of phosphate concentration, *J. Bacteriol.*, 146, 819, 1981.
86. **Carbon, J. and Fleck, E. W.,** Genetic alteration of structure and function in glycine transfer RNA of *Escherichia coli*: mechanism of suppression of the tryptophan synthetase A78 mutation, *J. Mol. Biol.*, 85, 371, 1974.
87. **Yarus, M.,** Translational efficiency of transfer RNAs: uses of an extended anticodon, *Science*, 218, 646, 1982.
88. **Schon, A., Bock, A., Ott, G., Sprinzl, M., and Soll, D.,** The selenocysteine-inserting opal suppressor serine tRNA from *E. coli* is highly unusual in structure and modification, *Nucleic Acid Res.*, 17, 7159, 1989.
89. **Hanyu, N., Kuchino, Y., and Nishimura, S.,** Dramatic events in ciliate evolution: alteration of UAA and UAG termination codons to glutamine codons due to anticodon mutations in two *Tetrahymena* $tRNAs^{Gln}$, *EMBO J.*, 5, 1307, 1986.
90. **Prather, N. E., Murgola, E. J., and Mims, B. H.,** Nucleotide insertion in the anticodon loop of a glycine transfer RNA causes missense suppression, *Proc. Natl. Acad. Sci. U.S.A.*,78, 7408, 1981.
91. **Yanofsky, C. and Soll, L.,** Mutations affecting $tRNA^{Trp}$ and its charging and their effect on regulation of transcription termination at the attenuator of the tryptophan operon, *J. Mol. Biol.*, 113, 663, 1977.
92. **Gefter, M. L. and Russell, R. L.,** Role of modifications in tyrosine transfer RNA: a modified base affecting ribosome binding, *J. Mol. Biol.*, 39, 145, 1969.
93. **Furuichi, Y., Wataya, Y., Hayatsu, H., and Ukita, T.,** Chemical modification of $tRNA^{tyr}$ yeast with bisulfite. A new method to modify isopentenyl-adenosine residue, *Biochem. Biophys. Res. Commun.*, 41, 1185, 1970.
94. **Petrullo, L. A. and Elseviers, D.,** Effect of a 2-methylthio-N^6-isopentenyladenosine deficiency on peptidyl-tRNA release in *Escher-ichia coli*, *J. Bacteriol.*, 165, 608, 1986.
95. **Vacher, J., Grosjean, H., Houssher, C., and Buckingham, R. H.,** The effect of point mutations affecting *Escherichia coli* tryptophan tRNA on anticodon–anticodon interactions and on UGA suppression, *J. Mol. Biol.*, 177, 329, 1984.
96. **Wilson, R. K. and Roe, B. A.,** Presence of the hypermodified nucleotide N^6-(Δ^2-isopentenyl)-2-methylthioadenosine prevents codon misreading by *Escherichia coli* phenylalanyl-transfer RNA, *Proc. Natl. Acad. Sci. U.S.A.*, 86, 409, 1989.

97. **Grosjean, H. and Houssier, C.,** Codon recognition: evaluation of the effects of modified bases on the anticodon loop of tRNA using the temperature-jump relaxation method, in *Chromatography and Modification of Nucleosides*, Gehrke, C. W. and Kuo, K. C. T., Eds., Elsevier, Amsterdam, 1990, A255.
98. **Bouadloun, F., Srichaiyo, T., Isaksson, L. A., and Bjork, G. R.,** Influence of modification next to the anticodon in tRNA on codon context sensitivity of translational suppression and accuracy, *J. Bacteriol.*, 166, 1022, 1986.
99. **Ericson, J. U. and Bjork, G. R.,** tRNA anticodons with the modified nucleoside 2-methylthio-N^6-(4-hydroxyisopentenyl) adenosine distinguish between bases 3′ of the codon, *J. Mol. Biol.*, 218, 509, 1991.
100. **Diaz, I., Pedersen, S., and Kurland, C. G.,** Effects of *mia*A on translation and growth rates, *Mol. Gen. Genet.*, 208, 373, 1987.
101. **Diaz, I. and Ehrenberg, M.,** ms^{2}i^{6}A deficiency enhances proofreading in translation, *J. Mol. Biol.*, 222, 1161, 1991.
102. **Kimball, M. E. and Soll, D.,** Phenylalanine transfer-RNA from *Mycoplasma* sp. (Kid) — Transfer-RNA lacking hypermodified nucleosides functional in protein synthesis, *Nucleic Acid Res.*, 1, 1713, 1974.
103. **Diaz, I., Ehrenberg, M., and Kurland, C. G.,** How do combinations of *rpsL*$^-$ and *miaA*$^-$ generate streptomycin dependence? *Mol. Gen. Genet.*, 202, 207, 1986.
104. **Laten, H., Gorman, J., and Bock, R. M.,** Isopentenyladenosine deficient tRNA from an antisuppressor mutant of *Saccharomyces cerevisiae*, *Nucleic Acid Res.*, 5, 4329, 1978.
105. **Janner, F., Vogeli, G., and Fluri, R.,** The antisuppressor strain *sin*1 of *Schizosaccharomyces pombe* lacks the modification isopentenyladenosine in transfer RNA, *J. Mol. Biol.*, 139, 207, 1980.
106. **Gowrishankar, J. and Pittard, J.,** Regulation of phenylalanine biosynthesis in *Escherichia coli* K-12: control of transcription of the *phe*A operon, *J. Bacteriol.*, 150, 1130, 1982.
107. **Buck, M. and Griffiths, E.,** Regulation of aromatic amino acid transport by tRNA: role of 2-methylthio-N^6-(Δ^2-isopentenyl)-adenosine, *Nucleic Acid Res.*, 9, 401, 1981.
108. **Buck, M. and Griffiths, E.,** Iron mediated methylthiolation of tRNA as a regulator of operon expression in *Escherichia coli*, *Nucleic Acid Res.*, 10, 2609, 1982.
109. **Blum, P. H.,** Reduced *leu* operon expression in a *mia*A mutant of *Salmonella typhimurium*, *J. Bacteriol.*, 170, 5125, 1988.
110. **Griffiths, E., Humphreys, J., Leach, A., and Scanlon, L.,** Alterations in the tRNAs of *Escherichia coli* recovered from lethally infected animals, *Infect. Immun.*, 22, 312, 1978.
111. **Landick, R., Yanofsky, C., Choo, K., and Phung, L.,** Replacement of the *Escherichia coli trp* operon attenuation control codons alters operon expression, *J. Mol. Biol.*, 216, 25, 1990.

Chapter 9

Cytokinin Metabolism and Compartmentation

Paula E. Jameson

TABLE OF CONTENTS

I. CHEMICAL DIVERSITY OF METABOLITES AND PATHWAYS OF METABOLISM

The true extent of the ability of plants to metabolize cytokinins was first revealed by Letham's group in the middle 1970s following the feeding of labeled cytokinins to blue lupin plants (*Lupinus angustifolius* L.) and the separation of numerous labeled, but uncharacterized, compounds. Ensuing metabolic studies resulted in the identification of 16 metabolites,[1-4] including new groups of zeatin derivatives, all of which have subsequently been shown to occur as endogenous compounds. Comprehensive reviews on cytokinin metabolism have been presented by Letham and Palni,[5] McGaw et al.,[6] and McGaw and Horgan.[7]

The elucidation of the metabolic pathways of the cytokinins has been largely based on the use of exogenous radio-labeled cytokinins. When applied to plant tissue, the cytokinins are usually rapidly distributed among the respective nucleotide, nucleoside, and base forms, but are ultimately either broken down by side chain cleavage to the corresponding adenine derivatives (with irreversible loss of biological activity) or are conjugated into storage or inactive forms. Structural modifications of zeatin can be grouped, therefore, into those reactions involving:

1. Modification of the purine ring (Figure 1)
2. Modification of the side chain (Figure 2)

A. MODIFICATIONS OF THE PURINE RING

1. Interconversions Between Base, Riboside, and Nucleotide Forms

When labeled cytokinins (e.g., ^{3}H-zeatin, ^{3}H-zeatin riboside, ^{3}H-dihydrozeatin riboside, ^{3}H-6-benzylaminopurine) are applied to derooted soybean

0-8493-6252-0/94/$0.00+$.50

plants,[8] to germinating *Lupinus luteus* L. seeds,[9] to isolated pieces of bark,[10] or to *Arabidopsis thaliana* plantlets,[11] and time-course studies are performed, the cytokinin nucleotides are the prominent metabolites formed immediately following uptake. The nucleotides are then, however, rapidly metabolized. Confirmation that the cytokinin nucleotides are converted into other cytokinins was shown for the first time by Singh et al.,[8] who supplied ^{3}H-zeatin nucleotide to soybean leaf blades. After 24 h the main metabolites were adenosine, adenine, and dihydrozeatin-*O*-glucoside. Since these were also the major metabolites of ^{3}H-zeatin riboside in the blades after a similar time period, Singh et al. (1988) suggested that in the explant system, zeatin nucleotide may be a metabolite of zeatin riboside and zeatin formed to facilitate uptake, after which it is rapidly metabolized.

The enzyme systems catalyzing the interconversions of cytokinin base–riboside–nucleotide appear to be the same as those for adenine–adenosine–adenosine monophosphate interconversions.[12-15] Burch and Stuchbury showed that the distribution and activity of these enzymes changed markedly during development and suggested that these enzymes may be important in the regulation of cytokinin activity.[16] Further, they suggested that the enzyme adenine phosphoribosyltransferase (APRT) may convert cytokinin bases to their nucleotide form and that this may facilitate uptake of cytokinin. Recently, Moffatt et al. showed that APRT was the main enzyme that converted 6-benzyl-aminopurine to its nucleotide form in young *A. thaliana* plants.[11] However, in addition they selected APRT-deficient plants and showed that the formation of 6-benzylaminopurine monophosphate from 6-benzylaminopurine was strongly and specifically reduced in these mutant plants. However, as 6-benzylaminopurine was rapidly absorbed by these plants, it appears that ribophosphorylation of

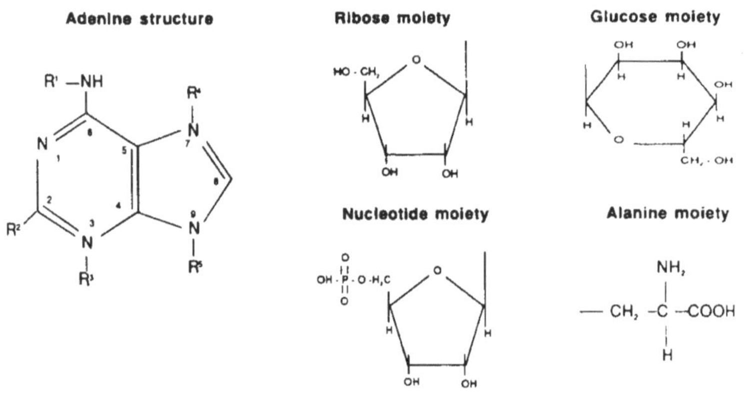

Modification	Form	Abbreviation	Substituents	
Purine ring	Base	iP	R1 = $-CH_2-CH=CH-CH_3$ CH_3	
		Z	R1 = $-CH_2-CH=C-CH_2-OH$ CH_3	
	Riboside	iPA	R1 = $-CH_2-CH=CH-CH_3$ CH_3	R5 = ribose moiety
		ZR	R1 = $-CH_2-CH=C-CH_2-OH$ CH_3	R5 = ribose moiety
	Nucleotide	iPNT	R1 = $-CH_2-CH=CH-CH_3$ CH_3	R5 = nucleotide moiety
		ZRNT	R1 = $-CH_2-CH=C-CH_2-OH$ CH_3	R5 = nucleotide moiety
	N-glucose conjugate		Glucose moiety attached at R3, R4 or R5	
	N-alanine conjugate		R5 = alanine moiety	

Figure 9-1 Structures of cytokinin metabolites: modifications of the purine ring.

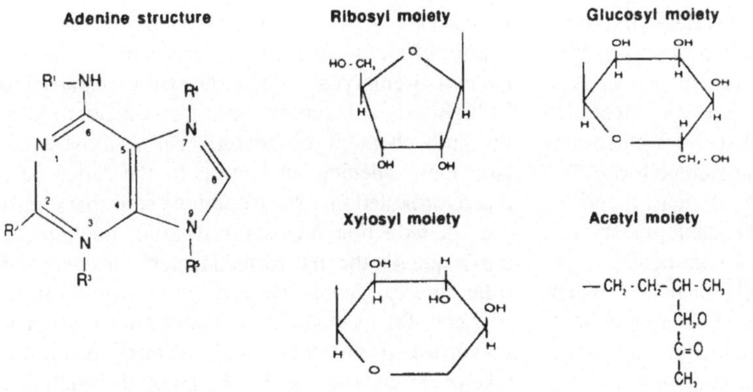

Modification	Form	Abbreviation	Substituents
Side chain	Dihydro-derivatives	DZ	$R1 = -CH_2-CH_2-CH(CH_3)-CH_2-OH$
		DZR	$R1 = -CH_2-CH_2-CH(CH_3)-CH_2-OH$ R5 = ribose moiety
	O-glucosyl derivatives	ZOG	$R1 = -CH_2-CH{=}C(CH_3)-CH_2-O$-Glucose
	O-acetyl derivatives	OAcZR	$R1 = -CH_2-CH{=}C(CH_3)-CH_2-O$-Acetyl R5 = ribose moiety
	O-xylosyl derivatives	OXZ	$R1 = -CH_2-CH{=}C(CH_3)-CH_2-O$-Xylose

Figure 9-2 Structures of cytokinin metabolites: modification of the side chain.

6-benzylaminopurine was not a prerequisite for its absorption by *A. thaliana* plantlets,[11] which is in contrast to the earlier suggestions that nucleotides were metabolites formed to facilitate uptake.

As the formation of the free cytokinins begins with the biosynthesis of the cytokinin nucleotides, it is clear that a considerable amount of interconversion of free bases, ribosides, and nucleotides is possible. The enzymes responsible are discussed more fully in Chapter 10.

2. Glucosylation

Glucosylation of cytokinins can occur at the 3, 7, and 9 positions of the purine moiety.[5] The 7-glucoside of zeatin, which occurs as an endogenous cytokinin,[17] was first detected following the application of ^{3}H-zeatin to derooted radish seedlings[18] where *N*-glucosylation was the predominant form of metabolism. When dihydrozeatin was supplied to derooted radish seedlings, it was similarly glucosylated, and dihydrozeatin 7-glucoside was the major metabolite.[19] In contrast, when ^{3}H-zeatin was supplied to roots of intact *Zea mays* seedlings, to derooted seedlings, and to a culture of embryonic tissue, the 9-glucoside was detected as a major metabolite, with the 7-glucoside being only of minor significance.[20,21] Zeatin 9-glucoside has also been found as a metabolite of zeatin in poplar leaves,[2,22] immature apple seeds,[23] and oat leaf segments.[24] Dihydrozeatin 9-glucoside was formed from dihydrozeatin in derooted radish seedlings,[19] and the 9-glucoside of isopentenyladenine has been identified as a naturally occurring compound in young seedlings and tubers of sweet potato.[25,26] The 3-glucoside of 6-benzylaminopurine was detected as a minor metabolite of 6-benzylaminopurine in derooted radish seedlings[27] and in oat leaf segments,[24] and dihydrozeatin 3-glucoside was formed from dihydrozeatin in derooted radish seedlings.[19]

Entsch et al.[28] have partially purified two glucosyltransferases from radish cotyledons where *N*-glucosylation is the predominant form of metabolism following application of zeatin (for more details see Chapter 10).

The 3-glucosides of zeatin and 6-benzylaminopurine can be hydrolyzed by β-glucosidase both *in vitro*[27,29] and *in vivo*,[29-31] to release the corresponding active bases that would account for their relatively high activity in cytokinin bioassays. As a consequence of their *in vivo* turnover, van Staden and Drewes have suggested that the 3-glucosides should be regarded as cytokinin storage forms.[29]

The 7- and 9-glucosides of zeatin and 6-benzylaminopurine appear to be stable metabolites in the tissues in which they are formed,[3,18,31] are relatively inactive in bioassays,[32,33] and are resistant to hydrolysis by β-glucosidase.[27,29] They are, however, potential substrates for cytokinin oxidase,[34] and some degradation of these compounds to the corresponding adenine derivatives has been observed in immature radish cotyledons[30] and in soybean callus.[35] It is believed that they are generally protected from this enzyme by compartmentalization in cells, which may account for their high stability in plant tissues. As a consequence of their low turnover *in vivo*, it is unlikely that they are storage forms. Further, their formation does not appear to be coupled with cytokinin action.[31,36] Consequently, these metabolites are generally regarded as detoxification or deactivation products.[3,6] However, it is of interest that the RolC locus of the Ti plasmid of *Agrobacterium rhizogenes* codes for a β-glucosidase capable of metabolizing cytokinin *N*-glucosides.[37]

3. *N*-Alanyl Derivatives

Conjugation of the cytokinins may also occur with alanine at the 9 position of the purine nucleus to yield lupinic acid [L-3-(zeatin-9-yl)alanine] and dihydrolupinic acid. Lupinic acid was first detected as a metabolite of zeatin in derooted blue lupin seedlings,[1,4,38] and later the corresponding 9-alanyl derivatives were identified as metabolites of 6-benzylaminopurine in bean seedlings[39] and dihydrozeatin in yellow lupin seeds.[23] Lupinic acid and dihydrolupinic acid were unequivocally identified in lupins[40] and have subsequently only been identified in members of the Fabaceae. The enzyme responsible, a β-(9-cytokinin)alanine synthase, has been purified from lupin seeds (see also Chapter 10). While the alanine conjugates exhibited enhanced metabolic stability in yellow lupin leaves[3] and were much less active than the corresponding bases in several bioassay systems,[32] the observation of the release of free zeatin from lupinic acid indicates that alanine conjugates may be potential storage forms rather than deactivation products.[35]

B. MODIFICATIONS OF THE SIDE CHAIN

1. Side Chain Reduction

When zeatin and zeatin riboside are supplied, metabolites exhibiting side chain reduction (the dihydroderivatives) are usually identified. Recently, Martin et al. isolated a reductase from *Phaseolus coccineus* embryos.[41] The reductase was shown to be highly specific for *trans*-zeatin; it did not modify the side chain of *cis*-zeatin, zeatin riboside, or isopentenyladenine, indicating in the latter case that hydroxylation of the isoprenoid side chain must precede reduction. As discussed earlier, nucleotides are frequently the first-formed metabolites detected following exogenous application of cytokinin. In soybean, for example, both zeatin nucleotide and dihydrozeatin nucleotide were formed during uptake of ^{3}H-zeatin riboside.[8] Singh et al. concluded that the rapid formation of dihydrozeatin nucleotide indicated that the reduction of the double bond in the isoprenoid side chain may occur at the nucleotide level.[8] Consequently, it remains to be established whether reductases from other sources are as specific as the *P. coccineus* zeatin reductase or whether, in fact, other reductases can reduce the nucleotide and riboside as well as the free base.

In many cases the reduced cytokinins are conjugated on the side chain and accumulate as dihydrozeatin-*O*-glucoside (for example, in primary leaves of *P. vulgaris*[42]), as dihydrozeatin riboside-*O*-glucoside (for example, in *Lupinus angustifolius* pod walls and seeds[43]), and as the 3-, 7-, and 9-glucosides in *Rhaphanus*.[44] By incubating ^{14}C-zeatin and ^{14}C-zeatin-*O*-xylose with enzyme extracts obtained from *P. coccineus* embryos, Mok et al. showed that while zeatin-*O*-xylose was not converted to its dihydroderivative, zeatin was converted to dihydrozeatin.[45] They concluded that reduction of the side chain seemed to precede *O*-xylosylation in *P. coccineus* embryos.

In vitro experiments have established that the dihydroderivatives are not usually substrates for cytokinin oxidase (but see below), and hence reduction enhances structural stability. Conjugation thus provides a mechanism by which the biological activity of these compounds may be controlled.[44]

2. Side Chain Cleavage

Oxidative side chain cleavage to produce adenine, adenosine, and adenine nucleotides is the predominant fate of exogenously supplied zeatin, zeatin riboside, and isopentenyladenine, in various plant tissues and results in irreversible destruction of cytokinin activity. The enzymes involved, cytokinin oxidases, are discussed in detail in Chapter 11. The enzymes are able to utilize zeatin, zeatin riboside, isopentenyladenine, and isopentenyladenosine as substrates, as well as the 7- and 9-glucosides and lupinic acid. The saturated dihydroderivatives are resistant to cytokinin oxidase obtained from *Zea mays* kernels and from

Vinca rosea crown gall tissue[34,46] (i.e., activity of these enzymes requires a Δ^2 double bond in the N^6 side chain), and the *O*-glucosyl derivatives are also resistant to oxidation.[34]

There is, however, some circumstantial evidence for the presence of other oxidative system(s) in plants. The side chain of 6-benzylaminopurine can be cleaved when 6-benzylaminopurine is fed to soybean and tobacco callus,[47,48] seedlings of *L. luteus*,[48] and radish,[27] and similarly, the kinetin side chain can be cleaved in *Acer* cells[49] and in decapitated tomato plants.[50] The dihydrozeatin-type cytokinins are subject to side chain cleavage in lupin seedlings,[3,51] in suspension cultures of *Chenopodium rubrum*,[52] and in soybean leaves.[8] While dihydrozeatin-*O*-glucoside was the major metabolite of dihydrozeatin riboside in soybean leaf blades, adenine and adenosine were also formed. Since no conversion of dihydrozeatin riboside to zeatin or zeatin riboside was detected, the formation of adenine and adenosine suggests that these tissues contain an uncharacterized enzyme system that has the ability to cleave a saturated side chain at N^6. The tentative identification of adenine as a minor metabolite of ^{3}H-dihydrozeatin (1.4% of total ^{3}H) has also been observed in detached leaves of *Phaseolus vulgaris*.[53]

3. Conjugation of the Side Chain

a. *O-Glucosylation*

Glucosylation of the isoprenoid side chain to form the *O*-β-D-glucopyranosyl derivatives of zeatin was first reported following the feeding of ^{3}H-zeatin to derooted lupin seedlings[3,38] and to soybean callus tissue.[54] Four *O*-glucosides (zeatin-*O*-glucoside, zeatin riboside-*O*-glucoside, dihydrozeatin-*O*-glucoside, dihydrozeatin riboside-*O*-glucoside) were subsequently identified in mature leaves of poplar as metabolites of supplied zeatin or zeatin riboside.[2,22] Recently, Dixon et al.[55] reported the isolation of an *O*-glucosyltransferase from *Phaseolus* embryos (also see Chapter 10).

The *O*-glucosides appear to be ubiquitous endogenous compounds present in developing, mature, and senescent tissues and have been shown to accumulate to what have been described as "surprisingly high levels" in some.[5] They have been identified in developing, mature, and senescent leaves[22,42,56-58] as well as in lupin pod walls and seeds,[40,43,59,60] sweet corn kernels,[61] *V. rosea* and *Datura innoxia* crown gall tissues,[62,63] and immature embryos of *P. lunatus*.[45] A novel *O*-glucoside of zeatin-9-glucoside was identified recently in wheat spikes, by Sayavedra-Soto et al.[64] The *O*-glucosides are not substrates for cytokinin oxidase,[34] but are readily degraded by β-glucosidases and, unlike the *N*-glucosyl conjugates, are readily metabolized in some systems. These observations come both from monitoring endogenous changes and from experiments where labeled *O*-glucosides were fed to plant tissues. For example, Wang et al.[56] showed that primary leaves of decapitated, disbudded bean plants accumulated large quantities of dihydrozeatin riboside-*O*-glucoside (later shown to be approximately 50 ng d^{-1} per leaf to about 500 ng g^{-1}) after 40 days from decapitation.[42] When secondary laterals were allowed to grow out, the levels of dihydrozeatin riboside-*O*-glucoside in the primary leaves rapidly declined. An analysis of detached leaves showed a similar decline of dihydrozeatin riboside-*O*-glucoside, which was presumed to be due to metabolism and not to export. Consequently, Palmer et al. concluded that in leaves of intact plants, the decrease of dihydrozeatin riboside-*O*-glucoside was due to metabolism rather than to export per se.[42]

The *O*-glucosides have also been shown to decline following induction of other phases of plant development, e.g., fertilization of the wheat ovule,[65] seed germination,[66] and bud growth.[67] Further, they show greater activity than the 7- or 9-glucosides or lupinic acid, in various bioassays.[32,56] Letham[68] suggested that as the very large and stable N^6-substituent would be expected to abolish cytokinin activity, zeatin-*O*-glucoside and related *O*-glucosides are unlikely to be active per se. That the *O*-glucoside moiety can be cleaved was shown clearly by McGaw et al.,[44] when dihydrozeatin, zeatin-*O*-glucoside, and dihydrozeatin riboside-*O*-glucoside were fed to derooted *Rhaphanus* seedlings. While substantial amounts of the *O*-glucosides remained unmetabolized, zeatin 7-glucoside and dihydrozeatin 7-glucoside were isolated from *Rhaphanus*, but *O*-glucosides of zeatin 7-glucoside and dihydrozeatin 7-glucoside were not detected, indicating the conversion of the *O*-glucosides to their corresponding bases prior to *N*-glucosylation, a conclusion supported by Letham and Gollnow.[31]

The glucosyl moiety of exogenous labeled *O*-glucosides has been shown to be cleaved in other plant tissues, presumably by β-glucosidase enzymes. In detached leaves of *L. luteus*, ^{3}H-zeatin-*O*-glucoside was converted predominantly to zeatin riboside-*O*-glucoside, but the formation of smaller quantities of adenosine, zeatin, and dihydrozeatin were reported.[3] Detached primary leaves of *P. vulgaris* metabolized both ^{3}H-zeatin-*O*-glucoside and ^{3}H-dihydrozeatin riboside-*O*-glucoside, but here cleavage of the side chain, rather than ribosylation, was the primary form of metabolism.[53] Van Staden and Papaphilippou suggested that the growth of soybean callus on medium supplied with high levels of zeatin-*O*-glucoside was due to the slow release of zeatin at nontoxic levels,[69] and certainly ^{14}C-zeatin-*O*-glucoside was shown to be rapidly degraded in *V. rosea* crown gall tissues.[70] The activity of these metabolites is therefore almost certainly a consequence of the cleavage of the glucosyl moiety, making the *O*-glucosides prime candidates for storage forms.[32]

b. O-Xylosylation

A species-specific side chain conjugation has been characterized by Mok and co-workers (see Chapter 10), following metabolism of ^{14}C-zeatin by immature embryos of *P. vulgaris* and *P. lunatus*. Rapid conversion of ^{14}C-zeatin to zeatin-*O*-glucoside occurred in *P. lunatus* embryos, whereas embryos of *P. vulgaris* converted the zeatin to *O*-β-D-xylopyranosylzeatin and its riboside.[71] Subsequently, the *O*-xylosyl derivatives of zeatin, zeatin riboside, and dihydrozeatin were shown to be endogenous compounds.[45,72] The inability of *P. lunatus* to form the xylose derivatives appears to be the exception in *Phaseolus* species, as incubation of the embryos of *P. coccineus* and *P. acutifolius* with radio-labeled zeatin also led to the formation of *O*-xylosylzeatin and its dihydroderivative.[73] However, it appears that xylosylation of zeatin and dihydrozeatin occurs only in the immature embryos.[73] The implications of selective expression of *O*-xylosylation with regard to embryo development are unknown. The enzyme responsible for *O*-xylosylation in *P. vulgaris* embryos has been partially purified and is discussed elsewhere (Chapter 10). The biological activity of *O*-xylosylzeatin is slightly lower than that of zeatin in the *P. lunatus* callus bioassay, but is about 100 times higher than that of zeatin in the *P. vulgaris* callus bioassay where zeatin is much less active.[74] This has led to the suggestion that *O*-xylosylzeatin may be an active form, since the hydrolysis of it to zeatin in *P. vulgaris* does not increase the activity. This was supported by the insignificant conversion of *O*-xylosylzeatin to zeatin as observed in short-term incubation studies.[73] However, a storage role, or even a transport role, for this group of compounds should not be ruled out.

c. O-Acetylation

In studies of the translocation and metabolism of ^{3}H-zeatin riboside in blue lupin (*L. angustifolius*) shoots, unknown nucleotide-like metabolites were detected in stem bark, pod walls, developing lateral shoots, and developing inflorescences.[75] In the case of at least bark and pod walls, the metabolites were transient. When dihydrozeatin was supplied to isolated segments of lupin bark, *O*-acetyl-9-β-D-ribofuranosyldihydrozeatin-5′-monophosphate was identified as a transient metabolite.[10,76] A metabolite of zeatin riboside formed in excised inflorescences was the corresponding unsaturated compound (*O*-acetyl zeatin nucleotide). Simultaneously with the report (from Letham's group) of the identification of this new group of cytokinin metabolites,[76] Laloue and Pethe reported the occurrence of *O*-acetyl metabolites of cytokinins in plant tumor tissue.[77] Hence, they may prove to be ubiquitous cytokinins.

Letham and Zhang noted that the *O*-acetyl group is slowly cleaved by dilute aqueous ammonia,[10] which is normally used to elute cytokinins from cellulose phosphate columns. Hence, this purification is not suitable for preparation of the *O*-acetyl bases and ribosides.

4. Methylation

Novel cytokinins are produced by cultures of the phytopathogen *Pseudomonas syringae* pv *savastanoi*. Two of these have a modified side chain and have been identified as ribosyl-1^{4}-methyl zeatin[78] and its riboside.[79,80]

Methylzeatin and methylzeatin riboside were shown, by Evidente et al.,[81] to be more active in stimulating chlorophyll synthesis in etiolated cucumber cotyledons than either zeatin or zeatin riboside. Whether the high biological activity of these cytokinins is a direct consequence of their altered side chain or is due to enhanced metabolic stability remains to be determined.

5. Aromatic Side Chains

An unusual cytokinin was isolated and identified from leaves of *Populus robusta*[82,83] as 6-(*O*-hydroxybenzylamino)-9-β-D-ribofuranosyl-purine. Its isolation represented the first report of a naturally occurring purinyl cytokinin having an aromatic side chain. More recently, 6-benzylaminopurine itself was reported as an endogenous cytokinin in primary crown gall tumors from the stems of tomato plants.[84]

C. NOVEL CONJUGATES OF THE SIDE CHAIN AND PURINE RING

A standard procedure for the detection of the *O*-glucosides of zeatin, dihydrozeatin, zeatin riboside, and dihydrozeatin riboside has been via β-glucosidase treatment and subsequent analysis of the aglycones. However, recently Singh et al. noted that β-glucosidase hydrolysis released more zeatin riboside than direct analyses of the intact *O*-glucosides indicated was present.[8] They concluded that other glucosides of zeatin riboside may exist in soybean leaves. Previously, both Davey and van Staden[59] and Dekhuijzen[85] commented on the fact that treatments of a sample with β-glucosidase released compounds that did not always completely coelute with zeatin and/or zeatin riboside. Conjugates that release cytokinin bases or ribosides by β-glucosidase hydrolysis include a novel ribosyl zeatin glycoside that was isolated from buds of *Pinus radiata* in which the hexosyl moiety is attached to the ribosyl moiety rather than to the side chain hydroxyl.[86] A similar metabolite was identified following the metabolism of ^{14}C-6-benzylaminopurine by shoot cultures of *Gerbera*.[87] In this case β-glucosidase treatment of the metabolite released 6-benzylaminopurine riboside. The suggested structure is similar to the *P. radiata* glucoside, with the glucose unit attached to the ribosyl moiety.[87]

Brassica campestris tissues infected with *Plasmodiophora brassicae* exhibit club-root disease and contain elevated levels of zeatin and zeatin riboside, as well as a novel cytokinin that appears to be a conjugate of zeatin and glucose-6-phosphate, linked through the side chain.[85] Zeatin is released after β-glucosidase treatment of the active fraction.

II. VARIATION BETWEEN SPECIES

The metabolic picture is extremely complex, and there are obvious variations between species and within organs of the same plant. Extensive metabolic studies have been conducted on radish seedlings, primary leaves of *Phaseolus vulgaris*, *Zea mays* kernels, lupins, and soybeans, as well as in crown gall tissues.

In some plants metabolism is similar, irrespective of the supplied cytokinin, whereas in others it varies depending on the substrate. This variation is clearly seen when comparing radish seedlings and *P. vulgaris* primary leaves. Zeatin, dihydrozeatin, and isopentenyladenine fed to derooted radish seedlings are primarily metabolized to the 7-glucoside; both the 3- and 9-glucosides were also detected following metabolism of dihydrozeatin,[19,44] which is similar to the pattern of metabolism of 6-benzylaminopurine in radish tissues. There was little metabolism by side chain oxidation. However, detached *Phaseolus* primary leaves supplied with ^{14}C-zeatin conserved zeatin by *O*-glucosylation, but degraded a similar amount by side chain oxidation. Little zeatin remained unmetabolized. By contrast, there was little degradation of dihydrozeatin, but substantial *O*-glucosylation, and a considerable amount of the dihydrozeatin remained unmetabolized after 72 h.[53] The contrast is even more apparent in alder: detached leaves of alder (*Alnus glutinosa*) supplied with zeatin or dihydrozeatin degraded most of the zeatin, while the activity of dihydrozeatin was conserved by *O*-glucosylation. Only small amounts of zeatin were *O*-glucosylated. This is in contrast to the endogenous cytokinin where the bulk of the activity detected in mature leaves was probably zeatin-*O*-glucoside.[88]

An interesting interspecific difference in metabolism of zeatin by immature embryos of *Phaseolus* sp.[71] is described below and by Mok and Martin (Chapter 10). In the lupin[3,75] and soybean,[8] species with which extensive metabolism studies have been performed, both side chain degradation and various forms of conjugation occur, while in *Zea mays* kernels, some conjugation occurs, but the major metabolism is via side chain degradation.[89]

III. VARIATION BETWEEN ORGANS

The extensive studies carried out on several plant species have revealed that metabolic differences exist even at the organ and tissue levels. Space precludes inclusion of all relevant examples, and those selected come from studies conducted since the major reviews of metabolism in the middle 1980s.[5-7]

A. FRUITS OF *PHASEOLUS* spp.

When immature embryos, seed coats, and pod tissues of *P. vulgaris* and *P. lunatus* were dissected and incubated with ^{14}C-zeatin, the species-specific metabolism observed originally in immature embryos[71] was also detected in the seed coats,[90] i.e., the formation of the *O*-xylosylzeatin in *P. vulgaris* tissues and of *O*-glucosylzeatin in *P. lunatus* tissue. However, the level of these conjugates was considerably lower in the seed coats, where the predominant form of metabolism was side chain degradation (especially in *P. vulgaris*; significantly more zeatin was unmetabolized by the seed coats of *P. lunatus*). Pod tissues did not metabolize zeatin actively — a relatively high level of ^{14}C-zeatin was recovered — the major metabolite was zeatin nucleotide. The pronounced interspecific differences in the metabolism of zeatin observed in the seeds (i.e., embryos, endosperm, and seed coats) did not occur in the pods.[90]

B. *ZEA MAYS*

Metabolism of zeatin and zeatin riboside by tissues of *Zea mays* is predominantly via side chain degradation. When ^{3}H-zeatin was fed to derooted *Zea mays* seedlings and to kernels at commercial maturity, the metabolism of exogenous cytokinin consisted mainly of side chain cleavage, although small amounts of labeled *O*-glucosides, as well as dihydrozeatin and dihydrozeatin riboside, were observed in kernels.[21,89] In contrast, when ^{3}H-zeatin or ^{3}H-zeatin riboside were fed to radicles or roots of young seedlings, *N*-glucosylation occurred to produce zeatin-9-glucoside in significant quantities.[21,91] Hocart et al. concluded that 9-glucosylation is the principal form of zeatin and zeatin riboside metabolism only in the root tissue in *Zea mays*,[91] whereas side chain cleavage by cytokinin oxidase occurs actively throughout the plant. Within the kernels themselves there is variation between the embryo and endosperm, both in their endogenous complement and in their ability to metabolize applied cytokinins.

The cytokinins in dry seeds of *Zea mays* were analyzed for whole seed and for embryo and endosperm, separately. Whole-seed analysis masked differential distribution of cytokinins between the embryo and endosperm.[92] Conjugation of the endogenous cytokinins by *N*-glucosylation was a significant feature, with the 9-glucosyl derivatives of both zeatin and dihydrozeatin detected. However, *O*-glucosides were also present, with high levels of dihydrozeatin riboside-*O*-glucoside detected in the embryo. Free bases (zeatin, dihydrozeatin, isopentenyladenine) were generally higher in the embryo, whereas ribosides were higher in the endosperm. Nucleotides of zeatin, dihydrozeatin, and isopentenyladenine were detected in both tissues, but the predominant nucleotide was that of isopentenyladenine in the endosperm.[92]

When germinating seeds were supplied with ^{3}H-zeatin riboside, it was rapidly metabolized in both embryo and endosperm tissues.[91] Little or no radioactivity cochromatographed with the glucosides, and no radioactivity coeluted with dihydrozeatin or dihydrozeatin riboside. In the embryo after 6 h, nucleotides of adenine and zeatin predominated over their ribosides. In contrast, in the endosperm, adenine, adenosine, and zeatin riboside collectively accounted for about 82% of the extracted ^{3}H, while adenine nucleotides contributed only 7%, and the zeatin nucleotides were not detected. Over time, deribosylation of the exogenously supplied ^{3}H-zeatin riboside was more prominent in the embryo than in the endosperm,[91] reflecting the endogenous situation in the dry seed.

C. *LUPINUS ANGUSTIFOLIUS*

A recent study of the blue lupin revealed differential distribution of cytokinins: this time between the seed coat and embryo of the seeds.[51] The majority (92%) of the seed-contained cytokinins were located in the seed coat, but the relative proportions of particular cytokinins in the seed coats were different from those in the embryos. For example, dihydrozeatin riboside was the most abundant form in the seed coats, but in the embryos it was low relative to dihydrozeatin nucleotide, the major cytokinin in this tissue. The nucleotides of zeatin and dihydrozeatin were prominent in both seed coats and embryos. The *O*-glucosides of zeatin and dihydrozeatin, and their ribosides, were much more prominent in seed coats than in embryos. The level of each cytokinin with a saturated N^6 side chain was greater than that of the corresponding unsaturated compound.[51]

When high specific ^{3}H-zeatin riboside and ^{3}H-dihydrozeatin riboside were introduced into the xylem plus pith of pod-bearing lateral stems of otherwise-intact lupin plants, it was possible to characterize ^{3}H-metabolites in seed coats and embryos that contained 0.2 and 0.01%, respectively, of the supplied radioactivity. After supplying ^{3}H-zeatin riboside, a moderate percentage of ^{3}H in the seed coats was due to cytokinin (mostly dihydro-zeatin riboside), but no ^{3}H in the embryos could be definitively attributed to cytokinins. Even after supplying ^{3}H-dihydrozeatin riboside, only a small proportion of the ^{3}H extracted from the embryos was due to cytokinin (mostly dihydrolupinic acid, dihydrozeatin-*O*-glucoside, and dihydrozeatin nucleotide).[51]

The small proportion of xylem cytokinin that does enter the embryo is not able to account for the high endogenous level of the seed.[51] However, those endogenous cytokinins supplied via the xylem were converted to metabolites similar to those detected endogenously: nucleotides, dihydroderivatives, and *O*-glucosides.

In studies of the translocation and metabolism of ^{3}H-zeatin riboside in blue lupin shoots, unknown nucleotide-like metabolites (subsequently shown to be *O*-acetyl derivatives[100]) were detected in stem bark, pod walls, developing lateral shoots, and developing inflorescences.[75] In experiments where ^{3}H-zeatin riboside and ^{3}H-dihydrozeatin riboside were supplied via a wick to the xylem, cytokinin metabolites were detected in the bark. Well-developed ray cells were present in the lupin plants used, and subsequently it was shown, using girdled stems, that ^{3}H-zeatin riboside and ^{3}H-dihydrozeatin riboside moved rapidly from xylem to the bark tissues. Metabolism of these ribosides in excised stem tissues indicated that it was the supplied ribosides that were translocated to the bark per se (and possibly as nucleotides) and that *O*-acetyl zeatin riboside and *O*-acetyl dihydrozeatin riboside were formed exclusively in the bark and were probably the precursors of the respective *O*-acetyl nucleotides. Further ribose cleavage and side chain reduction were largely confined to the bark.[76] This tissue-specific metabolism may represent stabilization of the cytokinin activity in living phloem tissue.

IV. VARIATION DURING DEVELOPMENT

A. LEAVES

Qualitative changes in the endogenous cytokinin complement during leaf development have been reported on numerous occasions. The changes described in earlier papers report the accumulation of *O*-glucoside-like compounds in mature and senescing leaves (e.g., *Xanthium strumarium*,[93] *Salix babylonica*,[94] and *Phaseolus vulgaris*[42]). Leaves of *Nicotiana rustica*, however, contained *O*-glucosides at similar levels at all stages of development.[58]

In several species the metabolism of applied cytokinins has also been observed to change with age: enhanced cytokinin deribosylation occurred in senescing leaves of *Populus nigra*,[22] and decreased conjugation and enhanced isoprenoid side chain cleavage was reported in both senescing radish cotyledons[31] and in blue lupin leaves (which exhibit gradual sequential leaf senescence).[75]

In contrast, in soybeans explants in which leaf senescence is monocarpic and related to pod development, metabolism of ^{3}H-zeatin riboside and ^{3}H-dihydrozeatin riboside in blades from explants from prepodfill to late podfill did not differ appreciably: dihydrozeatin-*O*-glucoside was a prominent foliar metabolite of xylem-supplied ^{3}H-zeatin riboside and ^{3}H-dihydrozeatin riboside in blades of all ages, and the predominant endogenous cytokinin in leaf blades.[8]

The major xylem cytokinins of *Nicotiana rustica* (zeatin riboside and dihydrozeatin riboside) were supplied directly to the xylem.[95] Surprisingly, zeatin-7-glucoside was the major metabolite of ^{3}H-dihydrozeatin riboside in leaf laminae, whereas adenine and adenosine were the main metabolites of ^{3}H-zeatin riboside. No *O*-glucosides were detected. Notably, leaves of differing maturity (*N. rustica* exhibits sequential leaf senescence) did not differ appreciably in the level of radioactivity or metabolism of supplied cytokinin, indicating that younger leaves were not diverting xylem cytokinin away from senescing leaves, nor were senescing leaves inactivating cytokinin at a different rate.[95]

B. SEEDS

Cytokinins have been implicated in seed development, germination, and early postgermination events, and there are a number of recent studies investigating the metabolism of cytokinins during these stages.[9,91,96-98] Biosynthesis of cytokinins in germinating seeds is addressed in detail by Letham (Chapter 5).

In a novel approach, green pods of intact *Lupinus luteus* plants were painted three to four times over an 18-day time period, with a solution of ^{3}H-zeatin riboside.[9] When the pods were dry and brown (about 20 days after the last treatment), they were separated into seed and pod wall, and the seeds were subsequently germinated in distilled water. The radioactivity from the pod wall was translocated to maturing seeds, and metabolites with pronounced cytokinin activity were detected. The majority of the activity was from zeatin riboside and dihydrozeatin riboside, with considerably lesser quantities of zeatin, dihydrozeatin and its *O*-glucoside, and the nucleotides of zeatin and dihydrozeatin. However, when the dry seeds were imbibed for 30 h, only 30% of the radioactivity that had been detected in the air-dry seeds remained, and radioactivity remaining as cytokinins was virtually nil, thus showing that endogenous cytokinins had been intensively metabolized in the phase of activation of germination in lupin seeds.[9] This metabolism, however, only reflects turnover of stored cytokinin, and recent information indicates biosynthesis is occurring during germination. Nandi et al. demonstrated the incorporation of ^{14}C-adenine into dihydrozeatin riboside and its nucleotide by intact seeds (dihydrozeatin was detected in both embryo axes and cotyledons).[96] It is noteworthy that previously Summons et al. had shown that dihydrozeatin riboside was one of the major endogenous cytokinins in lupin seed,[40] and bioassay work from van Staden's group shows, not a major decrease in cytokinin activity during germination, but fluctuating levels.[99] In addition, Nandi et al. showed that the ^{14}C-incorporation occurred in excised embryos, but not excised cotyledons.[96] They concluded that the ^{14}C-dihydrozeatin riboside that they detected in the intact cotyledon fraction was a consequence of transport from the embryo into the cotyledon.

Since dihydrozeatin riboside was a major endogenous cytokinin in lupin seeds[9] and was the cytokinin actually synthesized during germination from ^{14}C-adenine,[96] the most relevant metabolic studies are therefore those utilizing dihydrozeatin riboside.[97] ^{3}H-dihydrozeatin riboside was applied selectively to the embryonic axes or to the cotyledons of germinating lupin seeds 6 h following the start of imbibition. A highly polarized movement of ^{3}H-dihydrozeatin riboside from the embryonic axis to the cotyledons was detected. In contrast to the rapid metabolism observed in the labeled "endogenous" cytokinins described above, exogenously supplied ^{3}H-dihydrozeatin riboside was found to be very stable. The only significant metabolite was the nucleotide, which was predominantly in the embryo axes and which the authors suggest may facilitate transport to the cotyledons. They further suggest that the accumulation of the stable dihydroderivative of zeatin riboside in the cotyledon can be expected to elicit a prolonged hormonal response.[97]

V. COMPARTMENTATION

Some of the discrepancies observed between the nature of endogenous cytokinins and the metabolic fate of exogenous cytokinins may be explained by invoking compartmentation of either the cytokinins or their metabolic enzymes. The susceptibility of the 7- and 9-glucosyl- or 9-alanyl-conjugated cytokinins to cytokinin oxidase[44] came as a surprise, considering the apparent metabolic stability of these compounds. Zeatin 7-glucoside is the major metabolite of zeatin feeds to derooted radish seedlings[18] and radish roots,[100] and it was earlier suggested that the 7-glucosyl group may render the cytokinins insensitive to cytokinin oxidase. However, as the 7-glucoside is susceptible to cytokinin oxidase *in vitro*, the apparent metabolic stability of zeatin 7-glucoside may be due to either compartmentation or to the absence or low level of cytokinin oxidase, and as noted by McGaw et al.,[44] the activity of this enzyme does appear low in radish tissue. The different metabolic patterns in the axes (i.e., no oxidation[101]) and stems (oxidation[53]) of *Phaseolus* also point to organ-specific localization of cytokinin oxidase. However, in tobacco cell suspension cultures, 6-benzyl-aminopurine 7-glucoside was the major metabolite formed following 6-benzylaminopurine application, and no oxidation occurred.[102] Here, as suggested by McGaw et al.,[6] compartmentation must play a role, since this tissue is apparently rich in cytokinin oxidase.

As mentioned in the previous section, cytokinin-*O*-glucosides accumulate to high levels in senescing tissues of some species, indicating that these conjugates are ineffective in delaying senescence. As dihydrozeatin riboside-*O*-glucoside was shown to be more active than zeatin in preventing senescence in the *Phaseolus* disc bioassay (Wang et al., 1977, cited in *Planta*[56]), the endogenous *O*-glucosides were possibly compartmentalized to prevent metabolism by β-glucosidases, whereas the cytokinins applied to leaf discs were subject to the action of β-glucosidases. As exogenously applied *O*-glucosides are readily metabolized it would appear that compartmentation is preventing the endogenous *O*-glucosides from coming into contact with β-glucosidase.

In addition, in many tissues zeatin and/or zeatin riboside accumulate to relatively high levels, which, because of their marked susceptibility to cytokinin oxidase, is indicative of some form of compartmentation. This is exemplified in *Vinca* crown gall tissue and corn kernels, where zeatin and zeatin riboside are abundant[61,62] in tissues noted for their high cytokinin oxidation potential.[34]

There is, however, only one study in which compartmentation of cytokinins has been directly addressed, and this was based on the metabolism and compartmentation of ^{3}H-dihydrozeatin that was supplied to photoautotropic suspension cultures of *Chenopodium rubrum*.[52] ^{3}H-dihydrozeatin was selected because in other tissues it was known to be converted preferentially to *O*- and N-glucosides,[19,42,53,56] and suspension cultures of *C. rubrum* were chosen because they had been shown to be well suited to the study of intracellular compartmentation, in particular with respect to the vacuole.[103] In these cultures the predominant metabolites in extracts of the cells were dihydrozeatin-*O*-glucoside and dihydrozeatin riboside-*O*-glucoside. Moreover, as with the glucoside conjugates of abscisic acid, the gibberellins and 2,4-dichlorophenoxy acetic acid, these two cytokinin *O*-glucosides were shown to be compartmented within the vacuole, whereas dihydrozeatin and dihydrozeatin riboside (which were relatively minor metabolites in the extracts of cells) were localized predominantly outside the vacuole. In addition, the presence of dihydrozeatin riboside was noted in the culture medium, but similar efflux was not observed for the *O*-glucosides of dihydrozeatin or dihydrozeatin riboside. After 36 h only about 2% of the recovered radioactivity could be attributed to dihydrozeatin nucleotide-like compounds.[52]

Fusseder and Ziegler concluded that the vacuole may represent a subcellular storage compartment for glucosidic conjugates of plant hormones in general:[52] their findings provide a firm basis for many of the conflicting observations between metabolism of exogenously applied cytokinins and the levels of endogenous cytokinins.

VI. CONCLUSIONS

Studies of exogenously applied cytokinins have revealed that not only are there metabolic differences between species, but there are differences between organs and even tissues. However, considerable knowledge has been acquired concerning pathways of metabolism (Figure 3). Although the physiological significance of at least some forms of metabolism (side chain degradation, and conjugation of both side chain and purine ring) can be assigned with respect to regulating the pool size of "active" cytokinins, what in fact constitutes the "active" form(s) remains to be ascertained.

Metabolic studies have provided a firm foundation on which to base studies of cytokinin translocation (Chapter 5) and from which to probe for biosynthetic and metabolic enzymes. Such cytokinin-specific enzymes are currently being used as targets to isolate genes that can then be used as new tools by which to study the mechanisms regulating cytokinin levels (Chapters 10, 11, and 21).

The challenge for future studies is to relate the wealth of information on the metabolism of externally applied compounds to the dynamics of endogenous cytokinins. The essentially unphysiological nature of many feeding experiments (e.g., excised tissue, derooted plants, wick feeding) and the problems associated with the phenomena of uptake, transport, and compartmentation (both inter- and intracellular) have led to situations where the exogenous cytokinin metabolism is quite different from the endogenous picture. Ideally, we need to manipulate the levels of endogenous cytokinins by more subtle and less-invasive means. Transgenic plants provide such a means: the incorporation of a cytokinin biosynthesis gene whose expression is controlled in a spatial or temporal manner, or the incorporation of genes involved in cytokinin metabolism, should provide new ways by which we can probe the relationship between the cytokinins and specific developmental events.

ACKNOWLEDGMENTS

I wish to thank Professor D. S. Letham and Dr. C. H. Hocart for their constructive criticism and advice concerning this contribution, and Dr. J. Eason for Figures 1–3.

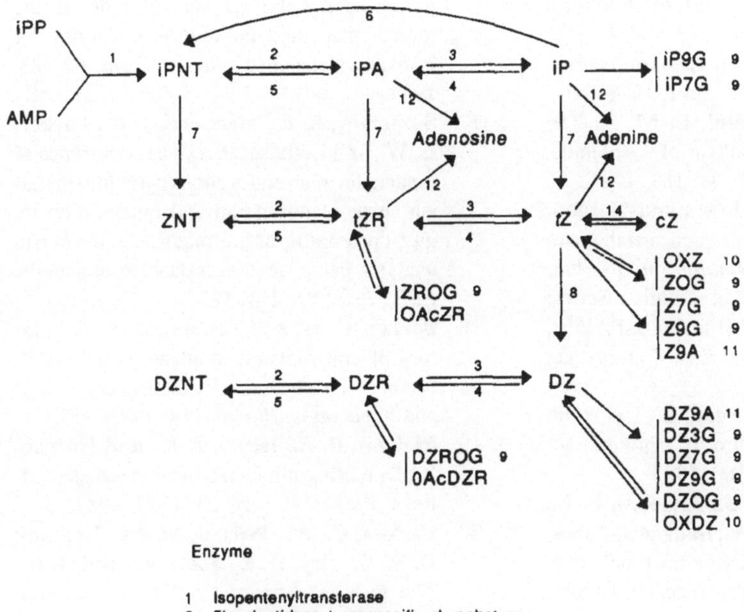

Enzyme

1 Isopentenyltransferase
2 5'nucleotidase / nonspecific phosphatases
3 adenine nucleotidase
4 adenine phosphorylase
5 adenine kinase
6 adenine phosphoribosyltransferase
7 microsomal mixed function oxidases
8 zeatin reductase
9 *O*-glucosyltransferase
10 *O*-xylosyltransferase
11 *B*-(-9-cytokinin alanine)synthase
12 cytokinin oxidase
13 *B*-glucosidase
14 zeatin *cis-trans* isomerase

From: McGaw and Horgan 1985
Letham and Palni 1983
Mok et al. 1988
Kaminek 1992

Figure 9-3 Biosynthesis and metabolism of cytokinins.

REFERENCES

1. **MacLeod, J. K., Summons, R. E., Parker, C. W., and Letham, D. S.,** Lupinic acid, a purinyl amino acid and a novel metabolite of zeatin, *J. Chem. Soc. Commun.*, 809, 1975.
2. **Letham, D. S., Parker, C. W., Duke, C. C., Summons, R. E., and MacLeod, J. K.,** *O*-glucosylzeatin and related compounds — a new group of cytokinin metabolites, *Ann. Bot.*, 41, 261, 1976.
3. **Parker, C. W., Letham, D. S., Gollnow, B. I., Summons, R. E., Duke, C. C., and MacLeod, J. K.,** Regulators of cell division in plant tissue. XXV. Metabolism of zeatin by lupin seedlings, *Planta*, 142, 239, 1978.
4. Duke, **C. C., MacLeod, J. K., Summons, R. E., Letham, D. S., and Parker, C. W.,** The structure and synthesis of cytokinin metabolites. II. Lupinic acid and *O*-β D-glucopyra-nosylzeatin from *Lupinus angustifolius*, *Aust. J. Chem.*, 31, 1291, 1978.
5. **Letham, D. S. and Palni, L. M. S.,** The biosynthesis and metabolism of cytokinins, *Ann. Rev. Plant Physiol.*, 34, 163, 1983.
6. **McGaw, B. A., Scott, I. M., and Horgan, R.,** Cytokinin biosynthesis and metabolism, in *Biosynthesis and Metabolism of Plant Hormones*, Soc. Exp. Biol. Seminar Series, Vol. 23, Crozier, A. and Hillman, J. P., Eds., Cambridge University Press, Cambridge, 1984, chap. 5.
7. **McGaw, B. A. and Horgan, R.,** Cytokinin metabolism and the control of cytokinin activity, *Biol. Plant.*, 27, 180, 1985.
8. **Singh, S., Letham, D. S., Jameson, P. E., Zhang, R., Parker, C. W., Badenoch-Jones, J., and Noodén, L. D.,** Cytokinin biochemistry in relation to leaf senescence. IV. Cytokinin metabolism in soybean explants, *Plant Physiol.*, 88, 788, 1988.
9. **Knypl, J. S., Letham, D. S., and Palni, L. M. S.,** Cytokinins in maturing and germinating *Lupinus luteus* L. seeds, *Biol. Plant.*, 27, 188, 1985.
10. **Letham, D. S. and Zhang, R.,** Cytokinin translocation and metabolism in lupin species. II. New nucleotide metabolites of cytokinins, *Plant Sci.*, 64, 161, 1989.
11. **Moffatt, B., Pethe, C., and Laloue, M.,** Metabolism of benzyladenine is impaired in a mutant of *Arabidopsis thaliana* lacking adenine phosphoribosyltransferase activity, *Plant Physiol.*, 95, 900, 1991.
12. **Chen, C. M. and Eckert, R. L.,** Phosphorylation of cytokinin by adenosine kinase from wheat germ, *Plant Physiol.*, 59, 443, 1977.
13. **Chen, C. M. and Kristopeit, S. M.,** Metabolism of cytokinin: dephosphorylation of cytokinin ribonucleotide by 5′-nucleotidase from wheat germ cytosol, *Plant Physiol.*, 67, 494, 1981.
14. **Chen, C. M. and Kristopeit, S. M.,** Metabolism of cytokinin: deribosylation of cytokinin ribonucleoside by adenine nucleosidase from wheat germ cells, *Plant Physiol.*, 68, 1020, 1981.
15. **Chen, C. M.,** Cytokinin biosynthesis in cell-free systems, in *Plant Growth Substances 1982*, Wareing, P. F., Ed., Academic Press, London, 1982, 155.
16. **Burch, L. R. and Stuchbury, T.,** Activity and distribution of enzymes that interconvert purine bases, ribosides and ribotides in the tomato plant and possible implications for cytokinin metabolism, *Physiol. Plant.*, 69, 283, 1987.
17. **Summons, R. E., MacLeod, J. K., Parker, C. W., and Letham, D. S.,** The occurrence of raphanatin as an endogenous cytokinin in radish seed: identification and quantitation by gas chromatographic-mass spectrometric analysis using deuterium-labeled standards, *FEBS Lett.*, 82, 211, 1977.
18. **Parker, C. W. and Letham, D. S.,** Regulators of cell division in plant tissues. XVI. Metabolism of zeatin by radish cotyledons and hypocotyls, *Planta*, 114, 199, 1973.
19. **McGaw, B. A., Heald, J. K., and Horgan, R.,** Dihydrozeatin metabolism in radish seedlings, *Phytochemistry*, 23, 1373, 1984.
20. **Parker, C. W., Wilson, M. M., Letham, D. S., Cowley, D. E., and MacLeod, J. K.,** The glucosylation of cytokinins, *Biochem. Biophys. Res. Commun.*, 55, 1370, 1973.
21. **Parker, C. W. and Letham, D. S.,** Regulators of cell division in plant tissues. XVIII. Metabolism of zeatin in *Zea mays* seedlings, *Planta*, 115, 337, 1974.
22. **Duke, C. C., Letham, D. S., Parker, C. W., MacLeod, J. K., and Summons, R. E.,** The complex of *O*-glucosylzeatin derivatives formed in *Populus* species, *Phytochemistry*, 18, 819, 1979.
23. **Entsch, B., Letham, D. S., Parker, C. W., and Summons, R. E.,** Metabolites of cytokinins, in *Plant Growth Substances 1979*, Skoog, F., Ed., Springer-Verlag, Berlin, 1980, 109.

24. **Tao, G. Q., Letham, D. S., Palni, L. M. S., and Summons, R. E.,** Cytokinin biochemistry in relation to leaf senescence. I. The metabolism of 6-benzylaminopurine and zeatin in oat leaf segments, *J. Plant Growth Regul.*, 2, 89, 1983.
25. **Hashizume, T., Suye, S.-I., Soeda, T., and Sugiyama, T.,** Isolation and characterisation of a new glucopyranosyl derivative of 6-(3-methyl-2-butenylamino)purine from sweet potato tubers, *FEBS Lett.*, 144, 25, 1982.
26. **Sugiyama, T., Suye, S.-I., and Hashizume, T.,** Mass spectrometric determination of cytokinins in young sweet potato plants using deuterium-labeled standards. *Agric. Biol. Chem.*, 47, 315, 1983.
27. **Letham, D. S., Wilson, M. M., Parker, C. W., Jenkins, I. D., MacLeod, J. K., and Summons, R. E.,** Regulators of cell division in plant tissues. XXIII. The identity of an unusual metabolite of 6-benzylaminopurine, *Biochim. Biophys. Acta*, 399, 61, 1975.
28. **Entsch, B., Parker, C. W., Letham, D. S., and Summons, R. E.,** Preparation and characterization, using high-performance liquid chromatography, of an enzyme forming glucosides of cytokinins, *Biochim. Biophys. Acta*, 570, 124, 1979.
29. **Van Staden, J. and Drewes, F. E.,** The stability and metabolism of benzyladenine glucosides in soybean callus, *J. Plant Physiol.*, 140, 92, 1992.
30. **Letham, D. S., Tao, G. Q., and Parker, C. W.,** An overview of cytokinin metabolism, in *Plant Growth Substances 1982*, Wareing, P. F., Ed., Academic Press, London, 1982, 143.
31. **Letham, D. S. and Gollnow, B. I.,** Regulators of cell division in plant tissues. XXX. Cytokinin metabolism in relation to radish cotyledon expansion and senescence, *J. Plant Growth Regul.*, 4, 129, 1985.
32. **Letham, D. S., Palni, L. M. S., Tao, G.-Q., Gollnow, B. I., and Bates, C. M.,** Regulators of cell division in plant tissues. XXIX. The activities of cytokinin glucosides and alanine conjugates in cytokinin bioassays, *J. Plant Growth Regul.*, 2, 103, 1983.
33. **Van Staden, J. and Drewes, F. E.,** The stability and metabolism of benzyladenine glucosides in soybean callus, *J. Plant Physiol.*, 140, 92 1992.
34. **McGaw, B. A. and Horgan, R.,** Cytokinin oxidase from *Zea mays* kernels and *Vinca rosea* crown-gall tissue, *Planta*, 159, 30, 1983.
35. **Palni, L. M. S., Palmer, M. V., and Letham, D. S.,** The stability and biological activity of cytokinin metabolites in soybean callus tissue, *Planta*, 160, 242, 1984.
36. **Laloue, M.,** Cytokinins: 7-glucosylation is not a prerequisite of the expression of their biological activity, *Planta*, 134, 273, 1977.
37. **Estruch, J. J., Chriqui, D., Grossmann, K., Schell, J., and Spena, A.,** The plant oncogene *rolC* is responsible for the release of cytokinins from glucoside conjugates, *EMBO J.*, 10, 2889, 1991.
38. **Parker, C. W., Letham, D. S., Wilson, M. M., Jenkins, J. D., MacLeod, J. K., and Summons, R. E.,** The identity of two new cytokinin metabolites, *Ann. Bot.*, 39, 375, 1975.
39. **Letham, D. S., Summons, R. E., Parker, C. W., and MacLeod, J. K.,** Regulators of cell division in plant tissues. XXVII. Identification of an amino-acid conjugate of 6-benzyl-aminopurine formed in *Phaseolus vulgaris* seedlings, *Planta*, 146, 71, 1979.
40. **Summons, R. E., Letham, D. S., Gollnow, B. I., Parker, C. W., Entsch, B., Johnson, L. P., MacLeod, J. K., and Rolfe, B. G.,** Cytokinin translocation and metabolism in species of Leguminoseae: studies in relation to shoot and nodule development, in *Metabolism and Molecular Activities of Cytokinins*, Guern, J. and Péaud-Lenoël, L., Eds., Springer-Verlag, Berlin, New York, 1981, 69.
41. **Martin, R. C., Mok, M. C., Shaw, G., and Mok, D. W. S.,** An enzyme mediating the conversion of zeatin to dihydrozeatin in *Phaseolus* embryos, *Plant Physiol.*, 90, 1630, 1989.
42. **Palmer, M. V., Horgan, R., and Wareing, P. F.,** Cytokinin metabolism in *Phaseolus vulgaris* L. I. Variations in cytokinin levels in leaves of decapitated plants in relation to lateral bud outgrowth, *J. Exp. Bot.*, 32, 1231, 1981.
43. **Summons, R. E., Entsch, B., Parker, C. W., and Letham, D. S.,** Mass spectrometric analysis of cytokinins in plant tissues. III. Quantitation of the cytokinin glycoside complex of lupin pods by stable isotope dilution, *FEBS Lett.*, 107, 21, 1979.
44. **McGaw, B. A., Horgan, R., and Heald, J. K.,** Cytokinin metabolism and the modulation of cytokinin activity in radish, *Phytochemistry*, 24, 9, 1985.
45. **Mok, D. W. S., Mok, M. C., Shaw, G., Dixon, S. C., and Martin, R. D.,** Genetic differences in the enzymatic regulation of zeatin metabolism in *Phaseolus* embryos, in *Plant Growth Substances 1988*, Pharis, R. P. and Rood, S. B., Eds., Springer-Verlag, Berlin, 1990, 267.

46. **Whitty, C. D. and Hall, R. H.,** A cytokinin oxidase in *Zea mays*, *Can. J. Biochem.*, 52, 781, 1974.
47. **Dyson, W., Fox, J. E., and McChesney, J. D.,** Short term metabolism of urea and purine cytokinins, *Plant Physiol.*, 49, 506, 1972.
48. **Fox, J. E., Comette, J., Deleuze, G., Dyson, W., Giersak, C., Niu, P., Zapata, J., and McChesney, J.,** The formation, isolation and biological activity of a cytokinin 7-glucoside, *Plant Physiol.*, 52, 627, 1973.
49. **Dorée, M. and Guern, J.,** Short-term metabolism of some exogenous cytokinins in *Acer pseudoplatanus* cells, *Biochim. Biophys. Acta*, 304, 611, 1973.
50. **Forsyth, C. and van Staden, J.,** Cytokinin metabolism in tomato plants. II. Metabolites of kinetin and benzyladenine in decapitated roots, *Plant Growth Regul.*, 6, 277, 1987.
51. **Zhang, R. and Letham, D. S.,** Cytokinin translocation and metabolism in lupin species. III. Translocation of xylem cytokinin into the seeds of lateral shoots of *Lupinus angustifolius*, *Plant Sci.*, 70, 65, 1990.
52. **Fusseder, A. and Ziegler, P.,** Metabolism and compartmentation of dihydrozeatin exogenously supplied to photoautrotrophic suspension cultures of *Chenopodium rubrum*, *Planta*, 173, 104, 1988.
53. **Palmer, M. V., Scott, I. M., and Horgan, R.,** Cytokinin metabolism in *Phaseolus vulgaris* L. II. Comparative metabolism of exogenous cytokinins by detached leaves, *Plant Sci. Lett.*, 22, 187, 1981.
54. **Horgan, R.,** A new cytokinin metabolite, *Biochem. Biophys. Res. Commun.*, 65, 358, 1975.
55. **Dixon, S. C., Martin, R. C., Mok, M. C., Shaw, G., and Mok, D. W. S.,** Zeatin glycosylation enzymes in *Phaseolus*. Isolation of *O*-glucosyltransferase from *P. lunatus* and comparison to *O*-xylosyltransferase from *P. vulgaris*, *Plant Physiol.*, 90, 1316, 1989.
56. **Wang, T. L., Thompson, A. G., and Horgan, R.,** A cytokinin glucoside from the leaves of *Phaseolus vulgaris* L., *Planta*, 135, 285, 1977.
57. **Van Staden, J.,** Endogenous cytokinins in Bougainvillea 'San Diego red'. II. The complex in mature leaves, *Bot. Gaz.*, 141, 245, 1980.
58. **Singh, S., Letham, D. S., and Palni, L. M. S.,** Cytokinin biochemistry in relation to leaf senescence. VII. Endogenous cytokinin levels and exogenous applications of cytokinins in relation to sequential leaf senescence of tobacco, *Physiol. Plant.*, 86, 388, 1992.
59. **Davey, J. E. and van Staden, J.,** Cytokinin activity in *Lupinus albus*. III. Distribution in fruits, *Physiol. Plant.*, 43, 87, 1978.
60. **Davey, J. E. and van Staden, J.,** Cytokinin activity in *Lupinus albus* L. IV. Distribution in seeds, *Plant Physiol.*, 63, 873, 1979.
61. **Summons, R. E., Duke, C. C., Eichholzer, J. V., Entsch, B., Letham, D. S., MacLeod, J. K., and Parker, C. W.,** Mass spectrometric analysis of cytokinins in plant tissues. II. Quantitation of cytokinins in *Zea mays* kernels using deuterium labelled standards, *Biomed. Mass Spectrosc.*, 6, 407, 1979.
62. **Scott, I. M., Martin, G. C., Horgan, R., and Heald, J. K.,** Mass spectrometric measurement of zeatin glucoside levels in *Vinca rosea* L. crown gall tissue, *Planta*, 154, 273, 1982.
63. **Palni, L. M. S., Tay, S. A. B., and MacLeod, J. K.,** Determination by gas chromatography-mass spectrometry of [$^{15}N_5$]adenine incorporation into endogenous cytokinins and the effect of tissue age on cytokinin biosynthesis in *Datura innoxia* crown gall tissue, *Plant Physiol.*, 84, 1158, 1987.
64. **Sayavedra-Soto, L. A., Durley, R. C., Trione, E. J., and Morris, R. O.,** Identification of cytokinins in young wheat spikes (*Triticum aestivum* cv. Chinese Spring), *J. Plant Growth Regul.*, 7, 169, 1988.
65. **Jameson, P. E., McWha, J. A., and Wright, G. J.,** Cytokinins and changes in their activity during the development of grains of wheat (*Triticum aestivum* L.), *J. Plant Physiol.*, 106, 27, 1982.
66. **Smith, A. R. and van Staden, J.,** Changes in endogenous cytokinin levels in kernels of *Zea mays* during imbibition and germination, *J. Exp. Bot.*, 29, 1067, 1978.
67. **Van Staden, J. and Dimalla, G. G.,** Endogenous cytokinins and the breaking of dormancy and apical dominance in potato tubers, *J. Exp. Bot.*, 29, 1077, 1978.
68. **Letham, D. S.,** Cytokinins, in *Phytohormones and Related Compounds — A Comprehensive Treatise*, Letham, D. S., Goodwin, P. B., and Higgins, T. J. V., Eds., Elsevier/North Holland, New York, 1978, 205.
69. **Van Staden, J. and Papaphilippou, A.P.,** Biological activity of *O*-β-D-glucopyranosylzeatin, *Plant Physiol.*, 60, 649, 1977.
70. **Horgan, H., Palni, L. M. S., Scott, I., and McGaw, B.,** Cytokinin biosynthesis and metabolism in *Vinca rosea* crown gall tissue, in *Metabolism and Molecular Activities of Cytokinins*, Guern, J. and Péaud-Lenoël, C., Eds., Springer-Verlag, Berlin, 1981, 56.

71. **Lee, Y. H., Mok, M. C., Mok, D. W. S., Griffin, D. A., and Shaw, G.,** Cytokinin metabolism in *Phaseolus* embryos: genetic difference and the occurrence of novel zeatin metabolites, *Plant Physiol.*, 77, 635, 1985.
72. **Shaw, G., Mok, M. C., and Mok, D. W. S.,** Purines, pyrimidines, and imidazoles. Part 63. Synthesis of *O*-β-D-xylopyranosyl-(*E*)-zeatin, a new cytokinin metabolite, *J. Chem. Res., (S)*, 110, 1987.
73. **Mok, M. C. and Mok, D. W. S.,** Metabolism of ^{14}C-zeatin in *Phaseolus* embryos: occurrence of *O*-xylosyldihydrozeatin and its ribonucleoside, *Plant Physiol.*, 84, 596, 1987.
74. **Mok, M.C., Mok, D. W. S., Marsden, K. E., and Shaw, G.,** The biological activity and metabolism of a novel cytokinin metabolite, *O*-xylosylzeatin, in callus tissue of *Phaseolus vulgaris* and *P. lunatus*, *J. Plant Physiol.*, 130, 423, 1987.
75. **Jameson, P. E., Letham, D. S., Zhang, R., Parker, C. W., and Badenoch-Jones, J.,** Cytokinin translocation and metabolism in lupin species. I. Zeatin riboside introduced into the xylem at the base of *Lupinus angustifolius* stems, *Aust. J. Plant Physiol.*, 14, 695, 1987.
76. **Letham, D. S., Parker, C. W., Zhang, R., Singh, S., Upadhyaya, M. N., Dart, P. J., and Palni, L. M. S.,** Xylem-translocated cytokinin: metabolism and function, in *Plant Growth Substances 1988*, Pharis, R. P. and Rood, S. B., Eds., Springer-Verlag, Berlin, 1990, 275.
77. **Laloue, M. and Pethe, C.,** Some novel aspects of cytokinin biosynthesis and metabolism, in *Abstracts of 13th Int. Conf. on Plant Growth Substances*, Pharis, R. P. and Rood, S. B., Eds., International Plant Growth Substances Association, Springer-Verlag, Canada, 1988, 93.
78. **Evidente, A., Surico, G., Iacobellis, N. S., and Randazzo, G.,** 1′-methyl-zeatin, an additional cytokinin from *Pseudomonas syringae* pv. *savastanoi*, *Phytochemistry*, 25, 525, 1986.
79. **Surico, G., Evidente, A., Iacobellis, N. S., and Randazzo, G.,** A new cytokinin from the culture filtrate of *Pseudomonas syringae* pv. *savastanoi*, *Phytochemistry*, 24, 1499, 1985.
80. **MacDonald, E. M. S., Powell, G. K., Regier, D. A., Glass, L., Kosuge, T., and Morris, R. O.,** Secretion of zeatin, ribosylzeatin and ribosyl-1″-methylzeatin by *Pseudomonas savastanoi*: plasmid-coded cytokinin biosynthesis, *Plant Physiol.*, 82, 742, 1986.
81. **Evidente, A., Fujii, T., Iacobellis, N. S., Riva, S., Sisto, A., and Surico, G.,** Structure–activity relationships of zeatin cytokinins produced by plant pathogenic *Pseudomonades*, *Phytochemistry*, 30, 3505, 1991.
82. **Horgan, R., Hewett, E. W., Horgan, J. M., Purse, J., and Wareing, P. F.,** A new cytokinin from *Populus x Robusta*, *Tet. Lett.*, No. 30, 2827, 1973.
83. **Horgan, R., Hewett, E. W., Horgan, J. M., Purse, J., and Wareing, P. F.,** A new cytokinin from *Populus x Robusta*, *Phytochemistry*, 14, 1005, 1975.
84. **Nandi, S. K., Letham, D. S., Palni, L. M. S., Wong, O. C., and Summons, R. E.,** 6-Benzylaminopurine and its glycosides as naturally occurring cytokinins, *Plant Sci.*, 61, 189, 1989.
85. **Dekhuijzen, H. M.,** The occurrence of free and bound cytokinins in clubroots and *Plasmodiophora brassicae* infected turnip tissue cultures, *Physiol. Plant.*, 49, 169, 1980.
86. **Taylor, J. S., Koshioka, M., Pharis, R. P., and Sweet, G. B.,** Changes in cytokinins and gibberellin-like substances in *Pinus radiata* buds during lateral shoot initiation and the characterization of ribosyl zeatin and a novel ribosyl zeatin glycoside, *Plant Physiol.*, 74, 626, 1984.
87. **Horgan, R.,** Cytokinin biosynthesis and metabolism, in *Plant Growth Substances 1985*, Bopp, M., Ed., Springer-Verlag, Berlin, 1986, 92.
88. **Henson, I. E.,** Types, formation and metabolism of cytokinins in leaves of *Alnus glutinosa* (L.) Gaertn, *J. Exp. Bot.*, 29, 935, 1978.
89. **Summons, R. E., Entsch, B., Letham, D. S., Gollnow, B. I., and MacLeod, J. K.,** Regulation of cell division in plant tissues. XXVIII. Metabolites of zeatin in sweet corn kernals: purifications and identifications using high performance liquid chromatography and chemical ionization mass spectrometry, *Planta*, 147, 422, 1980.
90. **Turner, J. E., Mok, M. C., and Mok, D. W. S.,** Zeatin metabolism in fruits of *Phaseolus*, comparison between embryos, seedcoat, and pod tissues, *Plant Physiol.*, 79, 321, 1985.
91. **Hocart, C. H., Letham, D. S., and Parker, C. W.,** Metabolism and translocation of exogenous zeatin riboside in germinating seeds and seedlings of *Zea mays*, *J. Exp. Bot.*, 41, 1517, 1990.
92. **Hocart, C. H., Badenoch-Jones, J., Parker, C. W., Letham, D. S., and Summons, R. E.,** Cytokinins of dry *Zea mays* seed: quantification by radioimmunoassay and gas chromatography-mass spectrometry, *J. Plant Growth Regul.*, 7, 179, 1988.

93. **Henson, I. E. and Wareing, P. F.,** Cytokinins in *Xanthium strumarium* L. Distribution in the plant and production in the root system, *J. Exp. Bot.*, 27, 1768, 1976.

94. **Van Staden, J.,** Seasonal changes in cytokinin content of leaves of *Salix babylonica*, *Physiol. Plant.*, 40, 296, 1977.

95. **Singh, S., Letham, D. S., and Palni, L. M. S.,** Cytokinin biochemistry in relation to leaf senescence. VIII. Translocation, metabolism and biosynthesis of cytokinins in relation to sequential leaf senescence of tobacco, *Physiol. Plant.*, 86, 398, 1992.

96. **Nandi, S. K., Palni, L. M. S., Letham, D. S., and Knypl, J. S.,** The biosynthesis of cytokinins in germinating lupin seeds, *J. Exp. Bot.*, 39, 1649, 1988.

97. **Nandi, S. K. and Palni, L. M. S.,** Transport and metabolism of dihydrozeatin riboside in germinating lupin seeds, *J. Exp. Bot.*, 40, 615, 1989.

98. **Hocart, C. H. and Letham, D. S.,** Biosynthesis of cytokinin in germinating seeds of *Zea mays*, *J. Exp. Bot.*, 41, 1525, 1990.

99. **Van Staden, J., Davey, J. E., and Brown, N. A. C.,** Cytokinins in seed development and germination, in *The Physiology and Biochemistry of Seed Development, Dormancy and Germination*, Khan, A. A., Ed., Elsevier, New York, 1982, 137.

100. **Gordon, M. E., Letham, D. S., and Parker, C. W.,** The metabolism and translocation of zeatin in intact radish seedlings, *Ann. Bot.*, [illegible] 809, 1974.

101. **Sondheimer, E. and Tzou, D.-S.,** The metabolism of hormones during seed germination and dormancy. II. The metabolism of 8-^{14}C-zeatin in bean axes, *Plant Physiol.*, 47, 516, 1971.

102. **Gawer, M., Laloue, M., Terrine, C., and Guern, J.,** Metabolism and biological significance of benzyladenine-7-glucoside, *Plant Sci. Lett.*, 8, 262, 1977.

103. **Ziegler, P. and Beck, E.,** Exoamylase activity in vacuoles isolated from pea and wheat leaf protoplasts, *Plant Physiol.*, 82, 1119, 1986.

Chapter 10

Cytokinin Metabolic Enzymes

David W. S. Mok and Ruth C. Martin

TABLE OF CONTENTS

I. INTRODUCTION

Cytokinin metabolites are structurally similar to, and presumably derived from, free bases. Metabolites were discovered in feeding studies with radiolabeled free cytokinins and were named in reference to the parent compound, such as zeatin riboside or dihydrozeatin riboside. Extensive metabolism studies led to the identification of a comprehensive array of derivatives, including nucleosides, nucleotides, *N*-glucosides, *O*-glycosides, lupinic acid, and reduced forms of parent compounds. Analytical methods were designed to detect and measure various forms of endogenous cytokinins. Traditionally, quantitative variations in cytokinin content have been regarded as having a dominant effect on growth and development. However, as the mode of cytokinin action is still unknown, it is prudent to take into consideration qualitative differences of various cytokinin derivatives. It is possible that these specific metabolic structural modifications are required for cytokinin activity and that switching between distinct metabolic pathways is correlated with particular developmental events. In *Mercurialis*, for example, fertile and male sterile flowers are associated with the occurrence of *trans*- or *cis*-zeatin pathways, respectively (see Chapter 22). Genes governing the switching of metabolic pathways may, in turn, regulate development. Identification of enzymes mediating interconversion of cytokinins may lead to the isolation of such genes. This chapter reviews key properties of known enzymes involved in cytokinin metabolism and discusses the possible significance of these enzymes in cytokinin action and plant development. Enzymes mediating the conversion of zeatin in *Phaseolus* are used as examples to illustrate their relevance in molecular and developmental studies. Cytokinin degradation, primarily via the action of cytokinin oxidases, is reviewed by Armstrong in Chapter 11.

II. ENZYMES INVOLVED IN THE CONVERSION OF CYTOKININ BASES, NUCLEOSIDES, AND NUCLEOTIDES

Certain enzymes produce modificationssimilar to the metabolism of adenine, i.e., the interconversion between free base, nucleoside, and nucleotides (Table 1). One such enzyme, adenosine phosphorylase, which catalyzes the ribosylation of N^6-(Δ^2-isopentenyl)adenine (i^{6}Ade), was identified from wheat germ.[1] This enzyme had a higher affinity for adenine (Ade) (Km 32 μM) than for i^{6}Ade (Km 57 μM), and mediated ribosylation of kinetin. It was also shown to be distinct from the inosine-guanosine phosphorylase, based on chromatographic properties and on the competitive inhibition of the two enzymes by adenosine and guanosine-inosine, respectively.

Another enzyme, mediating the opposite reaction (deribosylation), was also isolated from the same tissue. The adenosine nucleosidase was capable of deribosylating both N^6-(Δ^2-isopentenyl)-adenosine (i^{6}Ado) and adenosine (Ado) to their respective free bases. Its affinity for adenosine was about twice that of i^{6}Ado, with Kms of 1.4 and 2.4 μM, respectively.[2] The molecular weight of the enzyme was 59 kDa, and the reaction was inhibited by adenosine and other cytokinin nucleosides, but was insensitive to guanosine, uridine, and 3′-deoxyadenosine.

0-8493-6252-0/94/$0.00+$.50

Table 10-1 **Key Properties of General Cytokinin Metabolic Enzymes**

Name	Source	Substrate[a]	Km (μ*M*)	Optimal pH	Mol. Mass
Adenosine phosphorylase	Wheat germ	i^{6}Ade	57	6.5–7.5	NA[b]
		kinetin	46		
		adenine	32		
Adenosine nucleosidase	Wheat germ	i^{6}Ado	2.4	4.7	59 kDa
		Ado	1.4		
Adenosine kinase	Wheat germ	i^{6}Ado	31	6.8–7.4	NA
		Ado	8.7		
Adenosine kinase	Lupine	Ado	1.5	7.0–7.5	38 kDa
Adenine phosphoribosyl-transferase	Wheat germ	i^{6}Ade	130	7.5	23 kDa
		Ade	74		
5′-Nucleotidase I	Wheat germ	i^{6}AMP	3.5	7.0	57 kDa
		AMP	3.2		
5′-Nucleotidase II	Wheat germ	i^{6}AMP	12.8	7.0	110 kDa
		AMP	11.5		
Cytokinin-7-glucosyl-transferase	Radish	*t*-io^{6}Ade	150	7.0	46 kDa
		c-io^{6}Ade			
		ipn^{6}Ade			
		bzl^{6}Ade			
		Ade			
β-(9-cytokinin)-alanine synthase	Lupine	*t*-io^{6}Ade	800	NA	NA
		i^{6}Ade			
		dihydrozeatin			
		bzl^{6}Ade			
		Ade	26,000		

[a] i^{6}Ade: N^6-(Δ^2-isopentenyl)adenine; i^{6}Ado: N^6-(Δ^2-isopentenyl)adenosine; i^{6}AMP: N^6-(Δ^2-isopentenyl)adenosine monophosphate; *t*-io^{6}Ade: *trans*-zeatin; *c*-io^{6}Ade: *cis*-zeatin; ipn^{6}Ade: N^6-(Δ^2-isopentyl)adenine; bzl^{6}Ade: N^6-benzyladenine; [b] NA: not available.

With respect to the formation of nucleotides, phosphorylation of adenosine and i^{6}Ado by an adenosine kinase from wheat germ was reported.[3] The reaction product of i^{6}Ado (Km of 31 μ*M*) appeared to be primarily the corresponding monophosphate, while adenosine (Km of 8.7 μ*M*) as the substrate resulted in the formation of 55% adenosine monophosphate (AMP), 45% adenosine diphosphate (ADP), and traces of adenosine triphosphate (ATP). The property of the enzyme is very similar to an adenosine kinase isolated from lupine.[4] The lupine enzyme utilized ATP as the phosphate donor. The two enzymes are probably the same as the ATP:adenosine-5′-phosphotransferase (EC 2.7.1.20) isolated from animal systems,[5,6] which utilized N^6-substituted adenosines, including N^6-(Δ^2-isopentenyl)adenosine and N^6-benzyladenosine, as substrates.

In the course of studying adenosine kinase, crude extracts from wheat germ and tobacco callus tissues were found to contain enzyme activity catalyzing the phosphoribosylation of adenine and i^{6}Ade. Presumably an adenine phosphoribosyltransferase that uses phosphoribosyl pyrophosphate (PRPP) as a phosphoribosyl donor was present in such preparations. Subsequently, the enzyme was purified about 200-fold from crude extracts.[7] The affinity of the enzyme for adenine (Km 74 μ*M*) was higher than for i^{6}Ado (Km 130 μ*M*). These results indicated that cytokinin nucleotides may be formed directly from cytokinin bases or via their ribosyl derivatives. It is expected that most adenine metabolic enzymes will have an effect on cytokinin metabolism, as indicated by the observation that the formation of N^6-benzyladenine nucleotides were impaired in an *Arabidopsis* mutant defective in adenine phosphoribosyltransferase.[8]

Dephosphorylation of these nucleotides by nucleotidases from wheat germ also favored adenosine monophosphate as the substrate.[9] Two forms of 5′-nucleotidases (57 and 110 kDa, respectively) were compared, and both were able to dephosphorylate AMP and N^6-(Δ^2-isopentenyl)-adenosine monophosphate (i^{6}AMP). The rate of hydrolysis of the cytokinin nucleotide by the 57- and 110-kDa enzymes was 72 and 86% of that of AMP.

These studies indicate that the interconversion of cytokinin bases, nucleosides, and nucleotides are most likely related to the adenine or purine metabolic pathways.[10] As the configuration of the N^6-side chain does not confer specificity in the conversion, the enzymatic controls of such steps are probably shared with those regulating adenine and adenosine metabolism.

III. ENZYMES MODIFYING THE PURINE RING OF CYTOKININS

Next to ribosylation and phosphoribosylation, the most frequently observed modifications of the purine ring involve glucosylation and the formation of alanine conjugates. *N*-glucosylation at the 3, 7, or 9 position of zeatin and dihydrozeatin have been studied extensively in tobacco, radish, and lupine.[11-13] Two peaks of enzyme activity mediating the formation of both N^6-benzyladenine 7- and 9-*N*-glucosides were separated on a DEAE cellulose column.[14] The enzyme from radish present in larger quantity was examined further.[15,16] The molecular weight of the enzyme, designated as "cytokinin-7-glucosyltransferase," was 46 kDa and catalyzed the formation of both 7- and 9-glucosides, but favored the 7 position. Both UDPG (Km of 1.9×10^{-4} *M*) and TDPG served as the glucosyl donor. Cytokinin substrates included *trans*-zeatin (Km of 1.5×10^{-4} *M*), *cis*-zeatin, dihydrozeatin, N^6-benzyladenine, and a number of chemical variants. The structural requirements for substrates *in vitro* appeared to be similar to those needed for high cytokinin activity: an intact purine ring with aliphatic or cyclic N^6-substituents with at least a 3-carbon chain. Although the requirement for a purine ring seemed stricter than the configuration of the side chain, the relative amounts of the product varied with the substrate: most notably, the formation of appreciable quantities of 9-glucosides of zeatin and N^6-benzyladenine vs. only trace amounts when dihydrozeatin and zeatin-*O*-glucoside were used as the substrates. The possibility of other adenine derivatives serving as substrates for this enzyme was not tested due to the difficulty in extracting sufficient amounts of the enzyme.

There were also differences between *in vitro* assays and *in vivo* feeding experiments, in radish tissues. For example, feeding-labeled zeatin yielded only 7-glucoside, while dihydrozeatin gave 3- and 9-glucosides.[17] These differences may suggest the presence of additional *N*-glucosylation enzymes that have not been identified or the interconversion of the product *in vivo*. *N*-glucosides of cytokinins are biologically inactive and very stable. It is assumed that the *N*-glucosylation may be important in regulating levels of active cytokinin, possibly through inactivation. Nevertheless, the precise function of these compounds is not known. Although the affinity of the *N*-glucosylation enzyme *in vitro* showed some correlation with the cytokinin activity of the substrates, the structural requirement is broad. Inhibition by an inhibitor of zeatin, 3-methyl-7-(n-pentylamino)pyrazolo[*4,3-d*]-pyrimidine, was probably related to the structural similarity of the two compounds and may not indicate specificity of the enzymes to cytokinins.

The formation of 9-alanine conjugates of zeatin and benzyladenine was first observed in *Lupinus* and *Phaseolus*.[16,18-20] The zeatin conjugate was named lupinic acid. An enzyme, β-(9-cytokinin)-alanine synthase, was analyzed in lupine. The Km for *trans*-zeatin was 0.8 m*M*, and *O*-acetylserine (Km of 47 m*M*) served as the alanine donor. A large number of purines were also substrates for the enzyme. The relative affinity of the enzyme to some cytokinins was i^{6}Ade $\gg$ *trans*-zeatin = N^6-(Δ^2isopentyl)adenine $\gg$ dihydrozeatin $\gg$ *cis*-zeatin.[16] Benzyladenine, *O*-glucosylzeatin, adenine, methyladenine, hydro-xyethyladenine, propyladenine, and hydro-xylhexyladenine were also substrates. Similar to *N*-glycosylation of cytokinins, the alanyl conjugation probably represents a process of cytokinin inactivation, since the alanine conjugates are also biologically inactive and stable.

IV. ENZYMES MODIFYING THE N^6-SUBSTITUTED SIDE CHAIN

Cytokinins with a hydroxylated N^6-side chain are often glycosylated to form the *O*-glucoside and *O*-xyloside.[21-26] The *O*-glucosyl derivatives are biologically active in bioassays.[27] The activity was suggested to be related to the ubiquitous β-glucosidases, resulting in conversion to cytokinin bases. The enzymes mediating the *O*-glycosylation were isolated much later than the discovery of *O*-glucosylzeatin, and the work was centered on zeatin and dihydrozeatin in *Phaseolus* species. Two enzymes, zeatin *O*-xylosyltransferase and zeatin *O*-glucosyltransferase, have been isolated from immature seeds of *Phaseolus vulgaris* and *P. lunatus*, respectively.[28,29] The enzymes are similar in molecular mass (50 kDa), but can be separated on anion exchange columns. In addition, they can be distinguished by their substrate specificity (Table 2). The *O*-xylosyltransferase isolated from *P. vulgaris* utilizes *trans*-zeatin as the main substrate and UDPX as the donor of the xylosyl moiety. The enzyme

Table 10-2 **Zeatin Metabolic Enzymes Isolated from *Phaseolus***

Zeatin *O*-xylosyltransferase of *P. vulgaris*			
Reaction	*t*-zeatin + UDPX → *O*-xylosylzeatin		
	Km	2 μ*M*	3 μ*M*
Reaction	dihydrozeatin + UDPX → *O*-xylosyldihydrozeatin		
	Km	10 μ*M*	3 μ*M*
Optimal pH	8.5		
Mol. Mass	50 kDa		
Zeatin *O*-glucosyltransferase of *P. lunatus*			
Reaction	*t*-zeatin + UDPG → *O*-glucosylzeatin		
	Km	28 μ*M*	200 μ*M*
Reaction	*t*-zeatin + UDPX → *O*-xylosylzeatin		
	Km	28 μ*M*	2700 μ*M*
Optimal pH	8.0		
Mol. Mass	50 kDa		
Zeatin reductase			
Reaction	*t*-zeatin → dihydrozeatin		
	Km	70 μ*M*, high molecular weight (HMW) form	
	100 μ*M*, low molecular weight (LMW) form		
Optimal pH	7.5		
Mol. Mass	55 kDa (HMW), 25 kDa (LMW)		
Zeatin *cis-trans* isomerase			
Reaction	*c*-zeatin ⇄ *t*-zeatin		

also accepts dihydrozeatin as a substrate, but not *cis*-zeatin or ribosylzeatin. The *O*-gluco-syltransferase occurring in *P. lunatus* seeds recognizes *trans*-zeatin, but not dihydrozeatin, *cis*-zeatin, or ribosylzeatin. In addition, both UDPG and UDPX serve as glycosyl donors for the enzyme, although UDPG is the favored substrate. The products of the enzymatic reaction have been found to be endogenous in *Phaseolus*.[24] In terms of cytokinin action and plant development, it is interesting to note that the occurrence of the enzymes is species specific, and the activity is pronounced, in immature seeds. These observations may indicate genetic, as well as developmental, regulation of cytokinin metabolism.[30]

Similar to earlier findings regarding *O*-glucosylzeatin, *O*-xylosylzeatin is also biologically active in callus bioassays.[31] Moreover, short-term feeding experiments with labeled *O*-xylosylzeatin did not result in substantial conversion to zeatin. Therefore, it is possible that some of the *O*-glycosylzeatin derivatives exert biological action independent of the free base.

Conversion of *trans*-zeatin to dihydrozeatin, mediated by a reductase, was also discovered in *Phaseolus*.[32] The enzyme requires nicotinamide-adenine dinucleotide phosphate (NADPH). Two forms of the enzyme, 50 and 25 kDa, have been identified with Kms for *trans*-zeatin of 70 and 100 μ*M*, respectively (Table 2). However, *cis*-zeatin, ribosylzeatin, and N^6-(Δ^2-isopentenyl)adenine are not substrates for the enzyme, indicating a high substrate specificity. The level of zeatin reductase was high in *P. vulgaris*, a species known to have high cytokinin oxidase activity.[33,34] As cytokinin oxidases selectively cleave unsaturated N^6-side chains of cytokinins (see Chapter 12), reduction of the side chain may be a mechanism to maintain the level of active cytokinins.

An isomerase mediating the interconversion of *cis*- and *trans*-zeatin was also detected in immature seeds of *Phaseolus*.[35,36] The balance of the interconversion favors *trans*-zeatin formation. The reaction requires flavin adenine dinucleotide (FAD) or flavin mononucleotide (FMN), and light. The enzyme was purifed over 2000-fold using a series of chromatographic steps, including molecular sieving, ion exchange and hydrophobic interaction FPLC, and Concanavalin A columns.[36] Retention on a Co column suggests that it is a glycoprotein. The molecular weight of the enzyme was estimated to be 68 kDa, by gel filtration. As the *cis*-isomer of zeatin is not, or only weakly, active, isomerization is a direct and efficient mechanism for increasing the level of active cytokinins. More importantly, it may provide a link between nucleic acid metabolism and cytokinin biosynthesis.

V. THE SIGNIFICANCE OF CYTOKININ METABOLIC ENZYMES

Both naturally occurring (e.g., zeatin) and synthetic (e.g., N^6-benzyladenine) cytokinins have been used as the parent compounds to trace metabolic products. Greater progress has been made in the chemical identification, than in the enzymology or functional analysis, of the metabolites. The importance of those enzyme-catalyzing steps in common with adenine metabolism is more difficult to fathom, since they have broad substrate affinity and are expected to utilize adenine cytokinins as substrates. For example, of the five zeatin metabolic enzymes described above, the *N*-gluco-syltransferase recognizes a wide range of adenyl compounds as substrates, including zeatin, N^6-benzyladenine, and others. The product of the enzymatic reaction is not biologically active, and no functional role has been demonstrated. Therefore, it is likely that the enzyme is involved in the general metabolism of purines. The other four enzymes have high substrate specificity to cytokinins; thus the properties of these enzymes and their products may be more relevant to the understanding of cytokinin metabolism.

O-glucosylzeatin is active in bioassays. Its ready conversion to the corresponding free bases may account for the biological activity. In addition to possibly having a storage role, it may also be involved in transport between and within cells. In *Chenopodium*, *O*-glucosides of dihydrozeatin and ribosyldihydrozeatin are localized in the vacuoles, while dihydrozeatin and its riboside are present in the cytosol and in the medium.[37] These findings suggest that *O*-glucosylation may result in intracellular compartmentalization. The *O*-glucosyltransferase obtained from *Phaseolus lunatus* appears to be specific to *trans*-zeatin. In *in vitro* assays dihydrozeatin is not a substrate for the enzyme. Thus, it may be argued that the specificity of this enzyme does not reside in the presence of the hydroxyl moiety per se. This observation is puzzling, since the *O*-glucoside of dihydrozeatin has been reported in leaves of *P. vulgaris* plants.[38] It is possible that other *O*-glucosyltransferases exist, but have not been identified. The *O*-glucosyltransferase of *P. lunatus* can utilize both UDPG and UDPX for the formation of the corresponding *O*-glycosides of *trans*-zeatin. It is likely that this enzyme and the *O*-xylosyltransferase are variants of a family of enzymes. Since the *O*-xylosytransferase recognizes *trans*-zeatin and dihydrozeatin, but only uses UDPX as the glycosyl donor, it is conceivable that other enzymes adapted to catalyzing the formation of *O*-glucosyldihydrozeatin, using UDPG, may also occur. The expression of individual enzymes may be tissue specific, which would account for the presence of *O*-glucosyldihydrozeatin in *P. vulgaris* leaves. In any event, the information on the *O*-glycosyltransferases in *Phaseolus* indicates that the control of zeatin metabolism is precise and is likely to be species- and organ-specific. The evolutionary relationship between these enzymes may be further clarified by analyzing the genes encoding these enzymes.

It is generally accepted that cytokinin biosynthesis can occur via either a *de novo* pathway or indirect synthesis via tRNAs.[39,40] The *de novo* synthesis involves the formation of i^6AMP from AMP and iPP, mediated by an isopentenyl transferase (perhaps similar to the *ipt* enzyme of *Agrobacterium*), followed by hydroxylation of the side chain. The isopentenyl transferase activity has been detected from cytokinin autonomous tobacco callus,[41] but isolation of the enzyme mediating side chain hydroxylation has not been reported, although such conversion may occur in *Actinidia* tissue culture.[42,43] Moreover, no sequence homology to the *ipt* genes of microbial systems could be detected in plant DNA preparations,[44,45] and the occurrence of zeatin in *E. coli* harboring the *ipt* gene is attributed to hydroxylation enzymes of the host.[46] Incubation of maize embryos with labeled adenine led to the formation of radiolabeled zeatin, without the recovery of isopentenyl derivatives or nucleotides.[47] Therefore, the *de novo* pathway (or pathways) of cytokinin biosynthesis in plants remains to be defined. Another pathway suggested for zeatin (and N^6-(Δ^2-isopentenyl)adenine) biosynthesis involves breakdown of cytokinin-bearing tRNAs.[48] A major challenge to this model is the finding that plant tRNAs contain *cis*-zeatin derivatives that are not biologically active.[49] The occurrence of the isomerase mediating the conversion from *cis*-zeatin to its *trans*-isomer removes such a theoretical constraint for the indirect pathway of zeatin biosynthesis.

VI. MOLECULAR ANALYSES

The molecular mechanism regulating cytokinin function is unknown. However, Mendelian factors related to cytokinins have been reported. These include genes controlling cytokinin autonomy in *Phaseolus* and *Nicotiana*,[50-52] cytokinin resistance in *Nicotiana* and *Arabidopsis*, and cytokinin overproduction in moss.[53-57] It is difficult to isolate genes involved in these genotypic variations, since the gene products are not known. Zeatin- or other cytokinin-specific enzymes identified from *Phaseolus* and other species can be used more readily in cloning cytokinin genes.

The utilization of zeatin metabolic enzymes in developmental and molecular analysis of cytokinins was tested initially using the *O*-xylosyltransferase due to its stability and high substrate affinity. Monoclonal antibodies were generated using purified enzyme preparations. A hybridoma line, designated as XZT-1, producing antibody recognizing both the *O*-xylosyltransferase from *P. vulgaris* and the *O*-glucosyltransferase from *P. lunatus*, was identified.[58] The MAb was employed in a number of developmental studies and gene cloning experiments. The occurrence of the enzymes in other tissues was examined. No immunopositive proteins were detected in the cytosolic (soluble) fractions of leaves, roots, pedicels, and callus cultures of *Phaseolus*,[59] but immuno-positive proteins were detected in soluble fractions from seeds of maize and lupin. The amount of the enzyme is highest in very small seeds, and the quantity decreases as the seeds mature.[60] It appears that zeatin *O*-glycosyltransferases occur in dicots as well as in monocots and are likely be important in seed development.

Immunolocalization was performed in immature seeds of *P. vulgaris,* using tissue printing, light, and EM sections.[61] Immunogenic protein was localized primarily in the endosperm. The amount of the enzyme in the endosperm is approximately 200 times that of the cotyledonary tissues. Northern hybridization using a cDNA probe indicated that endosperm tissues also accumulate the corresponding mRNA. Interestingly, light-microscope samples of endosperm nuclei were found to be immunopositive. Subsequently, the nuclear location of the enzyme was also confirmed by EM sections and studies of isolated nuclei.[61] It is likely that the enzymes are associated with the chromatin or the nuclear matrix. These findings suggest that some zeatin-specific metabolic enzymes may be bifunctional proteins; in addition to metabolizing zeatin, they may also interact directly with DNA or be involved in cytokinin movement between the nucleus and cytoplasm, especially in rapidly dividing tissues such as the endosperm.

Isolation of cDNAs of zeatin *O*-xylosyltransferase was accomplished by the immunoscreening of expression libraries constructed in lambda gt22. Positive plaques containing insert encoding immunopositive fusion proteins we selected.[58] Full-length cDNAs were subsequent cloned into a number of plasmids, for purification of the protein and sequencing of the inserts. The work represents the first isolation of cytokinin genes from plants and demonstrates the feasibility of using zeatin-specific enzymes to study the genetics of cytokinin metabolism.

VII. CONCLUDING REMARKS

Our knowledge of the regulation of cytokinin metabolism and biosynthesis remains speculative. The proposed pathways of biosynthesis in plants require confirmation. Similarly, metabolism studies have dealt primarily with the identification of metabolites. Perhaps additional attention should be focused on the demanding, but effective, approach of isolating cytokinin-specific enzymes to provide more definitive answers. The discovery of the isomerase and the nuclear localization of the *O*-xylosyltransferase are examples of the unexpected, but potentially interesting, findings of such endeavors. More importantly, these enzymes confer a realistic opportunity for isolating cytokinin genes in plants.

ACKNOWLEDGMENT

Research described was supported by grants from the National Science Foundation (DCB-8818620 and IBN-9116490).

REFERENCES

1. **Chen, C. M. and Petschow, B.,** Metabolism of cytokinin: ribosylation of cytokinin bases by adenosine phosphorylase from wheat germ, *Plant Physiol.*, 62, 871, 1978.
2. **Chen, C. M. and Kristopeit, S. M.,** Metabolism of cytokinin: deribosylation of cytokinin ribonucloside by adenine nucleosidase from wheat germ cells, *Plant Physiol.*, 68, 1020, 1981.
3. **Chen, C. M. and Eckert, R. L.,** Phosphorylation of cytokinin by adenosine kinase from wheat germ, *Plant Physiol.*, 59, 443, 1977.
4. **Guranowski, A.,** Plant adenosine kinase: purification and some properties of the enzyme from *Lupinus luteus* seeds, *Arch. Biochem. Biophys.*, 196, 220, 1979.
5. **Lindberg, B., Klenow, H., and Hansen, K.,** Some properties of partially purified mammalian adenosine kinase, *J. Biol. Chem.*, 242, 350, 1967.
6. **Divekar, A. Y. and Hakala, M.,** Adenosine kinase of Sarcoma 180 cells, N^6-substituted adenosine as substrates and inhibitors, *Mol. Pharmacol.*, 7, 663, 1971.

7. **Chen, C. M., Melitz, D. K., and Clough, F. W.,** Metabolism of cytokinin: phosphoribosilation of cytokinin bases by adenine phos-phoribosyltransferase from wheat germ, *Arch. Biochem. Biophys.*, 214, 634, 1982.
8. **Moffatt, B., Pethe, C., and Laloue, M.,** Metabolism of benzyladenine is impaired in a mutant of *Arabidopsis thaliana* lacking adenine phosphoribosyltransferase activity, *Plant Physiol.*, 95, 900, 1991.
9. **Chen, C. M. and Kristopeit, S. M.,** Metabolism of cytokinin: dephosphorylation of cytokinin ribonucleotide by 5′-nucleotidase from wheat germ cytosol, *Plant Physiol.*, 67, 494, 1981.
10. **Burch, L. R. and Stuchbury, T.,** Activity and distribution of enzymes that interconvert purine bases, ribosides and ribotides in the tomato plant and possible implications for cytokinin metabolism, *Physiol. Plant.*, 69, 1987, 283.
11. **McGaw, B. A., Heald, J. K., and Horgan, R.,** Dihydrozeatin metabolism in radish seedlings, *Phytochemistry*, 23, 1373, 1984.
12. **Fox, J. E., Cornette, J., Deleuze, G., Dyson, W., Giersak, G., Niu, P., Zapata, J., and McChesney, J.,** The formation, isolation and biological activity of a cytokinin 7-glucoside, *Plant Physiol.*, 52, 627, 1973.
13. **Cowley, D. E., Duke, C. C., Liepa, A. J., MacLeod, J. K., and Letham, D. S.,** The structure and synthesis of cytokinin metabolites. I. The 7- and 9-β-D-glucofuranosides and pyranosides of zeatin and 6-benzylaminopurine, *Aust. J. Chem.*, 31, 1095, 1978.
14. **Entsch, B. and Letham, D. S.,** Enzymatic glucosylation of the cytokinin 6-benzylaminopurine, *Plant Sci. Lett.*, 14, 205, 1979.
15. **Entsch, B., Parker, C. W., Letham, D. S., and Summons, R. E.,** Preparation and characterization, using high performance liquid chromatography, of an enzyme forming glucosides of cytokinins, *Biochim. Biophys. Acta*, 570, 124, 1979.
16. **Entsch, B., Letham, D. S., Parker, C. W., Summons, R. E., and Gollnow, B. I.,** Metabolites of cytokinins, in *Plant Growth Substances 1979*, Skoog, F., Ed., Springer-Verlag, Berlin, 1980, 109.
17. **McGaw, B. A., Horgan, R., and Heald, J. K.,** Cytokinin metabolism and the modification of cytokinin activity in radish, *Phytochemistry*, 24, 9, 1985.
18. **Summons, R. E., Letham, D. S., Gollnow, B. I., Parker, C. W., Entsch, B., Johnson, L. P., MacLeod, J. K., and Rolfe, B. G.,** Cytokinin translocation and metabolism in species of the *Leguminoseae*: studies in relation to shoot and nodule development, in *Metabolism and Molecular Activities of Cytokinins*, Guern, J. and Péaud-Lenoël, C., Eds., Springer-Verlag, Berlin, 1981, 69.
19. **MacLeod, J. K., Summons, R. E., Parker, C. W., and Letham, D. S.,** Lupinic acid, a purinyl amino acid and a novel metabolite of zeatin, *J. Chem. Soc. Chem. Commun.*, 809, 1975.
20. **Letham, D. S., Summons, R. E., Parker, C. W., and MacLeod, J. K.,** Regulators of cell division in plant tissues. XXVII. Identification of an amino-acid conjugate of 6-benzylaminopurine formed in *Phaseolus vulgaris* seedlings, *Planta*, 146, 71, 1979.
21. **Duke, C. C., Letham, D. S., Parker, C. W., MacLeod, J. K., and Summons, R. E.,** The complex of *O*-glucosylzeatin derivatives formed in *Populus* species, *Phytochemistry*, 18, 819, 1979.
22. **Lee, Y. H., Mok, M. C., Mok, D. W. S., Griffin, D. A., and Shaw, G.,** Cytokinin metabolism in *Phaseolus* embryos: genetic difference and the occurrence of novel zeatin metabolites, *Plant Physiol.*, 77, 635, 1985.
23. **Letham, D. S., Parker, C. W., Duke, C. C., Summons, R. E., and MacLeod, J. K.,** *O*-glucosylzeatin and related compounds. A new group of cytokinin metabolites, *Ann. Bot.*, 41, 261, 1977.
24. **Mok, M. C. and Mok, D. W. S.,** Metabolism of ^{14}C-zeatin in *Phaseolus* embryos: occurrence of *O*-xylosyldihydrozeatin and its ribonucleoside, *Plant Physiol.*, 84, 596, 1987.
25. **Morris, R. O.,** Mass spectroscopic identification of cytokinins. Glucosylzeatin and glucosyl ribosylzeatin from *Vinca rosea* crown gall, *Plant Physiol.*, 59, 1029, 1977.
26. **Parker, C. W., Wilson, M. M., Letham, D. S., Cowley, D. E., and MacLeod, J. K.,** The glucosylation of cytokinins, *Biochem. Biophys. Res. Commun.*, 55, 1370, 1973.
27. **McGaw, B. A., Scott, I. M., and Horgan, R.,** Cytokinin biosynthesis and metabolism, in *The Biosynthesis and Metabolism of Plant Hormones*, Crozier, A. and Hillman, J. R., Eds., Cambridge University Press, Cambridge, 1984, 105.

28. **Turner, J. E., Mok, D. W. S., Mok, M. C., and Shaw, G.,** Isolation and partial purification of the enzyme catalyzing the formation of *O*-xylosylzeatin in *Phaseolus vulgaris* embryos, *Proc. Natl. Acad. Sci. U.S.A.*, 84, 3714, 1987.

29. **Dixon, S. C., Martin, R. C., Mok, M. C., Shaw, G., and Mok, D. W. S.,** Zeatin glycosylation enzymes in *Phaseolus*: isolation of *O*-glucosyltransferase from *P. lunatus* and comparison to *O*-xylosyltransferase from *P. vulgaris.*, *Plant Physiol.*, 90, 1316, 1989.

30. **Mok, D. W. S., Mok, M. C., Shaw, G., Dixon, S. C., and Martin, R. C.,** Genetic differences in the enzymatic regulation of zeatin metabolism in *Phaseolus* embryos, in *Plant Growth Substances 1988*, Pharis, R. P. and Rood, S. B., Eds., Springer-Verlag, Berlin, 1990, 267.

31. **Mok, M. C., Mok, D. W. S., Marsden, K. E., and Shaw, G.,** The biological activity and metabolism of a novel cytokinin metabolite, *O*-xylosylzeatin, in callus tissue of *Phaseolus vulgaris* and *P. lunatus.*, *J. Plant Physiol.*, 130, 423, 1987.

32. **Martin, R. C., Mok, M. C., Shaw, G., and Mok, D. W. S.,** An enzyme mediating the conversion of zeatin to dihydrozeatin in *Phaseolus* embryos, *Plant Physiol.*, 90, 1630, 1989.

33. **Mok, M. C., Mok, D. W. S., Dixon, S. C., Armstrong, D. J., and Shaw, G.,** Cytokinin structure-activity relationships and the metabolism of N^6-(Δ^2-isopentenyl)adenosine -8-^{14}C in *Phaseolus* callus tissues, *Plant Physiol.*, 70, 173, 1982.

34. **Chatfield, J. M. and Armstrong, D. J.,** Regulation of cytokinin oxidase activity in callus tissues of *Phaseolus vulgaris* L. cv. Great Northern, *Plant Physiol.*, 80, 493, 1986.

35. **Mok, M. C., Martin, R. C., Mok, D. W. S., and Shaw, G.,** Cytokinin activity, metabolism and function in *Phaseolus*, in *Physiology and Biochemistry of Cytokinins in Plants*, Kamínek, M., Mok, D. W. S., and Zazímalová, E., Eds., SPB Academic Publishing, The Hague, 1992, 41.

36. **Bassil, N. V., Mok, D. W. S., and Mok, M. C.,** Partial purification of a *cis-trans*-isomerase of zeatin from immature seed of *Phaseolus vulgaris* L., *Plant Physiol.*, 102, 867, 1993.

37. **Fusseder, A. and Ziegler, P.,** Metabolism and compartmentation of dihydrozeatin exogenously supplied to photoautotrophic suspension cultures of *Chenopodium rubrum*, *Planta*, 173, 104, 1988.

38. **Wang, T. L., Thompson, A. G., and Horgan, R.,** A cytokinin glucoside from the leaves of *Phaseolus vulgaris* L., *Planta*, 135, 285, 1977.

39. **Maaß, H. and Klambt, D.,** On the biogenesis of cytokinins in roots of *Phaseolus vulgaris*, *Planta*, 151, 353, 1981.

40. **Klambt, D.,** The biogenesis of cytokinins in higher plants: our present knowledge, in *Physiology and Biochemistry of Cytokinins in Plants*, Kamínek, M., Mok, D. W. S., and Zazímalová, E., Eds., SPB Academic Publishing, The Hague, 1992, 25.

41. **Chen, C. M. and Melitz, D. K.,** Cytokinin biosynthesis in a cell-free system from cytokinin-autotrophic tobacco tissue cultures, *FEBS Lett.*, 107, 15, 1979.

42. **Einset, J. W.,** Zeatin biosynthesis from N^6-(Δ^2-isopentenyl)adenine in *Actinidia* and other woody plants, *Proc. Natl. Acad. Sci. U.S.A.*, 83, 972, 1986.

43. **Einset, J. W.,** Biosynthesis of zeatin from N^6-(Δ^2-isopentenyl)adenine in *Actinidia*: site and seasonal changes in activity, *Proc. Natl. Acad. Sci. U.S.A.*, 83, 7751, 1986.

44. **Morris, R. O.,** Genes specifying auxin and cytokinin biosynthesis in prokaryotes, in *Plant Hormones and Their Role in Plant Growth and Development*, Davies, P. J., Ed., 1987, 636.

45. **Smigocki, A. C. and Owens, L. D.,** Cytokinin gene fused with a strong promoter enhances shoot organogenesis and zeatin levels in transformed plant cells, *Proc. Natl. Acad. Sci. U.S.A.*, 85, 5131, 1991.

46. **Heinemeyer, W., Buchmann, I., Tonge, D. W., Windass, J. D., Alt-Moerbe, J., Weiler, E. W., Botz, T., and Schroder, J.,** Two *Agrobacterium tumefaciens* genes for cytokinin biosynthesis: Ti plasmid-coded isopen-tenyltransferases adapted for function in prokaryotic or eukaryotic cells, *Mol. Gen. Genet.*, 210, 156, 1987.

47. **van Staden, J. and Drewes, F. E.,** Cytokinin biosynthesis in germinating maize embryos, *J. Expl. Bot.*, 44, 265, 1993.

48. **Skoog, F. and Armstrong, D. J.,** Cytokinins, *Ann. Rev. Plant Physiol.*, 21, 359, 1970.

49. **Vreman, H. J., Schmitz, R. Y., Skoog, F., Playtis, A. J., Frihart, C. R., and Leonard, N. J.,** Synthesis of 2-methylthio-*cis*- and *trans*-ribosylzeatin and their isolation from *Pisum* tRNA, *Phytochemistry*, 13, 31, 1974.

50. **Mok, M. C., Mok, D. W. S., Armstrong, D. J., Rabakoarihanta, A., and Kim, S. G.,** Cytokinin autonomy in tissue cultures of *Phaseolus*: a genotype-specific and heritable trait, *Genetics*, 94, 673, 1980.

51. **Meins, F., Jr., Foster, R., and Lutz, J. D.,** Evidence for a Mendelian factor controlling the cytokinin requirement of cultured tobacco cells, *Develop. Genet.*, 4, 129, 1983.
52. **Hansen, C. E., Meins, F., Jr., and Aebi, R.,** Hormonal regulation of zeatin riboside accumulation by cultured tobacco cells, *Planta*, 172, 520, 1987.
53. **Ashton, N. W., Grimsley, N. H., and Cove, D. J.,** Analysis of gametophytic development in the moss *Physcomitrella patens* using auxin and cytokinin resistant mutants, *Planta*, 144, 427, 1979.
54. **Blonstein, A. D., Parry, A. D., Horgan, R., and King, P. J.,** A cytokinin resistant mutant of *Nicotiana plumbaginifolia* is wilty, *Planta*, 183, 244, 1991.
55. **Su, W. and Howell, S. H.,** A single genetic locus, *Ckr1*, defines *Arabidopsis* mutants in which root growth is resistant to low concentrations of cytokinin, *Plant Physiol.*, 99, 1569, 1992.
56. **Futers, T. S., Wang, T. L., and Cove, D. J.,** Characterization of a temperature sensitive gametophore over-producing mutant of the moss *Physcomitrella patens*, *Mol. Gen. Genet.*, 203, 529, 1986.
57. **Wang, T. L., Futers, T. S., McGeary, F., and Cove, D. J.,** Moss mutants and the analysis of cytokinin metabolism, in *The Biosynthesis and Metabolism of Plant Hormones*, Crozier, A. and Hillman, J. R., Eds., Cambridge University Press, Cambridge, 1984, 135.
58. **Martin, R. C., Martin, R. R., Mok, M. C., and Mok, D. W. S.,** A monoclonal antibody specific to zeatin *O*-glycosyltransferase of *Phaseolus*, *Plant Physiol.*, 94, 1290, 1990.
59. **Mok, D. W. S., Mok, M. C., Martin, R. C., Bassil, N. V., and Lightfoot, D. A.,** Zeatin metabolism in Phaseolus: enzymes and genes, in *Progress in Plant Growth Regulation*, Karssen, C.M., van Loon, L. C., and Vreugdenhil, D., Eds., Kluwer Academic Publishers, Dordrecht, 1992, 597.
60. **Mok, D. W. S., Mok, M. C., Martin, R. C., Bassil, N. V., and Shaw, G.,** Immunoanalyses of zeatin metabolic enzymes of *Phaseolus*, in *Physiology and Biochemistry of Cytokinins in Plants*, Kamínek, M., Mok, D. W. S., and Zazímalová, E., Eds., SPB Academic Publishing, The Hague, 1992, 17.
61. **Martin, R. C., Mok, M. C., and Mok, D. W. S.,** Cytolocalization of zeatin *O*-xylosyltransferase in *Phaseolus*, *Proc. Natl. Acad. Sci. U.S.A.*, 90, 953, 1993.

Chapter 11

Cytokinin Oxidase and the Regulation of Cytokinin Degradation

Donald J. Armstrong

TABLE OF CONTENTS

I. INTRODUCTION: DISCOVERY OF CYTOKININ OXIDASE

A. *IN VIVO* DEGRADATION OF CYTOKININS

Cytokinin degradation is an important component process in the metabolic network that controls the levels of cytokinin metabolites and their distribution in plant systems. The existence in plant tissues of enzymatic machinery capable of degrading cytokinins to inactive products was demonstrated in the earliest studies of cytokinin metabolism. In 1962 McCalla et al. observed that the metabolic products formed when the cytokinin N^6-benzyladenine-8-^{14}C was supplied to senescing leaves of cocklebur (*Xanthium pensylvanicum*) included the degradation products adenosine, guanosine, and inosine monophosphates (AMP, GMP, IMP), urea, and several ureides.[1] In the 30 years that have elapsed since this initial study, the metabolism of cytokinins has been examined in a wide variety of plant systems, using a number of different cytokinins as radioactive tracers. As illustrated in Table 1, degradation has been observed to be a significant component of cytokinin metabolism in many plant tissues. Adenine, adenosine, and AMP have been the degradation products most frequently identified.

0-8493-6252-0/94/$0.00+$.50

B. DEMONSTRATION OF CYTOKININ DEGRADATION IN CELL-FREE SYSTEMS

The degradation of cytokinins by a cell-free enzyme system was first reported by Paces et al. in 1971.[2] These investigators observed that N^6-(Δ^2-isopentenyl)adenosine (i^{6}Ado) was degraded to adenosine by an enzyme preparation from tobacco callus tissue. The term cytokinin oxidase was introduced by Whitty and Hall in 1974 to describe a similar enzyme activity obtained from corn kernels.[3] This term may be applied to any enzyme that oxidatively cleaves the N^6-side chain of cytokinin bases and nucleosides to yield adenine or adenosine. As shown in Figure 1, an aldehyde corresponding to the N^6-side chain of cytokinin substrates is the other product of cytokinin oxidase action.[4] As originally demonstrated by Whitty and Hall,[3] molecular oxygen is required for the reaction.

Cytokinin oxidase activities have now been isolated from a number of higher-plant sources, including *Zea mays* kernels and seedlings,[3-6] *Vinca rosea* crown gall tumor tissue, and normal *V. rosea* callus tissue,[7,8] *Phaseolus* callus tissues,[9,10] wheat germ,[11] tobacco (*Nicotiana tabacum* cv Wisconsin 38) callus tissue,[2,12,13] and poplar (*Populus x euroamericana* cv Robusta) callus tissue.[13] Some of the properties reported for the enzymes from different plant sources are summarized in Table 2. At this time the only plant enzymes known to catalyze the degradation of cytokinins to inactive products that lack the N^6-side chain are cytokinin oxidases.

Table 11-1 **Plant Systems Reported to Exhibit Significant Cytokinin Degradation**

Cytokinin Tested[a]	Plant Material Tested	References
N^6-benzyladenine	*Nicotiana tabacum* (Callus Tissue)	32
	Glycine max (Callus Tissue; Excised Leaves)	33,38
	Xanthium pensylvanicum (Senescing Leaves)	1
Kinetin	*Acer pseudoplatanus* (Cell Suspensions)	29
	Funaria hygrometrica (Moss Protonema)	55
	Glycine max (Callus Tissue)	30,78
	Lactuca sativa (Germinating Seeds)	79
i^{6}Ade	*Nicotiana tabacum* (Cell Suspensions)	80
	Pisum sativum (Excised Organs)	83
	Vinca rosea (Crown Gall Tumor Tissue)	81,82
i^{6}Ado	*Nicotiana tabacum* (Cell Suspensions)	35,80
	Phaseolus vulgaris (Callus Tissue)	37
Zeatin	*Alnus glutinosa* (Detached Leaves)	84
	Glycine max (Callus Tissue; Stem Cuttings)	30,78,85,86
	Lupinus angustifolius (Seeds and Shoots)	87,88
	Malus species (Seeds)	88
	Phaseolus lunatus (Developing Fruits)	77
	Phaseolus vulgaris (Developing Fruits and Detached Leaves)	27,77
	Populus alba (Leaves)	88,89
	Vinca rosea (Crown Gall Tumor Tissue)	81,82
	Zea mays (Shoots, Roots, and Kernels)	88,90,91
Zeatin riboside	*Glycine max* (Callus Tissue)	36,92
	Helianthus tubersos (Tuber Tissue)	93
	Nicotiana rustica (Leaf Discs)	94
	Nicotiana suaveolens x *N. langsdorffii* and *Nicotiana glauca* x *N. langsdorffii* (Normal and Tumor Tissue)	28
	Zea mays (Kernels, Seeds, and Seedlings)	88,95

[a] Abbreviations: i^{6}Ade, N^6-(Δ^2-isopentenyl)adenine; i^{6}Ado, N^6-(Δ^2-isopentenyl)adenosine.

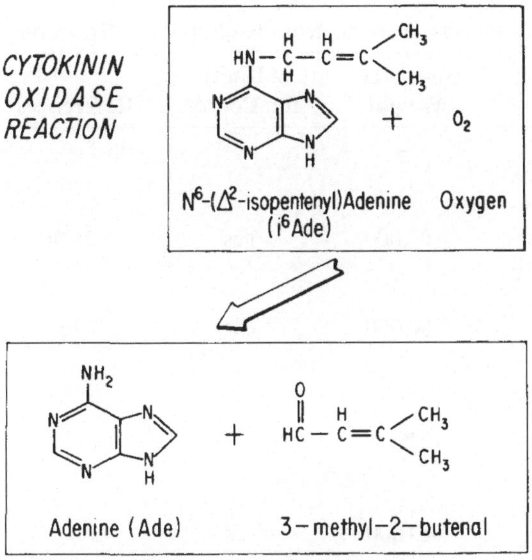

Figure 11-1 Degradation of N^6-(Δ^2-isopentenyl)adenine by cytokinin oxidase.

II. CATALYTIC PROPERTIES OF CYTOKININ OXIDASE

A. ASSAY OF ACTIVITY

The catalytic properties of cytokinin oxidase have usually been investigated using ring-labeled N^6-(Δ^2-isopentenyl)adenine (i^{6}Ade) or i^{6}Ado as a substrate in assays in which the adenine or adenosine product is separated from unreacted cytokinin by an appropriate chromatographic procedure.[2,7,9,11] As indicated in Table 2, the reported pH optima for assays with enzymes from different plant sources vary from pH 6.0 to 9.0. A spectrophotometric assay, based on a change in UV absorption associated with cytokinin degradation, was utilized by Whitty and Hall in their early study of the cytokinin oxidase activity from immature corn kernels.[3] However, the biochemical basis of the observed change in the spectrum is still uncertain.[14]

B. SUBSTRATE SPECIFICITY

The cytokinin oxidase activities isolated from different higher-plant sources appear to have very similar substrate specificities.[2,3,7,9-11,13,15] The information derived from studies of substrate specificity may be summarized as follows:

1. The cytokinins i^{6}Ade and i^{6}Ado are preferred substrates for all higher-plant cytokinin oxidase activities for which data are available, including the enzymes from tobacco callus cultures,[2,13,15] *Zea mays* kernels,[3,7] *V. rosea* callus cultures,[7] *P. vulgaris* and *P. lunatus* callus cultures,[9,10] wheat germ,[11] and poplar callus cultures.[13]
2. Zeatin and zeatin riboside are also substrates for all of the cytokinin oxidase activities examined to date. The enzymes from *Z. mays* kernels,[7] *V. rosea* crown gall tissue,[7] and wheat germ[11] appear to exhibit little substrate preference in attacking zeatin, i^{6}Ade, and the corresponding nucleosides. However, the results of substrate competition tests with the cytokinin oxidase activities from *Phaseolus* callus tissues,[9,10] tobacco callus tissues,[13,15] and poplar callus tissues[13] indicate that the enzymes from these plant sources more readily cleave the side chain of i^{6}Ade than the hydroxylated side chain of zeatin. On the basis of substrate competition tests, both the *trans*- and the *cis*-isomers of zeatin appear to be cleaved by cytokinin oxidase, although sometimes at different rates.[9,10,13,15]
3. The double bond in the isoprenoid side chains of zeatin and i^{6}Ade is important to substrate activity. Reduction of the double bond (forming dihydrozeatin or N^6-isopentyladenine) confers resistance to the attack of the cytokinin oxidase activities isolated from higher-plant tissues.[3,7,9,10,13] A shift in the position of the double bond in the side chain of i^{6}Ade (Δ^2 to Δ^3) also eliminates substrate activity.[3]
4. *O*-glucosylation of the side chain of zeatin appears to confer resistance to the *in vitro* attack of cytokinin oxidase. *O*-glucosylzeatin was not degraded by cell-free cytokinin oxidase preparations from *Z. mays* kernels and *V. rosea* callus tissues.[7]
5. *N*-glucosylation or other forms of ring substitution may increase Km values for the substrates of cytokinin oxidase, but do not necessarily eliminate substrate activity. For example, *N*-glucosyl derivatives of zeatin (substituted at

Table 11-2 **Properties of Cytokinin Oxidase Activities Isolated from Higher-Plant Sources**

Plant source	pH Optima	Michaelis Constant[a]	Molecular Weight	Affinity for Con A[a]	References
Nicotiana tabacum callus tissue	9.0	4.0 μM i^{6}Ade	—	—	13
Phaseolus vulgaris callus tissue	6.5	<0.1 μM i^{6}Ade	67,000	Yes	9,24,96
Phaseolus lunatus callus tissue	8.4	—	60,000	No	10,96
Populus x *euroamericana* callus tissue	8.5	4.0 μM i^{6}Ade	—	—	13
Triticum aestivum wheat germ	7.5	0.3 μM i^{6}Ado	40,000	Yes	5,11
Vinca rosea crown gall tumor tissue	7.0	27.7 μM i^{6}Ade 25.6 μM zeatin 28.6 μM zeatin riboside	25,100	—	7,8
Zea mays kernels	6.0	19.2 μM i^{6}Ade 33.3 μM i^{6}Ado 28.6 μM zeatin 33.3 μM zeatin riboside	78,000 94,000 88,000	Yes	3,5,7,25
Zea mays seedlings	—	—	<67,000	Yes	5

[a] Abbreviations: i^{6}Ade, N^6-(Δ^2-isopentenyl)adenine; i^{6}Ado, N^6-(Δ^2-isopentenyl)adenosine; Con A, Concanavalin A.

the 7 or 9 position of the purine ring) as well as 9-alanylzeatin (lupinic acid) served as substrates for the cytokinin oxidase activities from *Z. mays* kernels and *V. rosea* callus tissues.[7]

6. Cytokinin nucleotides do not appear to be substrates for cytokinin oxidase. The nucleotide form of i^{6}Ade (i^{6}AMP) was not degraded by partially purified preparations of the cytokinin oxidase from wheat germ.[11]
7. The cytokinin N^6-benzyladenine is resistant to the *in vitro* attack of cytokinin oxidase activities from higher-plant sources.[3,7,9,10,11,13] Laloue and Fox observed slow degradation of N^6-benzyladenine by cell-free preparations of the cytokinin oxidase from wheat germ,[11] but the rates were only 1/40 of that observed using i^{6}Ado as a substrate.
8. Kinetin (N^6-furfuryladenine) also appears to be resistant to degradation by cell-free preparations of cytokinin oxidase activities from higher-plant sources, as exemplified by the enzymes from *Phaseolus* callus cultures[9,10] and *Z. mays* kernels.[3] However, kinetin is reported to be a preferred substrate for a cytokinin oxidase activity obtained from protonemal cells of the moss *Funaria hygrometrica*,[16] and the inability of cytokinin oxidase activities from higher plants to degrade kinetin should not be assumed for all sources of the enzyme.
9. The specificity of cytokinin oxidase for substitutions at the N^6 position of adenine is indicated by the inability of the cytokinin oxidase from *V. rosea* crown gall tissue to use the alkaloid triacanthine, N^3-(Δ^2-isopentenyl)-adenine, as a substrate.[7]

C. MICHAELIS CONSTANTS

The Michaelis Constant (Km) values that have been reported for cytokinin oxidase vary by two orders of magnitude for the same substrate tested with enzyme preparations from different plant sources (Table 2). For example, the Km values reported for i^6Ado vary from 0.3 μM for the cytokinin oxidase from wheat germ[11] to over 30 μM for the cytokinin oxidase activity from *Z. mays* kernels.[3,7] The Km values reported for i^6Ade exhibit a similar range (Table 2). Considering the difficulties inherent in the manipulations required to assay for cytokinin oxidase activity, the Km values listed in Table 2 are rough estimates at best.

D. INTERACTION WITH PHENYLUREA DERIVATIVES

The activity of cytokinin oxidase appears to be inhibited by cytokinin-active phenylurea derivatives.[9-11,13] The cytokinin thidiazuron (*N*-phenyl-*N'*-1,2,3-thiadiazol-5-ylurea) inhibited the degradation of i^6Ade by cell-free preparations of cytokinin oxidase from *Phaseolus* callus tissue, by roughly 50% when supplied at a concentration (10 μM) equal to that of the i^6Ade substrate.[9,10] Similarly, the highly active cytokinin *N*-(2-chloro-4-pyridyl)-*N'*-phenylurea (CPPU) strongly inhibited the cell-free activity of cytokinin oxidase isolated from wheat germ[11] and from *Z. mays* kernels.[5] The results with thidiazuron have subsequently been confirmed with cytokinin oxidase preparations from tobacco and poplar callus tissues.[13] The possibility that cytokinin-active phenylurea derivatives may themselves serve as substrates for cytokinin oxidase cannot be entirely excluded. However, Burch and Horgan examined the kinetics of CPPU inhibition of the cytokinin oxidase activity from *Z. mays* and concluded that the mechanism of inhibition appeared to be noncompetitive.[5] This result is consistent with the observation by Mok and Mok that the cytokinin thidiazuron is very slowly metabolized in *Phaseolus* callus tissues,[17] and it may explain the rise in endogenous cytokinins observed by Thomas and Katterman in soybean callus grown on thidiazuron.[18] It should be noted that cytokinin-active adenine derivatives that are inactive as substrates of cytokinin oxidase do not inhibit the activity of the enzyme to any significant extent. For example, N^6-benzyladenine and kinetin, even in tenfold excess, had little, if any, effect on the degradation of i^6Ade by cell-free preparations of the cytokinin oxidase from *P. vulgaris* callus cultures.[9]

E. REACTION MECHANISM

The reaction catalyzed by cytokinin oxidase has been the subject of several investigations. The reaction's requirement for molecular oxygen has been repeatedly demonstrated,[3,7,9,11] and the products of the degradation of i^6Ade by the cytokinin oxidase from *Z. mays* kernels have been isolated and unambiguously characterized, by UV and mass spectral analysis, as adenine and the side-chain fragment 3-methyl-2-butenal.[4,14] Similar data have been obtained for the cytokinin oxidase from *V. rosea* crown gall tissue.[7] The involvement of an unstable intermediate in the reaction was postulated by Whitty and Hall,[3] on the basis of spectral data (a shift in γ_{max} from 269 nm to 248 nm) obtained in assays with the cytokinin oxidase from corn kernels. Laloue and Fox have subsequently obtained evidence that the reaction proceeds via an iminopurine intermediate that can be trapped (apparently as an adduct) by carrying out the reaction in the presence of β-mercaptoethanol.[11,19] Unfortunately, most of the pertinent analytic data, and details of the procedures used to characterize the iminopurine intermediate, have never been published in anything other than abstract form.

The identification of an imine intermediate in the cytokinin oxidase reaction suggests analogies to certain amine oxidases[20,21] that catalyze reactions of the type:

$$R-CH_2-NH_2+O_2+H_2O \rightarrow RCHO+NH_3+H_2O_2$$

In this reaction, transfer of electrons from the amine to oxygen results in the formation of an imine group. Hydrolysis of the imine double bond releases ammonia and generates an aldehyde. The amine oxidases that contain copper as a prosthetic group (rather than a flavin) catalyze their own inactivation if supplied aminoacetonitrile as a substrate.[22] Therefore, it is of interest that the cytokinin oxidases from wheat germ and from *Z. mays* kernels were inhibited approximately 30 and 60%, respectively, by exposure to 5-mM aminoacetonitrile.[5] Cyanide has also been reported to inhibit cytokinin oxidase activity, with 1-mM cyanide giving 60 to 80% inhibition of the enzymes from *Z. mays* kernels and *V. rosea* crown-gall tissue.[3,7] The inhibitory effects of cyanide and aminoacetonitrile are consistent with the properties expected of a copper-containing oxidase, but other types of prosthetic groups are certainly not excluded by these data.

The effects of a number of metals on the activity of cytokinin oxidase were tested by Chatfield and Armstrong,[23] using cell-free enzyme preparations from *P. vulgaris* callus cultures. Cytokinin oxidase activity from these cultures was found to be greatly enhanced (up to 50-fold increases in rate) in reaction mixtures that contained copper-imidazole complexes.[23] Similar results were obtained with cytokinin oxidase preparations from *P. lunatus* callus cultures.[10] Enzyme catalysis was required for the enhanced activity, and both copper and imidazole had to be present. The substrate specificity of the cytokinin oxidase reaction (including sensitivity to inhibition by phenylurea derivatives) was unaltered in the presence of copper-imidazole complexes,[10,23] and the enhanced activity was shown to chromatograph, on DEAE-cellulose, with cytokinin oxidase activity assayed using conventional reaction mixtures.[23] However, the pH optima for the cytokinin oxidase reaction was shifted to lower values, in the presence of copper-imidazole complexes,[10] and the enhanced reaction was not inhibited by the anaerobic conditions achieved by sparging sealed reaction mixtures with molecular nitrogen.[10,23] Either the reaction catalyzed in the presence of copper-imidazole does not require oxygen or the efficiency of utilization of traces of oxygen remaining in the reaction mixture (perhaps bound by the metal complex) is greatly enhanced in the presence of copper-imidazole complexes.

The enhanced rates of reaction observed when cytokinin oxidase is incubated in the presence of copper-imidazole complexes suggest that the enzyme may be capable of degrading cytokinins by alternative reactions that involve components (and reaction mechanisms) other than those utilized in conventional assays for the enzyme. The results of attempts in the author's laboratory to detect naturally occurring cofactors that might stimulate cytokinin oxidase activity are discussed below (Section IV.E).

III. MOLECULAR CHARACTERISTICS OF CYTOKININ OXIDASE

A. CHROMATOGRAPHIC PROPERTIES

The physical characterization of cytokinin oxidase proteins has been limited, to date, primarily to information defined by the chromatographic properties of the enzymes. All of the enzymes for which data are available bind to anion exchange materials.[5,6,10,11,24,25] Additional chromatographic techniques that have been employed with cytokinin oxidase activities from various higher-plant sources include hydroxylapatite chromatography,[3,7] hydrophobic interaction chromatography,[5,6,25] chromatofocusing,[11,24] gel filtration,[3,7,11] and lectin affinity chromatography.[5,6,10,24,25]

B. MOLECULAR WEIGHTS

The molecular weights of cytokinin oxidase activities from a number of plant sources have been estimated by gel filtration (Table 2). The reported values range from 25,100 for the enzyme from [illegible]. *rosea* crown gall tumor tissue[7] to approximately 90,000 for the enzyme from *Z. mays* kernels[3,7] and include an intermediate value of 40,000 reported for the enzyme from wheat germ.[11] The molecular weight of the cytokinin oxidase from corn kernels has been estimated by SDS-PAGE,[5] and the value obtained by this method (78 kDa) is in reasonable agreement with estimates derived from gel filtration (88 kDa and 94 kDa).[3,7]

C. GLYCOSYLATION AND ISOZYME VARIATION

The cytokinin oxidase proteins from several plant sources have been shown to be glycoproteins, as evidenced by the affinity of the enzymes for the lectin concanavalin A. Binding of the cytokinin oxidase from *P. vulgaris* cv Great Northern callus tissue to concanavalin A was demonstrated by Chatfield and Armstrong.[24] The cytokinin oxidases from *Z. mays* and wheat germ[5,25] were also found to bind to the lectin, and this property of the *Z. mays* enzyme has proven useful in its purification. However, affinity for concanavalin A is not a universal property of cytokinin oxidase. The cytokinin oxidase from *P. lunatus* cv Kingston callus tissue is not recognized by concanavalin A.[10] Thus, either the Kingston enzyme is unglycosylated or the pattern of glycosylation is significantly different from that of the other cytokinin oxidase activities examined to date.

The possibility of isozyme variation in cytokinin oxidase is suggested by the large variation in the molecular weights of cytokinin oxidase activities from different plant sources (Table 2) and by the genotypic differences in glycosylation observed with the enzymes isolated from *Phaseolus* callus cultures.[10] However, direct evidence for the presence of isozymes in cytokinin oxidase preparations from a single plant source is limited. Laloue and Fox resolved two peaks of cytokinin oxidase activity,[11] using high performance chromatofocusing columns to fractionate wheat germ enzyme preparations. The smaller peak (about one third of the total activity) eluted with an apparent pI value of 5.0 compared to a value of 4.75 for the larger peak. Whitty and Hall, in their study of the cytokinin oxidase activity from corn kernels,[3] observed that only part of the cytokinin oxidase activity from this source bound to hydroxylapatite, and the pH curve for the activity of crude enzyme preparations exhibited both a major peak at pH 6.0 and a shoulder peak at pH 8.5. However, no indication of isozyme variation was reported by Horgan and co-workers in their work on the corn enzyme.[5,6,25] Although the data published by the latter investigators suggest some difference in the molecular weights of the enzymes from kernels and from seedlings,[6,25] a direct comparison of the enzymes from the two sources was not made.

D. DEVELOPMENT OF MOLECULAR PROBES

The cytokinin oxidase activities from *Z. mays* kernels and seedlings have been purified to homogeneity by Horgan and co-workers.[5,6,25] The *N*-terminal amino residue of the enzyme from corn seedlings was found to be blocked.[6] Polyclonal antibodies prepared against the corn kernel enzyme cross-reacted with cytokinin oxidase activities from wheat germ and from corn seedlings and with an *in vitro* translation product generated with RNA from corn seedlings.[6] The cytokinin oxidase antibody preparation has been used to screen a cDNA expression library prepared, in lambda gt11, from messenger RNA obtained from corn seedlings. Positive plaques containing a 1-kilobase insert have been identified, and the cloned insert is being characterized.[6]

IV. REGULATION OF CYTOKININ OXIDASE AND CYTOKININ DEGRADATION

A. ROLE OF SUBSTRATE SPECIFICITY

The regulation of cytokinin degradation in plant tissues is, undoubtedly, dependent in part on the other metabolic transformations to which these molecules are subject and on the substrate specificity of cytokinin oxidase. Dihydrozeatin, which is not a substrate of cytokinin oxidase, might be expected to be more resistant to degradation than zeatin is, and this particular expectation appears to be supported by the available data from *in vivo* studies with labeled dihydrozeatin.[26,27] Moreover, the rates of cytokinin degradation observed in cultured genetic tumor tissues derived from interspecific hybrids of *Nicotiana* were correlated with the endogenous production of dihydrozeatin.[28] Labeled zeatin riboside was rapidly degraded by side chain cleavage in cultured *N. suaveolens* x *N. langsdorffii* tumor tissues that contained dihydrozeatin derivatives as their major endogenous cytokinins, but the cultured tumor lines derived from *N. glauca* x *N. langsdorffii* hybrids, which contained zeatin derivatives as their major endogenous cytokinins, degraded zeatin riboside more slowly.

The resistance of N^6-benzyladenine to attack by cytokinin oxidase is also correlated with relatively low rates of *in vivo* degradation, as observed in studies of the short-term metabolism of this cytokinin in *Acer pseudoplatanus* cell suspension cultures,[29] in soybean callus cultures,[30] and in cells of moss protonema.[31] However, the substrate specificity of cytokinin oxidase is not always a reliable guide to the *in vivo* stability of cytokinin molecules in particular plant tissues. In longer-term studies with senescing cocklebur leaves[1] and with soybean callus cultures,[32,33] a substantial fraction of labeled N^6-benzyladenine supplied to these tissues was degraded, either by cytokinin oxidase or by some other enzyme activity. Moreover, on the basis of pulse-chase experiments with labeled i^{6}Ado, Laloue and Pethe concluded that for this particular cytokinin (but not for N^6-benzyladenine), a major degradative pathway operated at the nucleotide level in tobacco cell suspension cultures.[34] Thus, the existence in plant tissues of additional pathways of cytokinin degradation (other than cytokinin oxidase) cannot be excluded and may complicate the regulation of cytokinin degradation.

B. REGULATION BY CYTOKININ SUPPLY

The regulatory mechanisms that control cytokinin degradation and the levels of cytokinin oxidase activity present in plant tissues are known to be sensitive to cytokinin supply. Terrine and Laloue found that the *in vivo* degradation of i^{6}Ado by cytokinin-autonomous tobacco (*Nicotiana tabacum* cv Wisconsin 38) cell suspension cultures was stimulated two- to threefold by 3- to 4-h exposure to exogenous sources of either i^{6}Ado or N^6-benzyladenine.[35] Chatfield and Armstrong subsequently demonstrated that transient increases in the cytokinin supply available to cytokinin-dependent *P. vulgaris* cv Great Northern callus tissues induced a rapid increase in cytokinin oxidase activity in the callus tissue.[9] The increase in cytokinin oxidase activity could be detected after 1 h and continued for about 8 h, reaching values two- to threefold higher than the controls. The cytokinin-induced increase in cytokinin oxidase activity was inhibited in tissues pretreated with cordycepin or cycloheximide, suggesting that RNA and protein synthesis were required for the response. All cytokinin-active compounds tested, including both substrates and nonsubstrates of cytokinin oxidase and both adenine and phenylurea structures, were effective in inducing elevated levels of cytokinin oxidase in Great Northern callus tissue. The results obtained with Great Northern callus tissues have been replicated with cytokinin-dependent callus tissues of *P. lunatus* cv Kingston[10] and *N. tabacum* cv Wisconsin 38.[12,13] The rates of zeatin riboside degradation in cytokinin-dependent soybean callus tissue have also been reported to be dependent on cytokinin concentration.[36] Thus, the regulatory mechanism that links cytokinin oxidase levels to cytokinin supply appears to be widely distributed in plant tissues.

C. ROLE OF GLYCOSYLATION

Quantitative adjustments in cytokinin oxidase activity are not the only means by which plant tissues regulate cytokinin degradation. Callus tissues of *P. vulgaris* cv Great Northern and *P. lunatus* cv Kingston exhibit pronounced differences in the rates at which they degrade exogenous i^6Ado-8-^{14}C.[37] However, the difference in rates of cytokinin degradation in these two callus lines are correlated with qualitative differences in the properties of the corresponding cytokinin oxidase proteins, rather than with quantitative differences in the levels of enzyme activity. Great Northern callus tissue (which rapidly degrades i^6Ado) contains a cytokinin oxidase that has a pH optimum of 6.5 and exhibits the properties of a glycoprotein, as evidenced by the affinity of the enzyme for the lectin concanavalin A.[24] Kingston callus tissue (which exhibits much lower rates of i^6Ado degradation) contains a cytokinin oxidase that does not bind to concanavalin A and has a pH optimum of 8.5.[10] On the basis of these results, Kamínek and Armstrong suggested that the compartmentation of cytokinin oxidase may differ in Great Northern and Kingston callus tissues.[10] The difference in rates of cytokinin degradation in the two tissues can be rationalized if the glycosylated cytokinin oxidase of Great Northern callus tissues is localized in the cell wall or plasmalemma and the Kingston enzyme is sequestered in an internal compartment that restricts access of the enzyme to exogenous cytokinin substrates. However, this hypothesis has not yet been tested, and no data on the compartmentation of cytokinin oxidase are currently available for any plant tissue.

D. REGULATORY LINKS TO OTHER PLANT HORMONES

Regulatory links between cytokinin metabolism and the other plant hormones undoubtedly exist in plant tissues. Several studies suggest that auxins, either directly or indirectly, participate in the regulation of cytokinin degradation. In soybean leaves, treatment with auxins increased side chain cleavage of externally supplied N^6-benzyladenine,[38] presumably by inhibiting the enzyme that forms alanine conjugates of cytokinins.[39] Palni et al. found that the degradation of zeatin riboside by tobacco pith explants could be doubled by increasing the auxin concentration in the medium.[40] In addition, the *in vitro* activity of a partially purified cell-free preparation of cytokinin oxidase from corn kernels was increased up to 30% when the enzyme was assayed in the presence of any one of four different synthetic auxins. The authors suggested that the observed regulatory effect of auxin on cytokinin degradation might explain both the unexpected increases in cytokinin accumulation observed when the auxin production associated with the *tms* locus in crown-gall tumors is reduced by inactivation of the locus,[41,42] and the fall in endogenous cytokinin levels observed when some lines of tobacco crown gall tissues were cultured on media containing auxin.[43] Similar reasoning might be applied to the finding (of Wyndaele et al.[44]) that the endogenous cytokinin content of soybean tumor tissues and habituated lines of soybean callus tissue was inversely correlated with the auxin content of these tissues. However, the stimulatory effects of auxins on cytokinin oxidase activity (reported by Palni et al.[40]) were rather modest, and rates of cytokinin degradation were not measured in the referenced studies of auxin effects on *in vivo* levels of cytokinins. Motyka and Kamínek found that transient increases in the auxin levels supplied to tobacco callus cultures failed to elicit any increase in the cytokinin oxidase activity that could be extracted from the callus tissue.[12,13] Although *in vivo* rates of cytokinin degradation were not measured, this result does suggest that auxin effects on cytokinin degradation may be mediated by mechanisms that do not involve increases in the levels of cytokinin oxidase protein.

The role of abscisic acid in regulating cytokinin degradation may also merit examination. Abscisic acid has been reported to suppress the degradation of kinetin-8-^{14}C (as well as other aspects of cytokinin metabolism) in Grand Rapids lettuce seeds[45] and to inhibit the conversion of zeatin to dihydrozeatin in embryonic axes of *P. vulgaris*.[46] Biological interactions of abscisic acid and cytokinins have been described in a large number of plant systems, and additional metabolic interactions of these hormones appear likely.

E. REGULATION BY PHENOLIC COMPOUNDS

The search for naturally occurring cofactors that might mimic the effects of copper-imidazole complexes (see Section II.E) in stimulating cytokinin oxidase activity has generated results that may have significance for the *in vivo* regulation of cytokinin degradation. Recent investigations in the author's laboratory have demonstrated that certain phenolic compounds greatly stimulate the activity of cell-free preparations of cytokinin oxidase. Selected results are shown in Table 3. The most effective compound identified to date is 2,6-dimethoxyphenol (DMP), which stimulates the activity of crude cell-free preparations of the cytokinin oxidase from *P. vulgaris* cv Great Northern callus tissue up to 50-fold. The ability to stimulate cytokinin oxidase activity appears to be a property of a limited range of phenolic structures that include naturally occurring compounds such as acetosyringone. The

Table 11-3 **Effects of Selected Phenolic Compounds on the Activity of a Cell-Free Preparation of Cytokinin Oxidase from *Phaseolus vulgaris* cv Great Northern Callus Tissue**[a]

Phenolic Compound Tested (300 μM)	Cytokinin Oxidase Activity[b]: i^{6}Ade Degraded (pmol/h/Assay)	Stimulation (Test/Control)
None (standard assay control)	18	—
Acetosyringone (3′,5′-dimethoxy-4′-hydroxyacetophenone)	199	11.1
Acetovanillone (3′-methoxy-4′-hydroxyacetophenone)	19	1.1
2,6-Dimethoxyphenol	901	50.1
4′-Hydroxyacetophenone	19	1.1
Syringic acid (3,5-dimethoxy-4-hydroxybenzoic acid)	29	1.6

[a] Armstrong, D. J. and Keener, W.; [b] Cytokinin oxidase activity was assayed as described by Chatfield and Armstrong,[9] using 10 μM N^6-(Δ^2-isopentenyl)adenine-2,8-^{3}H (100 μCi/μmol) as a substrate in 50 μl assay volumes containing 0.1 *M* phosphate buffer (pH 7.6), the test compound (300 μM), and enzyme equivalent to 100 mg of callus tissue (fresh weight). The assays were incubated at 35°C for 30 min.

activity of acetosyringone (3,5-dimethoxy-4-hydroxy-acetophenone), which has a pattern of ring substitution similar to that of DMP, is of particular interest because this compound is reported to serve as a signal compound in the interaction of the plant pathogen *Agrobacterium tumefaciens* with its higher-plant hosts.[47] Acetosyringone and related phenolics released by host plants activate the *vir* region of the *Agrobacterium* Ti plasmid[48] and stimulate the cytokinin production associated with the *tzs* gene of the *vir* region in nopaline strains of *Agrobacterium*.[49,50] Thus, the cytokinin production controlled by the *vir* region of *Agrobacterium* is regulated by a compound that may play a role in cytokinin degradation by the host plant.

The biochemical basis of the stimulatory effect of particular phenolic structures on cytokinin oxidase activity remains to be determined. To date, these compounds have been tested using relatively crude preparations of cytokinin oxidase, and it is possible that an additional enzyme involved in cytokinin degradation is being activated or that coupled reactions are occurring that in some way enhance cytokinin degradation by cytokinin oxidase. However, an interesting parallel to the results reported here is provided by the known ability of particular phenolic compounds to stimulate or inhibit (depending on structure) the degradation of auxins by indoleacetic acid oxidase and peroxidase activities.[51,52] An association of cytokinin oxidase with peroxidase activity has occasionally been suggested. Whitty and Hall reported that a peroxidase activity migrated with cytokinin oxidase activity during gel electrophoresis of a partially purified cytokinin oxidase fraction from *Z. mays* kernels,[3] and Van Staden and Forsyth obtained evidence suggesting that a low level of cytokinin oxidase activity was associated with horseradish peroxidase.[53] However, Chatfield and Armstrong found that the cytokinin oxidase activity in an enzyme preparation from *Phaseolus* callus tissues could be separated from two major peaks of peroxidase activity by chromatofocusing.[24] The possibility that cytokinin oxidase may be capable of catalyzing some type of mixed-function oxidase reaction involving phenolic compounds merits consideration, but further data are required to resolve the questions generated by the observed effects of phenolic structures on cytokinin degradation.

V. CYTOKININ DEGRADATION IN SYSTEMS OTHER THAN HIGHER PLANTS

A. MOSS PROTONEMA

The ability of cytokinins to stimulate gametophore development in moss protonema has made the cytokinin metabolism of this lower-plant system of special interest. Gerhauser and Bopp have shown that protonemal cells of the moss *Funaria hygrometrica* contain a cytokinin oxidase activity that is capable of rapidly degrading kinetin (N^6-furfuryladenine-8-^{14}C) to adenine,[16,31] both *in vivo* in intact protonemal cells[31] and *in vitro* in cell-free enzyme preparations.[16] The cytokinin i^6Ade also serves as a substrate for the cytokinin oxidase from protonemal cells,[54] but N^6-benzyladenine is only slowly degraded by the moss enzyme.[16] The resistance of N^6-benzyladenine to attack by the moss cytokinin oxidase is correlated with the *in vivo* persistence of this cytokinin in cells of the moss protonema.[55] Pretreatment of moss protonema with kinetin stimulated the subsequent degradation of this cytokinin by the protonemal cells, suggesting that cytokinin oxidase activity is regulated in this system (as in higher-plant tissues) by cytokinin supply.[56]

B. CELLULAR SLIME MOLDS

The only organism, other than green plants, in which cytokinin oxidase activity has been unequivocally identified is the cellular slime mold *Dictyostelium discoideum.*[57] *Dictyostelium* is an interesting, and perhaps unique, example of a nonplant system in which cytokinins are used as regulatory signals in development. In this organism, i^6Ade and the cytokinin-active compound discadenine are synthesized during spore maturation, by enzymes that are developmentally regulated.[58-63] Discadenine is derived from i^6Ade by transfer of a 3-amino-3-carboxypropyl substituent from S-adenosylmethionine to the 3 position of the purine ring of i^6Ade.[63] Discadenine functions as a water-soluble inhibitor of spore germination in *Dictyostelium* and prevents precocious germination of spores prior to dispersal. Armstrong and Firtel demonstrated that *Dictyostelium* amoebae contain a cytokinin oxidase activity that is developmentally regulated.[57] The enzyme is present in the vegetative amoebae and falls to low levels when aggregation of the amoebae and the multicellular development that leads to spore formation are initiated by starvation. Thus, the activity of cytokinin oxidase decreases in *Dictyostelium* as the enzymes associated with discadenine synthesis begin to rise. As is the case for other developmentally regulated enzymes in *Dictyostelium*, cytokinin oxidase activity in starving amoebae is affected by cyclic-AMP treatment and by cell–cell contacts, which in this case accelerate the fall in enzyme activity. Although i^6Ade was the preferred substrate of the *Dictyostelium* cytokinin oxidase, the substrate specificity of the enzyme appeared to be somewhat broader than that of cytokinin oxidase activities from higher-plant sources. Discadenine was not available for testing as a substrate. The cytokinin oxidase activity in *Dictyostelium* may play a role in the regulation of spore germination. Alternatively, the presence of the enzyme in vegetative amoebae may signal an additional, and as yet unidentified, role of cytokinins in regulating vegetative growth and cell multiplication in *Dictyostelium.*

C. YEAST

Indirect evidence for the presence of a cytokinin oxidase activity in yeast (*Saccharomyces cerevisiae*) has been obtained by Van Kast and Laten.[64] These investigators found that kinetin and a number of other cytokinin-active bases (but not N^6-benzyladenine) would substitute for the nutritional requirements of adenine-requiring mutants of *S. cerevisiae*, including a mutant blocked in AMP biosynthesis at a position downstream from the formation of IMP. Hypoxanthine, the product that might be expected from an attack of cytokinins by a deaminase activity, does not support the growth of the latter mutant. The ability of kinetin, but not N^6-benzyladenine, to substitute for the adenine requirement of this yeast mutant indicates the presence of an enzyme activity similar to the cytokinin oxidase activity described in the moss *Funaria hygrometrica.*[16] The presence of a cytokinin oxidase activity in yeast suggests that cytokinins are being metabolized in yeast cells. Transfer RNA molecules are always a potential source of i^6Ado residues, but Laten and Zahareas-Doktor presented evidence for the existence of a tRNA-independent pathway for the biosynthesis of i^6Ado in yeast.[65]

D. MAMMALIAN SYSTEMS

The metabolism of cytokinins in mammalian systems was investigated during the 1970s, as a result of interest in possible therapeutic uses of i^6Ado and other cytokinin nucleosides as antitumor agents.[66] Chheda and Mittelman found that i^6Ado-8-^{14}C administered to human subjects, either orally or intravenously, was rapidly metabolized and degraded,[67] and a large portion of the label was excreted as compounds that did not absorb UV light. The latter observation indicated that the degradation of i^6Ado followed a path somewhat different from that involved in the degradation of the common purine nucleosides, which are excreted as uric acid (with the UV-absorbing purine ring intact) produced through the action of xanthine oxidase. Interestingly, a different cytokinin nucleoside, N^6-benzyladenosine, was found to be excreted as products with an intact purine ring.[66] A number of investigators have demonstrated that cytokinin bases are substrates for xanthine oxidase.[68-70] The products of the reaction are 8-hydroxy and 2,8-dihydroxy derivatives that retain the N^6-side chain.

The degradation of cytokinin nucleosides in animal systems may also involve adenosine deaminase (adenosine aminohydrolase) activities. The ability of cytokinin nucleosides to act as competitive inhibitors of the degradation of adenosine by a commercial preparation of adenosine deaminase from calf intestinal mucosa was reported by Terrine et al.[71] Although i^{6}Ado is a rather poor substrate for the enzyme from calf intestinal mucosa, Hall and co-workers subsequently demonstrated that i^{6}Ado was more readily degraded to inosine by adenosine deaminase activities present in chicken and rabbit bone marrow and in porcine and human plasma and blood cells.[72,73] A number of cytokinin-active N^6-substituted adenosine derivatives served assubstrates for these enzymes, but the free base forms of cytokinins as well as nucleosides bearing hydroxylated isoprenoid side chains (e.g., ribosylzeatin) were resistant to attack by the deaminase activities.[72] (It should be noted that McGaw and Horgan pointed out that the identification of inosine as the product of i^{6}Ado degradation by the chicken bone marrow enzyme was based solely on spectral data that could be ambiguous.[14]) The involvement of adenosine deaminase activities in the degradation of cytokinins in plant systems has not been demonstrated.

VI. PHYSIOLOGICAL SIGNIFICANCE OF CYTOKININ DEGRADATION

A. TISSUE CULTURE SYSTEMS

The physiological impacts of cytokinin degradation are most clearly evident in certain plant tissue culture systems. The pronounced differences in cytokinin degradation observed by Mok et al.[37] in callus cultures derived from the *Phaseolus* genotypes *P. vulgaris* cv Great Northern and *P. lunatus* cv Kingston were reflected in the structure–activity relationships exhibited by cytokinins in the two types of callus cultures.[74] Great Northern callus tissues rapidly degraded i^{6}Ado-8-^{14}C to adenosine and responded only weakly to zeatin and i^{6}Ade, requiring high concentrations of these cytokinins to support maximum growth. In Kingston callus tissue, the major short-term metabolite of i^{6}Ado was the corresponding nucleotide, rather than adenosine, and zeatin and i^{6}Ade were highly active as cytokinins in this callus system. In contrast to these results, the two lines of *Phaseolus* callus tissues did not differ significantly in their responses to dihydrozeatin and to other cytokinins resistant to the attack of cytokinin oxidase.[74] Results similar to those obtained with *Phaseolus* callus cultures have been reported by Horgan[75] in studies of two lines of soybean cell suspension cultures that differed markedly in their ability to grow on zeatin. The cell line that grew more slowly on zeatin also exhibited a significantly greater capacity to degrade zeatin-8-^{14}C to adenine and related metabolites. The cytokinin N^6-benzyladenine, which is resistant to the attack of cytokinin oxidase, was equally effective in promoting the growth of either soybean line. Thus, the responses of these particular systems to externally supplied cytokinins appear to be controlled by the rates of cytokinin degradation.

B. RELATION TO PLANT DEVELOPMENT

The role of cytokinin degradation in the regulation of specific processes and events in the growth and development of intact plants has received limited attention to date. Nonetheless, the available data indicate that rates of cytokinin degradation vary in different tissues and in the same tissue at different stages of development. Thus, in developing fruits of *Phaseolus* genotypes, the degradation of labeled zeatin was much more rapid in seed-coat tissue than in embryo and pod tissue.[76] The cytokinin oxidase activity associated with developing corn kernels increased as development progressed.[3,77] This increase appears to be correlated with a rise in cytokinin levels in the developing kernels,[77] suggesting that cytokinin oxidase activity is regulated by the endogenous cytokinin supply. The developmental increase in cytokinin oxidase activity may account for the subsequent fall in cytokinin levels as the kernels mature.[77] It will be of interest to determine whether increases in cytokinin degradation are among the changes in hormonal metabolism that are related to the onset of seed dormancy in this and other systems. On the basis of the limited studies to date, cytokinin degradation in plant tissues does appear to be developmentally regulated. This regulation suggests that cytokinin oxidase has a significant role in the developmental control of cytokinin levels in plant tissues.

VII. CONCLUSION: DIRECTIONS OF FUTURE RESEARCH

The developmental regulation of cytokinin oxidase and cytokinin degradation is an area of cytokinin research where the application of molecular techniques might be expected to yield interesting and significant results. Cloning of the cytokinin oxidase gene and the development of molecular probes for the protein and the corresponding messenger RNA are within reach in at least two plant systems. The availability of these molecular tools should greatly facilitate investigations of the developmental and cellular regulation of cytokinin oxidase. Information concerning the tissue-specific expression of the cytokinin oxidase gene, and the factors that control that expression, should enhance our understanding of the role of cytokinin degradation in specific processes and events in plant development. At the cellular level, the compartmentation of the cytokinin oxidase protein and the relationship of cytokinin degradation to the control of the cell cycle and cellular differentiation events await investigation. As the regulatory sites and sequences that control cytokinin oxidase are identified, the development of new strategies for the manipulation of *in vivo* rates of cytokinin degradation, for experimental and practical purposes, may be anticipated.

The molecular basis of the regulation of cytokinin oxidase activity is an important problem with many ramifications, but there also remain significant questions concerning cytokinin degradation that require physiological and biochemical investigation. The nature of the prosthetic group of cytokinin oxidase could have important regulatory implications for the enzyme. The reaction catalyzed by cytokinin oxidase appears to have received inadequate attention, particularly considering the complex interaction of alternative reaction mechanisms observed with some oxidative enzymes. Indeed, it is not certain, nor necessarily likely, that the cytokinin oxidase reaction represents the only degradative pathway for cytokinins available in any and all plant systems. The relationship of cytokinin degradation to the regulation of cytokinin biosynthesis and other aspects of cytokinin metabolism needs to be examined. The role of phenolic compounds in the *in vivo* regulation of cytokinin degradation has yet to be explored, and the interactions of cytokinin oxidase with other potential regulatory ligands has received limited attention to date.

The regulatory phenomena that have emerged in investigations of cytokinin oxidase hint at the underlying complexity of the regulatory networks that must link cytokinin degradation to other aspects of hormonal metabolism and to a larger framework of developmental signaling. The hypotheses generated from knowledge gained in studies of cytokinin oxidase provide a means of probing these networks. Thus, further advances in our understanding of the regulation of cytokinin degradation may serve to illuminate a number of problems and questions concerning cytokinin metabolism and function in plant development.

REFERENCES

1. **McCalla, D. R., More, D. J., and Osborne, D. J.,** The metabolism of a kinin, benzyladenine, *Biochim. Biophys. Acta*, 55, 522, 1962.
2. **Paces, V., Werstiuk, E., and Hall, R. H.,** Conversion of N^6-(Δ^2-isopentenyl)adenosine to adenosine by enzyme activity in tobacco tissue, *Plant Physiol.*, 48, 775, 1971.
3. **Whitty, C. D. and Hall, R. H.,** A cytokinin oxidase in *Zea mays*, *Can. J. Biochem.*, 52, 789, 1974.
4. **Brownlee, B. G., Hall, R. H., and Whitty, C. D.,** 3-Methyl-2-butenal: an enzymatic degradation product of the cytokinin N^6-(Δ^2-isopentenyl)adenine, *Can. J. Biochem.*, 53, 37, 1975.
5. **Burch, L. R. and Horgan, R.,** The purification of cytokinin oxidase from *Zea mays* kernels, *Phytochemistry*, 28, 1313, 1989.
6. **Burch, L. R. and Horgan, R.,** Cytokinin oxidase and the degradative metabolism of cytokinins, in *Physiology and Biochemistry of Cytokinins in Plants*, Kamínek, M., Mok, D. W. S., and Zazímalová, E., Eds., SPB Academic Publishing, The Hague, 1992, 30.
7. **McGaw, B. A. and Horgan, R.,** Cytokinin oxidase from *Zea mays* kernels and *Vinca rosea* crown-gall tissue, *Planta*, 159, 30, 1983.
8. **Scott, I. M., McGaw, B. A., Horgan, R., and Williams, P. E.,** Biochemical studies of cytokinins in *Vinca rosea* crown gall tissue, in *Plant Growth Substances 1982*, Wareing, P. F., Ed., Academic Press, London, 1982, 165.
9. **Chatfield, J. M. and Armstrong, D. J.,** Regulation of cytokinin oxidase activity in callus tissues of *Phaseolus vulgaris* L. cv Great Northern, *Plant Physiol.*, 80, 493, 1986.
10. **Kamínek, M. and Armstrong, D. J.,** Genotypic variation in cytokinin oxidase from *Phaseolus* callus cultures, *Plant Physiol.*, 93, 1530, 1990.
11. **Laloue, M. and Fox, J. E.,** Cytokinin oxidase from wheat: partial purification and general properties, *Plant Physiol.*, 90, 899, 1989.
12. **Motyka, V. and Kamínek, M.,** Regulation of cytokinin catabolism in tobacco callus cultures, in *Progress in Plant Cellular and Molecular Biology*, Nijkamp, H. J. J., Van Der Plas, L. H. W., and Van Aartrijk, J., Eds., Kluwer Academic Publishers, Dordrecht, 1992, 492.
13. **Motyka, V. and Kamínek, M.,** Characterization of cytokinin oxidase from tobacco and poplar callus cultures, in *Physiology and Biochemistry of Cytokinins in Plants*, Kamínek, M., Mok, D. W. S., and Zazímalová, E., Eds., SPB Academic Publishing, The Hague, 1992, 33.

14. **McGaw, B. A. and Horgan, R.,** Cytokinin catabolism and cytokinin oxidase, *Phytochemistry*, 22, 1103, 1983.
15. **Paces, V. and Kamínek, M.,** Effect of ribosylzeatin isomers on the enzymatic degradation of N^6-(Δ^2-isopentenyl)adenosine, *Nucleic Acids Res.*, 3, 2309, 1976.
16. **Gerhauser, D. and Bopp, M.,** Cytokinin oxidases in mosses. 2. Metabolism of kinetin and benzyladenine *in vitro*. *J. Plant Physiol.*, 135, 714, 1990.
17. **Mok, M. C. and Mok, D. W. S.,** The metabolism of [^{14}C]-thidiazuron in callus tissues of *Phaseolus lunatus*, *Physiol. Plant.*, 65, 427, 1985.
18. **Thomas, J. C. and Katterman, F. R.,** Cytokinin activity induced by thidiazuron, *Plant Physiol.*, 81, 681, 1986.
19. **Laloue, M. and Fox, J. E.,** Characterization of an imine intermediate in the degradation of isopentenylated cytokinins by a cytokinin oxidase from wheat, in *Abst. 12th Int. Conf. on Plant Growth Substances*, International Plant Growth Association, Heidelberg, 1985, 23.
20. **Kapeller-Adler, R.,** *Amine Oxidases and Methods for Their Study*, Wiley-Interscience, New York, 1970, 26.
21. **Hamilton, G. A.,** Oxidases with monocopper reactive sites. Galactose oxidase and copper-containing amine oxidases, in *Copper Proteins*, Spiro, T. G., Ed., Wiley-Interscience, New York, 1981, 193.
22. **Maycock, A. L., Suva, R. H., and Abeles, R. H.,** Novel inactivators of plasma amine oxidases, *J. Am. Chem. Soc.*, 97, 5613, 1975.
23. **Chatfield, J. M. and Armstrong, D. J.,** Cytokinin oxidase activity from *Phaseolus vulgaris* callus tissues. Enhanced *in vitro* activity of the enzyme in the presence of copper-imidazole complexes, *Plant Physiol.*, 84, 726, 1987.
24. **Chatfield, J. M. and Armstrong, D. J.,** Cytokinin oxidase activity from *Phaseolus vulgaris* callus tissues. Affinity for concanavalin A, *Plant Physiol.*, 88, 245, 1988.
25. **Horgan, R., Burch, L. R., and Palni, L. M. S.,** Cytokinin oxidase and the degradative metabolism of cytokinins, in *Plant Growth Substances 1988*, Pharis, R. P. and Rood, S. B., Eds., Springer-Verlag, New York, 1990, 282.
26. **Palmer, M. V., Horgan, R., and Wareing, P. F.,** Cytokinin metabolism in *Phaseolus vulgaris* L. Identification of endogenous cytokinins and metabolism of [8-^{14}C]-dihydrozeatin in stems of decapitated plants, *Planta*, 153, 297, 1981.
27. **Palmer, M. V., Scott, I. M., and Horgan, R.,** Cytokinin metabolism in *Phaseolus vulgaris* L. II. Comparative metabolism of exogenous cytokinins by detached leaves, *Plant Sci. Lett.*, 22, 187, 1981.
28. **Nandi, S. K., De Klerk, G. J. M., Parker, C. W., and Palni, L. M. S.,** Endogenous cytokinin levels and metabolism of zeatin riboside in genetic tumour tissues and non-tumorous tissues of tobacco, *Physiol. Plant.*, 78, 197, 1990.
29. **Doree, M. and Guern, J.,** Short-term metabolism of some exogenous cytokinins in *Acer pseudoplatanus* cells, *Biochim. Biophys. Acta*, 304, 611, 1973.
30. **Van Staden, J. and Mooney, P. A.,** The effect of cytokinin preconditioning on the metabolism of adenine derivatives in soybean callus, *J. Plant Physiol.*, 133, 466, 1988.
31. **Gerhauser, D. and Bopp, M.,** Cytokinin oxidases in mosses. 1. Metabolism of kinetin and benzyladenine *in vivo*, *J. Plant Physiol.*, 135, 680, 1990.
32. **Fox, J. E.,** Incorporation of a kinin N^6-benzyladenine into soluble RNA, *Plant Physiol.*, 41, 75, 1966.
33. **Dyson, W. H., Fox, J. E., and McChesney, J. D.,** Short term metabolism of urea and purine cytokinins, *Plant Physiol.*, 49, 506, 1972.
34. **Laloue, M. and Pethe, C.,** Dynamics of cytokinin metabolism in tobacco cells, in *Plant Growth Substances 1982*, Wareing, P. F., Ed., Academic Press, London, 1982, 185.
35. **Terrine, C. and Laloue, M.,** Kinetics of N^6-(Δ^2-isopentenyl)adenosine degradation in tobacco cells, *Plant Physiol.*, 65, 1090, 1980.
36. **Palmer, M. V. and Palni, L. M. S.,** Substrate effects on cytokinin metabolism in soybean callus tissue, *J. Plant Physiol.*, 126, 365, 1987.
37. **Mok, M. C., Mok, D. W. S., Dixon, S. C., Armstrong, D. J., and Shaw, G.,** Cytokinin structure-activity relationships and the metabolism of N^6-(Δ^2-isopentenyl)adenosine-8-^{14}C in *Phaseolus* callus tissues, *Plant Physiol.*, 70, 173, 1982.
38. **Zhang, R., Letham, D. S., Wong, O. C., Nooden, L. D., and Parker, C. W.,** Cytokinin biochemistry in relation to leaf senescence. II. The metabolism of 6-benzylaminopurine in soybean leaves and the inhibition of its conjugation, *Plant Physiol.*, 83, 334, 1987.

39. **Parker, C. W., Entsch, B., and Letham, D. S.,** Inhibitors of cytokinin metabolism. I. Inhibitors of two enzymes which metabolize cytokinins, *Phytochemistry*, 25, 303, 1986.
40. **Palni, L. M. S., Burch, L., and Horgan, R.,** The effect of auxin concentration on cytokinin stability and metabolism, *Planta*, 174, 231, 1988.
41. **Morris, R. O., Akiyoshi, D. E., MacDonald, E. M. S., Morris, J. W., Regier, D. A., and Zaerr, J. B.,** Cytokinin metabolism in relation to tumor induction by *Agrobacterium tumefaciens*, in *Plant Growth Substances 1982*, Wareing, P. F., Ed., Academic Press, London, 1982, 175.
42. **McGaw, B. A., Horgan, R., Heald, J. K., Wullems, G. J., and Schilperoort, R. A.,** Mass-spectrometric quantitation of cytokinins in tobacco crown-gall tumours induced by mutated octopine Ti plasmids of *Agrobacterium tumefaciens*, *Planta*, 176, 230, 1988.
43. **Hansen, C. E., Meins, F., Jr., and Milani, A.,** Clonal and physiological variation in the cytokinin content of tobacco cell lines differing in cytokinin requirement and capacity for neoplastic growth, *Differentiation*, 29, 1, 1985.
44. **Wyndaele, R., Christiansen, J., Horseele, R., Rudelsheim, P., and Van Onckelen, H.,** Functional correlation between endogenous phytohormone levels and hormone autotrophy of transformed and habituated soybean cell lines, *Plant Cell Physiol.*, 29, 1095, 1988.
45. **Miernyk, J. A.,** Abscisic acid inhibition of kinetin nucleotide formation in germinating lettuce seeds, *Physiol. Plant.*, 45, 63, 1979.
46. **Sondheimer, E. and Tzou, D.,** The metabolism of 8-^{14}C-zeatin in bean axes, *Plant Physiol.*, 47, 516, 1971.
47. **Stachel, S. E., Messens, E., Van Montagu, M., and Zambryski, P.,** Identification of the signal molecules produced by wounded plant cells that activate T-DNA transfer in *Agrobacterium tumefaciens*, *Nature*, 318, 624, 1985.
48. **Bolton, G. W., Nester, E. W., and Gordon, M. P.,** Plant phenolic compounds induce expression of the *Agrobacterium tumefaciens* loci needed for virulence, *Science*, 232, 983, 1986.
49. **Morris, R. O.,** Genes specifying auxin and cytokinin biosynthesis in phytopathogens, *Ann. Rev. Plant Physiol.*, 37, 509, 1986.
50. **John, M. C. and Amasino, R. M.,** Expression of an *Agrobacterium* Ti plasmid gene involved in cytokinin biosynthesis is regulated by virulence loci and induced by plant phenolic compounds, *J. Bacteriol.*, 170, 790, 1988.
51. **Fox, L. R. and Purves, W. K.,** Mechanism of enhancement of IAA oxidation by 2,4-dichlorophenol, *Plant Physiol.*, 43, 454, 1968.
52. **Lee, T. T., Starratt, A. N., and Jevnikar, J. J.,** Regulation of enzymic oxidation of indole-3-acetic acid by phenols: structure-activity relationships, *Phytochemistry*, 21, 517, 1982.
53. **Van Staden, J. and Forsyth, C.,** Peroxidase and zeatin stability, *J. Plant Physiol.*, 118, 367, 1985.
54. **Bopp, M., Gerhauser, D., and Kessler, U.,** On the hormonal system of mosses, in *Plant Growth Substances 1985*, Bopp, M., Ed., Springer-Verlag, New York, 1986, 263.
55. **Bopp, M. and Erichsen, U.,** Metabolism of cytokinins in mosses, in *Metabolism and Molecular Activities of Cytokinins*, Guern, J. and Péaud-Lenoël, C., Eds., Springer-Verlag, Berlin, 1981, 105.
56. **Bopp, M. and Gerhauser, D.,** Localization of cytokinin responses in the moss *Funaria hygrometrica*, *Biol. Plant.*, 27, 265, 1985.
57. **Armstrong, D. J. and Firtel, R. A.,** Cytokinin oxidase activity in the cellular slime mold, *Dictyostelium discoideum*, *Dev. Biol.*, 136, 491, 1989.
58. **Abe, H., Uchiyama, M., Tanaka, Y., and Saito, H.,** Structure of discadenine, a spore germination inhibitor from the cellular slime mold, *Dictyostelium discoideum*, *Tet. Lett.*, 42, 3807, 1976.
59. **Tanaka, Y., Abe, H., Uchiyama, M., Taya, Y., and Nishimura, S.,** Isopentenyladenine from *Dictyostelium discoideum*, *Phytochemistry*, 27, 543, 1978.
60. **Nomura, T., Yoshimasa, T., Abe, H., and Uchiyama, M.,** Cytokinin activity of discadenine: a spore germination inhibitor of *Dictyostelium discoideum*, *Phytochemistry*, 16, 1819, 1977.
61. **Ihara, M., Taya, Y., and Nishimura, S.,** Developmental regulation of cytokinin, spore germination inhibitor discadenine, and related enzymes in *Dictyostelium discoideum*, *Exp. Cell Res.*, 126, 273, 1980.
62. **Ihara, M., Taya, Y., Nishimura, S., and Tanaka, Y.,** Purification and some properties of Δ^2-isopentenylpyrophosphate:5′-AMP Δ^2-isopentenyl-transferase from the cellular slime mold, *Dictyostelium discoideum*, *Arch. Biochem. Biophys.*, 230, 652, 1984.
63. **Ihara, M., Tanaka, Y., Yanagisawa, K., Taya, Y., and Nishimura, S.,** Purification and some properties of discadenine synthase from *Dictyostelium discoideum*, *Biochim. Biophys. Acta*, 881, 135, 1986.

64. **Van Kast, C. A. and Laten, H.,** Cytokinin utilization by adenine-requiring mutants of the yeast *Saccharomyces cerevisiae*, *Plant Physiol.*, 83, 726, 1987.
65. **Laten, H. and Zahareas-Doktor, S.,** Presence and source of free isopentenyladenosine in yeasts, *Proc. Natl. Acad. Sci. U.S.A.*, 82, 113, 1985.
66. **Mittelman, A., Evans, J. T., and Chheda, G. B.,** Cytokinins as chemotherapeutic agents, *Ann. N. Y. Acad. Sci.*, 255, 225, 1975.
67. **Chheda, G. B. and Mittelman, A.,** N^6-(Δ^2-isopentenyl)adenosine metabolism in man, *Biochem. Pharmacol.*, 21, 27, 1972.
68. **Henderson, T. R., Skinner, C. G., and Eakin, R. E.,** Kinetin and kinins as substrates and inhibitors of xanthine oxidase, *Plant Physiol.*, 35, 552, 1962.
69. **McLennan, B. D. and Pater, A.,** Evidence for the oxidation of N^6-(Δ^2-isopentenyl)-adenine and N^6-(3-hydroxy-3-methylbutyl)-adenine by xanthine oxidase, *Can. J. Biochem.*, 51, 1123, 1973.
70. **Chen, C. M., O'Brien, D. S., and McChesney, J. D.,** Biosynthesis and cytokinin activity of 8-hydroxy and 2,8-dihydroxy derivatives of zeatin and N^6-(Δ^2-isopentenyl)adenine, *Biochemistry*, 14, 3088, 1975.
71. **Terrine, C., Sadorge, P., Gawer, M., and Guern, J.,** Etude de la degradation par voie enzymatique des analogues N^6-substitues de l'adenosine. I. Action de l'adenosine aminohydrolase *in vitro*, *Physiol. Veg.*, 7, 425, 1969.
72. **Hall, R. H., Alam, S. N., McLennan, B. D., Terrine, C., and Guern, J.,** N^6-(Δ^2-isopentenyl)adenosine: its conversion to inosine, catalyzed by adenosine aminohydro-lases from chicken bone marrow and calf intestinal mucosa, *Can. J. Biochem.*, 49, 623, 1971.
73. **Hall, R. H. and Mintsioulis, G.,** Enzymatic activity that catalyzes degradation of N^6-(Δ^2-isopentenyl)adenosine, *J. Biochem.*, 73, 739, 1973.
74. **Mok, M. C., Mok, D. W. S., and Armstrong, D. J.,** Differential cytokinin structure-activity relationships in *Phaseolus*, *Plant Physiol.*, 61, 72, 1978.
75. **Horgan, R.,** Plant growth regulators and the control of growth and differentiation in plant tissue cultures, in *Plant Tissue and Cell Culture*, Green, C. E., Somers, D. A., Hackett, W. P., and Biesboer, D. D., Eds., Alan R. Liss, New York, 1987, 135.
76. **Turner, J. E., Mok, M. C., and Mok, D. W. S.,** Zeatin metabolism in fruits of *Phaseolus*: comparison between embryos, seedcoat, and pod tissues, *Plant Physiol.*, 79, 321, 1985.
77. **Jones, R. J., Schreiber, B. M., McNeil, K., Brenner, M. L., and Foxon, G.,** Cytokinin levels and oxidase activity during maize kernel development, in *Physiology and Biochemistry of Cytokinins in Plants*, Kamínek, M., Mok, D. W. S., and Zazímalová, E., Eds., SPB Academic Publishing, The Hague, 1992, 235.
78. **Forsyth, C. and Van Staden, J.,** The metabolism and cell division activity of adenine derivatives in soybean callus, *J. Plant Physiol.*, 124, 275, 1986.
79. **Miernyk, J. A. and Blaydes, D. F.,** Short term metabolism of radioactive kinetin during lettuce seed germination, *Physiol. Plant.*, 39, 4, 1977.
80. **Laloue, M., Terrine, C., and Guern, J.,** Cytokinins: metabolism and biological activity of N^6-(Δ^2-isopentenyl)adenosine and N^6-(Δ^2-isopentenyl)adenine in tobacco cells and callus, *Plant Physiol.*, 59, 478, 1977.
81. **Horgan, R., Palni, L. M. S., Scott, J. M., and McGaw, B. A.,** Cytokinin biosynthesis and metabolism in *Vinca rosea* crown gall tissue, in *Metabolism and Molecular Activities of Cytokinins*, Guern, J. and Péaud-Lenoël, C., Eds., Springer-Verlag, Berlin, 1981, 56.
82. **Palni, L. M. S. and Horgan, R.,** Cytokinin biosynthesis in crown gall tissue of *Vinca rosea*: metabolism of isopentenyladenine, *Phytochemistry*, 22, 1597, 1983.
83. **King, R. A. and Van Staden, J.,** The metabolism of N^6-(Δ^2-isopentenyl)[^{3}H]adenine by isolated organs of *Pisum sativum*, *Plant Physiol.*, 131, 181, 1987.
84. **Henson, I. E.,** Types, formation, and metabolism of cytokinins in leaves of *Alnus glutinosa* (L.) Gaertn., *J. Exp. Bot.*, 29, 935, 1978.
85. **Van Staden, J.,** Short term metabolism of different concentrations of zeatin applied to soybean callus, *Z. Pflanzenphysiol.*, 109, 163, 1983.
86. **Nooden, L. D. and Letham, D. S.,** Cytokinin control of monocarpic senescence in soybean, in *Plant Growth Substances 1985*, Bopp, M., Ed., Springer-Verlag, Berlin, 1986, 324.
87. **Parker, C. W., Letham, D. S., Gollnow, B. I., Summons, R. E., Duke, C. C., and MacLeod, J. K.,** Regulators of cell division in plant tissues. XXV. Metabolism of zeatin by lupin seedlings, *Planta*, 142, 239, 1978.
88. **Entsch, B., Letham, D. S., Parker, C. W., Summons, R. E., and Gollnow, B. I.,** Metabolites of cytokinins, in *Plant Growth Substances 1979*, Skoog, F., Ed., Springer-Verlag, New York, 1979, 109.

89. **Duke, C. C., Letham, D. S., Parker, C. W., MacLeod, J. K., and Summons, R. E.,** The complex of *O*-glucosylzeatin derivatives formed in *Populus* species, *Phytochemistry*, 18, 819, 1979.
90. **Summons, R. E., Entsch, B., Letham, D. S., Gollnow, B. I., and MacLeod, J. K.,** Regulators of cell division in plant tissues. XXVIII. Metabolites of zeatin in sweet-corn kernels: purification and identification using high performance liquid chromatography and chemical-ionization mass spectrometry, *Planta*, 147, 422, 1980.
91. **Parker, C. W. and Letham, D. S.,** Regulators of cell division in plant tissues. XVIII. Metabolism of zeatin in *Zea mays* seedlings, *Planta*, 115, 337, 1974.
92. **Palni, L. M. S., Palmer, M. V., and Letham, D. S.,** The stability and biological activity of cytokinin metabolites in soybean callus tissue, *Planta*, 160, 242, 1984.
93. **Palmer, M. V., Letham, D. S., and Gunning, B. E. S.,** Cytokinin metabolism in nondividing and auxin-induced dividing explants of *Helianthus tuberosus* L. tuber tissue, *J. Plant Growth Regul.*, 2, 289, 1984.
94. **Singh, S., Palni, L. M. S., and Letham, D. S.,** Cytokinin biochemistry in relation to leaf senescence. V. Endogenous cytokinin levels and metabolism of zeatin riboside in leaf discs from green and senescent tobacco (*Nicotiana rustica*) leaves, *J. Plant Physiol.*, 139, 279, 1992.
95. **Hocart, C. H., Letham, D. S., and Parker, C. W.,** Metabolism and translocation of exogenous zeatin riboside in germinating seeds and seedlings of *Zea mays*, *J. Exp. Bot.*, 41, 1517, 1990.
96. **Huang, B. and Armstrong, D. J.,** Unpublished data, 1990.

Chapter 12

Cytokinins and Plant Development — An Overview

Machteld C. Mok

TABLE OF CONTENTS

I. INTRODUCTION

The discovery of cytokinins by Skoog and Miller (see Chapter 1) marked the beginning of a decade of intensive studies to elucidate the effects of cytokinins on plant growth and development. Although cytokinins were defined as cell-division factors,[1,2] they influence a wide array of important biological processes. Numerous investigations showed that cytokinins play a role — major or minor — throughout development, from seed germination to leaf and plant senescence, and modulate physiological processes important throughout the life of a plant, including photosynthesis and respiration (see Chapters 13 and 14).

It is clear that plant development is determined by a number of internal and external factors; and cytokinins are only one of these factors. The different types of plant hormones can act synergistically or antagonistically;[3] for instance, cytokinins delay leaf senescence, but so do auxins and gibberellins, while abscisic acid and ethylene promote senescence. Moreover, environmental stimuli such as light and temperature have pronounced effects on plant development. Even though it may be difficult to separate the influence of cytokinins from that of other hormones or the environment, a discussion of the factors other than cytokinins would be outside the scope of this volume; besides, the literature on developmental processes such as seed germination and senescence is extensive, and entire volumes have been devoted to each of them. In this chapter a summary will be given of the involvement of cytokinins in seed germination, *de novo* bud formation, release of buds from apical dominance, leaf expansion, reproductive development, and senescence. In addition, since many different bioassays have been developed based on the various biological effects of cytokinins, the use of these bioassays for assessing cytokinin activity will be discussed.

II. CYTOKININS AND PLANT DEVELOPMENT

A. SEED DORMANCY AND GERMINATION

Abscisic acid and gibberellins play essential roles in regulating seed dormancy and germination. Abscisic acid accumulates in embryos during the middle to late stages of development, whereas its concentration decreases during seed desiccation and dormancy.[4,5] Studies with viviparous mutants have provided the most definitive evidence for involvement of abscisic acid in seed dormancy; these mutants were shown to be defective in the synthesis of, or response to, abscisic acid.[6] Gibberellins, on the other hand, stimulate germination. The level of the active gibberellin, GA_1, is low in dry cereal grains, but increases during germination, causing expression of hydrolases.[4] Several recent studies have focused on the molecular mechanisms by which abscisic acid and gibberellin modulate the expression of genes encoding key proteins in late embryogenesis or seed germination.[7-12]

0-8493-6252-0/94/$0.00+$.50

The effects of cytokinins on seed germination are not as pronounced; however, numerous studies have shown that cytokinins can influence germination, particularly when conditions are suboptimal for germination. Examples in earlier literature are reviewed by Khan;[13] they include germination of seeds at lower light intensities in the presence of cytokinin than in its absence, and cytokinin reversal of inhibition of germination by abscisic acid, coumarin, xanthatin, some unidentified inhibitors, and red light. The "permissive" role of cytokinin is further supported by studies with lettuce, in which inhibition of seed germination by darkness or high temperature could be overcome by gibberellin alone, in some genotypes, but required the simultaneous application of cytokinin in others.[14,15] Application of kinetin, but not gibberellin, could alleviate the inhibitory effects of cycloheximide on the germination of freshly harvested lettuce seeds, leading the authors to postulate direct triggering of germination by cytokinin.[16] However, kinetin was not effective when applied to older seeds,[16] indicating that its effect is limited to precocious germination.

Endogenous cytokinins have been measured in germinating seeds of various species. The level of cytokinin is generally low in dry seeds and initially decreases even further, but then increases during germination.[17-20] Delaying chickpea seed germination by increasing the temperature resulted in a concomitant shift in the fall and rise in cytokinin concentrations.[21] During light-stimulated celery seed germination, the levels of n-butanol-soluble cytokinins, such as zeatin and ribosylzeatin, increased, but those of the water-soluble nucleotides decreased, suggesting conversion of nucleotides rather than *de novo* synthesis.[22] However, germinating lupin embryos were shown to incorporate radiolabeled adenine into cytokinins,[23] indicating *de novo* biosynthesis of cytokinins. A comparison of cotyledons and embryo axes demonstrated that synthesis took place in the embryo axes.[23]

To explain the permissive role of cytokinins, Thomas proposed that cytokinins control the redistribution of gibberellins between subcellular sites;[15] however, this has yet to be demonstrated. The results of several studies, taken together, point to a role of cytokinin in the utilization of storage proteins in the cotyledons of dicots. First, cytokinins are synthesized in the embryo axes of germinating seeds[23] and transported to the cotyledons.[23-25] Second, the embryo axis was found to be important in mobilizing storage proteins in cotyledons of some species.[25-28] Third, exogenous cytokinins could replace the axis.[29,30] For example, application of dihydrozeatin to chickpea cotyledons accelerated protein degradation and increased carboxypeptidase and caseinase activity.[30] The mechanisms by which these enzyme activities are modulated have not yet been identified; possibly, induction mechanisms similar to those involved in gibberellin stimulation of α-amylase activity are involved. Thus, cytokinins may not be the trigger of germination, but may be important for seedling growth as well as for mobilization of storage proteins once germination has begun.

B. *DE NOVO* BUD FORMATION

The classic studies by Skoog and Miller demonstrated the dramatic effects of cytokinins on shoot differentiation in tobacco callus.[31] These studies established the interaction between cytokinin and auxin in stimulating redifferentiation of shoots and roots, with high cytokinin-to-auxin ratios favoring shoot formation and low cytokinin-to-auxin ratios favoring root development. Other compounds such as casein hydrolysate (or individual amino acids, e.g., tyrosine and phenylalanine) could increase the number of buds formed, but the presence of cytokinin was always required. Subsequently, regeneration by organogenesis was demonstrated in a number of other species;[32] generally, cytokinins are essential for shoot bud formation from callus tissues as well as directly from cultured plant tissues, such as leaf discs, hypocotyls, or internode sections. In some species, extremely high cytokinin concentrations are required for adventitious bud formation; therefore, the stable and highly active phenylurea derivatives, thidiazuron[33] and pyridylphenylureas,[34] are often used for those species.

The endogenous cytokinins as well as the metabolism of exogenous cytokinins were determined in relation to adventitious bud regeneration. The concentrations of zeatin, isopentenyladenine, and their ribosides were higher in begonia leaves exposed to environmental conditions conducive to bud formation than in leaves exposed to conditions inhibiting bud formation.[35] However, when radiolabeled benzyladenine was supplied to petunia leaf discs and the fate of the radiolabel determined at commitment to shoot formation, glucosyl derivatives were found to be the major, and nucleotides the minor, metabolites.[36] No free base or riboside could be detected. The latter findings could be the consequence of the high cytokinin levels in the medium and, consequently, in the tissues, resulting in conjugation to glucosylated products assumed to be storage forms of cytokinins.

Regeneration techniques have been crucial for the development of plant transformation systems. Without the possibility of regeneration, either by organogenesis or somatic embryogenesis, gene transfer would be limited to the few successful cases of foreign DNA incorporation into meristems or pollen. In spite of the importance of organogenesis to biotechnology, our understanding of how cytokinins bring about the changes in organization is very limited. Several systems have been used to study protein patterns associated with organogenesis: the thin-cell-layer system of tobacco,[37,38] which can be manipulated to give shoots, roots, flowers, or callus; stem segments of *Torenia*[39] and cotyledons of Douglas fir[40] and *Picea abies*,[41] on which adventitious buds are induced by cytokinins; and regenerating *Petunia* calli.[42] Specific polypeptides were associated with bud induction in tobacco, *Torenia*, and Douglas fir. However, no proteins unique to bud induction could be identified in *Picea abies* cotyledons and *Petunia* calli; rather, several proteins were less abundant during bud induction in these systems. Thus, results have varied considerably between species. It should be noted that proteins were analyzed at different times in these studies, from a few hours after inductive treatment to just before the morphological changes became visible; therefore, modifications in protein profiles may not have been related to the initial changes in gene expression.

In differentiating callus tissues or when adventitious buds are formed on cultured plant organs, only a small portion of the cells give rise to meristems. Why only particular cells display morphogenetic competence is not easy to answer. An obvious prerequisite is the ability of cells to divide, and thus the involvement of cytokinins in cell division should be relevant to shoot formation as well (see Chapter 15). However, the signal transduction chain leading to the initial changes in developmental patterns is unknown. Once the first developmental changes have been induced, sufficient levels of cytokinins and other hormones may be produced in the developing meristem, to sustain further growth and development.

C. RELEASE OF BUDS FROM APICAL DOMINANCE

Outgrowth of lateral buds is generally prevented by a correlative signal from the apical bud; removal of the apical bud eliminates this signal and allows the growth of lateral shoots. The importance of auxin in apical dominance was first demonstrated by Thimann and Skoog;[43] whereas decapitation of *Vicia faba* plants allowed lateral bud growth, treatment of the apical stump with auxin prevented it. In contrast, cytokinin application to lateral buds often overcomes apical dominance,[44,45] as shown originally by Wickson and Thimann.[46]

Among attempts to relate endogenous cytokinins with branching patterns, the clearest correlation has been obtained with mutants and trans-formants. The Torosa-2 (nonbranching) tomato mutant contained lower levels of cytokinins than did the normal branching wild type.[47] Similar results have been obtained more recently with another nonbranching mutant, Craigella Lateral Suppressor.[48] Transformants of tobacco and petunia, containing high cytokinin levels due to overexpression of the isopentenyltransferase (*ipt*) gene, were more branching than the nontransformed controls (see Chapter 21). Further evidence implicating cytokinins in the release from apical dominance has been derived from studies of the cytokinin concentrations in different types of buds; terminal buds of tomato[48] and a water fern[49] contained more zeatin and ribosylzeatin than did quiescent lateral buds. Moreover, the levels of zeatin and zeatin riboside decreased in lateral buds, proportional to their distance to the apical bud.[49] Surprisingly, the isopentenyladenosine levels were tenfold higher in the subapical bud,[49] leading to the hypothesis that conversion of isopentenyladenosine to zeatin riboside may be necessary for bud and shoot growth. In support of this hypothesis, isopentenyladenine was much less effective than zeatin in releasing pea buds from apical dominance.[50] However, no direct evidence showing conversion of isopentenyladenosine to zeatin riboside, in buds released from apical dominance, has been provided.

Cytokinins in buds may originate from the roots, by transport through the xylem, or they may be synthesized in the buds themselves. As evidence for the former origin, radiolabeled benzyladenine supplied to the roots of pea seedlings was transported to the dominant bud.[51] Transport of cytokinin was influenced by auxin, since decapitation of potato shoots combined with the application of auxin to the apical end directed radiolabeled benzyladenine to the decapitated stump; without auxin application, more label was channeled to the axillary buds.[52] Cytokinin synthesis in the axillary buds may be essential, as well, for shoot development, as treatment of buds with the purine inhibitor hadacidin prevented growth, while subsequent treatment of these buds with cytokinin led to shoot growth.[53] However, the role of cytokinins is, most likely, secondary; the pivotal force in apical dominance is the basipetal transport of auxin from the dominant bud.[44,45] Although it was initially assumed that auxin moves into the lateral buds, this does not seem to be the case.[45] Recent studies[45,54,55] have implicated the inhibition of polar auxin transport in the axillary buds, by polar auxin transport from the dominant bud through the shoot, possibly mediated by auxin efflux carrier randomization in the cells of the inhibited buds.[54] Within hours after decapitation, the auxin levels in the subapical bud increase,[56] abscisic acid levels decrease,[56] and transport of nutrients and cytokinins from the roots to this bud is enhanced.[52] The primary action of cytokinins in bud growth is unknown, although it may be envisioned that stimulation of cell division, Ca^{2+} distribution, changes in cell wall extensibility (see Section II.D), and synthesis of proteins[57] could be involved.

D. LEAF EXPANSION

The ability of cytokinins to stimulate leaf expansion was first shown by Kuraishi and Okumura[58] in 1956 and subsequently confirmed in a number of other studies.[59-61] Competition experiments demonstrated that the effects of cytokinins were restricted to the treated leaf and that leaf expansion was even diminished when cytokinin was applied to the adjacent leaf.[59] Analyses of leaf cytokinins established a correlation between endogenous cytokinins and leaf expansion.[62] Higher concentrations of zeatin and ribosylzeatin occurred in the basal part of pepper leaves, the region where leaf expansion takes place, than in the distal portion; older, fully expanded leaves had a more uniform distribution of cytokinins.[62]

Studies focusing on the possible action of cytokinins in leaf expansion have implied that cytokinins enhance sink strength for the allocation of assimilates. Treatment with kinetin resulted in the movement of nutrients from the untreated to the treated part of the tobacco leaf,[59,63] and application of cytokinins raised the osmotic concentration in radish[64] and watermelon[65] cotyledons. This interpretation is also consistent with the results of the bean leaf competition experiments mentioned above.[59] However, more recent studies indicate that the change in nutrient allocation may be an indirect result of the increase in cytokinin concentration. Pepper leaf expansion stimulated by application of benzyladenine was not associated with significant changes in sucrose or leucine uptake, partitioning of photosynthates, or activities of enzymes such as acid invertase or pyrophosphate-dependent phosphofructokinase, in the leaves.[62,66] In addition, determination of the parameters contributing to the cell expansion equation[67,68] have shown that initiation of growth promotion by cytokinin (as well as by gibberellin and light) is the result of an increase in wall extensibility and that changes in osmotic concentration are secondary to initiation of growth.[69]

E. REPRODUCTIVE DEVELOPMENT

The shift from vegetative to reproductive development is generally evoked by exposure of the plant to environmental cues, such as the proper daylength or temperature. In plants responsive to the photoperiod, the photoreceptive organ is the leaf,[70] indicating the presence of a flower-inducing compound. However, although the "florigen" concept was proposed by Chailakhyan as early as 1936,[71] no such compound has been identified, in spite of numerous research endeavors.

Each group of plant hormones affects flowering in some species, but none of them qualify as florigen. Most interesting is the role of gibberellin, since it can induce flowering, under noninductive conditions, in a broad range of species,[70] yet it can suppress flower formation in a number of fruit tree species.[72] Cytokinins stimulate flowering in several plant species, but generally only under favorable (inductive) conditions.[72] Cultured pedicel tissues of tobacco have been used by several groups to establish a connection between cytokinins and flowering, since these tissues can be programmed to give vegetative buds, flower buds, roots, or callus, depending on the concentrations of auxin and cytokinin.[73] Although studies did not agree on the relative efficacies of exogenous cytokinins,[74-77] the more stable cytokinins, such as benzyladenine and dihydrozeatin, were most effective. Cytokinins were required during the first 4 to 8 days only, indicating a role early in initiation.[76,78] Early cytokinin treatment as well as the site of treatment were important in *Pharbitis nil*, where application of benzyladenine to the cotyledons at the start of, or before, the inductive dark period increased the number of flowers, but benzyladenine treatment of the plumule had no effect.[79] In *Pharbitis* as well as most other systems, from cultured rape inflorescences[80] to conifers,[81,82] the effects of exogenous cytokinins were merely quantitative, increasing the number of buds.

A correlation has been sought between flowering and the endogenous cytokinin present in plant tissues. Higher cytokinin concentrations were found in root and leaf exudates of the long-day plant *Sinapis alba* after 16 h of light.[83] Cytokinin levels were also linked to flowering in *Chenopodium*[84] and *Perilla frutescens*.[85] In contrast, cytokinins in the short-day species *Xanthium strumarium* decreased under inductive conditions.[86] Elevated levels of cytokinins (determined by radioimmunoassay with antibody to zeatin riboside) were found in tobacco pedicel explants cultured on medium containing benzyladenine,[78] but no such increase was observed in another study.[87] Thus, there does not seem to be general agreement on the relationship between flowering and endogenous cytokinins.

The mode of cytokinin action in flowering is unclear. Changes in gene expression have been implicated in the tobacco pedicel explants.[88] The studies by Bernier and co-workers (see Chapter 15) indicate that cytokinins stimulate the early mitotic wave that invariably occurs in evoked meristems, and support the contention that the floral stimulus consists of several components, one of which is cytokinin. Nevertheless, Metzger has argued against the induction of flowering by the cumulative effects of concentration changes in a number of compounds,[70] based, among other reasons, on the failure to mimic inductive conditions by application of combinations of hormones and the low likelihood that a specific ratio of hormones reaches the apex under all conditions that induce flowering. If a specific flower-inducing compound exists, the role of cytokinins may be related to their enhancement of assimilate flow to the shoot apex,[79] thereby increasing translocation of the "floral stimulus."

For a discussion of the involvement of cytokinin in sex expression and male sterility, the reader is referred to Chapter 22.

F. DELAY OF SENESCENCE

A role of cytokinins in delaying leaf senescence was first implied in 1957 by Richmond and Lang's demonstration that immersion of petioles in water with kinetin could significantly slow the loss of protein and chlorophyll in *Xanthium* leaves, as compared to the controls without kinetin.[89] Further studies by Mothes and co-workers showed that the direct application of cytokinin to parts of tobacco leaves resulted in chlorophyll retention, but only in cytokinin-treated areas.[90,91] These early observations have been confirmed by numerous studies involving diverse plant species and an array of cytokinins.[92-94] Only a few exceptions were found; for instance, cytokinin was not effective in delaying senescence in *Taraxacum*, although gibberellin was effective in this species.[95] Senescence could be delayed by cytokinin application to intact plants as well; however, the effects of cytokinins on intact plants were generally less clear than those on detached leaves, probably due to correlative influences of other plant parts.[93]

Leaf senescence as well as monocarpic senescence have been the subject of extensive reviews with numerous citations.[96-100] The evidence supporting cytokinin involvement in delaying senescence is convincing and includes, besides results of exogenous cytokinin treatments, comparisons of endogenous cytokinins in young, maturing, and senescing tissues, and correlations of cytokinin contents with various senesence-inducing and -delaying treatments.[92,93] Although some of the older studies concerning the determination of endogenous cytokinins were inconclusive — analyses were often limited to a few forms of cytokinins, such as zeatin and zeatin riboside, or involved bioassays — leaf senescence was generally found to be accompanied by a decrease in endogenous cytokinins. In a recent study with tobacco, nonsenescent leaves contained two to eight times higher cytokinin levels than senescent leaves near the stem base.[94] Application of ammonium nitrate retarded senescence and increased zeatin and dihydrozeatin levels, in excised tobacco leaves.[101] In addition, radiolabeled adenine was incorporated into cytokinins in young expanding leaves, but not in mature and senescing leaves.[102]

Since senescence involves a loss of chlorophyll, and cytokinins stimulate synthesis of chloroplast DNA and photosynthetic enzymes, as well as grana formation and chloroplast replication,[93] it is tempting to conclude that cytokinins retard leaf senescence through promotion of chlorophyll synthesis or preservation of the chloroplast structure. However, cytokinins affect a number of other important processes associated with senescence. Protein and nucleic acid levels decline during senescence,[103] although this decline seems to affect only certain proteins, while others are unaffected or even increased. Among the proteins with higher activity during senescence are proteolytic enzymes. Cytokinins prevent this increase in proteolytic activity[104,105] and decrease lipase[106] and lipoxygenase,[107] which are involved in membrane breakdown. The effect of cytokinin on membrane integrity may be important in delaying senescence, since it maintains compartmentation and prevents leakage of nutrients and enzymes.

Senescence is accompanied by a decrease in stomatal aperture. A number of compounds that enhance stomatal closure also promote senescence, while those that limit stomatal closure, including the cytokinins, delay senescence.[92] As a consequence of stomatal aperture, transpiration and nutrient flow through the xylem, as well as transport of cytokinins from the roots, are enhanced.[92]

Cytokinins can inhibit oxidation[108] and prevent the sharp rise in respiration usually occurring during senescence (see Chapter 13). The depression of cyanide-resistant respiration by cytokinins is particularly interesting, even though rather high cytokinin concentrations are required to exert inhibition. The effects of cytokinins on the early physiological and cellular changes are likely to be most important, although the exact chain of events is difficult to determine. Senescence is a complex problem, and cytokinin is just one influencing factor. Confounding the situation even more is the fact that cytokinins stimulate the biosynthesis of ethylene,[109] one of the major factors promoting senescence. This underscores the intricacy of interactions between plant hormones, including synergistic or antagonistic action and influence on each other's biosynthesis and metabolism.

III. BIOASSAYS

Bioassays have been indispensable for the detection of cytokinin activity; in fact, the discovery of cytokinins depended on the tobacco pith bioassay devised by Skoog and co-workers. In the first two decades after the discovery of cytokinins, the tobacco callus bioassay served to evaluate the activities of numerous compounds. These studies delineated structure–activity relationships, among others the optimal length of the N^6-side chain and the effects of ring-structure modifications.[110] The need for an optimized culture medium resulted in the development of the Murashige and Skoog medium,[111] which has since been the most widely used medium for plant tissue culture. Other callus bioassays have also been utilized, such as the soybean,[112] carrot,[113] and *Phaseolus*[114] bioassays.

A number of bioassays have been devised based on the various biological effects of cytokinins. They include, but are not limited to, the lettuce seed germination assay;[115] *Funaria* protonemata[116] and pea lateral bud[117] bioassays, related to the ability of cytokinin to promote formation of new buds and release buds from apical dominance, respectively; leaf or cotyledon expansion assays, such as the etiolated bean leaf disc,[118] *Spirodela* frond expansion,[119] and radish cotyledon[60] bioassays; senescence bioassays with various plant species;[119,120] the *Amaranthus* bioassay[121,122] measuring the formation of betacyanin; and the cucumber cotyledon assay[123] depending on the formation of chlorophyll.

Bioassays have been used to assess cytokinin activity of new compounds as well as to determine the cytokinin activity in plant extracts. For the former use, they are indispensable, since they offer the only means of determining whether a compound, either directly or indirectly, exerts cytokinin activity. For the latter use, bioassays were essential in the earlier period of cytokinin research, but in recent years they have been replaced by more precise methods of cytokinin analyses, e.g., chromatography followed by immunoassays or quantitative mass spectra analyses using internal standards.

There are a number of essential attributes that determine the efficacy of a bioassay.[110,124] A bioassay should be specific to cytokinins and highly sensitive, detecting minute quantities of the more active cytokinins. This is important for detection of cytokinins in plant extracts, which are difficult to obtain in large quantities, as well as for evaluation of compounds that are only weakly active. Moreover, bioassays should be quantitative, allowing comparisons to the activity of known standards. Although not absolutely necessary, aseptic conditions could enhance the reliability of the assays, by eliminating cytokinins present in microorganisms. Repeatability, not only in one experimenter's hands, but also between groups, is essential for assigning relative activities to the large array of cytokinins. Furthermore, a short assay time is desirable in order to avoid long delays in experiments and to obtain more accurate measurements of activity. Most cytokinins, particularly the naturally occurring compounds, are rapidly metabolized or broken down in plant tissues (see Chapters 9–11). Bioassay results cannot distinguish between direct cytokinin action of the compound and indirect effect, i.e., having activity after being metabolized or even by inducing a change in endogenous cytokinin biosynthesis and metabolism, or altering the sensitivity of tissues to cytokinins. Thus, an ideal bioassay would last only hours or even minutes.

All bioassays have their intrinsic shortcomings. Since other groups of hormones influence the same biological processes, they are often not cytokinin specific. For instance, lettuce seed germination is affected by a number of compounds, most notably the gibberellins, and the *Amaranthus* assay responds to the presence of jasmonic acid.[122] Leaf disc expansion assays are greatly influenced by the age of the leaves. Interestingly, the *Funaria* bioassay does not detect *trans*-zeatin riboside, while it is sensitive to *trans*-zeatin and even detects *cis*-zeatin.[125] The cucumber chlorophyll retention bioassay suffers from variability, since it is dependent on light and influenced by the presence of salts;[123] moreover, cytokinin treatment increases chlorophyll only up to fivefold over the control. Callus bioassays are generally specific, sensitive, and quantitative, but require a long assay time (3 to 5 weeks). The long assay time allows significant metabolic and catabolic activity, in some instances leading to significant differences in cytokinin structure–activity relationships between some systems. For instance, cytokinins with unsaturated side chains, such as zeatin and isopentenyladenine, were only weakly active in the *P. vulgaris* callus bioassay, although highly active in the *P. lunatus* bioassay, due to rapid degradation in the former system.[114,126] Therefore, stable synthetic cytokinins are more likely to display similar relative activities in the various bioassay systems.

Recently, an interesting new bioassay has been described[127] that utilizes leaf mesophyll protoplasts from tobacco plants transformed with a chimeric gene consisting of the gene-5 promoter (Pg5) of

Agrobacterium tumefaciens T-DNA fused to the *gus* gene (encoding β-glucuronidase). The gene-5 promoter responds to the presence of cytokinin and auxin, and therefore the bioassay can be used for detection of both types of compounds. For detection of auxin, cytokinin is included in the assay medium, and vice versa, cytokinin detection requires addition of auxin. The assay seems sufficiently sensitive for detection of cytokinin activity; however, the auxin requirement renders the assay less effective for detection of cytokinins in plant extracts, since the presence of auxins or auxin inhibitors could conceivably alter responses to cytokinins. The use of protoplasts ensures rapid uptake of cytokinins, and the use of multiwell dishes allows for large numbers of assays. Although the assay time of 3 days is short compared to the callus bioassays, it is still relatively long for a promoter response assay. In the future, with further characterization of plant genes responding to cytokinins, additional promoters may become available that show rapid induction (hours or even minutes).

The particular bioassay used for the evaluation of cytokinin activity has depended on the familiarity of the researcher with the assay and the ready availability of plant material or cultures. Although some assays are better than others, none has emerged as clearly superior in all aspects. Only when a bioassay assay with extremely fast response (minutes) becomes available will we know which compounds are "true" cytokinins, i.e., those exerting cytokinin activity directly.

IV. CONCLUSIONS

A substantial amount of literature has accumulated in support of a role for cytokinins at the various phases of plant development. Most evidence has been based on determinations of the effects of exogenous cytokinins and analyses of endogenous cytokinins. Positive results with exogenous cytokinins do not necessarily imply the involvement of endogenous cytokinins, but form starting points for further studies. Analyses of endogenous cytokinins have tended to affirm the involvement of cytokinins in some of the developmental processes; however, the cytokinin analyses were frequently flawed by inadequate analytic techniques, reliance on bioassays, or determination of only a few cytokinin metabolites. With the recent advances in analytical techniques, more accurate determinations of the various cytokinin metabolites should be feasible. Research focusing on mutants and transformants has provided the most convincing evidence linking endogenous cytokinins to plant development. Expression of the *ipt* gene of *Agrobacterium* in plants has led to alterations in development patterns (e.g., a decrease in apical dominance, a delay in senescence, an increase in chlorophyll formation), with accompanying increases in endogenous cytokinins (see Chapter 21). With the availability of additional genes involved in cytokinin biosynthesis or metabolism for plant transformation, the link between cytokinin concentrations and plant development may be strengthened. Although it has been proposed that changes in sensitivity to growth substances may regulate plant development, rather than levels of these compounds,[128] concrete evidence in support of this theory has yet to be provided.

Most interesting, although more difficult to answer, is the question of how cytokinins evoke physiological and developmental changes. Changes in gene expression must precede or accompany developmental transitions, and cytokinins have been shown to regulate expression of a number of genes of known and unknown function (see Chapter 17). However, it is not known how far the effects on gene expression are removed from the primary action of cytokinins; moreover, there may be multiple cellular targets of cytokinins. Thus, it is important to start to unravel the signal transduction pathways leading to cytokinin-induced changes at the cellular, tissue, and organ levels. In-depth reviews of the progress toward elucidation of these mechanisms will be presented in the following chapters.

REFERENCES

1. **Miller, C. O., Skoog, F., Von Saltza, M. H., and Strong, F. M.,** Kinetin, a cell division factor from deoxyribonucleic acid, *J. Am Chem. Soc.*, 78, 1392, 1955.
2. **Skoog, F., Strong, F. M., and Miller, C. O.,** Cytokinins, *Science*, 148, 532, 1965.
3. **Thimann, K. V.,** Antagonisms and similarities between cytokinins, abscisic acid and auxin (mini review), in *Physiology and Biochemistry of Cytokinins in Plants*, Kamínek, M., Mok, D. W. S., and Zazímalová, E., Eds., SPB Academic Publishers, The Hague, 1992, 395.
4. **Jacobsen, J. V. and Chandler P. M.,** Gibberellin and abscisic acid in germinating cereals, in *Plant Hormones and their Role in Plant Growth and Development*, Davies, P. J., Ed., Martinus Nijhoff, Dordrecht, 1987, 164.
5. **Quatrano, R. S.,** The role of hormones during seed development, in *Plant Hormones and their Role in Plant Growth and Development*, Davies, P. J., Ed., Martinus Nijhoff, Dordrecht, 1987, 494.

6. **Karssen, C. M. and van Loon, L. C.,** Probing hormone action in developing seeds by ABA-deficient and -insensitive mutants, in *Progress in Plant Growth Regulation*, Karssen, C. M., van Loon, L. C., and Vreugdenhill, D., Eds., Kluwer, Dordrecht, 1992, 43.
7. **Marcotte, W. R., Jr, Russell, S. H., and Quatrano, R. S.,** Abscisic acid-responsive sequences from the Em gene of wheat, *Plant Cell*, 1, 969, 1989.
8. **Guiltinan, M. J., Marcotte, W. R., and Quatrano, R. S.,** A plant leucine zipper protein that recognizes an abscisic acid response element, *Science*, 250, 267, 1990.
9. **Skriver, K., Olsen, F. L., Rogers, J. C., and Mundy, J.,** *cis*-Acting DNA elements responsive to gibberellin and its antagonistic abscisic acid, *Proc. Natl. Acad. Sci. U.S.A.*, 88, 7266, 1991.
10. **Lanahan, M. B., Ho, T.-H. D., Rogers, S. W., and Rogers, J. C.,** A gibberellin response complex in cereal α-amylase gene promoters, *Plant Cell*, 4, 203, 1992.
11. **Gubler, F. and Jacobsen, J. V.,** Gibberellin-responsive elements in the promoter of a barley high-pI α-amylase gene, *Plant Cell*, 4, 1435, 1992.
12. **Rogers, J. C. and Rogers, S. W.,** Definition and functional implications of gibberellin and abscisic acid *cis*-acting hormone response complexes, *Plant Cell*, 4, 1443, 1992.
13. **Khan, A. A.,** Cytokinins: permissive role in seed germination, *Science*, 171, 853, 1971.
14. **Thomas, T. H.,** Cytokinin involvement in the control of seed dormancy, in *Physiology and Biochemistry of Cytokinins in Plants*, Kamínek, M., Mok, D. W. S., and Zazímalová, E., Eds., SPB Academic Publishers, The Hague, 1992, 323.
15. **Biddington, N. L. and Thomas, T. H.,** Influence of different cytokinins on the germination of lettuce (*Lactuca sativa*) and celery (*Apium graveolens*) seeds, *Physiol. Plant.*, 37, 12, 1976.
16. **Schultz, C. and Small, J. G. C.,** Inhibition of lettuce seed germination by cycloheximide and chloramphenicaol is alleviated by kinetin and oxygen, *Plant Physiol.*, 97, 836, 1991.
17. **Webb, D. P., van Staden, J., and Wareing, P. F.,** Changes in endogenous cytokinins, gibberellins and germination inhibitors during the breaking of dormancy in *Acer saccharum* Marsh., *J. Exp. Bot.*, 24, 105, 1973.
18. **Pinfield, N. J. and Davies, H. V.,** Hormonal changes during after-ripening of *Acer platanoides* L. seeds, *Z. Pflanzenphysiol.*, 90, 171, 1975.
19. **Pinfield, N. J. and Sanches-Torres, J. O.,** The involvement of endogenous cytokinin-like and gibberellin-like substances in germination and early seedling growth of *Cucurbita pepo* L., *J. Plant Physiol.*, 116, 293, 1984.
20. **Martin, L., Diez, A., Nicolas, G., Legaz, M. E., and Villalobos, N.,** Cytokinins in chick-pea seeds: identification and transformations during germination in seedling growth, *J. Plant Physiol.*, 128, 133, 1987.
21. **Martin, L., Diez, A., Nicolas, G., and Villalobos, N.,** Variation of the levels and transport of cytokinins during germination of chick-pea seeds, *J. Plant Physiol.*, 128, 141, 1987.
22. **Thomas, T. H., Biddington, N. L., and Palevitch, D.,** The role of cytokinins in the phytochrome-mediated germination of dormant, imbibed celery (*Apium graveolens* L.) seeds, *Photochem. Photobiol.*, 27, 231, 1978.
23. **Nandi, S. K., Palni, L. M. S., Letham, D. S., and Knypl, J. S.,** The biosynthesis of cytokinins in germinating lupin seeds, *J. Expl. Bot.*, 39, 1649, 1988.
24. **Van Onckelen, H. A., Caubergs, R., and Degreef, J. A.,** Effect of light treatment and endogenous growth hormones on α- and β-amylase activities in cotyledons of *Phaseolus vulgaris* L., *Plant Cell Physiol.*, 18, 1029, 1977.
25. **Revilla, M. E., Martin, L., Nicolas, G., Legaz, M. E., and Villalobos, N.,** Effects of high temperature on the variation and transport of endogenous cytokinins during the germination of chick-pea seeds, *J. Plant Physiol.*, 132, 223, 1988.
26. **Penner, D. and Ashton, F. M.,** Hormonal control of proteinase activity in squash cotyledons, *Plant Physiol.*, 42, 791, 1967.
27. **Yomo, H. and Varner, J. E.,** Control of the formation of amylases and proteases in cotyledons of germinating peas, *Plant Physiol.*, 51, 708, 1973.
28. **Morohashi, Y.,** Control of development of aminolytic and proteolytic activities in cotyledons of germinating black gram seeds, *Physiol. Plant.*, 56, 189, 1982.
29. **Ilan, I. and Gepstein, S.,** Hormonal regulation of food reserve breakdown in germinating dicotyledonous seeds, *Israel J. Bot.*, 29, 193, 1981.

30. **Pino, E., Martin, L., Guerra, H., Nicolas, G., and Villalobos, N.,** Effect of dihydrozeatin on the mobilization of protein reserves in protein bodies during the germination of chick-pea seeds, *J. Plant Physiol.*, 137, 425, 1991.
31. **Skoog, F. and Miller, C. O.,** Chemical regulation of growth and organ formation in plant tissues cultured *in vitro*, *Symp. Soc. Expl. Biol.*, 11, 118, 1957.
32. **Flick, C. E., Evans, D. A., and Sharp, W. R.,** Organogenesis, in *Handbook of Plant Cell Culture*, Vol. 1, Evans, D. A., Sharp, W. R., Ammirato, P. V., and Yamada, Y., Eds., Macmillan, New York, 1983, 13.
33. **Mok, M. C., Mok, D. W. S., Turner, J. E., and Mujer, C. V.,** Biological and biochemical effects of cytokinin-active phenylurea derivatives in tissue culture systems, *Hortscience*, 22, 1194, 1987.
34. **Fellman, C. D., Read, P. E., and Hosier, M. A.,** Effects of Thidiazuron and CPPU on meristem formation and shoot proliferation, *Hortscience*, 22, 1197, 1987.
35. **Heide, O. M.,** Hormonal mediation of environmental regulation of organogenesis and flowering in *Begonia*, in *Physiology and Biochemistry of Cytokinins in Plants*, Kamínek, M., Mok, D. W. S., and Zazímalová, E., Eds., SPB Academic Publishers, The Hague, 1992, 347.
36. **Auer, C. A., Laloue, M., Cohen, J. D., and Cooke, T. J.,** Benzyladenine uptake and metabolism during *Petunia* shoot organogenesis, in *Physiology and Biochemistry of Cytokinins in Plants*, Kamínek, M., Mok, D. W. S., and Zazímalová, E., Eds., SPB Academic Publishers, The Hague, 1992, 69.
37. **Tran Thanh Van, K.,** *In vitro* control of *de novo*, flower, bud, root, and callus differentiation from excised epidermal tissues, *Nature*, 246, 44, 1973.
38. **Tran Thanh Van, K. and Trinh, T. H.,** Fundamental and applied aspects of differentiation *in vitro* and *in vivo*, in *Handbook of Plant Cell Culture*, Vol. 4, Evans, D.A., Sharp, W.R., and Ammirato, P.V., Eds., Macmillan, New York, 1986, 316.
39. **Tanimoto, S. and Harada, H.,** Protein synthesis during adventitious bud initiation in superficial layers of *Torenia* stem segments cultured in vitro, *Biochem. Physiol. Pflanzen*, 178, 391, 1983.
40. **Yasuda, T., Hasegawa, P. M., and Cheng, T.-Y.,** Analysis of newly synthesized proteins during differentiation of cultured Douglas fir cotyledons, *Physiol. Plant.*, 48, 83, 1980.
41. **Stabel, P., Eriksson, T., and Engström, P.,** Changes in protein synthesis upon cytokinin-mediated adventitious bud induction and during seedling development in Norway spruce, *Picea abies*, *Plant Physiol.*, 92, 1174, 1990.
42. **Renaudin, J. P., Tournaire, C., Traas, J., and Teyssendier de la Serve, B.,** Cellular and molecular markers of cytokinin-induced bud formation in *Petunia* calli, in *Physiology and Biochemistry of Cytokinins in Plants*, Kamínek, M., Mok, D. W. S., and Zazímalová, E., Eds., SPB Academic Publishers, The Hague, 1992, 361.
43. **Thimann, K. V. and Skoog, F.,** On the inhibition of bud development and other functions of growth substances in *Vicia faba*, *Proc. R. Soc. Lond. [Biol.]*, 114, 317, 1934.
44. **Cline, M. G.,** Apical dominance, *Bot. Rev.*, 57, 318, 1991.
45. **Tamas, I. A.,** Hormonal regulation of apical dominance, in *Plant Hormones and their Role in Plant Growth and Development*, Davies, P.J., Ed., Martinus Nijhoff, Dordrecht, 1987, 393.
46. **Wickson, M. and Thimann, K. V.,** The antagonism of auxin and kinetin in apical dominance, *Physiol. Plant.*, 11, 62, 1958.
47. **Mapelli, S. and Lombardi, L.,** A comparative auxin and cytokinin study in normal and to-2 tomato plants, *Plant Cell Physiol.*, 23, 751, 1982.
48. **Sossountzov, L., Maldiney, R., Sotta, B., Sabbagh, I., Habricot, Y., Bonnet, M., and Miginiac, E.,** Immunocytochemical localization of cytokinins in Craigella tomato and a sideshootless mutant, *Planta*, 175, 291, 1988.
49. **Pilate, G., Sossountzov, L., and Miginiac, E.,** Hormone levels and apical dominance in the aquatic fern *Marsilea drummondii* A. Br., *Plant Physiol.*, 90, 907, 1989.
50. **King, R. A. and Van Staden, J.,** Differential responses of buds along the shoot of *Pisum sativum* to isopentenyladenine and zeatin application, *Plant Physiol. Biochem.*, 26, 253, 1988.
51. **Prochazka, S. and Jacobs, W. P.,** Transport of benzyladenine and gibberellin A_1 from roots in relation to the dominance between the axillary buds of pea (*Pisum sativum* L.) cotyledons, *Plant Physiol.*, 76, 224, 1984.
52. **Wooley, D. J. and Wareing, P. F.,** The interaction between growth promoters in apical dominance. I. Hormonal interaction, movement and metabolism of a cytokinin in rootless cuttings, *New Phytol.*, 71, 781, 1972.

53. **Lee, P. K. -W., Kessler, B., and Thimann, K. V.,** The effect of hadacidin on bud development and its implications for apical dominance, *Plant Physiol.*, 31, 11, 1974.
54. **Morris, D. A. and Johnson, C. F.,** The role of auxin efflux carriers in the reversible loss of polar auxin transport in the pea (*Pisum sativum* L.) stem, *Planta*, 181, 117, 1990.
55. **Li, C. J. and Bangerth, F.,** The possible role of cytokinins, ethylene and indoleacetic acid in apical dominance, in *Progress in Plant Growth Regulation*, Karssen, C. M., van Loon, L. C., and Vreugdenhill, D., Eds., Kluwer, Dordrecht, 1992, 431.
56. **Gocal, G. F. W., Pharis, R. P., Yeung, E. C., and Pearce, D.,** Changes after decapitation in concentrations of indole-3-acetic acid and abscisic acid in the larger axillary bud of *Phaseolus vulgaris* L. cv. Tender Green, *Plant Physiol.*, 95, 344, 1991.
57. **Stafstrom, J. P. and Sussex, I. M.,** Expression of a ribosomal protein gene in pea axillary buds, *Planta* 176, 497, 1988.
58. **Kuraishi, S. and Okumura, F. S.,** The effect of kinetin on leaf growth, *Bot. Mag.*, 69, 817, 1956.
59. **Leopold, A. C. and Kawase, M.,** Benzyladenine effects on bean leaf growth and senescence, *Am. J. Bot.*, 51, 294, 1964.
60. **Letham, D. S.,** Regulators of cell division in plant tissues. XII. A cytokinin bioassay using excised radish cotyledons, *Physiol. Plant.*, 25, 391, 1971.
61. **Brock, T. G. and Cleland, R. E.,** Role of acid efflux during growth promotion of primary leaves of *Phaseolus vulgaris* L. by hormones and light, *Planta*, 177, 476, 1989.
62. **Ulstov, P., Nielsen, T. H., Seiden, P., and Marcussen, J.,** Cytokinins and leaf development in sweet pepper (*Capsicum annuum* L.). I. Spatial distribution of endogenous cytokinins in relation to leaf growth, *Planta*, 188, 70, 1992.
63. **Mothes, K.,** Die Beitrag der Kinetinforschung zum Verständnis pflanzlicher Korrelationen, *Ber. Dtsch. Bot. Ges.*, 74, 24, 1961.
64. **Longo, G. P., Longo, C. P., Rossi, G., Vitale, A., and Predetti, M.,** Variations in carbohydrate and in osmotic potential of watermelon cotyledons treated with benzyladenine, *Plant Sci. Lett.*, 12, 179, 1978.
65. **Norris, R.F.,** Expansion of radish cotyledons: an interaction between NaCl and cytokinins, *Plant Sci. Lett.*, 6, 63, 1976.
66. **Nielsen, T. H. and Ulvskov, P.,** Cytokinins and leaf development in sweet pepper (*Capsicum annuum*). II. Sink metabolism in relation to cytokinin-promoted leaf expansion, *Planta*, 188, 78, 1992.
67. **Lockhart, J. A.,** An analysis of irreversible plant cell elongation, *J. Theor. Biol.*, 8, 264, 1965.
68. **Ray, P. M., Green, P. B., and Cleland, R. E.,** Role of turgor in plant cell growth, *Nature*, 239, 163, 1972.
69. **Brock, T. G. and Cleland, R. E.,** Biophysical basis of growth promotion in primary leaves of *Phaseolus vulgaris* L. by hormones versus light. Solute accumulation and the growth potential, *Planta*, 182, 427, 1990.
70. **Metzger, J. D.,** Hormones and reproductive development, in *Plant Hormones and their Role in Plant Growth and Development*, Davies, P. J., Ed., Martinus Nijhoff, Dordrecht, 1987, 431.
71. **Chailakhyan, M. K.,** On the hormonal theory of plant development, *Dokl. Acad. Sci. USSR*, 12, 443, 1936 (in Russian).
72. **Zeevaart, J. A. D.,** Phytohormones and flower formation, in *Plant Hormones and Related Compounds*, Vol. II, Letham, D. S., Goodwin, P. B., and Higgins, T. J. V., Eds., Elsevier, Amsterdam, 1978, 291.
73. **Tran Thanh Van, K.,** Direct flower neoformation from superficial tissue of small explants of *Nicotiana tabacum* L., *Planta*, 165, 87, 1973.
74. **Tran Thanh Van, K.,** Control of morphogenesis in in vitro cultures, *Ann. Rev. Plant Physiol.*, 32, 291, 1981.
75. **Meeks-Wagner, D. R., Dennis, E. S., Tran Thanh Van, K., and Peacock, W.J.,** Tobacco genes expresssed during in vitro floral initiation and their expression during normal plant development, *Plant Cell*, 1, 25, 1989.
76. **Van der Krieken, W. M., Croes, A. F., Smulders, M. J. M., and Wullems, G. J.,** Cytokinins and flower bud formation *in vitro* in tobacco, *Plant Physiol.*, 92, 565, 1990.
77. **Heylen, C. and Vendrig, J. C.,** Further studies on the effects of different auxins and cytokinins on flower formation in thin cell layers of *Nicotiana tabacum*: the effects of zeatin riboside, dihydrozeatin and dihydrozeatin riboside, *Physiol. Plant.*, 83, 574, 1991.
78. **Heylen, C., Vendrig, J. C., and Van Onckelen, H.,** The accumulation and metabolism of plant growth regulators during organogenesis in cultures of thin cell layers of *Nicotiana tabacum*, *Physiol. Plant.*, 83, 578, 1991.

79. **Ogawa, Y. and King, R. W.,** Indirect action of benzyladenine and other chemicals on flowering of *Pharbitis nil* Chois, *Plant Physiol.*, 63, 643, 1979.
80. **Polowick, P. L. and Sawhney, V. K.,** In vitro floral development of oilseed rape (*Brassica napus* L.): the effect of pH and plant growth regulators, *J. Exp. Bot.*, 42, 1583, 1991.
81. **Ross, S. D. and Pharis, R. P.,** Promotion of flowering in the pinaceae by gibberellins. I. Sexually mature, non-flowering grafts of Douglas fir, *Physiol. Plant.*, 36, 182, 1976.
82. **Imbault, N., Tardieu, I., Joseph, C., Zaerr, J. B., and Bonnet-Masimbert, M.,** Possible role of isopentenyladenine in flowering of *Pseudotsugs menziesii*: endogenous variations and exogenous applications, *Plant Physiol. Biochem.*, 26, 3, 1988.
83. **Lejeune, P., Kinet, J.-M., and Bernier, G.,** Cytokinin fluxes during floral induction in the long day plant *Sinapis alba* L., *Plant Physiol.*, 86, 1095, 1988.
84. **Machackova, I. and Krekule, J.,** Cytokinins in photoperiodic flower induction in *Chenopodium* species, *Flowering Newslett.*, 12, 15, 1991.
85. **Grayling, A. and Hanke, D. E.,** Cytokinins in exudates from leaves and roots of red *Perilla*, *Phytochem.*, 31, 1863, 1992.
86. **Hensen, I. E. and Wareing, P. F.,** Cytokinins in *Xanthium strumarium*: a rapid response to short day treatment, *Physiol. Plant.*, 32, 185, 1974.
87. **Van der Krieken, W. M., Van Eldik, G., Nijtmans, L., Croes, A. F., and Wullems, G. J.,** Endogenous cytokinins in flower bud forming explants of tobacco, *J. Exp. Bot.*, 42, 77, 1991.
88. **Meeks-Wagner, D. R., Dennis, E. S., Tran Thanh Van, K., and Peacock, W. J.,** Tobacco genes expressed during in vitro floral initiation and their expression during normal plant development, *Plant Cell*, 1, 25, 1989.
89. **Richmond, A. E. and Lang, A.,** Effect of kinetin on protein content and survival of detached *Xanthium* leaves, *Science*, 125, 650, 1957.
90. **Mothes, K. and Baudisch, W.,** Untersuchungen über die Reversibilität der Ausbleichung grüner Blätter, *Flora*, 146, 521, 1958.
91. **Mothes, K. and Engelbrecht, L.,** Kinetin-induced directed transport of substances in excised leaves in the dark, *Phytochemistry*, 1, 58, 1961.
92. **Thimann, K. V.,** The senescence of leaves, in *Senescence in Plants*, Thimann, K. V., Ed., CRC Press, Boca Raton, FL, 1980, 85.
93. **Van Staden, J., Cook, E. L., and Noodén, L. D.,** Cytokinins and senescence, in *Senescence and Aging in Plants*, Noodén, L. D. and Leopold, A. C., Eds., Academic Press, San Diego, 1988, 281.
94. **Singh, S., Letham, D. S., and Palni, L. M. S.,** Cytokinin biochemistry in relation to leaf senescence. VII. Endogenous cytokinin levels and exogenous applications of cytokinins in relation to sequential leaf senescence in tobacco, *Physiol. Plant.*, 86, 388, 1992.
95. **Fletcher, R. A. and Osborne, D. J.,** Gibberellin as a regulator of protein and nucleic acid synthesis during senescence in leaf cells of *Taraxacum officinale*, *Can. J. Bot.*, 44, 739, 1966.
96. **Thimann, K. V.,** *Senescence in Plants*, CRC Press, Boca Raton, FL, 1980.
97. **Stoddart, J. L. and Thomas, H.,** Leaf Senescence, in *Encyclopedia of Plant Physiology, New Series*, Vol. 14A, Bouleter, D. and Parthier, B., Eds., Springer-Verlag, Berlin, 1982, 592.
98. **Lesham, Y. Y., Halevy, A. H., and Frenkel, C.,** *Processes and Control of Plant Senescence*, Elsevier, Amsterdam, 1986.
99. **Goldthwaite, J. J.,** Hormones and plant senescence, in *Plant Hormones and their Role in Plant Growth and Development*, Davies, P.J., Ed., Martinus Nijhoff, Dordrecht, 1987, 553.
100. **Noodén, L. D. and Leopold, A. C.,** *Senescence and Aging in Plants*, Academic Press, San Diego, 1988.
101. **Singh, S., Letham, D. S., Zhang, X., and Palni, L. M. S.,** Cytokinin biochemistry in relation to leaf senescence. VI. Effects of nitrogenous nutrients on cytokinin levels and senescence of tobacco leaves, *Physiol. Plant.*, 84, 262, 1992.
102. **Singh, S., Letham, D. S., and Palni, L. M. S.,** Cytokinin biochemistry in relation to leaf senescence. VIII. Translocation, metabolism and biosynthesis of cytokinins in relation to sequential leaf senescence in tobacco, *Physiol. Plant.*, 86, 398, 1992.
103. **Brady, C. J.,** Nucleic acid and protein synthesis, in *Senescence and Aging in Plants*, Noodén, L. D. and Leopold, A. C., Eds., Academic Press, San Diego, 1988, 147.
104. **Anderson, J. W. and Rowan, K. S.,** Activity of peptidase in tobacco leaf tissue in relation to senescence, *Biochem. J.*, 97, 741, 1965.
105. **Legocka, J. and Sweykowska, A.,** The role of cytokinins in the development and metabolism of barley leaves. VI. The effect on the protein metabolism in various cell compartments during leaf senescence, *Acta Physiol. Plant.*, 5, 11, 1983.

106. **Sodek, L. and Wright, S. T. C.,** The effect of kinetin on ribonucleic acid phosphatase, lipase and esterase levels in detached wheat leaves, *Phytochemistry*, 8, 1629, 1983.
107. **Grossman, S. and Lesham, Y.,** Lowering of endogenous lipoxygenase activity in *Pisum sativum* foliage by cytokinin as related to senescence, *Physiol. Plant.*, 43, 359, 1978.
108. **Lesham, Y.,** Interactions of cytokinins with lipid-associated oxy free radicals during senescence: a prospective mode of cytokinin action, *Can. J. Bot.*, 62, 2943, 1984.
109. **Yu, Y., Yang, S.F., Corse, J., Kuhnle, J. A., and Hua, S.,** Structures of cytokinins influence synergistic production of ethylene, *Phytochemistry*, 20, 1191, 1981.
110. **Skoog, F. and Armstrong, D. J.,** Cytokinins, *Ann. Rev. Plant Physiol.*, 21, 359, 1970.
111. **Murashige, T. and Skoog, F.,** A revised medium for rapid growth and bioassays with tobacco tissue cultures, *Physiol. Plant.*, 15, 473, 1962.
112. **Miller, C. O.,** A kinetin-like compound in maize, *Proc. Natl. Acad. Sci. U.S.A.*, 47, 170, 1961.
113. **Shaw, G., Smallwood, B. M., and Steward, F. C.,** The structure and physiological activity of some N^6-substituted adenines, *Phytochemistry*, 10, 2329, 1971.
114. **Mok, M. C., Mok, D. W. S., and Armstrong, D. J.,** Differential cytokinin structure–activity relationships in *Phaseolus*, *Plant Physiol.*, 61, 72, 1978.
115. **Miller, C. O.,** The relationship of the kinetin and red-light promotions of lettuce seed germination, *Plant Physiol.*, 33, 115, 1958.
116. **Hahn, H. and Bopp, M.,** A cytokinin test with high specificity, *Planta*, 83, 115, 1968.
117. **Thimann, K. V. and Sachs, T.,** The role of cytokinins in the fasciation disease caused by *Corynebacterium fasciens*, *Am. J. Bot.*, 53, 731, 1966.
118. **Miller, C. O.,** Kinetin and kinetin-like compounds, in *Modern Methods of Plant Analysis*, Linskens, H. F. and Tracey, M. V., Eds., Springer, Berlin, 1963, 194.
119. **Letham, D.S.,** Regulators of cell division in plant tissues. V. A comparison of the activities of zeatin and other cytokinins in five bioassays, *Planta*, 74, 228, 1967.
120. **Osborne, D. J. and McCalla, D. R.,** Rapid bioassay for kinetin and kinins using senescing leaf tissue, *Plant Physiol.*, 36, 219, 1961.
121. **Conrad, K.,** Ein sensibilisierter *Amaranthus*-Cytokinintest (AT 74), *Biochem. Physiol. Pflanzen*, 165, 531, 1974.
122. **Köhler, K.-H., Opitz, C., and Fritsch, G.,** Physiology and biochemistry of the *Amaranthus* cytokinin bioassay and its applications, *Biol. Plant.*, 29, 10, 1987.
123. **Fletcher, R. A., Kallidumbil, V., and Steele, P.,** An improved bioassay for cytokinins using cucumber cotyledons, *Plant Physiol.*, 69, 675, 1982.
124. **Letham, D. S.,** Chemistry and physiology of kinetin-like compounds, *Ann. Rev. Plant Physiol.*, 18, 349, 1967.
125. **Spiess, L. D.,** Cooperative activity of isomers of zeatin and ribosyl-zeatin on *Funaria hygrometrica*, *Plant Physiol.*, 55, 583, 1975.
126. **Mok, M. C., Mok, D. W. S., Dixon, S. C., Armstrong, D. J., and Shaw, G.,** Cytokinin structure–activity relationships and the metabolism of N^6-(Δ^2-isopentenyl)adenosine-8-^{14}C in *Phaseolus* callus tissues, *Plant Physiol.*, 70, 173, 1982.
127. **Boerjan, W., Genetello, C., Van Montagu, M., and Inze, D.,** A new bioassay for auxins and cytokinins, *Plant Physiol.*, 99, 1090, 1992.
128. **Trewawas, A.,** How do growth substances work? *Plant Cell Environ.* 4, 203, 1981.

Chapter 13

Cytokinins and Oxidative Processes

Mary E. Musgrave

TABLE OF CONTENTS

I. INTRODUCTION

A. RESPIRATION INHIBITION BY CYTOKININS DURING SENESCENCE

Soon after the identification of kinetin and its effects on cell division and callus growth, there followed a flurry of research effort in the early 1960s searching for alternative "rapid bioassays" that could be used to identify compounds with cytokinin activity. Such bioassays utilized tissue excised from plants and often relied on the ability of cytokinins to delay senescence, as a means of identifying active compounds. Typically, cytokinins had to occur at concentrations higher than those needed to elicit response in the callus bioassay, but the appeal of having a more rapid response may have outweighed the decreased sensitivity.

After reports that cytokinins delayed chlorophyll loss in senescing *Xanthium* leaves,[1,2] Dedolph et al. observed that postharvest treatment of asparagus spears and broccoli heads with benzyladenine diminished the high respiratory rates that accompanied postharvest deterioration and also maintained the visual appeal and fresh weight of the vegetables.[3,4] They went so far as to suggest that "many of the phenomena observed by others, such as chlorophyll retention, changes in nitrogen metabolism, and general inhibition in green plant tissues subsequent to treatment with kinetin and related compounds, may be a consequence of respiration inhibition."[4]

In the years following, many other reports linking cytokinin responses by tissues to the inhibition of respiration appeared. The plant materials studied were diverse (pea stem sections,[5] tobacco leaf discs,[6] beet roots,[7] *Rumex* leaves,[8] broccoli florets,[9] and potato tuber[10]), and while prevention of senescence predominated as the action exerted by cytokinins, the reports were not confined to senescent systems.[5]

In 1974 Tetley and Thimann opened a new phase in this area of research, with their report that in

0-8493-6252-0/94/$0.00+$.50

senescing oat leaf tissue, the respiration affected by cytokinin treatment was resistant to cyanide.[11] The idea that cytokinins might act as specific inhibitors of the cyanide-resistant respiratory pathway was explored in detail through the 1980s. This argument is especially intriguing given the enigmatic nature of this respiratory pathway[12] and its potential for influencing the energetics of the cell. As will be discussed below, a mechanism for controlling the contribution of the cyanide-resistant pathway to total respiration could be of prime importance in plant growth and development.

B. RESPIRATION STIMULATION BY CYTOKININS DURING DEVELOPMENT

The idea that cytokinins might exert a growth and development role, through effects on respiration, was also suggested by researchers working with developing tissues. In sunflower[13] and watermelon[14] seedling cotyledons, benzyladenine accelerated the development of mitochondria. Morohashi suggested that the stimulating effects of the axis in the development of cotyledon respiration might be mediated by cytokinins transported from the axis.[15] Seeds of black gram (*Vigna mungo* L.) treated with benzyladenine solution (3×10^{-4} *M*) had accelerated development of mitochondria and higher state-3 respiration rates (by almost 20%). Activities of respiratory enzymes increased: 32% higher succinate dehydrogenase activity, 38% higher malate dehydrogenase activity, and 32% higher cytochrome oxidase activity, than water-treated controls after 1 day of germination. Morohashi concluded that benzyladenine worked by enhancing the activation of mitochondria already existing in the cotyledons, rather than by stimulating biogenesis of mitochondria.[15] Moore and Miller also observed a short-term stimulation of respiration by kinetin (0.1 to 10 μ*M*) when soybean callus tissue was initially placed into suspension culture.[16] However, at concentrations above 10 μ*M*, kinetin inhibited respiration, probably through the mechanism to be described below.

II. CYANIDE-RESISTANT RESPIRATION

A. IMPLICATIONS FOR ENERGETICS

The first suggestion of a general role for cyanide-resistant respiration in plants developed from Bahr and Bonner's observation that in isolated mitochondria, cyanide-resistant respiration only operated when the cytochrome pathway was exceeded or restricted.[17] They suggested that the cyanide-resistant respiratory pathway may accept electrons not handled by the cytochrome pathway when the latter becomes saturated. Palmer extended this analysis, suggesting that the presence of a nonphosphorylating electron transport chain would enable the mitochondria to metabolize carbon skeletons even in the presence of a high level of ATP, since NADH could still be oxidized.[18] Cyanide-resistant respiration was lower in mitochondria isolated from roots and leaves of ten species than in the intact tissue, leading Lambers et al. to conclude that through adenylate control of glycolysis, the cytochrome pathway in intact tissues does not operate at the level achieved in state-3 respiration in isolated mitochondria.[19] Thus, while glycolysis is controlled *in vivo*, regulation of the cyanide-resistant pathway seems to be quite flexible, since the pathway apparently operates under any condition of energy overflow.[20] Because the pathway bypasses two energy conservation sites on the cytochrome pathway, it has been hypothesized that use of the cyanide-resistant pathway makes plants less energetically efficient.[12] Thus, Tetley and Thimann's finding that cytokinin controls the contribution of cyanide-resistant respiration to the respiratory climacteric in senescing oat leaves[11] raised the exciting possibility that this effect on respiration could represent a mode of action by the growth regulator.

B. INHIBITION BY EXOGENOUS CYTOKININS

1. *In Vitro*

The Tetley and Thimann data[11] on cytokinin inhibition of cyanide-resistant respiration were consistent with the later observation by Miller[21] that cytokinins are effective inhibitors (at 0.5 m*M*) of the cyanide-resistant pathway in cells and mitochondria of soybean callus. Miller subsequently showed that inhibition of cyanide-resistant respiration by cytokinins was a widespread phenomenon occurring in mitochondria from bush bean (*Phaseolus vulgaris* L.), mung bean (*Phaseolus aureus* Roxburgh), soybean (*Glycine max* L. Merrill), maize (*Zea mays* L.), pea (*Pisum sativum* L.), and wheat (*Triticum aestivum* L.).[22]

In a later study using sonicated mitochondria or submitochondrial particles from pea (*P. sativum* L.) and mung bean (*Vigna radiata* L.), Miller elucidated the nature of cytokinin modification of mitochondrial function.[23] Cytokinins inhibited the flow of electrons from internally generated NADH and also from succinate and externally supplied NADH when the cyanide-resistant respiratory pathway was operating (Figure 2E).

Dizengremel et al. found similar results with potato tuber mitochondria (Figure 1).[24] In these mitochondria the large cyanide-resistant respiratory pathway present (about 85% of total respiration) could be fully inhibited by 1 m*M* benzyladenine when using succinate as a substrate (Figure 1A, solid symbols). When malate was the substrate, only about 40% of the respiration could be inhibited by cytokinin, but the response saturated at a lower concentration of benzyladenine (200 μM) (Figure 1A, open symbols). Dizengremel et al. determined through kinetic analysis that the site of inhibition on the cyanide-resistant respiratory pathway was different from that of other specific inhibitors, SHAM (salicylhydroxamic acid) and disulfiram.[24]

The discovery by Keim et al. of a high-affinity ($K_d = 1 \times 10^{-8}$ M) cytokinin binding site in mung bean mitochondrial membranes has added another type of support to the idea that an early event in the response to cytokinins might be in an oxidative system.[25] Further support for the idea comes from the observation that in addition to the substituted adenine derivatives and the synthetic urea-derivative cytokinins, a new class of cytokinins, the nitroguanidines,[26] has been shown to inhibit cyanide-resistant respiration in isolated mitochondria.[27] However, in comparing the effectiveness of different adenine derivatives in inhibiting the cyanide-resistant respiration in potato tuber mitochondria, Dizengremel et al. noted that the relationship was poor relative to the cytokinin activities of the compounds.[24] Instead, their activity as respiration inhibitors seemed to be more closely predicted by their octanol:water partition coefficients (Figure 1B). Dizengremel et al. concluded that since the correlation of cytokinin activities in bioassays with their activity as respiratory inhibitors was poor, this provided evidence against the idea that cytokinins acted through an oxidative pathway. Miller also mentioned the poor performance by zeatin as a respiratory inhibitor in cells and mitochondria (Figures 2C and D), but proposed that it might be due to structural characteristics that prevent its accumulation into an active site quickly enough to give strong immediate effects.[21,22] He considered the situation analogous to the weak action of zeatin in some senescence tests, compared to its great effectiveness in other bioassays.

One of the difficulties in interpreting the relevance of these mitochondrial studies to how cytokinins work *in vivo* was the relatively high concentration of cytokinin (1 mM) needed to inhibit the cyanide-resistant pathway *in vitro*. Another drawback was that no causative relationship was shown between any physiologicalresponses to cytokinins and the inhibition of the cyanide-resistant pathway. Subsequent studies have addressed both of these concerns, as described below.

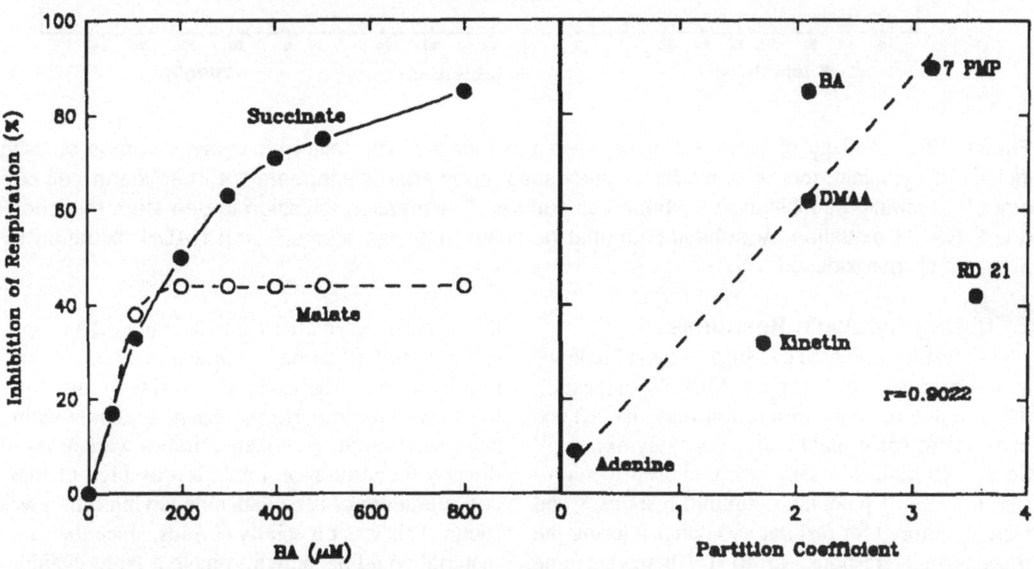

Figure 13-1 Inhibition of cyanide-resistant respiration by potato tuber mitochondria metabolizing succinate (filled circles) or malate (open circles), by increasing concentrations of benzyladenine. The right-hand panel shows the relationship between cyanide-resistant respiration inhibition by cytokinins and their octanol:water partition coefficients. The activity of the adenine derivative RD21 (6-benzoylamino-9-benzyl-purine) was considered to relate poorly to liposolubility, because of steric occupation of this compound, so it was not included in the regression estimate.[24] (DMAA: dimethyl-allyl-adenine.)

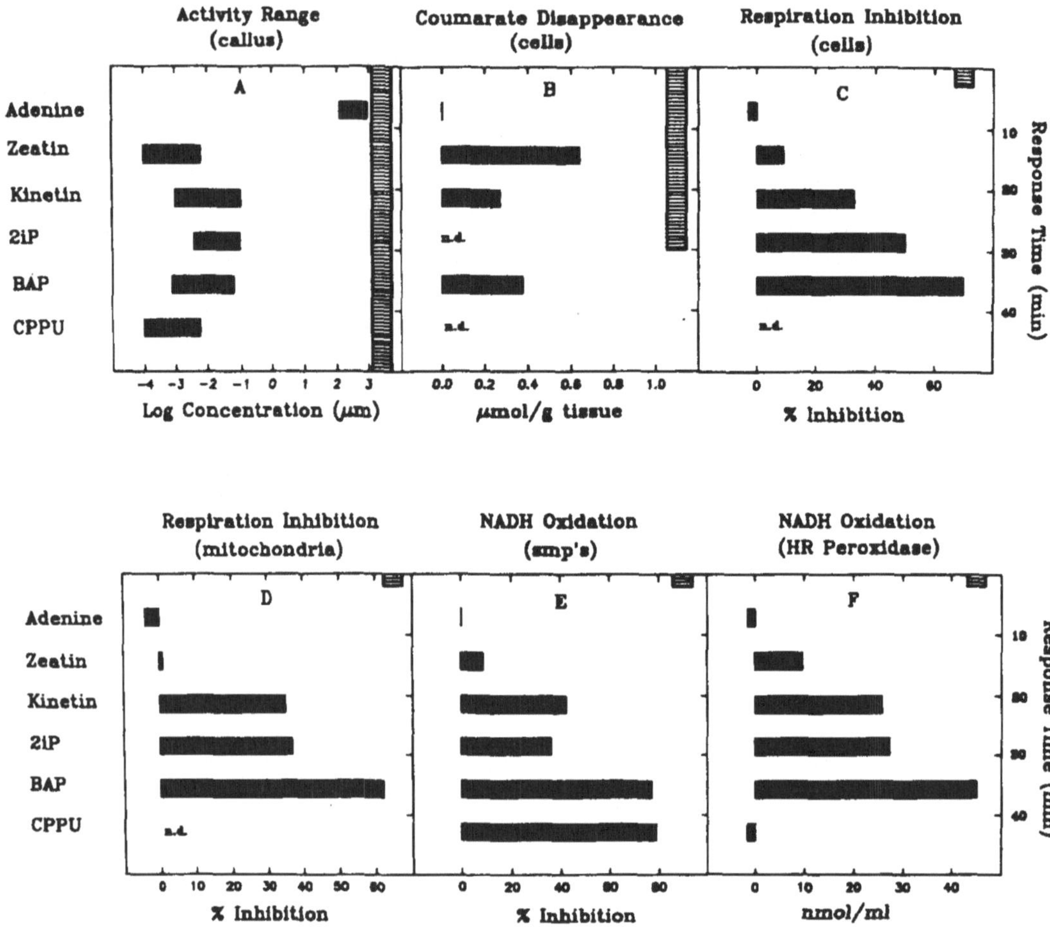

Figure 13-2 Activity of various cytokinins in the tobacco pith callus bioassay,[57] compared with indices of cytokinin control of oxidative processes: coumarate disappearance in soybean cell cultures,[47] respiration inhibition in soybean cell cultures,[21] respiration inhibition in pea stem mitochondria,[22] NADH oxidation by submitochondrial particles from pea stems,[23] and NADH oxidation by horseradish peroxidase.[51]

2. During Cytokinin Responses

The results presented by Musgrave and Siedow[28] and Musgrave et al.[29] support Miller's suggestion that cyanide-resistant respiration may, in fact, be involved in some plant responses to cytokinins.[22] Seven cytokinin bioassays were chosen to represent a range of potential cytokinin responses and were monitored for respiratory changes during the timecourse of response to 40 μ*M* benzyladenine (Table 1). All of the materials tested (with the exception of *Funaria*) showed about a 20% decline in the contribution of cyanide-resistant respiration to total respiration. The decline in cyanide-resistant respiration activity preceded the appearance of the cytokinin response, in each of these bioassays. Those bioassays that permitted early respiration determinations (*Amaranthus caudatus* betacyanin assay and *Glycine max* deoxyisoflavone assay) showed that as early as 6 h after benzyladenine (BA) application, a full effect had been achieved on respiration inhibition. The exception to this observation, the moss *Funaria hygrometrica*, was interesting because when the protonema tissues were assayed directly for respiration inhibition by high (1 m*M*) concentrations of benzyladenine, no inhibition was found. This was especially curious, since the same material could be shown to have a large cyanide-resistant respiratory component, through direct inhibition by SHAM in the presence of cyanide. This finding will be discussed in more detail in a later section. In further support of the idea that inhibition of cyanide-resistant respiration is an early part of a cytokinin response, Vankova et al. found that the apparent capacity for cyanide-resistant respiration in cytokinin-dependent cell suspensions of *Beta vulgaris* L. was diminished by the addition

Table 13-1 **Cyanide-Resistant Respiration Changes During Responses to 40 μ*M* BA in Bioassays Listed**[a]

Species	Tissue	Bioassay	Δ Respiration (% of total)
Amaranthus caudatus	Seedling	Betacyanin	−16
Avena sativa	Coleoptile	Carotenoids	−18
Cucumis sativus	Cotyledon	Expansion	−21
Funaria hygrometrica	Protonema	Bud formation	0
Glycine max	Cell culture	Deoxyisoflavone	−24
Pisum sativum	Seedling	Ethylene	−14
Xanthium pensylvanicum	Leaf disc	Chlorophyll	−22

[a] Respiration column expresses amount of cyanide-resistant respiration suppressed (as a percentage of total respiration rate) during the bioassay response period.[28,29]

of cytokinins to the liquid medium and that an inverse relationship existed between the degree of growth stimulation by the cytokinins and the amount of cyanide-resistant respiration.[30]

3. Anticytokinin Activity

Anticytokinins have commonly been identified as reversible inhibitors of cytokinin-induced growth of callus tissue in the tobacco pith bioassay.[31] In an attempt to use an anticytokinin to test the idea that cytokinins might exert their action through depression of cyanide-resistant respiratory pathway activity, Dizengremel et al. investigated the effect of a cytokinin antagonist, 7-(pentylamino)-3-methylpyrazolo (*4,3-d*) pyrimidine (7-PMP), on the cyanide-resistant pathway in potato tuber mitochondria.[24] They reasoned that if inhibition of alternative respiration *were* of importance in plant responses to cytokinins, then a cytokinin antagonist should not be inhibitory. When they found 7-PMP to be an effective inhibitor of the alternative pathway (Figure 1B), they concluded that the physiologicalmode of action of cytokinins was not through inhibition of the alternative pathway.

However, a growing body of literature suggests that compounds identified as anticytokinins in this way may, in fact, have *cytokinin* activity in other bioassays.[32,33] Dizengremel et al. did not demonstrate that 7-PMP was, indeed, a cytokinin antagonist in the potato tuber system they were using,[24] and results from another lab suggest that 7-PMP probably has *cytokinin* activity in potato tissue.[32] Thus, the question of whether an anticytokinin specific for the tissue being used can inhibit the alternative pathway remains unsettled.

C. CYANIDE-RESISTANT RESPIRATION INHIBITORS AS CYTOKININS

In following the premise that depression of cyanide-resistant respiration may be a mode of cytokinin activity, Musgrave et al. assayed several alternative pathway inhibitors for activity in various cytokinin bioassays.[29] In two antisenescence bioassays, the *Avena* coleoptile assay and the *Xanthium* leaf disc assay, SHAM (1 m*M*), an inhibitor of the alternative pathway, was as good as benzyladenine (40 μ*M*) in preventing loss of chlorophyll. SHAM also had some activity in the *Cucumis* cotyledon expansion bioassay. In the *Amaranthus* bioassay testing betacyanin production, SHAM was evidently toxic, since the response to BA by the seedlings was reduced when SHAM and benzyladenine were applied together. SHAM was also toxic in the soybean deoxyisoflavone assay. Propyl gallate (0.1 m*M*), another inhibitor of cyanide-resistant respiration, was also an effective "cytokinin" in the *Xanthium* antisenescence assay and in the *Cucumis* cotyledon expansion assay.[29] The interchangeability of these compounds in allaying senescence is further evidence that cytokinin effects on senescence are mediated through control of an oxidative system.

Miller noted that if cytokinin action in the mitochondria is important to plant growth and development, we would expect that some of the many lipophilic substances that inhibit oxidation of internally generated NADH by mitochondria would be able to produce "cytokinin" effects.[23] Satler and Thimann observed that certain alcohols delay leaf senescence, as do cytokinins.[34] The growth and chlorophyll levels of radish stimulated by adenine, isopentyl alcohol, and cytokinins released from the bacterium *Azotobacter chroococcum* may be attributable to alcohol and cytokinin effects on mitochondrial function.[35]

D. ENDOGENOUS CYTOKININS AND CYANIDE-RESISTANT RESPIRATION

1. Cell Suspension Cultures

The concern that responses seen upon addition of cytokinins may not relate to those occurring due to endogenous levels of the growth regulator was

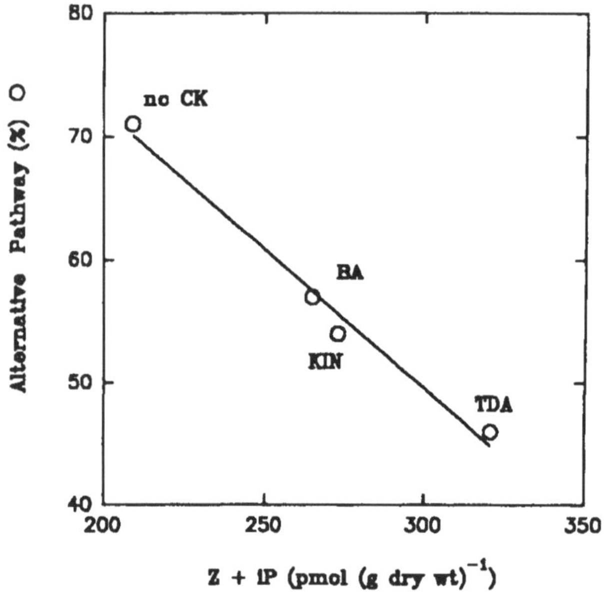

Figure 13-3 Addition of synthetic cytokinins (BA, benzyladenine; KIN, kinetin; TDA, thidiazuron) to beet cell suspension culture stimulated the intracellular content of natural cytokinins zeatin and iP and correlated well (r = .992) with decreases in alternative pathway.[30]

addressed in the study by Vankova et al.,[30] who examined the influence of exogenously applied synthetic cytokinins on levels of endogenous cytokinins and respiration in cytokinin-dependent cell suspensions of *Beta vulgaris* L. The addition of synthetic cytokinins to the cell suspension medium induced an increase in endogenous cytokinins and cell dry weight. When cells were harvested in the linear-growth phase and assayed for respiration pathways, an inverse relationship was found between endogenous cytokinin levels and the percentage of cyanide-resistant respiration (Figure 3).

2. Male Sterile Tissues

A number of reports have linked endogenous levels of cytokinins to male sterility in plants.[36-38] Exogenous application of synthetic cytokinins has also been observed to greatly reduce the production of male flowers in cultured spinach apices[39-41] and hemp plants.[40,41] Musgrave et al. tested various male sterile tissues for cyanide-resistant respiration activity.[42] When compared with male fertile material, six of the seven lines tested exhibited smaller cyanide-resistant pathway capacities. For example, maize line cms-C had 26% less alternative respiration in the isolated mitochondria than those from the normal line. When assayed in intact tissues, cyanide-resistant respiration was not detectable at all in the cms-C line, while in the normal line 24% of the total respiration was cyanide resistant.

These data suggested that high endogenous cytokinin levels associated with male sterility may suppress activity of cyanide-resistant respiration. It was further hypothesized that the greater vigor seen in male sterile genetic lines might be attributable to the suppression of the contribution, by this energetically wasteful pathway, to total respiration.[42]

Connett and Hanson found that cells from cytoplasmic male sterile and fertile *Petunia* lines grown in suspension culture differed in their amounts of cyanide-resistant respiration.[43] After 3 days of culture, cells from the male sterile lines exhibited much lower cyanide-resistant respiration than did cells from the fertile lines. Furthermore, a cell line from a fertile plant carrying a nuclear fertility restorer gene and male sterile cytoplasm had higher cyanide-resistant respiration than did the male sterile lines. The role cytokinins may play in the relationship between cyanide-resistant respiration and male sterility in this system is an interesting topic for further study.

III. OTHER OXIDATIVE PROCESSES

A. CYTOKININS PREVENT OXIDATION AND DELAY SENESCENCE

1. Antioxidant Activity

The antisenescence activity of cytokinins is not limited to the effect of suppressing respiratory climacterics. In a study on the susceptibility of membranes to ozone damage, Pauls and Thompson compared the effects of cytokinins and various antioxidants.[44] They found that following ozone treatment, microsomal membranes from bean cotyledons suffered phase separation of lipids and that these effects could be prevented by adding cytokinins (10^{-4} *M*) or antioxidants such as propyl gallate. They suggested that cytokinins act as radical scavengers to protect plant tissue from ozone damage and suggested that the antisenescence properties of cytokinins are related to their antioxidant activity.

2. Lipoxygenase Inhibition

An additional method by which cytokinins suppress oxidative processes in the cell is through the control of lipoxygenase. Grossman and Leshem found that when cytokinins were applied to intact pea plants, endogenous levels of lipoxygenase were lowered.[45] The triggering of senescence, which they induced by leaf detachment, was accompanied by an increase in lipoxygenase activity. Cytokinin application decelerated the lipoxygenase increase. Swamy and Suguna found that 15 μ*M* benzyladenine reduced lipoxygenase activity in cowpea (*Vigna unguiculata* L.) leaf discs during progressive senescence in the dark.[46] Both groups concluded that the repression of lipoxygenase by cytokinin contributes to the overall antisenescence action of the growth regulator.

B. ACTION ON PEROXIDASE

Miller examined the rate of disappearance of coumarate from soybean cell suspensions, in an attempt to find quick metabolic effects of cytokinins that are not dependent upon the overall processes of growth.[47] Soybean cell suspensions produce the deoxyisoflavone daidzein about 8 h after being supplemented with cytokinins,[48,49] and in investigating the source of this stimulation, Miller tested the influence of added precursors (such as *p*-coumarate) of daidzein on its formation.[47] He observed that coumarate disappeared from the medium and that the simultaneous addition of cytokinins promoted this disappearance in concentrations ranging from 40 to 80 μ*M*, after the suspended cells had been aerobically exposed to the cytokinin for about 30 min. The compounds that elicited a promotive effect on coumarate disappearance corresponded well to those that promote cell division at the concentrations employed (Figure 2A and B). This cytokinin effect was very dependent on the concentration and timing of addition, however, and *inhibition* of coumarate disappearance was obtained in the 0.0005- to 1-μ*M* concentration range if the cells were incubated with cytokinin for 1 h prior to the addition of coumarate.

Although the disappearance of coumarate from the medium may result from several metabolic activities, one route is from peroxidation.[50] Miller suggested that cytokinins might modify reactions that produce hydrogen peroxide; control other peroxidations competing for the available peroxide and for catalase, which destroys it; or affect peroxidases directly.

In a subsequent paper,[51] a possible regulatory role for cytokinins in NADH oxidation by peroxide, involving a copper interaction, was described. Cytokinins promoted NADH oxidation by horseradish peroxidase in the presence of cupric ions after just 2 min, with cytokinin concentrations as low as 5 μ*M*. Most of the compounds tested had similar relative activities in cytokinin bioassays and the peroxidase test, with the exception of the substituted ureas (Figure 2A and F).

C. CONTROL OF DEHYDRODICONIFERYL ALCOHOL GLUCOSIDES

Recently, dehydrodiconiferyl alcohol glucosides (DCGs) were identified as a component of a cytokinin-mediated regulatory circuit controlling cell division in tobacco. These phenolics accumulate in cytokinin-autonomous and cytokinin-stimulated tobacco tissues and substitute for cytokinins in cell division assays.[52] Zeatin riboside (1 μ*M*) added to tobacco pith cultures stimulated the production of DCGs, over a 7-day period, to a far greater extent than did the base medium without cytokinin, and there was a good correlation between DCG concentration and cell divisions. Interestingly, although synthetic DCGs have cell division-promoting activity, they do not have biologicalactivity in other systems (*Amaranthus* betacyanin assay; shoot initiation on *Nicotiana tabacum* leaf explants) in which adenine-derived cytokinins are active.

The production of DCGs requires dimerization of coniferyl alcohol by a soluble intracellular peroxidase.[53] Peroxidase activity is required to convert coniferyl alcohol to the aglycone. The DCG pool is large and long lived; its size is determined not by breakdown, but by precursor supply. Thus, the cytokinin control of peroxidase activity reported by Miller may connect cytokinin effects on oxidative processes to the production of the cell division factor DCG (Figure 4).

Some further evidence has recently appeared in support of this idea. Phenotypic alterations that occurred at specific developmental times were observed in transgenic tobacco plants overexpressing the anionic peroxidase.[54] Also, in cultured cells of *Zinnia*, an isoperoxidase appeared more than a day before differentiating tracheary elements were observed or 4-coumarate:Co-A ligase activity was detected. In this system the expression of the peroxidase was inhibited by an anticytokinin treatment that also delayed or inhibited the differentiation.[55]

IV. CYTOKININS AND OXIDATIVE PROCESSES

A. RANGE OF OXIDATIVE PROCESSES UNDER CYTOKININ CONTROL

From the foregoing discussion it should be apparent that cytokinins exert control on a broad

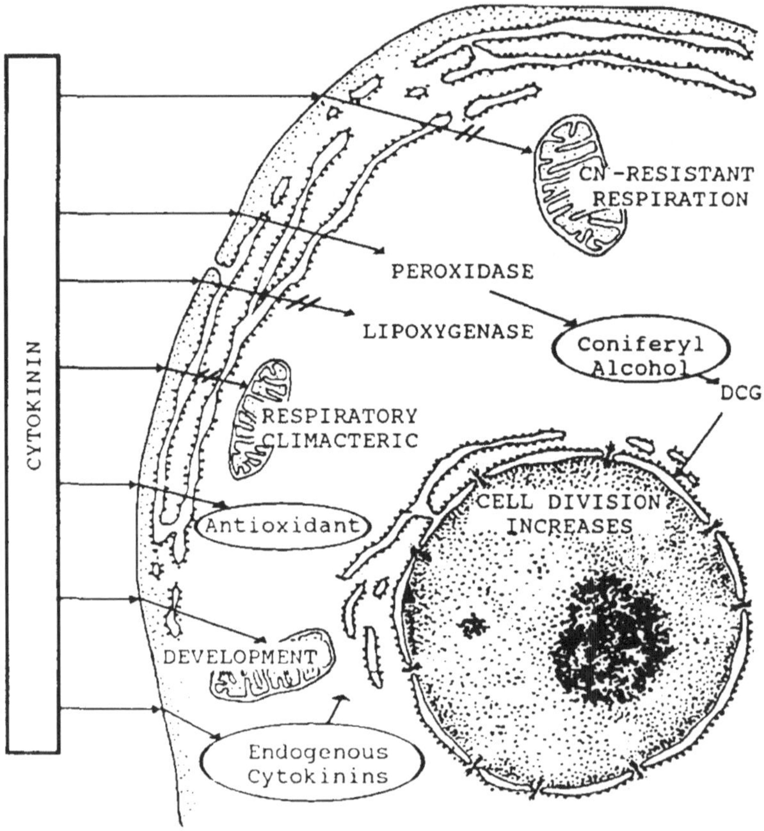

Figure 13-4 Pathways through which cytokinins may exert control on oxidative processes in the cell and give rise to "cytokinin" responses (Table 2).

range of oxidative processes. The many reports of cytokinin effects on oxidative processes can be divided into five main effects: senescence inhibition, cell growth, secondary-compound metabolism, respiration inhibition, and stimulation of respiration development (Table 2). Because of the many different lines of evidence supporting a role for cytokinins in the control of oxidative processes during their activity as cytokinins, it is possible to propose a general scheme for cytokinin activity, involving these processes (Figure 4).

B. SPECIFICITY QUESTIONS

While the correspondence between the activity of cytokinins as cell division factors and their activity as controllers of oxidative processes is good (Figures 2A–F), conceptual problems persist. Does the inactivity of zeatin in controlling oxidative processes connote more than possible structural exclusion from the active sites?[21] Why are the substituted urea cytokinins active in callus bioassays and most oxidative tests conducted, but not in controlling NADH oxidation by horseradish (HR) peroxidase?[51] How can the moss *Funaria* have a cyanide-resistant respiratory pathway that is insensitive to inhibition by benzyladenine if this respiratory inhibition is important for cytokinin responses?[29] Why do nitroguanidine compounds that perform equally well in cytokinin bioassays show an "all or nothing" response as inhibitors of cyanide-resistant respiration?[27]

After finding that many synthetic cytokinins, including both the substituted-purine and substituted-urea types, could modify mitochondrial function, Miller was surprised to find that rat liver mitochondria were also inhibited by cytokinins.[23] No definite role of cytokinins is known in rat growth and development, suggesting by analogy that there is no constructive influence of this inhibition by cytokinins. Is it just a coincidence that these lipid-soluble compounds inhibit electron transport? Miller suggested that the ineffectiveness of zeatin in inhibiting oxidative processes may have resulted from a modification of cytokinin by the plant in order to keep its natural growth regulator from activating sensitive systems while performing the "true" role(s) in growth and development.[23]

However, the evidence for a role in control of oxidative processes by cytokinins is too strong to be discounted because of these problems. Perhaps a more productive view of the situation is to consider cytokinin responses as being overlapping suites of effects that may be engaged by different taxa at different developmental junctures.

Table 13-2 **Processes Reported to be Influenced Through Cytokinin Control of Oxidative Events (Figure 4)**

Process	Species	References
Senescence inhibition		
Chlorophyll loss	*Xanthium pensylvanicum*	28, 29
Carotenoid loss	*Avena sativa*	11, 28, 29
Lipoxygenase control	*Pisum sativum*	45
	Vigna unguiculata	46
Antioxidant properties	*Phaseolus vulgaris*	44
Cell growth		
Division	*Nicotiana tabacum*	52, 53
	Beta vulgaris	30
Expansion	*Cucumis sativus*	28, 29
Secondary compound metabolism		
p-Coumarate	*Glycine max*	47
Betacyanin	*Amaranthus caudatus*	28, 29
Deoxyisoflavone	*Glycine max*	29, 49
Peroxidase control	*Zinnia elegans*	55
	Armoracia rusticana	51
Respiration inhibition		
Florets	*Brassica oleracea*	3, 4, 9
Stems	*Pisum sativum*	5
	Solanum tuberosum	10
Roots	*Beta vulgaris*	7
	Glycine max	42
	Helenium amarum	42
	Plantago lanceolata	42
	Zea mays	42
Coleoptiles	*Avena sativa*	11, 28, 29
Seedlings	*Amaranthus caudatus*	28, 29
	Pisum sativum	28
Cell cultures	*Daucus carota*	28
	Glycine max	16, 21, 29
	Beta vulgaris	30
Leaf discs	*Xanthium pensylvanicum*	28, 29
	Rumex crispus	8
	Nicotiana tabacum	6
	Cucumis sativus	28, 29
Cotyledons	*Zea mays*	22, 42
Mitochondria	*Glycine max*	21, 22
	Solanum tuberosum	24
	Pisum sativum	22, 23
	Vigna radiata	23
	Phaseolus aureus	22
	Triticum aestivum	22
	Phaseolus vulgaris	22
Respiration development		
Cotyledons	*Helianthus annuus*	13
	Phaseolus mungo	15
	Citrullus vulgaris	14
Cell culture	*Nicotiana tabacum*	56

This idea is supported by the observations of Teutonico et al.[52] that (1) cytokinins stimulate the production of DCGs, (2) these, in turn, control cell division, and (3) DCGs alone cannot elicit the same amount of cell division as the application of cytokinin in an amount needed to produce that quantity of DCGs. It is tantalizing to suggest that some other oxidative effect of cytokinin is needed, in combination with the cell division factor, in order to realize the full magnitude of response (Figure 4).

C. CONCLUSIONS

The data suggest that there is a suite of oxidative processes controlled by cytokinin levels (Figure 4). For some types of cytokinin responses involving antisenescence, the effects may be directly on respiration or senescence-related oxidative processes. In these tissues, inhibitors of cyanide-resistant respiration, lipoxygenase, or antioxidants, in general, should be considered as cytokinins, since the site and mode of activity is probably the same. For other assays, such as those involving the production of secondary compounds, cytokinins probably work through their direct action on peroxidases involved in the biosynthetic pathway of these compounds. Thus, betacyanin production, deoxy-isoflavone synthesis, and metabolism of *p*-coumarate may all be regulated by cytokinin control of enzymes contributing to their metabolism.

Cell division stimulation in tobacco by cytokinins, through production of the compound DCG, may also be controlled by a cytokinin effect on an oxidative system. Whether these DCGs will prove to be ubiquitous as components of cytokinin-mediated controls of cell division remains to be elucidated. This is an exciting area for future research.

Although the evidence for cytokinin activity through oxidative processes seems compelling enough, there is another reason why study of these effects has merit. As Miller pointed out,[23] synthetic cytokinins are often applied in high enough concentrations to make their effects on oxidative processes important. A broader understanding of cytokinin effects on oxidative processes will serve to strengthen our search for future knowledge about the mechanisms of cytokinin activities.

ACKNOWLEDGMENT

Approved for publication by the director of the Louisiana Agricultural Experiment Station as paper #92-38-6358.

REFERENCES

1. **Richmond, A. E. and Lang, A.,** Effect of kinetin on protein content and survival of detached *Xanthium* leaves, *Science*, 125, 650, 1957.
2. **Osborne, D. J. and McCalla, D. R.,** Rapid bioassay for kinetin and kinins using senescing leaf tissue, *Plant Physiol.*, 36, 219, 1961.
3. **Dedolph, R. R., Wittwer, S. H., and Tuli, V.,** Senescence inhibition and respiration, *Science*, 134, 1075, 1961.
4. **Dedolph, R. R., Wittwer, S. H., Tuli, V., and Gilbart, D.,** Effect of N^6-benzyl-aminopurine on respiration and storage behavior of broccoli (*Brassica oleracea* var. *italica*), *Plant Physiol.*, 37, 509, 1962.
5. **Katumi, M.,** Physiological effects of kinetin. Effect on the elongation, water uptake and oxygen uptake of etiolated sections, *Physiol. Plant.*, 16, 66, 1963.
6. **Sigiura, M.,** Inhibitory effect of kinetin on respiration of tobacco leaf discs, *Bot. Mag. (Tokyo)*, 76, 359, 1963.
7. **Palmer, J. M.,** The influence of growth regulating substances on the development of enhanced metabolic rates in thin slices of beetroot storage tissue, *Plant Physiol.*, 41, 1173, 1966.
8. **Goldthwaite, J.,** Energy metabolism of *Rumex* leaf tissue in the presence of senescence-regulating hormones and sucrose, *Plant Physiol.*, 54, 399, 1974.
9. **Halevy, A. H., Dilley, D. R., and Wittwer, S. H.,** Senescence inhibition and respiration induced by growth retardants and N^6-benzyladenine, *Plant Physiol.*, 41, 1085, 1966.
10. **Hourmant, A., Pradet, A., and Penot, M.,** Action de la benzylaminopurine sur l'absorption du phosphate et le metabolisme des composes phosphoryles des disques de tubercule de Pomme de terre en survie, *Physiol. Végétale*, 17, 484, 1979.
11. **Tetley, R. M. and Thimann, K. V.,** The metabolism of oat leaves during senescence. I. Respiration, carbohydrate metabolism, and the action of cytokinins, *Plant Physiol.*, 62, 232, 1974.
12. **Lambers, H.,** The physiological significance of cyanide-resistant respiration in higher plants, *Plant Cell Environ.*, 3, 293, 1980.
13. **Servettaz, O., Cortesi, F., and Longo, C. P.,** Effect of benzyladenine on some enzymes of mitochondria and microbodies in excised sunflower cotyledons, *Plant Physiol.*, 58, 569, 1976.

14. **Longo, G. P., Pedretti, M., Rossi, G., and Longo, C. P.,** Benzyladenine stimulates the development of mitochondria in watermelon cotyledons, *Plant Sci. Lett.*, 14, 213, 1979.
15. **Morohashi, Y.,** Effect of benzyladenine on the development of mitochondrial activities in imbibed black gram cotyledons, *J. Plant Physiol.*, 116, 235, 1984.
16. **Moore, T. S., Jr. and Miller., C. O.,** Effects of cytokinins on the respiration of soybean callus tissue, *Plant Physiol.*, 50, 594, 1972.
17. **Bahr, J. T. and Bonner, W. D., Jr.,** Cyanide-insensitive respiration. I. The steady state of skunk cabbage spadix and bean hypocotyl mitochondria, *J. Biol. Chem.*, 248, 3441, 1973.
18. **Palmer, J. M.,** The organization and regulation of electron transport in plant mitochondria, *Ann. Rev. Plant Physiol.*, 27, 113, 1976.
19. **Lambers, H., Day, D. A., and Azcon-Bieto, J.,** Cyanide-resistant respiration in roots and leaves. Measurements with intact tissues and isolated mitochondria, *Physiol. Plant.*, 58, 148, 1983.
20. **Day, D. A. and Lambers, H.,** The regulation of glycolysis and electron transport in roots, *Physiol. Plant.*, 58, 155, 1983.
21. **Miller, C. O.,** Cytokinin inhibition of respiration by cells and mitochondria of soybean, *Glycine max* (L.) Merrill, *Planta*, 146, 503, 1979.
22. **Miller, C. O.,** Cytokinin inhibition of respiration by cells and mitochondria from six plant species, *Proc. Natl. Acad. Sci. U.S.A.*, 77(8), 4731, 1980.
23. **Miller, C. O.,** Cytokinin modification of mitochondrial function, *Plant Physiol.*, 69, 1274, 1982.
24. **Dizengremel, P., Chauveau, M., and Roussaux, J.,** Inhibition by adenine derivatives of the cyanide-insensitive electron transport pathway of plant mitochondria, *Plant Physiol.*, 70, 585, 1982.
25. **Keim, P., Erion, J., and Fox, J. E.,** The current status of cytokinin-binding moieties, in *Metabolism and Molecular Activities of Cytokinins*, Guern, J. and Péaud-Lenoël, C., Eds., Springer-Verlag, New York, 1981, 179.
26. **Rodaway, S. and Lutz, A. W.,** Nitroguanidines: a new class of synthetic cytokinins, *Plant Physiol.*, 77, (Suppl.) 21, 1985.
27. **Musgrave, M. E.,** Alternative respiration and plant responses to cytokinins, in *Physiology and Biochemistry of Cytokinins in Plants*, Kamínek, M., Mok, D. W. S., and Zazímalová, E., Eds., SPB Academic Publishing, The Hague, 1992, 289.
28. **Musgrave, M. E. and Siedow, J. N.,** A relationship between cyanide-resistant respiration and plant responses to cytokinins, *Physiol. Plant.*, 64, 161, 1985.
29. **Musgrave, M. E., Miller, C. O., and Siedow, J. N.,** Do some plant responses to cytokinins involve the cyanide-resistant respiratory pathway? *Planta*, 172, 330, 1987.
30. **Vankova, R., Hsiao, K.-C., Bornman, C. H., and Gaudinova, A.,** Effects of synthetic cytokinins on levels of endogenous cytokinins and respiration patterns of *Beta vulgaris* cells in suspension, *J. Plant Growth Regul.*, 10, 197, 1991.
31. **Hecht, S. M.,** Probing the cytokinin receptor site(s), in *Plant Growth Substances 1979*, Skoog, F., Ed., Springer-Verlag, New York, 1979, 144.
32. **Iwamura, H., Masuda, N., Koshimizu, K., and Matsubara, S.,** Cytokinin-agonistic and antagonistic activities of 4-substituted-2-methylpyrrolo(2,3-d)pyrimidines, 7-deaza analogs of cytokinin-active adenine derivatives, *Phytochemistry*, 18, 217, 1979.
33. **Skoog, F. and Abdul Ghani, A. K. B.,** Relative activities of cytokinins and antagonists in releasing lateral buds of *Pisum* from apical dominance compared with their relative activities in the regulation and growth of tobacco callus, in *Metabolism and Molecular Activities of Cytokinins*, Guern, J. and Péaud-Lenoël, C., Eds., Springer-Verlag, New York, 1981, 140.
34. **Satler, S. O. and Thimann, K. V.,** The influence of aliphatic alcohols on leaf senescence, *Plant Physiol.*, 66, 395, 1980.
35. **Nieto, K. F. and Frankenberger, W. T., Jr.,** Influence of adenine, isopentyl alcohol and *Azotobacter chroococcum* on the growth of *Raphanus sativus*, *Plant Soil*, 127, 147, 1990.
36. **Ahokas, H.,** Cytoplasmic male sterility in barley: evidence for the involvement of cytokinins in fertility restoration, *Proc. Natl. Acad. Sci. U.S.A.*., 79, 7605, 1982.
37. **Kahlem, G., Champault, A., Louis, J.P., Bazin, M., Chabin, A., Delaigue, M., Dauphin, B., Durand, R., and Durand, B.,** Determination génétique et regulation hormonale de la differenciation sexuelle chez *Mercurialis annua* L., *Physiol. Végétale*, 13, 763, 1975.
38. **Louis, J. P. and Durand, B.,** Studies with the dioecious angiosperm *Mercurialis annua* L. (2n = 16)., *Mol. Gen. Genet.*, 165, 309, 1978.

39. **Culafic, L. and Neskovic, M.,** Effect of growth substances on flowering and sex expression in isolated apical buds of *Spinacea oleracea*, *Physiol. Plant.*, 48, 588, 1980.
40. **Chailakhyan, M. K. and Khryanin, V. N.,** The influence of growth regulators absorbed by the root on sex expression in hemp plants, *Planta*, 138, 181, 1978.
41. **Chailakhyan, M. K. and Khryanin, V. N.,** The role of roots in sex expression in hemp plants, *Planta*, 138, 185, 1978.
42. **Musgrave, M. E., Antonovics, J., and Siedow, J. N.,** Is male sterility in plants related to lack of cyanide-resistant respiration in tissues?, *Plant Sci.*, 44, 7, 1986.
43. **Connett, M. B. and Hanson, M. R.,** Differential mitochondrial transport through the cyanide-sensitive and cyanide-insensitive pathways in isonuclear lines of cytoplasmic male sterile, male fertile, and restored *Petunia*, *Plant Physiol.*, 93, 1634, 1990.
44. **Pauls, K. P. and Thompson, J. E.,** Effects of cytokinins and antioxidants on the susceptibility of membranes to ozone damage, *Plant Cell Physiol.*, 23, 821, 1982.
45. **Grossman, S. and Leshem, Y. Y.,** Lowering of endogenous lipoxygenase activity in *Pisum sativum* foliage by cytokinin as related to senescence, *Physiol. Plant.*, 43, 359, 1978.
46. **Swamy, P. M. and Suguna, P.,** Influence of calcium chloride and benzyladenine on lipoxygenase of *Vigna unguiculata* leaf discs during senescence, *Physiol. Plant.*, 84, 467, 1992.
47. **Miller, C. O.,** Cytokinin modification of metabolism of *p*-coumaric acid by a cell suspension of soybean (*Glycine max* (L.) Merrill), *Planta*, 140, 193, 1978.
48. **Miller, C. O.,** Control of deoxyisoflavone synthesis in soybean tissue, *Planta*, 87, 26, 1969.
49. **Miller, C. O.,** Modification of the cytokinin promotion of deoxyisoflavone synthesis in soybean tissue, *Plant Physiol.*, 49, 310, 1972.
50. **Stafford, H. A.,** The metabolism of aromatic compounds, *Ann. Rev. Plant Physiol.*, 25, 459, 1974.
51. **Miller, C. O.,** Possible regulatory roles of cytokinins. NADH oxidation by peroxidase and a copper interaction, *Plant Physiol.*, 79, 908, 1985.
52. **Teutonico, R. A., Dudley, M. W., Orr, J. D., Lynn, D. G., and Binns, A. N.,** Activity and accumulation of cell division-promoting phenolics in tobacco tissue cultures, *Plant Physiol.*, 97, 288, 1991.
53. **Orr, J. D. and Lynn, D. G.,** Biosynthesis of dehydrodiconiferyl alcohol glucosides: implications for the control of tobacco cell growth, *Plant Physiol.*, 98, 343, 1992.
54. **Lagrimini, L. M., Bradford, S., and Rothstein, S.,** Peroxidase-induced wilting in transgenic tobacco plants, *Plant Cell*, 2, 7, 1990.
55. **Church, D. L. and Galston, A. W.,** 4-Coumarate: coenzyme A ligase and isoperoxidase expression in *Zinnia* mesophyll cells induced to differentiate into tracheary elements, *Plant Physiol.*, 88, 679, 1988.
56. **Hoefnagel, M. H. N., Van der Zall, B. J., Van Iren, F., and Libbenga, K. R.,** Increase in the capacities of the cytochrome and alternative respiratory pathways in tobacco cells caused by 2,4-D and kinetin, *J. Plant Physiol.*, 140, 264, 1992.
57. **Matsubara, S.,** Structure–activity relationships of cytokinins, *Phytochemistry*, 19, 2239, 1980.

Chapter 14

Plastid Genes and Chloroplast Biogenesis

Ralf Reski

TABLE OF CONTENTS

I. PLASTIDS ARE A HETEROGENOUS ORGANELLE CLASS

Plastids are the DNA-containing organelles with the widest range of different forms and functions. Interconversion between different plastid types implies marked changes in their architecture and physiology. Furthermore, these changes go hand in hand with differentiation processes of cells and organs during plant development (reviewed by Herrmann et al.).[1] Energy, organic substances, and oxygen are provided by photosynthesis, and most of the photosynthetic products are formed in chloroplasts. They also catalyze key reactions in the reductive nitrogen and sulfur pathway, in lipid synthesis, and they synthesize important secondary metabolites. Specialized modifications are leukoplasts, amyloplasts, elaioplasts, and chromoplasts, which are characteristic for different tissues, exhibit marked structural differences, and lack chlorophyll. These organelle forms are involved in various fundamental biological processes, such as carbohydrate or lipid storage and plant propagation. During leaf senescence in angiosperms, chloroplast structures are degraded, temporally forming the last plastid type, gerontoplasts.[2] In vascular plants chloroplast biogenesis from less-differentiated organelle forms occurs: in vegetation tips the differentiation process starts from colorless proplastids. Angiosperms exhibit an additional pathway for chloroplast biogenesis: grown in darkness, proplastids differentiate to pale-yellow etioplasts with characteristic paracrystalline internal structures. Upon illumination, etioplasts develop into chloroplasts. In addition to light, in dicotyledonous angiosperms the biogenesis of chloroplasts is also regulated by intrinsic developmental signals that control leaf differentiation.[3] However, in monocots the processes of primary leaf development and chloroplast development are uncoupled.[4] The common view is that in vascular plants only mesophyll cells house mature chloroplasts, indicating that cell-specific signals are also important for chloroplast biogenesis.[5] Nevertheless, recent findings demonstrate active, but not fully differentiated, chloroplasts even in pith cells.[6] Finally, the developmental stage of the chloroplast itself appears to regulate the expression of nuclear genes coding for plastid proteins.[3]

There are several established experimental systems to study chloroplast biogenesis and correlated cell and organ morphogenesis at the molecular level.

The light-controlled differentiation is studied during greening of etiolated seedlings. Morphogenetic changes occur within a few hours[7] and are thus amenable to molecular analysis. Much work on chloroplast biogenesis in higher plants, particularly on the photocontrol of gene expression, has therefore been carried out with this etioplast/chloroplast transition.[1] However, this system is restricted to angiosperms, and plastid gene regulation during etioplast/chloroplast transition appears to be different from the one during regular proplastid/chloroplast transition.[8]

The *Arabidopsis thaliana det1* mutant develops as a light-grown plant even in darkness. The recessive nuclear mutation affects a variety of light-regulated traits, including leaf development, plastid morphology, and gene expression. Most probably, the wild-type *det1* gene product is a negative regulator in a signal transduction pathway that couples leaf and chloroplast development to light perception. This implies a constitutive expression of light-regulated genes in cell types where these genes are normally silent. In line with this model is the observation that cell cultures of this particular mutant become green even in the absence of cytokinins, conditions where normally no chloroplast biogenesis takes place.[9]

0-8493-6252-0/94/$0.00+$.50

Leaves of monocots are widely used to study plastid differentiation processes. Cell division occurs only in a basal meristem, resulting in a gradient of cells, with increasing age, from the base to the tip.[10] In a diurnal light regime, a continuous transition of proplastids to chloroplasts occurs in parallel to the basipetal increase in cell age.

In cereals the formation of plastid ribosomes is prevented by growth temperatures above 32°C.[11] The ribosome-deficient plastids resemble proplastids in size and structure. They do not differentiate into chloroplasts, although carotinoids are synthesized, and enzymes for chlorophyll biosynthesis are functioning, within these plastids.[12-14] In contrast to treating plants with inhibitors of protein synthesis, the heat-induced deficiency of plastid ribosomes provides a system for studying interactions between plastids and the cytosol, with minimal risk of toxic side effects or damage by photodestruction (reviewed by Feierabend and Berberich[15]). Ribosome-deficient plastids can also be obtained from the barley nuclear mutant *albostrians*, with the advantage of ruling out side effects of elevated temperatures.[16]

II. PLASTIDS ARE SEMIAUTONOMOUS ORGANELLES

Plant cells are characterized by three different genomes that have their own machinery for gene expression: the prokaryotic ancestors of plastids and mitochondria lost their autonomy by losing most of their genes during the establishment of endosymbiosis.[17] Some of these prokaryotic genes were transferred into the eukaryotic background of the nucleus. The problems arising for a coordinated expression of these genes, and the subsequent transport of their products into the ancestral organelle, are obvious and require the coevolution of three different genomes in one single cell.[1,17] The eukaryon achieved its capability of photosynthesis by a process that started with the engulfment of autonomous prokaryotic cells, continued with the establishment of endosymbiosis between eukaryon and chloroplast, and culminated in the domestication of virile remnants (chloroplasts) to nongreen derivatives (proplastids, amyloplasts, gerontoplasts, chromoplasts, etc.) with specialized functions in higher plants. Thus, the phenomenon of plastid differentiation demonstrates the domestication of the prokaryon by the eukaryon and, finally, the evolutionary aspect of plant development.

The activity of genes involved in this differentiation of plastids has to be under temporal and spatial control. Furthermore, these genes have to respond to multiple extrinsic signals notifying changes in the environment of the plant. Therefore, the role and importance of a specific signal, whether extrinsic or intrinsic, in the complex regulatory network around chloroplast biogenesis and accompanying cell and organ morphogenesis is likely to differ between the major plant groups.

Based on results from the *aurea* mutant[18,19] and the *ghost* mutant[20] of tomato, the *albostrians* mutant[21] and the *albina f17* mutant[22] from barley, and carotinoid-deficient mutants in maize,[23] and from norflurazon treated plants,[24,25] the *consensus* is now that the expression of some light-inducible nuclear genes is modulated by a factor generated by differentiated chloroplasts. This plastidic factor seems to be responsible for the expression of a certain set of nuclear genes: genes coding for proteins that are plastid constituents or that are, at least indirectly, involved in chloroplast biogenesis.[26-31] These observations reveal a common *trans*-acting factor modulating the expression of those nuclear genes that originally were coded by the prokaryotic DNA of the ancestors of chloroplasts. Nevertheless, the nature of the postulated plastidic factor is unknown.[30,32]

A common view about gene regulation during light-induced chloroplast biogenesis is that it is transcriptional for nuclear genes, but translational or posttranslational for plastid genes.[33] Nevertheless, the situation seems to be more complex, as several plastid differentiation processes have been shown to be accompanied not only by changes in general transcriptional activity,[34] but also by changes in the relative transcription rates of individual genes, indicating the existence of gene-specific mechanisms for transcriptional regulation.[1] Examples are the development of proplastids to chloroplasts in *Hordeum* and *Sorghum*,[8,35] light-dependent transformation of mature barley etioplasts to chloroplasts,[36] amyloplast formation in cultured tobacco cells,[37] and chromoplast formation in tomatoes.[38]

Several mechanisms for changes in the transcriptional activity of plastid genes during plastid differentiation have been suggested; changes in the ptDNA copy number per cell can only confer to the nonspecific regulation of plastid genes and do not seem to play a vital role in chloroplast biogenesis.[39] Developmentally related changes in DNA methylation were observed in some studies,[38] but not in others.[8,40] Likewise, variations in the conformation of the plastid DNA template, and differences in the three-dimensional structure of plastid nucleoids,[41] have not been unequivocally proven yet. Differences in promoter strength and competition, changes in RNA polymerase to template ratios,[35] and different half-lives of mRNAs may influence transcription rates and cause accumulation rates that differ relative to each other as well.[42,43] However, such models are based on the hypothesis that all DNA circles in a given plastid are, to the same extent, physiologically active and that the *in vivo* situation is reflected by our *in vitro* data. In addition to these theories are two just emerging models for differential gene regulation.

The first model is that changes in the molecular composition of the transcriptional apparatus itself can cause differences in gene expression and the subsequent plastid differentiation processes. There are two possibilities: either these changes in gene expression can be caused by changes in the populations of several putative RNA-polymerases, or a varying specificity can be conferred, as in prokaryotes, by sigma factors associated with a core RNA-polymerase.

Plastid DNA encodes the *rpo* genes, with all polypeptides resembling the bacterial RNA-polymerase, and these genes are transcribed and translated.[44] Additionally, first evidence is available about sigma-like factors in the plastids of higher plants.[45-47] Nevertheless, plastid extracts contain two fractions with RNA-polymerase activity, in all plants examined so far, whether alga or weed. These fractions differ from each other in biochemical terms (soluble vs. bound, molecular weight, composition) and were reported to differ in their preference to transcribe certain plastid genes, although these reports have given conflicting results.[33,44,48-53] Whether the two fractions indeed contain different enzymes or just different states of one common core enzyme with various auxiliary proteins is still unknown.[54] However, in the last decade there have been hints for a nuclear-encoded plastidic RNA-polymerase.[55-57] Utilizing barley with ribosome-deficient plastids, evidence was solidified for a plastidic RNA-polymerase translated on extraplastidic ribosomes.[15] This nuclear-encoded enzyme seems to be constitutively active in proplastids, chloroplasts, and senescing plastids, whereas the soluble fraction, possibly encoded by the prokaryotic *rpo* genes, develops activity solely in differentiated chloroplasts,[58,59] indicating a differential function of two different RNA-polymerases, at least in barley. Nevertheless, transcript levels of plastid genes coding proteins of the photosynthetic apparatus only accumulate during chloroplast biogenesis, pointing to the importance of posttranscriptional regulation.

The second model is that RNA-editing may be a mechanism involved in the differential regulation of gene expession. Editing was observed with kinetoplast RNA, mitochondrial, and nuclear genes.[60-62] All groups of plastid genes can be affected as well.[63-65] The main function of RNA editing is the creation of an mRNA that can be translated into a biologically active protein either by creating an initiator codon or by restoring codons for conserved amino acids. However, at the moment there are only first hints that differential mRNA editing really regulates gene expression. If so, quantitative Northern blots as well as *in vitro* transcription assays would not necessarily reflect the respective *in vivo* situation. According to our present knowledge, all editing processes in higher plants seem to bring down the peptide sequences to a *consensus* that is represented by the liverwort *Marchantia polymorpha*. This bryophyte is the first plant in which the total nucleotide sequence of ptDNA, as well as mtDNA, has been determined.[66,67] Our analyses of mitochondrial and plastid genes of the moss *Phys-comitrella patens* reveal that moss and liverwort share significant gene homology different from higher plants.[68-72] Thus, RNA-editing of plastid genes might be a regulatory mechanism evolved during the "domestication of chloroplasts."

Analyzing plastid differentiation we find a complex informational network between at least two different genomes in a single cell across several membranes.[1,73] Therefore, we have to dismiss the sedative simplification that each plant species reacts the same way as any other, neglecting the ongoing coevolution of these genomes.[17] For example, the plastid genes of monocots and dicots diverged 100 to 200 million years ago.[74] Why should they use the same dialect in their communication with nuclear genes?

III. PLASTIDS ARE TARGETS FOR CYTOKININ ACTION

Since the pioneering work of Mothes,[75] plastids have seemingly been a primary target of cytokinin action; numerous papers reported on the hormone-stimulated synthesis of chloroplast components during etioplast/chloroplast transition in greening seedlings, cotyledons, or tissue cultures. In detached leaves and in cell culture, both depleted in endogenous cytokinin, exogenously applied hormone stimulates this transition in light and in darkness. In contrast, cytokinins do not influence structural or physiological plastid parameters in intact plants. Likewise, cytokinins retard senescence in detached leaves, including maintenance, or even promote the synthesis of plastid components, but have no effect on senescence retardation of albino barley leaves, or in N-deficient, chlorophyll-free detached leaves of tobacco (reviewed by Parthier[76]).

Cytokinin-induced, as well as light-induced, chloroplast biogenesis is at the molecular level characterized by the accumulation of transcripts from plastid genes. A number of polyadenylated transcripts exhibit enhanced steady-state levels in plants treated with naturally occurring cytokinins or with their nonphysiological substitutes kinetin or benzyladenine, as well. Interestingly, the hormone seems to act only on a certain set of nuclear genes: genes coding for proteins that are plastid constituents or that are at least indirectly involved in chloroplast biogenesis.[31,77-80] These are the same eukaryotic remnants of prokaryotic ancestors that were suspected to be regulated by a yet-unknown plastidic factor. Nevertheless, this factor is triggering the expression of the *rbc*S and *cab* genes to a much higher extent than the expression of the gene coding chlorophyll synthetase.[81] Furthermore, this factor seems not to interfere with light responsiveness of nuclear genes, at least in monocots.[82]

It is still a controversy whether the enhancement of steady-state transcript levels of cytokinin-responsive genes is due to enhanced transcription rates or posttranscriptional events, although posttranscriptional regulation[83] is favored. Translational control by cytokinins has been described as polyribosome formation,[84] modification of the secondary structure of poly(A^+)RNA, which leads to "unmasking" of the message,[85] phosphorylation of ribosomal proteins,[86] or regulation of tRNA activity.[87] Posttranslational control may be conferred by retarded protein degradation.[88] Stimulation of correct RNA editing might be another posttranscriptional mechanism modulated by cytokinin.

Already the early reports of Stetler and Laetsch[89] and Seyer et al.[90] describe a correlation between the two cellular processes promoted by cytokinin: the cell cycle and chloroplast biogenesis. In one case an inverse relationship between chloroplast differentiation and tissue growth was observed (either chloroplasts differentiated or cells divided); in the latter a restriction of cytokinin action on chloroplast biogenesis, to the cell growth phase, was found: cytokinin added during the stationary phase was inactive. A similar correlation was described by Suzuki et al.,[91] who found preferential synthesis of plastid DNA prior to that of nuclear DNA, suggesting that ptDNA is synthesized and stored in large amounts to prepare for the following multiple cell divisions without organelle DNA synthesis. Thus, a communication between genome and plastome does exist, and cytokinin presumably interferes with it.

Repeatedly, a coaction of cytokinin and light, in plastid differentiation and multiplication, has been observed.[76] Even in molecular terms, a coaction of light and cytokinin is detectable.[92-94] Molecular genetic analysis of the regulation of cytokinin-responsive nuclear genes is scarce, since, up to now, no gene expression has been known to be triggered specifically by cytokinin. Thus, the vast amount of data on light-regulated genes[95] has to serve as a guideline for further characterizing the cytokinin stimulus on gene expression. In dicots, the expression of several "photogenes" (e.g., *rbc*S, *rbc*L) is seen only after exposure of the plants to light, leading to the idea that light is required for the expression of these genes. Leaf and chloroplast development also require exposure to light, again suggesting a control mechanism.[4] In monocots the situation is quite different, as primary leaf development and chloroplast biogenesis are uncoupled. Nevertheless, leaf expansion and the expression of several "photogenes" still seem to be coupled; in contrast to dicots, they occur at near-light-grown levels, even in dark-grown plants.[34]

The most extensively studied light-responsive genes are *rbc*S and *cab*. Expression of these genes coding for the small subunit of ribulose-1,5-bisphosphate carboxylase/oxygenase (*rbc*S) and the chlorophyll a/b-binding proteins (*cab*), respectively, is regulated by light at the transcriptional level, by sets of *cis*-acting elements and *trans*-acting factors. Moreover, their expression is dependent on genome–plastome interaction as well as on cytokinin. The light signal is mediated via phytochrome, although the expression of many light-responsive genes is modulated by more than one wavelength, e.g., blue light. In fact, there is increasing evidence that blue light plays an important role in modulating plastid differentiation and the accompanying expression of plastome- as well as genome-coded genes for plastid polypeptides.[96-100] A plastid promoter has been characterized that specifically responds to blue light: the light-responsive promoter (LRP) of the *psb*D-*psb*C operon.[101] While genes such as *rbc*L, *atp*B, and 16 S rRNA are fully active in dark-grown barley, the LRP has evolved a light-regulated specificity unlike other chloroplast promoters. Interestingly, LRP is highly conserved in monocots, less in dicots, and was not found in liverwort chloroplast DNA, indicating differences in light-induced gene expression between the major plant groups. The same indication was given by analysis of the nuclear phytochrome gene of the moss *Ceratodon purpureus*. In contrast to all known phytochromes, the 3′ terminus of *phyCer* codes for a putative polypeptide with striking homology to protein kinases.[102] These results suggest that phytochrome in moss is a light-regulated protein kinase, revealing a simple transduction pathway for the light stimulus. Protein kinases are [illegible] candidates for members of the signal transduction pathway for several stimuli in all kinds of plants; e.g., Kulaeva et al. proposed that cytokinin-binding proteins (CBPs) in barley may have protein kinase activity.[103]

Both light-responsive genes *rbc*S and *cab* are not single-copy genes, but represent multigene families with members possessing individual responsiveness to light in an organ- and developmental-specific manner.[104-108] An interaction between light responsiveness and an endogenous developmental program has also been proven for the expression of plastome-encoded genes.[109,110] Furthermore, Eckes et al. described the organ-specific expression of three cDNA clones from potato, that are regulated by light and are correlated with chloroplast development.

Regulatory circuits are even more complex; in addition to being cytokinin modulated, developmentally programmed, and light responsive, an endogenous oscillator also controls the expression of the multigene families of *cab* and *rbc*S. Thus, mRNA levels of these genes exhibit distinct circadian rhythms,[112,113] as do other light-responsive genes such as those coding for nitrate reductase and phosphoenolpyruvate carboxylase.[114,115] Further extrinsic and intrinsic signals interfering with cytokinin action on chloroplast biogenesis and accompanying gene expression are temperature[116] and auxin.[117]

Bearing these complex parameters in mind, it is small wonder that reports on the molecular genetic analysis of cytokinin action are scarce. However, cytokinins seem to interfere with, or modulate, the cooperation between the two genomes, but the molecular mechanisms of this regulatory circuit are far from being understood. Nevertheless, the use of mutants with altered response to external cytokinin as well as *ipt*-transformants[118,119] should facilitate the molecular analysis of cytokinin action.

IV. THE MOLECULAR ANALYSIS OF CHLOROPLAST DIVISION

Since Schimper, 110 years ago, suggested the continuity of plastids,[120] studies on the number, orientation, and division of chloroplasts have been undertaken to trace the mechanisms that provide for their nearly equal distribution into daughter cells during cytokinesis.[121] It is now generally assumed that the different plastid types can multiply by constriction division.[122-124] Additionally, species- as well as organ-specific correlations between nuclear DNA content and plastid number per cell are well documented.[125] However, genes involved in plastid division have not been identified yet, although they have been known from prokaryotes for nearly 30 years.[126,127] From inhibitor studies it was postulated that these genes are encoded by nuclear DNA in higher plants,[128] an indication supported by a nuclear mutant of *Arabidopsis thaliana*, hampered in chloroplast division.[129] Obviously, these yet-unknown genes are members of the regulatory network around plastid differentiation, woven by nuclear and plastid genes.[29,130,131]

The genome–plastome interaction in chloroplast multiplication can be seen in electron micrographs, as division appears to be a two-step process: (1) it is preceded by events within the plastid, such as the disintegration of starch grains and thylakoid membranes at the prospective constriction site, and is then followed by (2) the occurence of a plastid dividing ring (PD ring) outside the plastid around this constriction. This cytoplasmic PD ring is composed of a bundle of actin-like filaments and can be found around dividing plastids of red, brown, and green algae, mosses and ferns (archegoniates), and gymnosperms. In angiosperms, the plastid dividing ring is a doublet that is composed of an outer ring and an inner ring. In any case plastid division occurs by contraction of the filament bundle (reviewed by Kuroiwa[132]).

We have concentrated our studies on the mode of action of the physiological cytokinin N^6-(Δ^2-isopentenyl)adenine on differentiation of the haploid protonema of the moss *Physcomitrella patens*.[133] Special interest is directed at the mutant PC22, which is defective in chloroplast division, and thus the majority of cells possesses only one chloroplast per cell (Figures 1 and 2). Nevertheless, these macrochloroplasts have a normal internal fine structure (Figure 3). Under white-light conditions, they are severed during cell division by the enlarging cell plate, this being a, so far, unknown mechanism of plastid division. The defect of this particular mutant can be partially compensated for by exogenous cytokinin[134] and by blue-light treatment.[31] Cytokinin-induced chloroplast division is accompanied by a transient accumulation of several plastid polypeptides and a dramatic increase in *rbc*L mRNA levels. Such molecular changes cannot be detected under blue-light conditions with already divided chloroplasts, indicating a coaction of cytokinin and light quality in chloroplast division at the molecular level.[135]

Restriction Fragment Length Polymorphism (RFLP) analyses revealed methylation of a putative regulatory gene (*zfp*A gene[136]) near the *rbc*L in chloroplast mutant PC22,[135] as a possible form of transcriptional control in plastid differentiation. We sequenced the 3′ region of the wild-type *rbc*L and found a tRNA gene that was, so far, only known from the ptDNA of the liverwort *Marchantia*,[68] and an open reading frame (ORF) of 948 bp with homology to the *zfp*A from *Pisum*. The deduced polypeptide of this ORF contains a sequence (–Cys–X2–Cys–X15–Cys–X2–Cys–) that

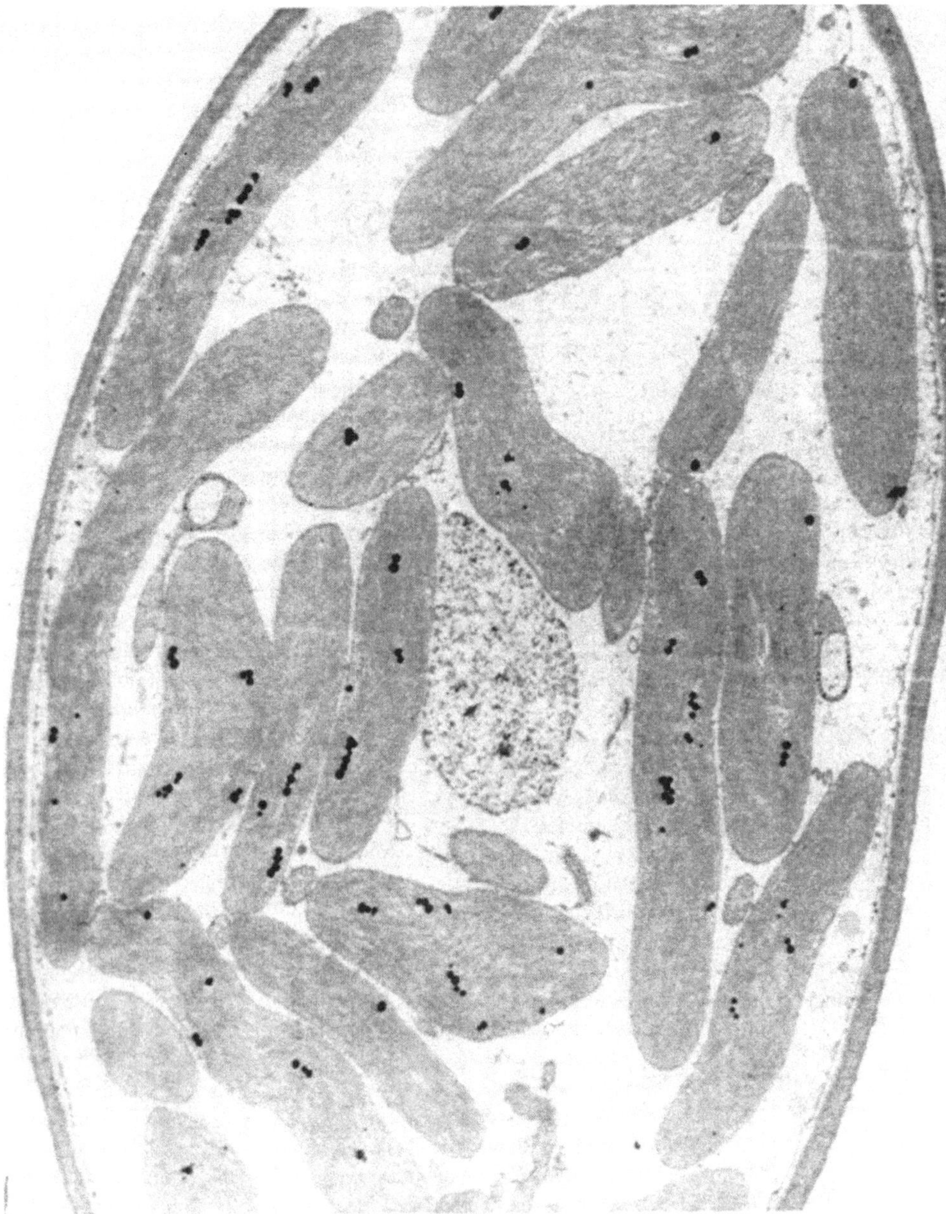

Figure 14-1 Electron micrographs of a cross-section through one protonema cell of (a) the wild type of *Physcomitrella patens* and (b) the chloroplast mutant PC22 derived from it. Note the cup-shaped giant chloroplast in the mutant.

may fold to form a zinc-finger element.[69] These elements can interact with nucleic acids, thereby revealing a regulatory function of the respective polypeptide. However, the function of this protein is still unknown and is a subject of debate.[137,138] Nevertheless, sequence comparison revealed that this protein contains two distinct domains: the hydrophilic zinc-finger element and a hydrophobic C-terminal region.[69] Moreover, *zfp*A may be a marker gene for plastid evolution, as it is highly homologous within the archegoniates and dicots, respectively, but differs significantly between the two groups.[139] Furthermore, it is mutilated in the plastid DNA of monocots. In the moss we found

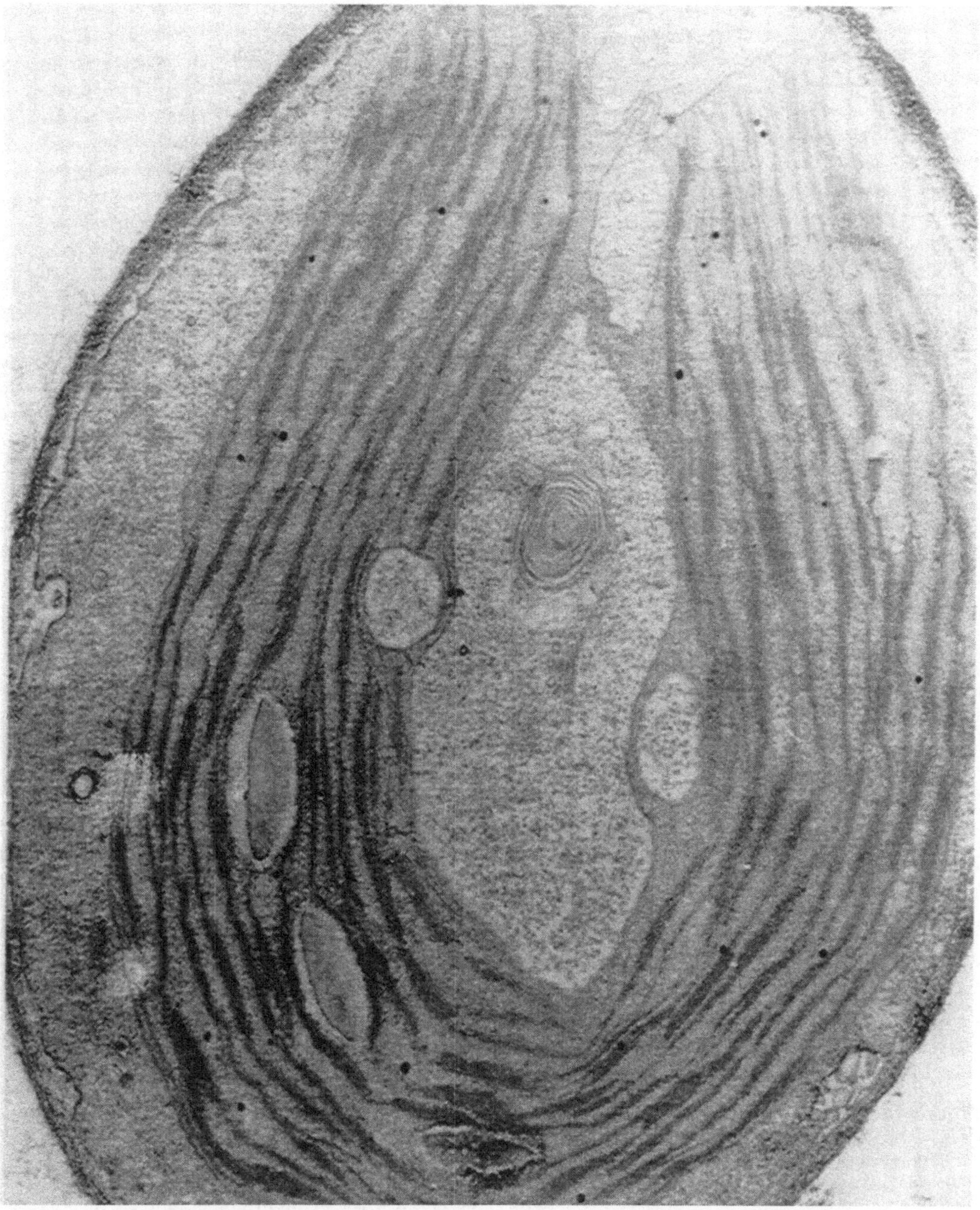

Figure 14-1 (continued).

prokaryotic promoter consensus sequences (-10, -35, Shine Dalgarno) as well as three TATA boxes that are known to be typical eukaryotic promoter sequences. Additionally, these elements are involved in the differential control of the *Sinapis alba* plastid *psb*A gene, depending on the state of plastid differentiation.[140] However, a multiplication of TATA boxes has not previously been reported for plastid promoters and reveals a complex mechanism of transcriptional control. A gene-specific hybridization probe detected three transcripts of low abundance with apparent molecular weights larger than the gene. RNA accumulation patterns differed between wild-type and chloroplast mutants. Additionally, under the influence of cytokinin, the high-molecular-weight transcript of the mutant decreased in amount (Figure 4). Using riboprobes we ensured that these transcripts were

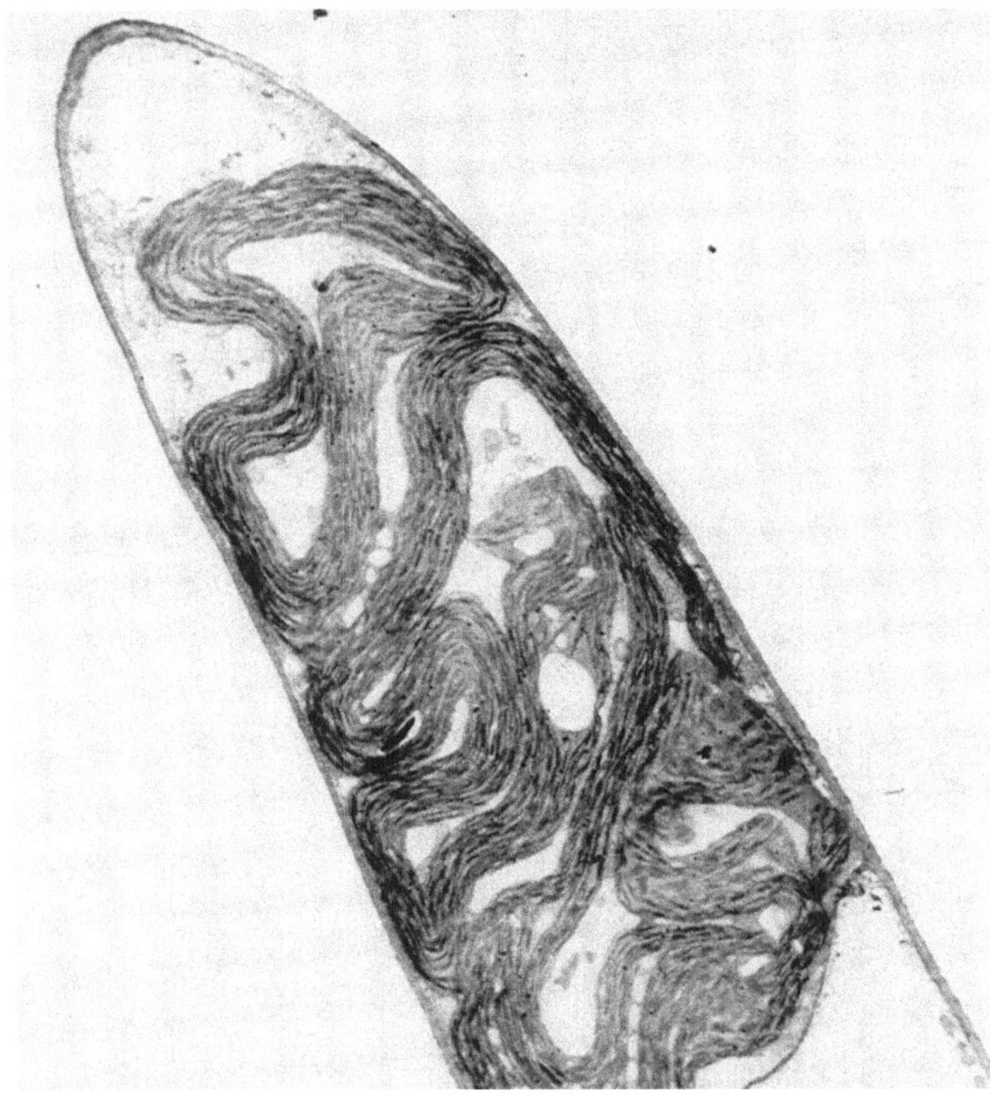

Figure 14-2 Electron micrograph of a longitudinal section through a PC22 protonema tip cell with one giant chloroplast.

encoded by the same ptDNA strand, thus being immature transcripts including *zfp*A sequences.

We performed Northern blots for several plastid genes with the *Physcomitrella* wild type, and two cytokinin-sensitive mutants derived from it (PC22 and P24), P24 not obviously being a chloroplast mutant. In every case we found an accumulation of immature plastid transcripts in both mutants (e.g., Figure 5). The amounts of high-molecular-weight transcripts were less in the wild type, as well as in mutant protonemata treated with cytokinin.[141] Reports on qualitative changes in the RNA processing patterns of plastid genes are infrequent, although most of them are organized in polycistronic transcription units, as in the ancestral bacterial operons.[142-146] Evidence came from light-induced chloroplast development in mustard,[109] maize,[147] and barley.[148] Likewise, in an *Oenothera* plastome, mutant processing of plastid proteins, whether encoded by ptDNA or by nucDNA, was affected.[149] Thus, processing of mRNAs and proteins might be one way for cytokinins to modulate plastid gene expression posttranscriptionally.

Obviously, Western analyses will need to be employed to prove changes in zfpA-protein amounts upon cytokinin treatment of the moss mutants, in

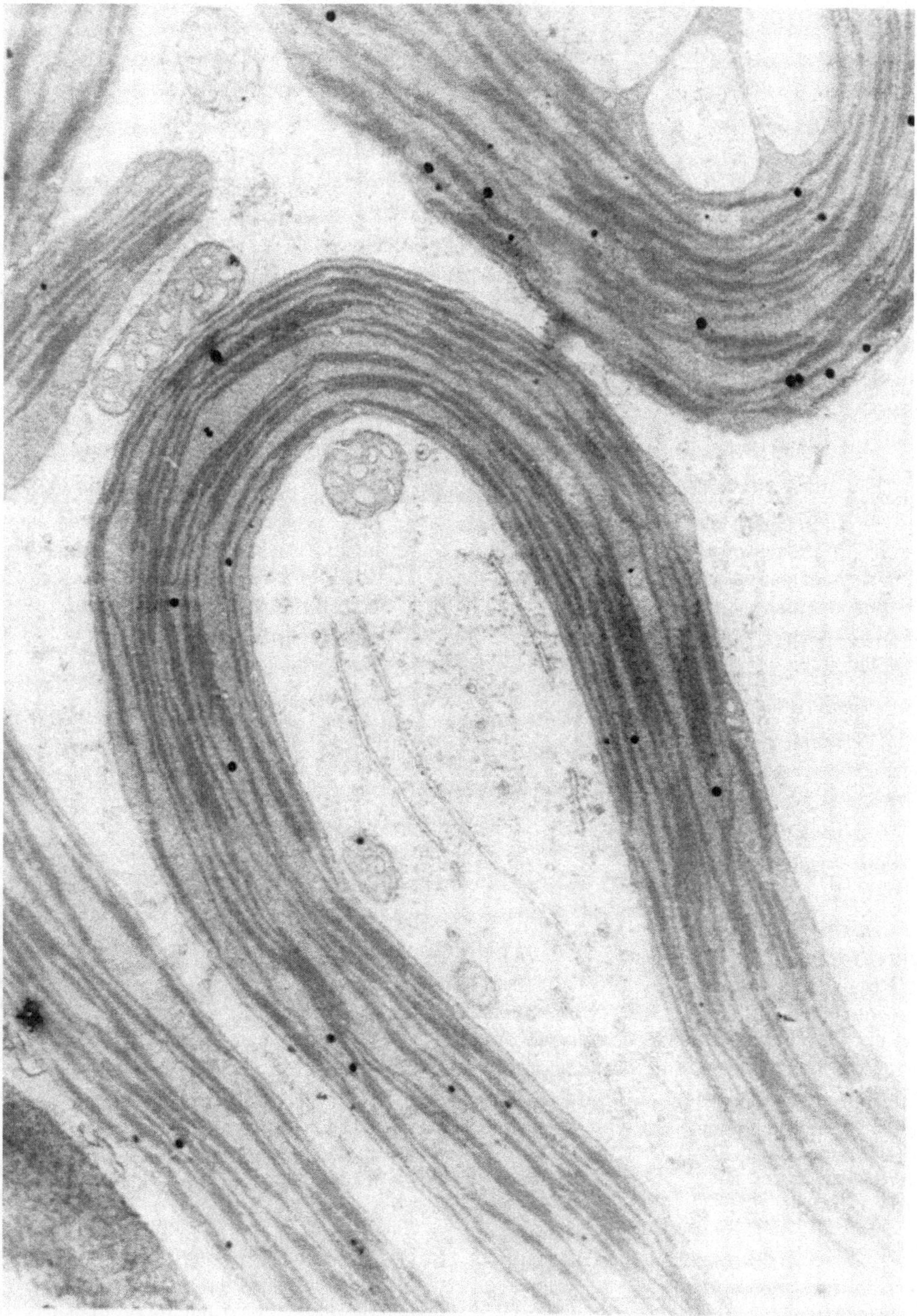

Figure 14-3 Electron micrograph of a PC22 protonema cell showing a normal internal fine structure of the mutant chloroplast.

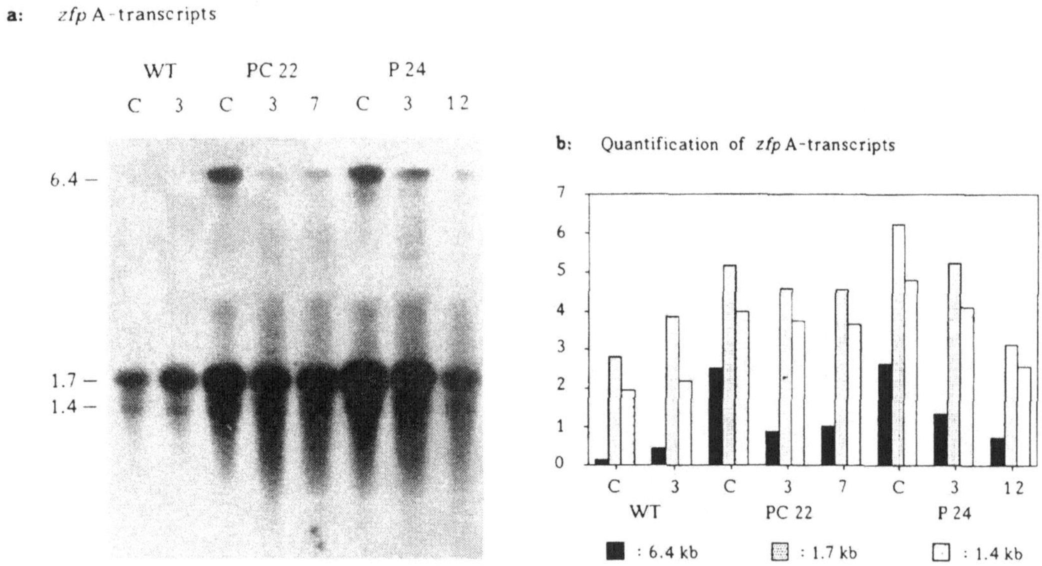

Figure 14-4 Northern analysis of *Physcomitrella zfp*A transcripts. [WT: wild-type; PC22: cytokinin-sensitive chloroplast mutant; P24: cytokinin-sensitive mutant; C: protonemata grown 12 days without cytokinin; 3: protonemata after 9 days of culture with additional 3 days in cytokinin; 7: protonemata after 5 days of culture with additional 7 days in cytokinin; 12: protonemata after 12 days in cytokinin (compare to References 135 and 154).] (a:) Northern blot using a radioactive *zfp*A-specific gene probe. (b:) Densitometric quantification of the autoradiogram shown in (a). (Unit: pixel × 1000. Transcript sizes are given in kilobases.)

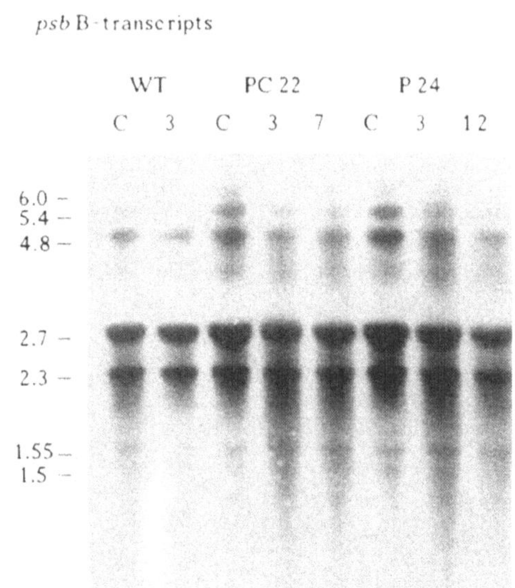

Figure 14-5 Northern analysis of *Physcomitrella psb*B transcripts. [WT: wild-type; PC22: cytokinin-sensitive chloroplast mutant; P24: cytokinin-sensitive mutant; C: protonemata grown 12 days without cytokinin; 3: protonemata after 9 days of culture with additional 3 days in cytokinin; 7: protonemata after 5 days of culture with additional 7 days in cytokinin; 12: protonemata after 12 days in cytokinin (compare to References 135 and 154).] (Transcript sizes are given in kilobases.)

order to correlate them to the proposed enhancing effect on transcript processing. Furthermore, the postulated interaction of zfpA protein with nucleic acids has to be proven. However, we do not yet know if the mutation affecting plastid division in this particular moss mutant is located on nucDNA or on ptDNA.[150] Furthermore, most of the data on plastid gene regulation during chloroplast development result from *in vitro* studies, and there is a need for plastid transformation systems, to study gene regulation *in vivo*. First reports on the successful transformation of the giant chloroplast of *Chlamydomonas*,[151] as well as of higher-plant chloroplasts,[152,153] demonstrate the feasibility of such an approach. The macrochloroplast of PC22 seems to be a prime candidate for chloroplast transformation of multicellular plants, as it is amenable for microinjection.[134] Thus, we will try to unravel the informational network around cytokinin-induced chloroplast division, with mutant curing experiments. Furthermore, we will attempt to transform moss mutants with *ipt* constructs, in order to manipulate endogenous cytokinin concentrations.

ACKNOWLEDGMENTS

Grants from the European Community, from the Deutsche Forschungsgemeinschaft, and from the city of Hamburg are gratefully acknowledged. Special thanks go to Prof. Dr. T. Börner and to Dr. K. Krupinska, who shared their unpublished results. I enjoyed stimulating discussions with Prof. Dr. W. O. Abel, Jon Falk, and the members of my group.

REFERENCES

1. **Herrmann, R. G., Westhoff, P., and Link, G.,** Biogenesis of plastids in higher plants, in *Plant Gene Research, Cell Organelles,* Herrmann, R. G., Ed., Springer-Verlag, New York, 1992, 275.
2. **Becker, W. and Apel, K.,** Differences in gene expression between natural and artificially induced leaf senescence, *Planta,* 189, 74, 1993.
3. **Mullet, J. E.,** Chloroplast development and gene expression, *Ann. Rev. Plant Physiol. Plant Mol. Biol.,* 39, 475, 1988.
4. **Dale, J. E.,** The control of leaf expansion, *Ann. Rev. Plant Physiol. Plant Mol. Biol.,* 39, 276, 1988.
5. **Dean, C. and Leech, R. M.,** Genome expression during normal leaf development, *Plant Physiol.,* 69, 904, 1982.
6. **Van Cleve, B., Forreiter, C., Sauter, J. J., and Apel, K.,** Pith cells of poplar contain photosynthetically active chloroplasts, *Planta,* 189, 70, 1993.
7. **Tobin, E. M. and Silverthorne, J.,** Light regulation of gene expression in higher plants, *Ann. Rev. Plant Physiol.,* 36, 569, 1985.
8. **Krupinska, K.,** Transcriptional control of plastid gene expression during development of primary foliage leaves of barley grown under a daily light-dark regime, *Planta,* 186, 294, 1992.
9. **Chory, J., Peto, C., Feinbaum, R., Pratt, L., and Ausubel, F.,** *Arabidopsis thaliana* mutant that develops as a light-grown plant in the absence of light, *Cell,* 58, 991, 1989.
10. **Boffey, S. A. and Leech, R. M.,** Chloroplast DNA levels and the control of chloroplast division in light-grown wheat leaves, *Plant Physiol.,* 69, 1387, 1982.
11. **Feierabend, J. and Mikus, M.,** Occurrence of a high temperature sensitivity of chloroplast ribosome formation in several higher plants, *Plant Physiol.,* 59, 863, 1977.
12. **Feierabend, J. and Schrader-Reichhardt, U.,** Biochemical differentiation of plastids and other organelles in rye leaves with a high-temperature-induced deficiency of plastid ribosomes, *Planta,* 129, 133, 1976.
13. **Rademacher, E. and Feierabend, J.,** Formation of chloroplast pigments and sterols in rye leaves deficient in plastid ribosomes, *Planta,* 129, 147, 1976.
14. **Hess, W. R., Schendel, R., Rüdiger, W., Fieder, B., and Börner, T.,** Components of chlorophyll biosynthesis in a barley *albina* mutant unable to synthesize δ-aminolevulinic acid by utilizing the transfer RNA for glutamic acid, *Planta,* 188, 19, 1992.
15. **Feierabend, J. and Berberich, T.,** Heat-induced ribosome-deficiency of plastids — mechanisms and applications, in *The Translational Apparatus of Photosynthetic Organelles,* Mache, R. and Subramanian, A. R., Eds., NATO ASI Series Vol. H55, Berlin, 1991, 215.
16. **Börner, T., Manteuffel, R., and Wellburn, A. R.,** Enzymes of plastid ribosome-deficient mutants. Chloroplast ATPase (CF1), *Protoplasma,* 98, 153, 1979.
17. **Gray, M. W.,** The endosymbiont hypothesis revisited, *Int. Rev. Cytol.,* 141, 233, 1992.
18. **Parks, B. M., Jones, A. M., Adamse, P., Korneef, M., Kendrick, R. E., and Quail, P. H.,** The *aurea* mutant of tomato is deficient in spectrophotometrically and immunochemically detectable phytochrome, *Plant Mol. Biol.,* 9, 97, 1987.
19. **Sharrock, R. A., Parks, B. M., Korneef, M., and Quail, P. H.,** Molecular analysis of the phytochrome deficiency in an *aurea* mutant of tomato, *Mol. Gen. Genet.,* 213, 9, 1988.
20. **Giuliano, G. and Scolnik, P. A.,** Transcription of two photosynthesis-associated nuclear gene families correlates with the presence of chloroplasts in leaves of the variegated tomato *ghost* mutant, *Plant Physiol.,* 86, 7, 1988.
21. **Hess, W. R., Schendel, R., Börner, T., and Rüdiger, W.,** Reduction of mRNA level for two nuclear encoded light regulated genes in the barley mutant *albostrians* is not correlated with phytochrome content and activity, *J. Plant Physiol.,* 138, 292, 1991.

22. **Batschauer, A., Mösinger, E., Krenz, K., Dörr, I., and Apel, K.,** The implication of a plastid-derived factor in the transcriptional control of nuclear genes encoding the light-harvesting chlorophyll a/b binding protein, *Eur. J. Biochem.*, 154, 625, 1986.
23. **Burgess, D. G. and Taylor, W. C.,** The chloroplast affects the transcription of a nuclear gene family, *Mol. Gen. Genet.*, 214, 89, 1988.
24. **Burgess, D. G. and Taylor, W. C.,** Chloroplast photooxidation affects the accumulation of cytosolic messenger RNA encoding chloroplast proteins in maize, *Planta*, 170, 520, 1987.
25. **Oelmüller, R., Schuster, C., and Mohr, H.,** Physiological characterization of a plastidic signal required for nitrate-induced appearance of nitrate and nitrite reductases, *Planta*, 174, 75, 1988.
26. **Stockhaus, J., Schell, J., and Willmitzer, L.,** Correlation of the expression of the nuclear photosynthetic gene ST-LS1 with the presence of chloroplasts, *EMBO J.*, 8, 2445, 1989.
27. **McHale, N. A., Kawata, E. E., and Cheung, A. Y.,** Plastid disruption in a thiamine requiring mutant of *Nicotiana sylvestris* blocks accumulation of specific nuclear and plastid mRNAs, *Mol. Gen. Genet.*, 221, 203, 1990.
28. **Börner, T.,** Chloroplast control of nuclear gene function, *Endocytobios, Cell Res.*, 3, 265, 1986.
29. **Taylor, W. C.,** Regulatory interactions between nuclear and plastid genomes, *Ann. Rev. Plant Physiol.*, 40, 211, 1989.
30. **Oelmüller, R.,** Photooxidative destruction of chloroplasts and its effect on nuclear gene expression and extraplastidic enzyme levels, *Photochem. Photobiol.*, 49, 229, 1989.
31. **Reski, R. and Abel, W. O.,** Analyzing cytokinins and plastid differentiation at the molecular level: interference with genome/plastome interaction, light quality, developmental state and endogenous oscillators (mini-review), in *Physiology and Biochemistry of Cytokinins in Plants*, Kamínek, M., Mok, D. W. S., and Zazímalová, E., Eds., SPB Academic Publishing, The Hague, 1992, 255.
32. **Susek, R. E. and Chory, J.,** A tale of two genomes: role of a chloroplast signal in coordinating nuclear and plastid genome expression, *Aust. J. Plant Physiol.*, 19, 387, 1992.
33. **Gruissem, W.,** Chloroplast gene expression: how plants turn their plastids on, *Cell*, 56, 161, 1989.
34. **Baumgartner, B. J., Rapp, J. C., and Mullet, J. E.,** Plastid transcription activity and DNA copy number increase early in barley chloroplast development, *Plant Physiol.*, 89, 1011, 1989.
35. **Schrubar, H., Wanner, G., and Westhoff, P.,** Transcriptional control of plastid gene expression in greening *Sorghum* seedlings, *Planta*, 183, 101, 1990.
36. **Klein, R. R. and Mullet, J. E.,** Light-induced transcription of chloroplast genes. *psbA* transcription is differentially enhanced in illuminated barley, *J. Biol. Chem.*, 265, 1895, 1990.
37. **Sakai, A., Kawano, S., and Kuriowa, T.,** Conversion of proplastids to amyloplasts in tobacco cultured cells is accompanied by changes in the transcriptional activities of plastid genes, *Plant Physiol.*, 100, 1062, 1992.
38. **Ngernprasirtsiri, J., Kobayashi, H., and Akazawa, T.,** DNA methylation as a mechanism of transcriptional regulation in nonphotosynthetic plastids in plant cells, *Proc. Natl. Acad. Sci. U.S.A.*, 85, 4750, 1988.
39. **Deng, X. W. and Gruissem, W.,** Control of plastid gene expression during development: the limited role of transcriptional regulation, *Cell*, 49, 379, 1987.
40. **Marano, M. R. and Carillo, N.,** Chromoplast formation during tomato fruit ripening. No evidence for plastid DNA methylation, *Plant Mol. Biol.*, 16, 11, 1991.
41. **Zaitlin, D., Hu, J., and Bogorad, L.,** Binding and transcription of relaxed DNA templates by fractions of maize chloroplast extracts, *Proc. Natl. Acad. Sci. U.S.A.*, 86, 876, 1989.
42. **Harpster, M. and Apel, K.,** The light-dependent regulation of gene expression during plastid development in higher plants, *Physiol. Plant.*, 64, 147, 1985.
43. **Deng, X. W., Tonkyn, J. C., Peter, G. F., Thornber, J. P., and Gruissem, W.,** Post-transcriptional control of plastid mRNA accumulation during adaptation of chloroplasts to different light quality environments, *Plant Cell*, 1, 645, 1989.
44. **Hu, J. and Bogorad, L.,** The chloroplast RNA polymerase: the 180-, 120-, and 38-kilodalton polypeptides are encoded in chloroplast genes, *Proc. Natl. Acad. Sci. U.S.A.*, 87, 1531, 1990.
45. **Jolly, S. O. and Bogorad, L.,** Preferential transcription of cloned maize chloroplast DNA sequences by maize chloroplast RNA polymerase, *Proc. Natl. Acad. Sci. U.S.A.*, 77, 822, 1980.

46. **Lerbs, S., Bräutigam, E., and Mache, R.,** DNA-dependent RNA polymerase of spinach chloroplasts: characterization of α-like and τ-like polypeptides, *Mol. Gen. Genet.*, 211, 459, 1988.
47. **Tiller, K., Eisermann, A., and Link, G.,** The chloroplast transcription apparatus from mustard (*Sinapis alba* L.). Evidence for three different transcription factors which resemble bacterial sigma factors, *Eur. J. Biochem.*, 198, 93, 1991.
48. **Hallick, R. B., Lipper, C., Richards, O. C., and Rutter, W. J.,** Isolation of transcriptionally active chromosome from chloroplasts of *Euglena gracilis, Biochemistry*, 15, 3039, 1976.
49. **Greenberg, B. M., Narita, J. O., DeLuca-Flaherty, C., Gruissem, W., Rushlow, K. A., and Hallick, R. B.,** Evidence for two RNA polymerase activities in *Euglena gracilis* chloroplasts, *J. Biol. Chem.*, 259, 14880, 1984.
50. **Narita, J. O., Rushlow, K. E., and Hallick, R. B.,** Characterization of a *Euglena gracilis* chloroplast RNA polymerase specific for ribosomal genes, *J. Biol. Chem.*, 260, 11194, 1985.
51. **Reiss, T. and Link, G.,** Characterization of transcriptionally active DNA-protein complexes from chloroplasts and etioplasts of mustard (*Sinapis alba* L.), *Eur. J. Biochem.*, 148, 207, 1985.
52. **Gruissem, W., Greenberg, B. M., Zurawski, G., and Hallick, R. B.,** Chloroplast gene expression and promoter identification in chloroplast extracts, *Methods Enzymol.*, 118, 253, 1986.
53. **Richter, G. and Ottersbach, N.,** Blue-light-dependent chloroplast differentiation in cultured plant cells: evidence for transcriptional control of plastid genes, *Bot. Acta*, 103, 168, 1990.
54. **Igloi, G. L. and Kössel, H.,** The transcription apparatus of chloroplasts, *Crit. Rev. Plant Sci.*, 10, 525, 1992.
55. **Bünger, W. and Feierabend, J.,** Capacity for RNA synthesis in 70 S ribosome deficient plastids of heat-bleached rye leaves, *Planta*, 149, 163, 1980.
56. **Siemenroth, A., Wollgiehn, R., Neumann, D., and Börner, T.,** Synthesis of ribosomal RNA in ribosome-deficient plastids of the mutant "albostrians" of *Hordeum vulgare* L., *Planta*, 153, 547, 1981.
57. **Lerbs, S., Bräutigam, E., and Parthier, B.,** Polypeptides of DNA-dependent RNA polymerase of spinach chloroplasts: characterization by antibody-linked polymerase assay and determination of sites of synthesis, *EMBO J.*, 4, 1661, 1985.
58. **Falk, J., Schmidt, A., and Krupinska, K.,** Characterization of plastid DNA transcription in ribosome deficient plastids of heat-bleached barley leaves, *J. Plant Physiol.*, 141, 176, 1993.
59. **Hess, W. R., Prombona, A., Fieder, B., Subramanian, A. R., and Börner, T.,** Chloroplast *rps*15 and the *rpo*B/C1/C2 gene cluster are strongly transcribed in ribosome-deficient plastids: evidence for a functioning non-chloroplast-encoded RNA polymerase, *EMBO J.*, 12, 563, 1993.
60. **Benne, R.,** RNA editing in trypanosomes: is there a message? *Trends Genet.*, 6, 177, 1990.
61. **Hiesel, R., Wissinger, B., Schuster, W., and Brennicke, A.,** RNA editing in plant mitochondria, *Science*, 246, 1632, 1989.
62. **Weissmann, C., Cattaneo, R., and Billeter, M. A.,** Sometimes an editor makes sense, *Nature*, 343, 697, 1990.
63. **Hoch, B., Maier, R. M., Appel, K., Igloi, G. L., and Kössel, H.,** Editing of a chloroplast mRNA by creation of an initiation codon, *Nature*, 353, 178, 1991.
64. **Kudla, J., Igloi, G. L., Metzlaff, M., Hagemann, R., and Kössel, H.,** RNA editing in tobacco chloroplasts leads to the formation of a translatable *psb*L mRNA by a C to U substitution within the initiation codon, *EMBO J.*, 11, 1099, 1992.
65. **Maier, R. M., Hoch, B., Zeltz, P., and Kössel, H.,** Internal editing of the maize chloroplast *ndhA* transcript restores codons for conserved amino acids, *Plant Cell*, 4, 609, 1992.
66. **Ohyama, K., Fukuzawa, H., Kohchi, T., Shrai, H., Sano, T., Sano, S., Shiri, H., Umesono, K., Shiki, Y., Takeuchi, M., Chang, Z., Aota, S., Inokuchi, H., and Ozeki, H.,** Chloroplast gene organization deduced from complete sequence of liverwort *Marchantia polymorpha* chloroplast DNA, *Nature*, 322, 572, 1986.
67. **Oda, K., Yamato, K., Ohta, E., Nakamura, Y., Takemura, M., Nozato, N., Akashi, K., Kanegae, T., Ogura, Y., Kohchi, T., and Ohyama, K.,** Gene organization deduced from the complete sequence of liverwort *Marchantia polymorpha* mitochondrial DNA. Primitive form of plant mitochondrial genome, *J. Mol. Biol.*, 223, 1, 1992.
68. **Kasten, B., Wehe, M., Reski, R., and Abel, W. O.,** *trnR-CCG* is not unique to the plastid DNA of the liverwort *Marchantia*: gene identification from the moss *Physcomitrella patens*, *Nucleic Acids Res.*, 19, 5074, 1991.

69. **Kasten, B., Wehe, M., Kruse, S., Reutter, K., Abel, W. O., and Reski, R.,** The plastome-encoded *zfpA* gene of a moss contains prokaryotic as well as eukaryotic promoter consensus sequences and its RNA abundance is modulated by cytokinin, *Curr. Genet.*, 22, 327, 1992.
70. **Marienfeld, J. R., Reski, R., and Abel, W. O.,** The first analysed archegoniate mitochondrial gene (*cox3*) exhibits extraordinary features, *Curr. Genet.*, 20, 319, 1991.
71. **Marienfeld, J. R., Reski, R., and Abel, W. O.,** Chondriom analysis of the moss *Physcomitrella patens*, *Crypt. Bot.*, 3, 23, 1992.
72. **Reutter, K., Wehe, M., Kasten, B., Abel, W. O., and Reski, R.,** Cloning plastid DNA of the moss *Physcomitrella patens, Mitt. Inst. Allg. Bot. Hamburg*, 24, 35, 1992.
73. **Keegstra, K. and von Heijne, G.,** Transport of proteins into chloroplasts, in *Plant Gene Research. Cell Organelles*, Herrmann, R. G., Ed., Springer-Verlag, New York, 1992, 353.
74. **Wolfe, K. H., Gouy, M., Yang, Y.-W., Sharp, P. M., and Li, W.-H.,** Data of the monocot-dicot divergence estimated from chloroplast DNA sequence data, *Proc. Natl. Acad. Sci. U.S.A.*, 86, 6201, 1989.
75. **Mothes, K.,** Über das Altern der Blätter und die Möglichkeit ihrer Wiederverjüngung, *Naturwissenschaften*, 47, 337, 1960.
76. **Parthier, B.,** The role of phytohormones (cytokinins) in chloroplast development, *Biochem. Physiol. Pflanzen*, 174, 173, 1979.
77. **Seyer, P. and Lescure, A.-M.,** Evidence for changes in plastid mRNA populations during cytokinin-induced chloroplast differentiation in tobacco cell suspensions, *Plant Sci. Lett.*, 36, 59, 1984.
78. **Lerbs, S., Lerbs, W., Klyachko, N. L., Romanko, E. G., Kulaeva, O. N., Wollgiehn, R., and Parthier, B.,** Gene expression in cytokinin- and light-mediated plastogenesis of *Cucurbita* cotelydons: ribulose-1,5-bisphosphate carboxylase/oxygenase, *Planta*, 162, 289, 1984.
79. **Teyssendier de la Serve, B., Axelos, M., and Péaud-Lenoël, C.,** Cytokinins modulate the expression of genes encoding the protein of the light harvesting chlorophyll a/b complex, *Plant Mol. Biol.*, 5, 155, 1985.
80. **Parthier, B.,** Hormone-induced alterations in plant gene expression, *Biochem Physiol. Pflanzen*, 185, 289, 1989.
81. **Hess, W. R., Blank-Huber, M., Fieder, B., Börner, T., and Rüdiger, W.,** Chlorophyll synthetase and chloroplast tRNA^{glu} are present in heat-bleached, ribosome-deficient plastids, *J. Plant Physiol.*, 139, 427, 1992.
82. **Börner, T. and Hess, W. R.,** Altered nuclear, mitochondrial and plastid gene expression in white barley cells containing ribosome-deficient plastids, in *Plant Mitochondria*, Brennicke, A. and Kück, U., Eds., VCI, Mannheim, 1993, in press.
83. **Flores, S. and Tobin, E.M.,** Cytokinin modulation of LHCP mRNA levels: the involvement of post-transcriptional regulation, *Plant Mol. Biol.*, 11, 409, 1988.
84. **Ohya, T. and Suzuki, H.,** Cytokinin-promoted polyribosome formation in excised cucumber cotyledons, *J. Plant Physiol.*, 133, 295, 1988.
85. **Jackowski, G., Jarmolowski, A., and Szweykowska, A.,** Kinetin modifies the secondary structure of poly(A)RNA in cucumber cotyledons, *Plant Sci.*, 52, 67, 1987.
86. **Yakovleva, L. A., Klueva, N. Y., and Kulaeva, O. N.,** Phosphorylation of ribosome proteins as a mechanism of phytohormone regulation of protein synthesis in plants, in *Physiology and Biochemistry of Cytokinins in Plants*, Kaminek, M., Mok, D. W. S., and Zazímalová, E., Eds., SPB Academic Publishing, The Hague, 1992, 169.
87. **Romanov, G. A.,** Cytokinins and tRNAs: a new insight into the old problem, *Sov. Plant Physiol.*, 37, 1196, 1990.
88. **Lamattina, L., Anchoverri, V., Conde, R. D., and Lezica, R. P.,** Quantification of the kinetin effect on protein synthesis and degradation in senescing wheat leaves, *Plant Physiol.*, 83, 497, 1987.
89. **Stetler, D. and Laetsch, W. M.,** Kinetin-induced chloroplast maturation in cultures of tobacco tissue, *Science*, 149, 1387, 1965.
90. **Seyer, P., Marty, D., Lescure, A. M., and Péaud-Lenoël, C.,** Effect of cytokinin on chloroplast cyclic differentiation in cultured tobacco cells, *Cell Differentiation*, 4, 187, 1975.
91. **Suzuki, T., Kawano, S., Sakai, A., Fujie, M., Kuroiwa, H., Nakamura, H., and Kuroiwa, T.,** Preferential mitochondrial and plastid DNA synthesis before multiple cell divisions in *Nicotiana tabacum*, *J. Cell Sci.*, 103, 831, 1992.
92. **Cotton, J. L. S., Ross, C. W., Byrne, D. H., and Colbert, J. T.,** Down-regulation of phytochrome mRNA abundance by red light and benzyladenine in etiolated cucumber cotyledons, *Plant Mol. Biol.* 14, 707, 1990.

93. **Jayabaskaran, C., Kuntz, M., Guillemaut, P., and Weil, J.-H.,** Variations in the levels of chloroplast tRNAs and aminoacyl-tRNA synthetase in senescing leaves of *Phaseolus vulgaris, Plant Physiol.*, 92, 136, 1990.
94. **Lu, J. L., Ertl, J. R., and Chen, C.-M.,** Cytokinin enhancement of the light induction of nitrate reductase transcript levels in etiolated barley leaves, *Plant Mol. Biol.*, 14, 585, 1990.
95. **Gilmartin, P. M., Sarokin, L., Memelink, J., and Chua, N.-H.,** Molecular light switches for plant genes, *Plant Cell*, 2, 369, 1990.
96. **Fluhr, R. and Chua, N.-H.,** Developmental regulation of two genes encoding ribulose-bisphosphate carboxylase small subunit in pea and transgenic tobacco plants: phytochrome response and blue-light induction, *Proc. Natl. Acad. Sci. U.S.A.*, 83, 2358, 1986.
97. **Richter, G., Dudel, A., Einspanier, R., Dannhauer, I., and Hüsemann, W.,** Blue-light control of mRNA level and transcription during chloroplast differentiation in photomixotrophic and photoautotrophic cell cultures (*Chenopodium rubrum* L.), *Planta*, 172, 79, 1987.
98. **Gamble, P. E. and Mullet, J. E.,** Blue light regulates the accumulation of two *psb*D-*psb*C transcripts in barley chloroplasts, *EMBO J.*, 8, 2785, 1989.
99. **Marrs, K. A. and Kaufman, L. S.,** Blue-light regulation of transcription in nuclear-encoded genes in pea, *Proc. Natl. Acad. Sci. U.S.A.*, 86, 4492, 1989.
100. **Oelmüller, R., Kendrick, R. E., and Briggs, W. R.,** Blue-light mediated accumulation of nuclear-encoded transcripts coding for proteins of the thylakoid membrane is absent in the labile phytochrome deficient *aurea* mutant of tomato, *Plant Mol. Biol.*, 13, 223, 1989.
101. **Christopher, D. A., Kim, M., and Mullet, J. E.,** A novel light-regulated promoter is conserved in cereal and dicot chloroplasts, *Plant Cell*, 4, 785, 1992.
102. **Algarra, P., Lindner, S., and Thümmler, F.,** Biochemical evidence that phytochrome of the moss *Ceratodon purpureus* is a light-regulated protein kinase, *FEBS Lett.*, 315, 69, 1993.
103. **Kulaeva, O. N., Novikova, G. V., Selivankina, Y., Karavaiko, N. N., and Moshkov, I. E.,** Protein kinases and cytokinin-binding proteins in plant cell response to cytokinin, in *Physiology and Biochemistry of Cytokinins in Plants*, Kamínek, M., Mok, D. W. S., and Zazímalová, E., Eds., SPB Academic Publishing, The Hague, 1992, 121.
104. **Kaufman, L. S., Thompson, W. F., and Briggs, W. R.,** Different red light requirements for phytochrome-induced accumulation of *cab* RNA and *rbc* RNA, *Science*, 226, 1447, 1984.
105. **Sugita, M. and Gruissem, W.,** Developmental, organ-specific, and light dependent expression of the tomato ribulose-1,5-bisphosphate carboxylase small subunit gene family, *Proc. Natl. Acad. Sci. U.S.A.*, 84, 7104, 1987.
106. **Silverthorne, J., Wimpee, C. F., Yamada, T., Rolfe, S. A., and Tobin, E. M.,** Differential expression of individual genes encoding the small subunit of ribulose-1,5-bisphosphate carboxylase in *Lemna gibba, Plant Mol. Biol.*, 15, 49, 1990.
107. **Hobbs, S. L. A., Pelcher, L. E., DeLong, C. M. O., Anderson, M., and Mahon, J. D.,** Genetic variability in the amount of ribulose-1,5-bisphosphate carboxylase/oxygenase and its small subunit mRNA in pea, *Planta*, 180, 510, 1990.
108. **Brusslan, J. A. and Tobin, E. M.,** Light-independent developmental regulation of *cab* gene expression in *Arabidopsis thaliana* seedlings, *Proc. Natl. Acad. Sci. U.S.A.*, 89, 7791, 1992.
109. **Dietrich, G., Detschey, S., Neuhaus, H., and Link, G.,** Temporal and light control of plastid transcript levels for proteins involved in photosynthesis during mustard (*Sinapis alba* L.) seedling development, *Planta*, 172, 393, 1987.
110. **Fiebig, C., Neuhaus, H., Teichert, J., Röcher, W., Degenhardt, J., and Link, G.,** Temporal and spatial pattern of plastid gene expression during crucifer seedling development and embryogenesis, *Planta*, 181, 191, 1990.
111. **Eckes, P., Schell, J., and Willmitzer, L.,** Organ-specific expression of three leaf/stem specific cDNAs from potato is regulated by light and correlated with chloroplast development, *Mol. Gen. Genet.*, 199, 216, 1985.
112. **Ernst, D., Apfelböck, A., Bergmann, A., and Weyrauch, C.,** Rhythmic regulation of the light-harvesting chlorophyll a/b protein and the small subunit of ribulose-1,5-bisphosphate carboxylase mRNA in rye seedlings, *Photochem. Photobiol.*, 52, 29, 1990.
113. **Riesselmann, S. and Piechulla, B.,** Effect of dark phases and temperature on the chlorophyll a/b binding protein mRNA level oscillations in tomato seedlings, *Plant Mol. Biol.*, 14, 605, 1990.
114. **Deng, M. D., Moureaux, T., Leydecker, M.-T., and Caboche, M.,** Nitrate-reductase expression is under the control of a circadian rhythm and is light inducible in *Nicotiana tabacum* leaves, *Planta*, 180, 257, 1990.

115. **Thomas, M., Cretin, C., Vidal, J., Keryer, E., Gadal, P., and Monsinger, E.,** Light-regulation of phosphoenolpyruvate carboxylase mRNA in leaves of C4 plants: evidence for phytochrome control on transcription during greening and for rhythmicity, *Plant Sci.*, 69, 65, 1990.

116. **Humbeck, K., Melis, A., and Krupinska, K.,** Chilling affects chloroplast development in barley primary foliage leaves, *J. Plant Physiol.*, in press, 1994.

117. **Dominov, J. A., Stenzler, L., Lee, S., Schwarz, J. J., Leisner, S., and Howell, H. H.,** Cytokinins and auxins control the expression of a gene in *Nicotiana plumbaginifolia* cells by feedback regulation, *Plant Cell*, 4, 451, 1992.

118. **Beinsberger, S. E. I., Valck, R. L. M., Clijsters, H. M. M., de Greef, J. A., and van Onckelen, H. A.,** Effects of enhanced cytokinin levels in *ipt*-transgenic tobacco, in *Physiology and Biochemistry of Cytokinins in Plants*, Kamínek, M., Mok, D. W. S., and Zazímalová, E., Eds., SPB Academic Publishing, The Hague, 1992, 77.

119. **Schmülling, T., Schell, J., and Spena, A.,** Construction of cytokinin overproducing mutants in higher plants, in *Physiology and Biochemistry of Cytokinins in Plants*, Kamínek, M., Mok, D. W. S., and Zazímalová, E., Eds., SPB Academic Publishing, The Hague, 1992, 87.

120. **Schimper, A. F.,** Über die Verteilung der Chlorophyllkörner und Farbkörper, *Bot. Ztg.*, 41, 105, 1883.

121. **Reski, R., Wehe, M., Kasten, B., Reutter, K., Marienfeld, J. R., and Abel, W. O.,** The molecular analysis of chloroplast division (mini-review), *Crypt. Bot.*, 3, 18, 1992.

122. **Chaley, N., Possingham, J. V., and Thompson, W. W.,** Chloroplast division in spinach leaves examined by scanning electron microscopy and freeze-etching, *J. Cell Sci.*, 46, 87, 1980.

123. **Whatley, J. M.,** The ultrastructure of plastids in roots, *Int. Rev. Cytol.*, 85, 175, 1983.

124. **Lawrence, M. E. and Possingham, J. V.,** Observations of microtubule-like structures within spinach plastids, *Biol. Cellulaire*, 52, 77, 1984.

125. **Butterfaß, T.,** Patterns of chloroplast reproduction. A developmental approach to protoplasmic plant anatomy, *Cell Biol. Monogr.*, 6, 1, 1979.

126. **Van de Putte, P. J., van Dillewijn, J., and Rörsch, A.,** The selection of mutants of *Escherichia coli* with impaired cell division at elevated temperature, *Mutat. Res.*, 1, 121, 1964.

127. **Lutkenhaus, J. F., Wolf-Watz, H., and Donachie, W. D.,** Organisation of genes in the *fts*A-*env*A region of the *Escherichia coli* genetic map and identification of a new *fts* locus: *fts*2, *J. Bacteriol*, 142, 615, 1980.

128. **Leonard, J. M. and Rose, R. J.,** Sensitivity of chloroplast division cycle to chloramphenicol and cycloheximide in cultured spinach leaves, *Plant Sci. Lett.*, 14, 159, 1979.

129. **Pyke, K. A. and Leech, R. M.,** Chloroplast division and expansion is radically altered by nuclear mutations in *Arabidopsis thaliana*, *Plant Physiol.*, 99, 1005, 1992.

130. **Parthier, B.,** The cooperation of nuclear and plastid genomes in plastid biogenesis and differentiation, *Biochem. Physiol. Pflanzen*, 177, 283, 1982.

131. **Hagemann, R.,** Plastid genetics in higher plants, in *Plant Gene Research. Cell Organelles*, Herrmann, R. G., Ed., Springer-Verlag, New York, 1992, 66.

132. **Kuroiwa, T.,** The replication, differentiation, and inheritance of plastids with emphasis on the concept of organelle nuclei, *Int. Rev. Cytol.*, 128, 1, 1991.

133. **Reski, R. and Abel, W. O.,** Induction of budding on chloronemata and caulonemata of the moss, *Physcomitrella patens*, using isopentenyladenine, *Planta*, 165, 354, 1985.

134. **Abel, W. O., Knebel, W., Koop, H.-U., Marienfeld, J. R., Quader, H., Reski, R., Schnepf, E., and Spörlein, B.,** A cytokinin-sensitive mutant of the moss, *Physcomitrella patens*, defective in chloroplast division, *Protoplasma*, 152, 1, 1989.

135. **Reski, R., Wehe, M., Hadeler, B., Marienfeld, J. R., and Abel, W. O.,** Cytokinin and light quality interact at the molecular level in the mutant PC22 of the moss *Physcomitrella*, *J. Plant Physiol.*, 138, 236, 1991.

136. **Sasaki, Y., Nagano, Y., Morioka, S., Ishikawa, H., and Matsuno, R.,** A chloroplast gene encoding a protein with one zinc finger, *Nucleic Acids Res.*, 17, 6217, 1989.

137. **Nagano, Y., Matsuno, R., and Sasaki, Y.,** Sequence and transcriptional analysis of the gene cluster *trn*Q-*zfp*A-*psa*I-ORF231-*pet*A in pea chloroplasts, *Curr. Genet.*, 20, 431, 1991.

138. **Smith, A. G., Wilson, R. M., Kaethner, T. M., Willey, D. L., and Gray, J. C.,** Pea chloroplast genes encoding a 4 kDa polypeptide of photosystem I and a putative enzyme of C1 metabolism, *Curr. Genet.*, 19, 403, 1991.

139. **Reski, R. and Kruse, S.,** Does a finger point to plastid evolution? Submitted.
140. **Eisermann, A., Tiller, K., and Link, G.,** In vitro transcription and DNA binding characteristics of chloroplast and etioplast extracts from mustard (*Sinapis alba*) indicate differential usage of the *psb*A promoter, *EMBO J.*, 9, 3981, 1990.
141. **Kruse, S., Quast, S., Abel, W. O., and Reski, R.,** Cytokinin-promoted processing of plastid transcripts in cytokinin-sensitive moss mutants, submitted.
142. **Westhoff, P. and Herrmann, R. G.,** Complex RNA maturation in chloroplasts: the *psb*B operon from spinach, *Eur. J. Biochem.*, 171, 551, 1988.
143. **Ohyama, K.,** Organization and expression of genes of plastid chromosomes from nonangiospermous land plants and green algae, in *Plant Gene Research. Cell Organelles*, Herrmann, R. G., Ed., Springer-Verlag, New York, 1992, 137.
144. **Palmer, J. D.,** Comparison of chloroplast and mitochondrial genome evolution in plants, in *Plant Gene Research. Cell Organelles*, Herrmann, R. G., Ed., Springer-Verlag, New York, 1992, 99.
145. **Whittier, R. F. and Sugiura, M.,** Plastid chromosomes from vascular plants — genes, in *Plant Gene Research. Cell Organelles*, Herrmann, R. G., Ed., Springer-Verlag, New York, 1992, 164.
146. **Kanno, A. and Hirai, A.,** A transcription map of the chloroplast genome from rice (*Oryza sativa*), *Curr. Genet.*, 23, 166, 1993.
147. **Crossland, L. D., Rodermel, S. R., and Bogorad, L.,** Single gene for the large subunit of ribulosebisphosphate carboxylase in maize yields two differentially regulated mRNAs, *Proc. Natl. Acad. Sci. U.S.A.*, 81, 4060, 1984.
148. **Gamble, P. E., Berends, T., Sexton, T., and Mullet, J. E.,** Light-dependent changes in *psb*D and *psb*C transcripts of barley chloroplasts: accumulation of two transcripts maintains *psb*D and *psb*C translation capability in mature chloroplasts, *EMBO J.*, 7, 1289, 1988.
149. **Johnson, E. M., Schnabelrauch, L. S., and Sears, B. B.,** A plastome mutation affects processing of both chloroplast and nuclear DNA-encoded plastid proteins, *Mol. Gen. Genet.*, 225, 106, 1991.
150. **Rother, S., Hadeler, B., Orsini, J. M., Abel, W. O., and Reski, R.,** Fate of a mutant macrochloroplast in somatic hybrids, *J. Plant Physiol.*, 143, 72, 1994.
151. **Boynton, J. E., Gillham, N. W., Newman, S. M., and Harris, E. H.,** Organelle genetics and transformation of *Chlamydomonas*, in *Plant Gene Research. Cell Organelles*, Herrmann, R. G., Ed., Springer-Verlag, New York, 1992, 4.
152. **Svab, Z., Hajdukiewitz, P., and Maliga, P.,** Stable transformation of plastids in higher plants, *Proc. Natl. Acad. Sci. U.S.A.*, 87, 8526, 1990.
153. **Spörlein, B., Streubel, M., Dahlfeld, G., Westhoff, P., and Koop, H.-U.,** PEG-mediated plastid transformation: a new system for transient gene expression assays in chloroplasts, *Theor. Appl. Genet.*, 82, 717, 1991.
154. **Ye, F., Gierlich, J., Reski, R., Marienfeld, J. R., and Abel, W. O.,** Isoenzyme analysis of cytokinin sensitive mutants of the moss *Physcomitrella patens*, *Plant Sci.*, 64, 203, 1989.

Chapter 15

Regulation of the Cell Cycle By Cytokinins

Annie Jacqmard, Claude Houssa, and Georges Bernier

TABLE OF CONTENTS

I. INTRODUCTION

Historically, the definition of cytokinins was based on their stimulation of cell division of plant tissue grown *in vitro*: hence, the generic name "cytokinins."[1] This property of cytokinins, and their essential role in organogenesis, were instrumental in the development of commercial micropropagation and somatic cell manipulation of crop plants. Subsequent studies indicated that cytokinins are also involved in the control of the cell cycle in meristems (apical meristems of roots and shoots, and secondary meristems of the cambia and phellogen) of intact plants. Thus, knowledge of the role of cytokinins in controlling cell division is of paramount importance for both fundamental and applied biology. Unfortunately, this topic has received relatively little attention, and the precise mode of action of cytokinin remains largely unknown.[2]

In this chapter we will discuss cytokinin regulation of the cell cycle in nonhabituated and nontumorous plant cells. As published results on this topic were often obtained from companion studies with other processes, such as apical dominance and flower initiation, or shoot/root regeneration and biomass production, separating the effects of cytokinin on cell division is difficult. In addition, the results were sometimes contradictory. Therefore, in this review, results obtained from our investigation of the shoot meristem of *Sinapis alba* will serve for comparison with other studies.

II. THE CELL CYCLE

The classical view of the cell cycle is illustrated in Figure 1. Plant cells that exit the cycle permanently, i.e., stop dividing, are usually displaced away from the meristematic area, as the result of differentiation. However, there are exceptions where cells leave the cycle temporarily while remaining within the meristems. These noncycling cells are found in active root[3,4] and shoot[5,6] meristems as well as in dormant meristems[7] and correlatively inhibited axillary buds.[8,9] These noncycling cells can resume proliferation when they receive an appropriate stimulus. Noncyling cells are often, but not always, arrested in G1. Lajtha proposed calling this phase G0,[10] where cell division is blocked, to distinguish it from the G1 of an active cycle. Recent work by Bergounioux and associates[11] revealed

0-8493-6252-0/94/$0.00+$.50

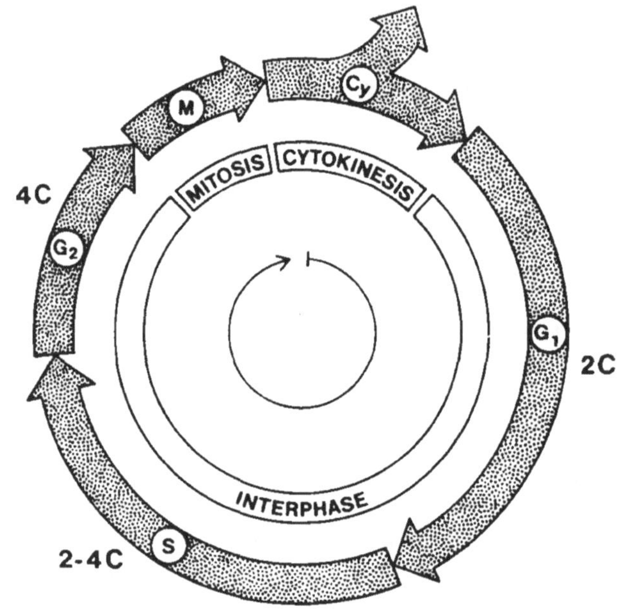

Figure 15-1 The five successive phases of the cell division cycle. After the nuclear division (mitosis, M) and the cytoplasmic division (cytokinesis, Cy), the two daughter cells begin interphase of a new cycle. Interphase consists of three successive phases: the presynthetic phase when the cell has a 2C nuclear DNA amount (G1); the DNA-synthetic phase (S), and the postsynthetic phase (G2) when the cell has 4C of DNA. DNA replication in S not followed by mitosis and cytokinesis gives rise to endopolyploid cells (endomitotic cycle). Mitosis not followed by cytokinesis leads to binucleate cells.

that G0 nuclei can indeed be distinguished from G1 nuclei, on the basis of their chromatin structure.

The percentage of cycling cells is called the "growth fraction."[12] It varies from 100% in some zones of active meristems[3,13,14] to 0% in dormant, or completely inhibited, axillary meristems as well as in quiescent embryos.[7,9,15,16] However, it should be noted that it is not always possible to distinguish between cells cycling very slowly and noncycling cells.

The classical catenary model of the cell cycle (Figure 1) is understandably an oversimplification. More complex models, including multisequential cycles, G0/G1 transition probability models, and oscillatory models, have been proposed for animal and yeast cells.[17-20] Whether these models apply to plant cells is unclear, and in this chapter we shall thus use only the scheme of Figure 1.

III. METHODS TO MEASURE THE RATE OF CELL DIVISION

Several techniques are available to study the cell division cycle. Some are more accurate than others, and some are better adapted for cultured cells than for cells present in the small and relatively inaccessible meristems of intact plants. Changes in the rate of division in a cell population are often inferred from changes in the mitotic index or the percentage of cells in mitosis. Determination of mitotic indices is simple and may seem to yield unambiguous data. Hence, this technique is commonly used. However, in order for the mitotic index to be a reliable indicator of the division rate, the duration of mitosis must be nearly identical for all cells. Although this is often the case, a meaningful interpretation of mitotic indices requires additional information, such as the duration of the M phase relative to other phases of the cell cycle. Similar comments can be made concerning the labeling index or percentage of radiolabeled nuclei after a short exposure of the tissue to [^{3}H]thymidine ([^{3}H]TdR), because the duration of S is more variable than that of mitosis. Microdensitometric measurements of the DNA content are often used to determine the relative numbers of 2C and 4C nuclei. These parameters provide rough estimates of the relative lengths of the G1, S, G2, and M phases of the cell division cycle.[21] Flow cytometry provides an alternative to obtain faster and more-reliable measurements of the proportions of cells in these phases.[22-25]

Direct counting of cells and comparison of cell numbers at successive intervals can be used to obtain an estimate of the cell doubling time. However, this technique requires the separation of tissues into single cells, a difficult task with either meristems of intact plants or cell suspension cultures.[26] As an increase in cell number often leads to an increase in biomass, many studies conducted with callus or cell suspension cultures equate changes in cell division with fresh- or dry-weight changes, but since the relationship between biomass and cell division is loose, these studies do not yield relevant data concerning the cell cycle. The most popular technique for estimating the durations of the cell cycle and its component phases in asynchronous cell population is the "pulse-labeled-mitoses" (PLM) technique,[27] which involves labeling subpopulations of cells at the S phase with [^{3}H]TdR and follows the subsequent fate of these labeled cells. In an ideal cell population where all cells divide and all have an identical cycle duration, this duration is equal to the cell doubling time. However, few, if any, cell populations are ideal. In addition, many are not perfectly asynchronous; i.e., cells are not distributed in the different phases of the cycle in proportion to the time they spend in each of these phases. Under these conditions, the results obtained with the PLM technique should be interpreted with caution.[28,29] A cell doubling time longer than the cell cycle duration is indicative of heterogeneity in the cell population, presumably a mixture of cycling and noncycling cells. The proportion of cycling cells, i.e., the growth fraction, can be estimated by various methods, as reviewed by Clowes,[3] and flow cytometry provides good measurement of such a parameter.[25]

Synchronized cell populations could be obtained by various experimental means, e.g., withdrawal, followed by resupply, of a critical regulator or the transient use of inhibitors. In such populations, peaks of cells in the G1 (cells with 2C of DNA), S (cells labeled by [^{3}H]TdR), G2 (cells with 4C of DNA), and M phase follow each other in sequence, and determination of their timing of occurrence thus provides information on the durations of the cell cycle and its phases.

IV. CYTOKININ STIMULATION OF CELL PROLIFERATION *IN VITRO* AND *IN VIVO*

The stimulatory effect of cytokinins on cell division was initially observed in *in vitro* cultured tobacco pith tissue[30,31] and confirmed with other tissues, cells, and protoplasts of a variety of plants.[32-40] Interestingly, cytokinins are also able to stimulate division of cultured haploid[41] and polyploid cells.[42] This effect is direct and not inductive; that is, they are not only required to initiate cell proliferation in nondividing tissues, but their continued presence is needed to sustain mitotic activity. In cell suspensions and protoplast cultures, cytokinin withdrawal and refeeding resulted in the corresponding arrest and partially synchronous recovery of cell divisions.[36,38,43,44] The general conclusion emerging from all these studies is that cytokinins are one of the major controlling factors of the cell cycle in cultured plant materials.

The question of whether this conclusion can be extended to intact plants is difficult to answer. Applications of cytokinins to intact plants, e.g., spraying whole plants, often do not promote their growth and may even inhibit it. This negative result might be due to the fact that plants normally contain optimal amounts of cytokinins, and an extra supply is unnecessary for, or even detrimental to, further growth.[45] This is especially relevant in the case of root tips, which are known as the major sites of cytokinin production during the phase of vegetative growth of the life cycle (see Chapter 5). Indeed, it has been found that exogenous cytokinins often inhibit growth, mitotic activity, and/or DNA replication, in root tips.[46-48]

Thus, the demonstration of cytokinins controlling the cell cycle in intact plants requires the use of either plants grown under conditions resulting in suboptimal cytokinin production or the analysis of tissues deficient in cytokinins.

In photoperiod-sensitive species, conditions that seem to prevent optimal cytokinin production are often also unfavorable for flowering. In plants exposed to nonpermissive photoperiods for flowering, an application of a cytokinin *directly* to the shoot apical meristem caused an increase of the mitotic and/or the labeling index. This phenomenon was observed in both long-day [*Sinapis alba* (Figure 2),[49,50] *Rudbeckia bicolor*,[51,52] *Lolium temulentum* (Table 1),[53] and *Arabidopsis thaliana*[54]] and short-day (*Chenopodium rubrum*[55] and *Perilla nankinensis*[56]) plants.

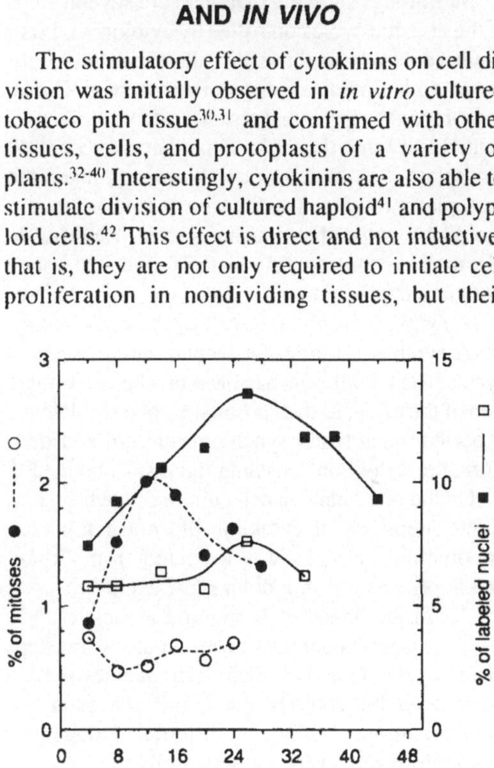

Figure 15-2 Mitotic (●- - -○) and labeling (■—□) indices in control and benzyladenine-treated *Sinapis alba* shoot meristems (control, open symbols; benzyladenine-treated, solid symbols) as a function of time after a single application of benzyladenine (4.5 × 10^{-5} *M*). Plants were grown for 2 months in 8-h short days, and their meristem was strictly vegetative. Benzyladenine and [^{3}H]TdR (specific activity 888 GBq/mmol^{-1}; concentration 370 kBq/ml^{-1}) were applied using a cotton plug inserted into the young leaves of the apical bud. (From Houssa, C. and Jacqmard, A.,[50] unpublished data.)

Table 15-1 **Effects of Benzyladenine on Labeling Index (%) in the Shoot Apex of *Lolium temulentum* L. cv Ceres**[a]

Hours after treatment	Control	Benzyladenine-treated
12	2.3	11.0
20	1.8	10.8
28	1.3	6.9
36	3.5	5.5

[a] From Jacqmard, A. and Houssa, C., unpublished data.[53] Plants were grown under short days (8 h) for 6 weeks and were vegetative. Benzyladenine ($4.5 \times 10^{-5} M$) and [^{3}H]TdR (specific activity 185 $GBq.mmol^{-1}$; concentration 370 $kBq\ ml^{-1}$) were injected into the leaf sheath cavity just above the apex.

Cells of tomato flowers ceased to divide, and the flowers were targeted for abortion, when plants were grown under short days at low irradiance.[57] The arrest was apparently due to a severe cytokinin deficiency, since exogenous cytokinin applied directly to the aborting flowers reactivated the cell division cycle in ovules and other floral parts.[57] Cytokinin was also limiting germination of excised embryos, as evidenced by the stimulation of cell division in benzyladenine-treated maize embryo axes.[58]

In normal plants some organs, such as the correlatively inhibited axillary buds, are deficient in cytokinins.[59] The inhibition can often be overcome by a local application of a cytokinin.[59,60] The growth promotion was found to be mediated, in part, by an activation of the cell division process, as reported in a variety of angiosperms such as soybean,[61] *Cicer arietinum,*[62] *Opuntia polyacantha,*[63] pea,[9] and the fern *Polypodium vulgare.*[64] Similarly, exogenous cytokinins induced cell division in dormant buds of potato tubers.[65] Limited evidence suggests that cytokinins also act on secondary meristems. Cytokinins appear to be needed to maintain activity in the basal intercalary meristem of the culm of *Cyperus alternifolius,*[66] and stimulation of cell division by benzyladenine in the procambium of soybean petioles was recently reported.[67] As for cultured materials, the cytokinin effect in intact plants is direct.[9] The observations cited above indicate that cytokinin is one of the major promoters of plant cell division *in vitro* as well as in intact plants.

V. DO CYTOKININS AFFECT ONE OR SEVERAL STEPS OF THE CYCLE?

Cytokinins may act at one or more specific steps of the cell cycle. Evidence on this central question, and identification of the controlled step(s), was at first controversial. In their pioneer work on cultured tobacco pith tissue, Skoog and associates concluded that cytokinins are required at three steps of the cycle: the G1/S and G2/M transitions and cytokinesis.[68] Contrasting results were later reported, by Jouanneau and Tandeau de Marsac[34] and Fosket and Short,[69] working on cytokinin-requiring cell cultures of tobacco (cv. Wisconsin) and soybean (cv. Biloxi), respectively. In tobacco cells partial synchronization of cell divisions could be induced by withdrawal followed by a resupply of cytokinin to the culture medium. This synchronization was not prevented by the presence of FUdR, an inhibitor of DNA synthesis. Moreover, it was found that omission of cytokinin from the medium did not markedly affect the amount of [^{14}C]TdR incorporated into nuclear DNA, at least during the first cycle.[34] Similarly, when soybean cells were transferred to a medium lacking cytokinin, cell proliferation ceased within 48 h, while DNA replication continued.[69] Apparently, cytokinin-deprived soybean cells were capable of successive rounds of DNA replication without intervening cell divisions (endomitosis). These results were taken to indicate a single step of the cycle requiring cytokinin and that this step occurs after the S phase, possibly at the G2/M transition.[34,70]

This interpretation has been challenged by more recent findings showing that there are several steps of the division cycle controlled by cytokinins. First, results contrary to those just described were obtained with similar experimental materials. Cooke and Meyer observed that absence of cytokinin prevented entry into the S phase in tobacco mesophyll protoplasts.[71] Addition of cytokinin stimulated the protoplasts to enter the S phase 5 to 6 h later. It should be noted that Jouanneau and Tandeau de Marsac[34] recognized that in their study, although cytokinin deprivation of tobacco cells did not affect DNA synthesis of the first cell cycle, it did reduce DNA synthesis during the second and subsequent cycles. Also, if all cells had been blocked at a single step of the cycle, as they postulated, one would have expected much better synchronization of cell division after cytokinin refeeding than was obtained.[43]

On the other hand, a cell culture of soybean (cv. Acme) deprived of cytokinin did not exhibit endopolyploid cells, as was expected from Fosket and Short's results, nor did it show exclusive arrest in G2, in the hands of Wang and associates.[36] In this case arrest apparently occurred at several steps of the cycle. Thus far, such discrepancies remain unresolved, but could be due to differences in culture conditions leading to different endogenous cytokinin levels in the cultured cells.

Second, in pea root cortical explants, kinetin induced two rounds of DNA replication prior to the first mitosis, the first round being apparently connected with the cells entering an endomitotic cycle.[72] Similarly, in primary leaves of intact bean plants treated with benzyladenine at a time when cell division in these leaves was terminated, cells entered a new round of nuclear DNA replication, but cell division was not resumed.[73-75] Clearly, in these two systems the cytokinin appeared to affect the G1/S transition. Third, in *Petunia* leaf protoplasts, cytokinin was found to promote the S/G2 transition.[39] Fourth, in the shoot apical meristem of intact *Sinapis* plants (Figure 2) and the ovules of aborting tomato flowers,[57] a single application of a cytokinin caused a nearly simultaneous increase in labeling and mitotic indices. This was interpreted as evidence that cytokinins act independently in these systems, on at least two steps of the cycle: the G1/S and G2/M transitions. Indeed, if cytokinins were acting only at one of these two transitions, the increase of one index would automatically occur several hours before the increase of the other index. Fifth, cytokinins also affect mitosis, as shown by the observation that in cytokinin-deprived tobacco cells the duration of the metaphase is considerably lengthened.[34]

Finally, convincing evidence supporting cytokinin action at more than one step of the cycle was obtained from analyzing cytokinin effects on the duration of component phases of the cell cycle, using the PLM technique. In root tips of *Vicia faba* seedlings,[76] MacLeod observed that kinetin application led to a reduction in cell cycle duration (T), as the result of shortening the G1 and S phases (Table 2). However, the duration of G2 was increased. The changes in the duration of the cell cycle and S phase were transient, whereas the changes in G1 and G2 were more persistent. Van't Hof was unable to detect similar effects of kinetin on cultured pea root tips;[77] the discrepancy may stem from differences in the endogenous cytokinin status between the *Vicia* and pea roots. Alternatively, it may have been caused by the long delay (72 h) before examining the effects of the cytokinins, in Van't Hof's study.

Using the vegetative shoot apical meristems of *Sinapis*, Houssa et al. found that a single application of benzyladenine caused a shortening of cell cycle time, via the reduction of S and G2 durations (Table 2).[78,79] The length of S was most affected: 2.5 times shorter in treated meristems. Interestingly, in both the *Vicia* root and *Sinapis* shoot meristems, the duration of M was also shortened in response to exogenous cytokinin (Table 2).

These studies suggest that cytokinins are essential at multiple steps of the cell cycle, for example, affecting, in the *Sinapis* shoot meristem, the G1/S transition (presumably by releasing noncycling cells arrested at the end of G1), the S phase (increase in the rate of DNA replication), the G2 phase, the G2/M transition, and mitosis. The effects on specific stages will be examined in more detail in the following sections.

VI. DO CYTOKININS INCREASE THE GROWTH FRACTION?

In some materials cytokinins stimulate growth and cell division by increasing the growth fraction, that is, by releasing cells arrested at a particular stage of the cell cycle. Depending on the cell

Table 15-2 **Duration (hours) of the Cell Cycle and its Component Phases in *Vicia faba* Root Tips and *Sinapis alba* Shoot Meristems[a,b]**

	T	G1	S	G2	M
Vicia faba					
Root tips					
Control	16.6[c]	3.6[c]	8.3[c]	2.8[c]	1.9[c]
Kinetin 10^{-5} *M*	13.2[c]	0.9[c]	5.4[c]	5.7[c]	1.2[c]
Sinapis alba					
Shoot meristems					
Control	66.0[e]	24.0[e]	15.0[e]	25.0[e]	2.1[c]
Benzyladenine 4.5×10^{-5} *M*	49.0[d]	24.4[d]	6.0[e]	17.4[e]	1.3[c]

[a] Compiled from Reference 76 for *Vicia faba*; and from Reference 79 for *Sinapis alba*; [b] Growing conditions for *Sinapis* plants and benzyladenine and [^{3}H]TdR treatments were as in Figure 2; [c] Estimated from the PLM curve;[27] in both *Vicia* and *Sinapis*, cytokinin was applied immediately after the pulse of [^{3}H]TdR; [d] Estimated from a formula of Cairnie, A. B. et al.[80]; [e] Estimated from a formula of MacLeod, R. D.[76]

population, the effects of cytokinins may be interpreted differently. For example, cytokinins were found to act at the G1/S transition in the cotyledonary bud of pea and the embryo axis of maize, since the majority of cells in these tissues were initially blocked at G1.[9,15,58] Cytokinins appeared to affect essentially the G2/M transition in the axillary bud of *Cicer*, since arrested cells were believed to be mostly in G2.[62] In the shoot meristem of *Sinapis*, cytokinin action on increasing the growth fraction was inferred from the increase in the labeling index (Figure 2) without changing the duration of G1 (Table 2). However, the growth fraction in this system cannot be measured consistently;[81] therefore, the effect of cytokinins on this parameter has not been conclusively determined.

VII. CYTOKININ ACTION ON THE S PHASE

A. THE PROCESS OF DNA REPLICATION

The chromosomal DNA of all eukaryotes is organized for the purpose of replication during the S phase into multiple discrete units, called the replicons.[82,83] Each replicon has an origin where replication starts, and two replication forks that diverge bidirectionally from the origin while replicating nascent DNA chains. Not all replicons replicate simultaneously, and usually families of replicons are activated sequentially throughout the S phase.[84,85] Therefore, the total duration of S varies considerably depending on three parameters:

1. The recruitment of replication origins, which determines the replicon size or interorigin distance
2. The fork progression rate
3. The degree of synchrony of replicon family activation[84]

These three parameters can be estimated using the technique of DNA-fiber autoradiography.[86] The tissue is labeled with high-specific-activity $[^3H]$TdR, and the DNA fibers from the labeled nuclei are spread on microscope slides and analyzed by autoradiography. Figure 3 illustrates the kind of autoradiographic preparations obtained, and Figure 4 shows the corresponding diagrams. The replicon size, R, is the midpoint-to-midpoint distance between pairs of labeled DNA segments. The fork progression rate, F, is calculated from the mean length of labeled segments for various labeling durations. The degree of synchronicity of replicon family activation is the ratio of the length of time, R_s, needed to replicate the DNA of an average replicon to the length of time required to replicate the whole chromosomal DNA or S-phase duration, T_s.

The molecular details of the process of DNA replication are very complex and beyond the scope of this review. (For the state of the art in this field, the reader is referred to reviews by Laskey et al.,[87] Diffley and Stillman,[88] and Held and Heintz.[89]) Two points will be emphasized here. First, the process of DNA replication is performed by a so-called "replication complex" including several specific proteins acting cooperatively and sequentially, like the DNA polymerases, DNA helicase, primase, topoisomerase, and accessory proteins.[90,91] The earliest step of this process is the recognition of replicon origins, which is achieved by the direct binding to the DNA sequence of these origins of a specific protein, called the initiator. The activity of this initiator protein is regulated by phosphorylation, which

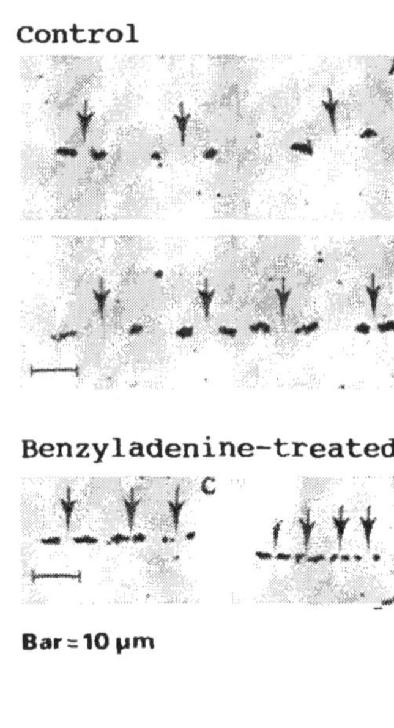

Figure 15-3 Autoradiograms of DNA fibers isolated from cells of untreated control (A, B) and benzyladenine-treated (C, D) vegetative shoot meristems of *Sinapis alba* pulsed with $[^3H]$TdR (specific activity 3.29 TBq/mmol^{-1}; concentration 37 MBq/ml^{-1}) for 4 h. Growing conditions for plants and benzyladenine treatment were as in Figure 2. $[^3H]$TdR was applied directly on the meriste[illegible] using a plastic collar fitted around the apex. The arrows indic[illegible] midpoints of pairs of labeled DNA segments. Replicon size[illegible] are seen as the distances between pairs of adjacent arrows. (Ba[illegible] 10 µM.) (From Houssa, C., Jacqmard, A., and Bernier, G., [illegible]anta, 181, 324, 1990. With permission.)

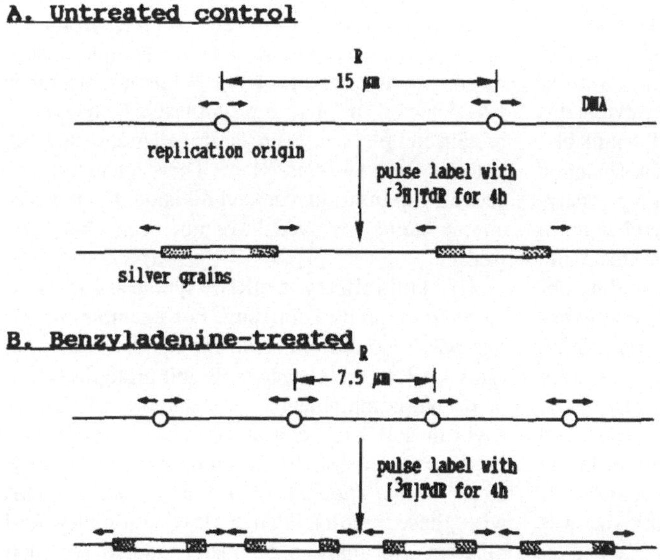

Figure 15-4 Diagrammatic representation of a DNA duplex isolated from shoot meristems of *Sinapis alba.* Equally spaced origins (open circles) and labeled DNA segments (dotted bars) are illustrated on autoradiograms of fibers isolated from cells of control (A) and benzyladenine-treated (B) vegetative meristems pulsed with [^{3}H]TdR for 4 h. The original autoradiograms are presented in Figure 3. R is the replicon size.

is essential for its ability to bind to single-stranded DNA at replicon origins. Evidence from various eukaryotes, including higher plants,[92] indicates that specific DNA sequences normally serve as replication origins. One critical, unresolved question is whether there is only one class of replicon origins in a large taxonomic group of eukaryotes, such as the higher plants. This is an important point because in the case of a single class of origins, changes in the recruitment of these origins (and thus in replicon size) can be due simply to the availability of a single initiator protein.[93] If there is more than one origin class and, correspondingly, more than one initiator molecule, the situation is then bound to be more complex.

Second, another level of control of origin activation is determined by the degree of synchrony of replicon family activation. This parameter can be very different in different cell types of the same organism.[94,95] In meristems of higher plants, the R_s/T_s ratios commonly range from 0.11 to 0.71,[96] indicating that replicon family activation is neither perfectly synchronous ($R_s/T_s=1$) nor totally asynchronous (R_s/T_s almost equal to 0). The basis for such differences in the degree of synchrony is essentially unknown, but some studies point to topological factors within the nuclear structure as being essential. As Diffley and Stillman[88] put it, "It is possible that one important role for nuclear structure in chromosomal replication may be organizational, pulling together colinear arrays of replication origins into clusters that function synchronously." If this is true, large changes in the R_s/T_s parameter should be associated with detectable alterations of the nuclear structure.

B. CYTOKININ ACTIVATION OF REPLICON ORIGINS

The effect of a single application of benzyladenine on the parameters of DNA replication in the vegetative shoot meristem of *Sinapis* plants was investigated by Houssa et al.,[78] using the technique of DNA-fiber autoradiography. Recall here that benzyladenine in this system caused a marked shortening of the S phase (Table 2). As seen in Table 3, this shortening was caused by two distinct mechanisms: (1) a decrease in the replicon size, R, from 15 to 7.5 μm, indicating that the cytokinin is capable of recruiting latent replication origins in this system; and (2) an increase in the degree of synchronicity of replicon family activation, R_s/T_s, from 0.39, a value that is typical of meristems in higher plants (see above), to near unity, the highest value ever found in these plants. This observation indicates that all replicon origins, whether they belong to a single class or multiple classes, are activated synchronously by cytokinin at the start of the S phase in *Sinapis* meristems. Therefore, cytokinin may exert its action at the start of the S phase, by increasing the availability of one key protein component common to all replication complexes.

As discussed above, large changes in the degree of synchrony of replicon family activation are expected to be accompanied by alterations of nuclear structure. Although the ultrastructure of nuclei in benzyladenine-treated *Sinapis* meristems, examined for quantitative changes by electron microscopy and stereological techniques, did not exhibit marked changes,[97] it should be noted that earlier studies of Piesco and Alvarez[98] found that the thermal stability of chromatin was decreased, while Feulgen dye binding was enhanced, after a kinetin treatment, in onion root tips. These changes seem to reflect alterations in chromatin configuration. Similar changes should be sought in systems where cytokinins are known to stimulate cell division. Presumably, techniques with greater resolution, such as scanning cytophotometry, flow cytometry, and image analysis, will be required to identify molecular changes in chromatin structure affected by hormone treatment.

We attempted to confirm the initial finding obtained with shoot meristem of *Sinapis* that exogenous cytokinin is an activator of latent replication origins. Two distinct experimental systems were selected: the ovule of tomato and vegetative apex of *Lolium temulentum*. In both systems it was found that a single application of benzyladenine led to increases in the labeling and mitotic indices (Table 1).[57] As observed in *Sinapis*, the replicon size was halved (unpublished). Thus, it is likely that activation of latent replicon origins could be a *general* effect of cytokinins in controlling the duration of the S phase in meristems of angiosperms. Whether cytokinin also synchronizes the initiation of DNA replication in all replicons of the tomato ovule or *Lolium* meristem remains to be investigated.

VIII. STIMULATION OF PROTEIN AND mRNA SYNTHESIS BY CYTOKININS IN RELATION TO THE CELL CYCLE

A number of studies have established that protein synthesis is required for progress of plant cells through the successive steps of the cell cycle.[99-106] Numerous attempts were made to correlate the effects of cytokinins on the cell cycle, via their effects on protein and/or mRNA synthesis. Unfortunately, available data obtained thus far do not allow any firm conclusions. In some cytokinin-dependent experimental systems, e.g., tobacco cells and protoplasts or pea root cortical explants, the rate of protein synthesis was not different with or without cytokinins,[71,107,108] while in other materials, such as soybean or apple cell suspensions, cytokinins were found to stimulate protein synthesis and polysome formation.[35,40]

Conflicting results were also obtained in studies concerning cytokinin effects on the complement of recently synthesized proteins. While no difference was detected in tobacco protoplasts,[109] changes in protein complements were described in several other cultured cell systems.[40,44,107] These changes preceded cytokinin-induced cell division, but the specific proteins involved have not been characterized.

Cytokinin effects on mRNA synthesis and translation are equally confusing. For example, no effect was reported either on the synthesis of total [poly A]$^+$RNA in tobacco cells and protoplasts[71,110] or on the complement of translatable mRNAs in soybean and bean cell suspensions.[111] However, more recent work, by Bergounioux and associates,[25,39] on *Petunia* leaf protoplasts was at odds with these results. Using flow cytometry and biparametric analysis of nuclei stained for total DNA and total RNA, they found that benzyladenine increased the RNA levels of G1, S, and G2 nuclei. In addition to an increase in rRNA transcripts, specific mRNAs were increased, including the mRNA of $p34^{cdc2}$ kinase (see Section IX). It is likely that the introduction of a more sensitive technique may allow the detection of cytokinin-stimulated transcription of cell-cycle-specific genes.

IX. THE MOLECULAR CONTROL OF THE G1/S AND G2/M TRANSITIONS. ARE CYTOKININS INTERACTING WITH THESE MECHANISMS?

The recent genetic and molecular studies on the regulation of the division cycle in yeast and animal cells have started to unravel the complex mechanisms that regulate the G1/S and G2/M transitions.[20,112-123] Briefly, these two critical transitions require the action of a protein kinase, $p34^{cdc2}$. The steady-state level of $p34^{cdc2}$ does not change during the division cycle in cycling cells. The key events triggering the two transitions are dependent on the interactions between the $p34^{cdc2}$ and specific cyclins (S and M forms). The latter accumulate transiently

Table 15-3 **The Effects of Benzyladenine on the Parameters of DNA Replication in Vegetative Shoot Meristematic Cells of *Sinapis alba*[a]**

	T_s (h)	R (μm)	F (μm/h^{-1})	R_s (h)	R_s/T_s
Control	18.0	15.0	1.05	7.14	0.39
Benzyladenine-treated	5.0	7.5	0.76	4.93	0.99

[a] From Reference 78, with permission. Growing conditions for plants and benzyladenine treatments were as in Figure 2; [^{3}H]TdR pulse was as in Figure 3.

in advance of these transitions. As a result of these interactions, specific cyclins activate the p34^{cdc2} kinase. Activation of p34^{cdc2} kinase is also controlled by the phosphorylation and dephosphorylation of specific threonine and tyrosine residues of the protein. Thus, active p34^{cdc2} becomes available at, or near, G1/S and G2/M transition points.

The cyclins that are acting at the two transition steps are different, although they share some structural homologies. In yeast, for example, the cyclins acting at the G1/S and G2/M transitions are called the G1 (S form) and mitotic (M form) cyclins, respectively. G1 cyclins accumulate during G1 and are then degraded during S, whereas mitotic cyclins accumulate during G2 and are destroyed during M. Inactivation of p34^{cdc2} kinase activity is believed to be the consequence of cyclin degradation during S and M.

When active, the p34^{cdc2} kinase phosphorylates a number of other proteins essential for progression through the cell cycle. At the G1/S transition, a possible substrate for p34^{cdc2} is the initiator protein(s) whose specific role is to recognize replicon origins, as described above. Another possible substrate is the CDC7 kinase, which when activated by phosphorylation, may, in turn, phosphorylate one or more target substrates essential for DNA replication.[124] Thus, it appears that the G1/S shift is likely to be controlled by a cascade of coupled protein phosphorylations, and several of the phosphorylated proteins are required to initiate DNA replication. In this regard it is important to stress that the p34^{cdc2} molecule has been shown to be a component of the so-called replication complex mentioned above.[125,126]

Apparently, the G2/M transition is controlled by similar processes, i.e., a series of phosphorylation events initiated by the p34^{cdc2} kinase. Nuclear envelope and nucleolus disassembly at prophase are brought about by phosphorylation of the lamin component of the nuclear envelope and nucleolin and other proteins of the nucleolus, chromosome condensation by phosphorylation of histone H1, cytoskeletal reorganization by phosphorylation of tyrosine kinase src, etc.[127] Exit from M requires p34^{cdc2} kinase inactivation, which presumably will allow all the forementioned cytological changes to be reversed at telophase.

Many of these regulatory proteins have been conserved during eukaryote evolution. For example, p34^{cdc2} and cyclins have been isolated from various organisms, including higher plants[128-137] (see reviews by Francis[138] and Jacobs[139]). The plant p34^{cdc2} was shown to be a *functional* homolog of the p34^{cdc2} kinase of yeast.[130,131,134,135,140] Thus, the regulation in higher plants of the onset of both the S and M phases may be similar to that in yeasts and animals. Mineyuki et al.[141] reported that p34^{cdc2} accumulates in the preprophase band (PPB) of microtubules in onion root tip cells. Since the plane of the cell division is predicted, if not predetermined, by the PPB, this suggests the possibility that p34^{cdc2} may also be involved in the determination of the division plane. Given the prominent role of cytokinins in the control of cell division in plants, an interesting question is whether this hormone interacts with the known regulatory proteins of the cell cycle. As both cytokinins and p34^{cdc2} act at multiple and some similar control points of the division cycle, i.e., G1/S and G2/M shifts and S and M phases, it may be speculated that cytokinins exert their control via the p34^{cdc2} kinase.

X. CELL PROLIFERATION AND ENDOGENOUS CYTOKININ LEVELS

The endogenous cytokinin level is usually high in mitotically active tissues, including root and shoot meristems, procambium, and cambium (Figure 5).[142-146] In tissues where the cell cycle is arrested, such as mature-plant parts and the quiescent center of root tips, the level of cytokinin is very low.[142,144-146]

Certain tissues in plants are thought to contain suboptimal levels of cytokinin. Examples include shoot apices of photoperiod-sensitive species grown under a nonpermissive daylength, aborting tomato flowers due to low light intensity, and axillary buds inhibited by apical dominance (see above). In all these materials, the progress of cells through the cell cycle is either slowed or arrested. Exogenously supplied cytokinins stimulate cell division. Interestingly, other treatments that lead to mitotic activity invariably increase the level of endogenous cytokinin. Transferring photoperiodic plants from a nonpermissive daylength to a favorable daylength for flowering was accompanied by an increase in cytokinins in *Sinapis alba*,[147,148] *Chenopodium murale*,[149] *Chenopodium rubrum*,[149] *Perilla frutescens*,[150] and *Xanthium strumarium*.[151] Flower bud abortion in tomato correlated with cytokinin deficiency, as compared with normal flowers under high light intensisty.[152] On the other hand, correlative inhibition of axillary buds by the shoot apex is believed to be mediated by cytokinin distribution;[59,144,153] removal of the shoot apex results in redirection of root-derived cytokinins to axillary buds, leading to bud break.

Changes in cytokinin content as cells progress through the cell cycle were elegantly demonstrated, by Nishinari and Syono,[154] in tobacco suspension cultures synchronized using aphidicolin, an inhibitor of DNA replication. Withdrawing the inhibitor

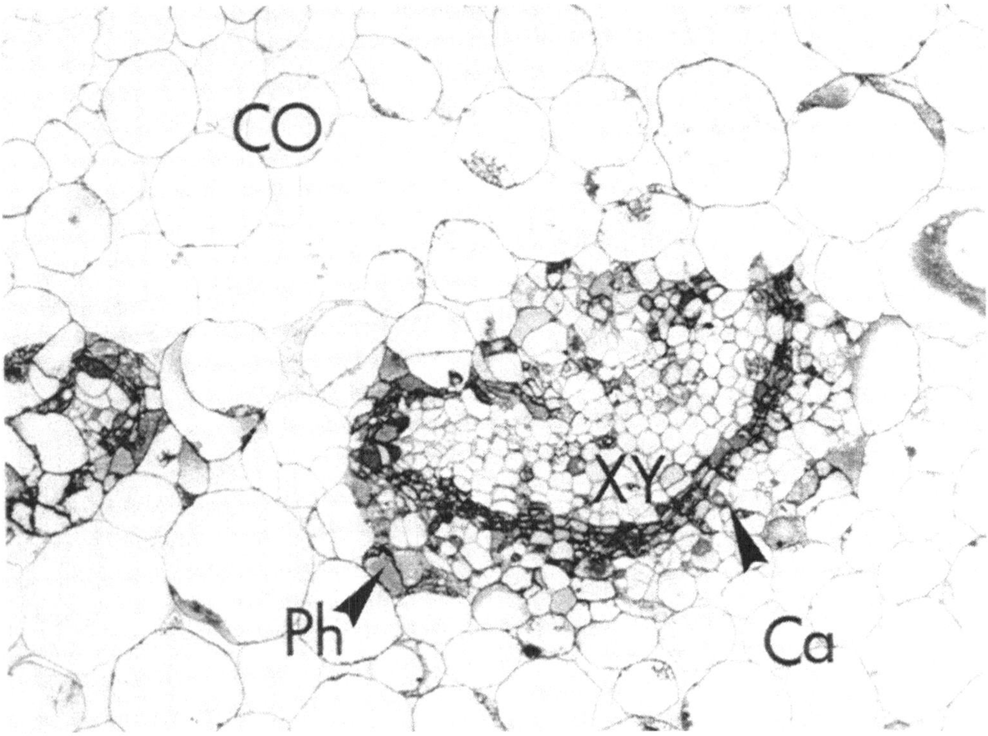

Figure 15-5 *Immunocytolocalization of cytokinin in cambial layers of tobacco petiole. The section was incubated with a rabbit antiribosylzeatin serum followed by goat anti-rabbit antibody conjugated to peroxidase. Location of ribosylzeatin in the cambium is marked by dark staining. Ca, cambium; Ph, phloem; XY, xylem; and Co, cortex. (Unpublished photograph, Stroobants,[145] kindly provided by Prof. E. Miginiac.)*

allowed cells arrested at G1 to enter S phase, and the synchrony was maintained for at least two cycles (Figure 6). A doubling of cytokinin content was detected in S, and a fourfold increase at the G2/M transition. The major cytokinins accumulating in this system at the G2/M shift were identified as zeatin and ribosylzeatin.[155] It remains to be determined if a qualitative difference in the cytokinin components is associated with specific steps of the cell cycle.

XI. INTERACTION OF CYTOKININS WITH OTHER REGULATORY FACTORS

Many factors regulate cell division in plants.[156] It is expected that cytokinins act in concert with these factors, in regulating cell division. The mode of action of such interactions is unknown. Nevertheless, results concerning selected factors will be reviewed briefly.

A. CARBOHYDRATES

Carbohydrates are known as major regulators of the plant cell division cycle.[157] In cultured root meristems starved of carbohydrates, division ceased, and cells were arrested preferentially in G1 and G2. This observation led Van't Hof and Kovacs to propose that the cell cycle is governed by two independent "Principal Control Points" (PCP):[157] one in G1 that regulates the G1/S transition, and another in G2 that controls the G2/M transition. A similar situation presumably prevails in shoot meristems.[5] As cytokinins also act at these two transitions, they may cooperate with carbohydrates as PCP factors. Unfortunately, no study aimed at cytokinin–carbohydrate interaction in relation to the cell cycle has been undertaken.

B. AUXINS

Since the initial work, of Skoog and associates, on cultured tobacco pith tissue, it has been known that not only a cytokinin, but also an auxin, is required for cells to complete the cell cycle.[31,68] Das et al. reported that a major feature of cytokinin action was an enhancement of cytokinesis in binucleate cells resulting from the presence of auxin alone.[31] This absolute requirement for two hormonal

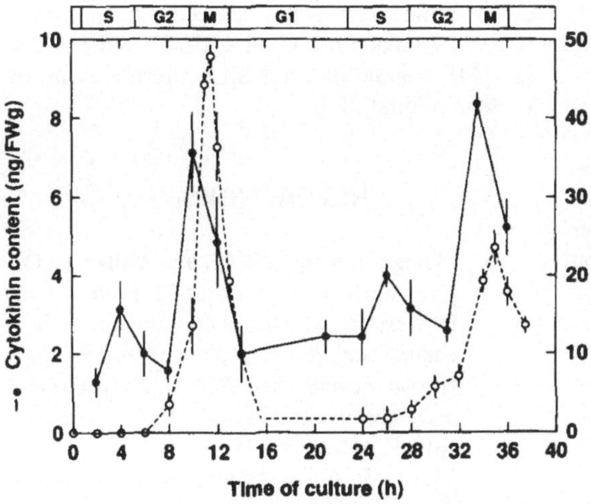

Figure 15-6 Changes in the mitotic index (MI) and cytokinin content after aphidicolin treatment of tobacco XD6S cells. The cells were precultured for 24 h in aphidicolin (5 μg/ml). The successive steps of the first two synchronous cell cycles, following removal of aphidicolin at time 0, are shown in the bar above the graph. (From Nishinari, N. and Syono, K., *Plant Cell Physiol.*, 27, 147, 1986. With permission.)

regulators was later confirmed in many other studies using a variety of cultured plant materials, including tobacco and soybean suspension cultures,[34,69] pea root cortical explants,[158] and tobacco and *Petunia* protoplasts.[38,39] In tobacco pith tissue and tuber tissue of the Jerusalem artichoke, the stimulatory effects of cytokinin on DNA replication and mitosis were only detectable in freshly excised explants, but not in explants that were previously grown on a hormone-free medium or were prewashed.[159,160] Growth on a hormone-free medium or prewashing are believed to eliminate a number of solutes, resulting in cytokinin insensitivity. Results of Simard[159] and Palmer and Gunning[160] suggest that endogenous auxins are among these leached-out solutes. Indeed, sensitivity to exogenous cytokinins, leading to increased DNA replication and mitosis, was restored by the addition of auxin to the culture medium. Although other solutes might also contribute to the tissue sensitivity to cytokinins, e.g., Ca^{2+} (see Chapter 19), it was apparent that auxin–cytokinin interaction was essential in the regulation of the cell cycle. A similar conclusion was reached in experiments where the timing of cytokinin supplied to cultured materials was changed relative to the timing of auxin supplied to the materials. In pea root cortical explant,[161] tobacco cell suspensions,[43] and protoplasts,[38] it was found that the presence of auxin in the culture medium for some time *before* the cytokinin was supplied was a prerequisite for the latter regulator to affect cell-cycle processes. We do not yet understand the nature of the auxin–cytokinin interaction. The fact that auxin seems to precede cytokinin action in different systems argues in favor of the hypothesis that there is at least one step early in the cycle, where auxin acts independently of cytokinins.

It is noteworthy that Hansen and co-workers observed that ribosylzeatin accumulation in cytokinin-autotrophic tobacco cells was blocked by the addition of auxin to the culture medium.[162] This is presumably related to the stimulatory action of auxin on cytokinin oxidase activity.[163]

C. OTHER REGULATORS

There is evidence to suggest that other hormones, gibberellins,[164-166] abscisic acid,[167,168] ethylene,[169,170] and polyamines[171-174] participate in the regulation of the division cycle in plants, at least in some experimental systems. As far as we know, there have been no studies on the possible interactions between cytokinins and these compounds. A clear interaction was demonstrated between cytokinins and trigonelline, a cotyledon-derived chemical that induced cell arrest at G2 in root meristems of legume seedlings.[175] The mechanism of the interaction, however, is not known.[176] Calcium, perhaps serving as a second messenger of hormones, was shown to affect the cell cycle (see Chapter 18).

XII. CONCLUSIONS

Cytokinins are intimately involved in the regulation of the cell cycle in both cultured cells and meristems of intact plants. Cytokinins are essential for advancing cells beyond several specific steps of the cycle. In the shoot meristem of *Sinapis alba,* cytokinins act at five steps: G1/S transition, S, G2, G2/M transition, and M. In other systems, S/G2 transition or cytokinesis are also shown to be controlled by cytokinins.

Cytokinin shortens the duration of the S phase as the result of the increased recruitment of replication origins, leading to a decrease in replicon size and synchronous activation of all replicon origins.

The effects at the start of the S phase are critical, since G1/S transition is known to be the activating step in DNA replication. Once activated, replication usually proceeds to completion.

The mechanism of action of cytokinins on the division cycle is largely unknown. A central question is whether cytokinins affect specific processes at different steps of the cycle or only a master process common to all controlled steps. A direction for future research is proposed based on the action of p34^{cdc2} kinase, which is the primary controlling factor of the cell cycle in all eukaryotes investigated thus far. As common transitional steps are affected by cytokinins and p34^{cdc2} kinase, it is tempting to speculate that cytokinins may in some fashion interact with this kinase. If p34^{cdc2} protein kinase is the target of cytokinin action, analysis of the mechanisms of action and the signal transduction pathway would follow.

Transgenic plants containing the isopentenyl transferase gene from the T-DNA of *Agrobacterium tumefaciens* have been engineered.[177,178] Under the control of appropriate promoters, increasing cytokinin in specific tissues and cell types is now possible.[179] It may be of interest to examine cell cycles in such tissues. Finally, identification of cytokinin mutants[180-182] with alteration in the cell cycle, and research on the interaction between cytokinins and other regulatory factors of cell division, are of paramount importance.

Since writing this chapter in late 1992, two contributions have appeared, showing that cytokinins can be stimulatory to p34^{cdc2} activity,[183,184] as we suggested. So far, however, the situation seems complex since cytokinins were found to regulate the transcription of the *cdc2* gene in some root tissues of *A. thaliana* and cultured tobacco protoplasts,[184] whereas they seemed to act directly on the catalytic activity of the p34^{cdc2} protein, i.e., at the posttranslational level, in cultured tobacco pith tissue.[183] In all these materials, cytokinins were shown to cooperate with auxins in the control of p34^{cdc2} activity.[183,184]

ACKNOWLEDGMENTS

We are greatly indebted to Prof. E. Miginiac, Université P. and M. Curie, Paris, for supplying unpublished photographic material and to copyright owners for their permission to reproduce published materials. We acknowledge M. Deberg and A. Pieltain for assistance in preparing autoradiograms of sections and DNA fibers, and M. Dejace for typing the manuscript.

Research in the author's laboratory has been supported by grants from Région Wallonne (2100), Université de Liège (Action de Recherche Concertée 88/93-129) and F.R.F.C. (2.4507-87 and 2.4529-92). C.H. is grateful to I.R.S.I.A. for the award of a research fellowship.

REFERENCES

1. **Skoog, F., Strong, F. M. and Miller, C. O.,** Cytokinins, *Science*, 148, 532, 1965.
2. **Davies, P. J.,** The plant hormones: their nature, occurrence, and functions, in *Plant Hormones and Their Role in Plant Growth and Development*, Davies, P. J., Ed., Martinus Nijhoff, Dordrecht, 1987, 1.
3. **Clowes, F. A. L.,** The proportion of cells that divide in root meristems of *Zea mays*, *Ann. Bot.*, 35, 249, 1971.
4. **Clowes, F. A. L.,** Size and activity of quiescent centres of roots, *New Phytol.*, 96, 13, 1984.
5. **Langenauer, H. D., Davis, E. L., and Webster, P. L.,** Quiescent cell populations in apical meristems of *Helianthus annuus*, *Can. J. Bot.*, 52, 2195, 1974.
6. **Gonthier, R., Jacqmard, A., and Bernier, G.,** Occurrence of two cell subpopulations with different cell-cycle durations in the central and peripheral zones of the vegetative shoot apex of *Sinapis alba* L., *Planta*, 165, 288, 1985.
7. **Cottignies, A.,** The blockage in the G1 phase of the cell cycle in the dormant shoot apex of ash, *Planta*, 147, 15, 1979.
8. **Naylor, J. M.,** Control of nuclear processes by auxin in axillary buds of *Tradescantia paludosa*, *Can. J. Bot.*, 36, 221, 1958.
9. **Nougarède, A., Rembur, J., and Rondet, P.,** Réactivation du bourgeon cotylédonaire du pois en réponse à la kinétine, *Can. J. Bot.*, 59, 590, 1981.
10. **Lajtha, L. G.,** Differential sensitivity of the cell life cycle, *J. Cell. Comp. Physiol.*, 62, 141, 1963.
11. **Bergounioux, C., Perennes, C., Brown, S. C., Sarda, C., and Gadal, P.,** Relation between protoplast division, cell-cycle stage and nuclear chromatin structure, *Protoplasma*, 142, 127, 1988.
12. **Mendelsohn, M. L.,** Autoradiographic analysis of cell proliferation in spontaneous breast cancer of C_3H mouse. III. The growth fraction, *J. Natl. Cancer Inst.*, 28, 1015, 1962.
13. **Nougarède, A. and Rembur, J.,** Determination of cell cycle and DNA synthesis duration in the shoot apex of *Chrysanthemum segetum* L. by double-labelling autoradiographic techniques, *Z. Pflanzenphysiol.*, 85, 283, 1977.

14. **Ormrod, J. C. and Francis, D.,** Effects of light on the cell cycle in the shoot apex of *Silene coeli-rosa* L. on the first day of floral induction, *Protoplasma*, 124, 96, 1985.
15. **Deltour, R. and Jacqmard, A.,** Relation between water stress and DNA synthesis during germination of *Zea mays* L., *Ann. Bot.*, 38, 329, 1974.
16. **Schatt, E., Landre, P., and Nougarède, A.,** Etat nucléaire des méristèmes du pois dans la graine sèche: imbibition et reprise du cycle, *Can. J. Bot.*, 63, 2200, 1985.
17. **John, P. C. L.,** *The Cell Cycle*, SEB Seminar Series, Cambridge University Press, Cambridge, 1981.
18. **Pardee, A. B.,** G1 events and regulation of cell proliferation, *Science*, 246, 537, 1989.
19. **Lewin, B.,** Driving the cell cycle: M phase kinase, its partners, and substrates, *Cell*, 61, 743, 1990.
20. **Forsburg, S. L. and Nurse, P.,** Cell cycle regulation in the yeasts *Saccharomyces cerevisiae* and *Schizosaccharomyces pombe*, *Ann. Rev. Cell Biol.*, 7, 227, 1991.
21. **Mak, S.,** Mammalian cell cycle analysis using microspectrophotometry combined with autoradiography, *Exp. Cell Res.*, 39, 286, 1965.
22. **Galbraith, D. W.,** Analysis of higher plants by flow cytometry and cell sorting, *Int. Rev. Cytol.*, 116, 165, 1989.
23. **Chiatante, D., Brusa, P., Levi, M., Sgorbati, S., and Sparvoli, E.,** A simple protocol to purify fresh nuclei from milligram amounts of meristematic pea root tissue for biochemical and flow cytometry applications, *Physiol. Plant.*, 78, 501, 1990.
24. **Houssa, C., Bomans, J., Greimers, R., and Jacqmard, A.,** High-yield isolation of protoplasts from microgram amounts of shoot meristematic tissues and rapid DNA content determination by flow cytometry, *Exp. Cell Res.*, 197, 153, 1991.
25. **Bergounioux, C., Brown, S. C., and Petit, P. X.,** Flow-cytometry and protoplast cell biology, *Physiol. Plant.*, 85, 374, 1992.
26. **Stepan-Sarkissian, G. and Grey, D.,** Growth determination and medium analysis, in *Methods in Molecular Biology*, Vol. 6, *Plant Cell and Tissue Culture*, Pollard, J. W. and Walker, J. M., Eds., Humana Press, Clifton, NJ, 1990, 13.
27. **Quastler, H. and Sherman, F. G.,** Cell population kinetics in the intestinal epithelium of the mouse. *Exp. Cell Res.*, 17, 420, 1959.
28. **Green, P. B.,** Growth and cell pattern on an axis: critique of concepts, terminology, and modes of study, *Bot. Gaz.*, 137, 187, 1976.
29. **Green, P. B.,** Organogenesis — a biophysical view, *Ann. Rev. Plant Physiol.*, 31, 51, 1980.
30. **Miller, C. O., Skoog, F., von Saltza, M. H., and Strong, F. M.,** Kinetin, a cell division factor from deoxyribonucleic acid, *J. Am. Chem. Soc.*, 77, 1392, 1955.
31. **Das, N. K., Patau, K., and Skoog, F.,** Initiation of mitosis and cell division by kinetin and indoleacetic acid in excised tobacco pith tissue, *Physiol. Plant.*, 9, 640, 1956.
32. **Tisserat, B.,** Embryogenesis, organogenesis and plant regeneration, in *Plant Cell Culture, a Practical Approach*, Dixon, R. A., Ed., IRL Press, Oxford, 1985, 79.
33. **Krikorian, A. D., Kelly, K., and Smith, D. L.,** Hormones in tissue culture and micropropagation, in *Plant Hormones and Their Role in Plant Growth and Development*, Davies, P. J., Ed., Martinus Nijhoff, Dordrecht, 1987, 593.
34. **Jouanneau, J. P. and Tandeau de Marsac, N.,** Stepwise effects of cytokinin activity and DNA synthesis upon mitotic cycle events in partially synchronized tobacco cells, *Exp. Cell Res.*, 77, 167, 1973.
35. **Short, K. C., Tepfer, D. A., and Fosket, D. E.,** Regulation of polyribosome formation and cell division in cultured soybean cells by cytokinin, *J. Cell Sci.*, 15, 75, 1974.
36. **Wang, T. L., Everett, N. P., Gould, A. R., and Street, H. E.,** Studies on the control of the cell cycle in cultured plant cells. III. The effects of cytokinin, *Protoplasma*, 106, 23, 1981.
37. **Zelcer, A. and Galun, E.,** Culture of newly isolated tobacco protoplasts: precursor incorporation into protein, RNA and DNA, *Plant Sci. Lett.*, 7, 31, 1976.
38. **Meyer, Y. and Cooke, R.,** Time course of hormonal control of the mitosis in tobacco mesophyll protoplasts cultivated *in vitro*, *Planta*, 147, 181, 1979.
39. **Bergounioux, C., Perennes, C., Brown, S. C., and Gadal, P.,** Cytometric analysis of growth-regulator-dependent transcription and cell-cycle progression in *Petunia* protoplast cultures, *Planta*, 175, 500, 1988.
40. **Ordás, R. J., Fernandéz, B., and Rodríguez, R.,** Benzyladenine-controlled protein synthesis and growth in apple cell suspensions, *Physiol. Plant.*, 84, 229, 1992.
41. **Blaschke, J. R., Forche, E., and Neumann, K.-H.,** Investigations on the cell cycle of haploid and diploid tissue cultures of *Datura innoxia* Mill and its synchronization, *Planta*, 144, 7, 1978.

42. **Torrey, J. G.,** Kinetin as trigger for mitosis in mature endomitotic plant cells, *Exp. Cell Res.*, 23, 281, 1961.
43. **Jouanneau, J. P.,** Contrôle par les cytokinines de la synchronisation des mitoses dans les cellules de tabac, *Exp. Cell Res.*, 67, 329, 1971.
44. **Fosket, D. E., Volk, M. J., and Goldsmith, J.,** Polyribosome formation in relation to cytokinin-induced cell division in suspension cultures of *Glycine max*, *Plant Physiol.*, 60, 554, 1977.
45. **Kende, H.,** The cytokinins, *Int. Rev. Cytol.*, 31, 301, 1971.
46. **Manos, G. E.,** The effect of growth substances on attached and detached root tips of *Pisum sativum* L., *Physiol. Plant.*, 14, 697, 1961.
47. **Goodwin, P. B.,** Phytohormones and growth and development of organs of the vegetative plant, in *Phytohormones and Related Compounds: A Comprehensive Treatise*, Letham, D. S., Goodwin, P. B., and Higgins, T. J. V., Eds., Elsevier/North Holland, Amsterdam, 1978, 31.
48. **Levi, M., Colombo, D., Sgorbati, S., Chiatante, D., and Sparvoli, E.,** Effect of benzyladenine on cell cycle in intact roots, *J. Plant Physiol.*, 131, 385, 1987.
49. **Bernier, G., Kinet, J.-M., Jacqmard, A., Havelange, A., and Bodson, M.,** Cytokinin as a possible component of the floral stimulus in *Sinapis alba*, *Plant Physiol.*, 60, 282, 1977.
50. **Houssa, C. and Jacqmard, A.,** unpublished data, 1990.
51. **Chailakhyan, M. Kh., Milyaeva, E. L., and Yanina, L. I.,** The influence of physiologically-active substances of growth, development of apices of *Rudbeckia bicolor* (in Russian), *Biol. Rev.* (Armenia), 23, 13, 1970.
52. **Jacqmard, A. and Bernier, G.,** unpublished data, 1978.
53. **Jacqmard, A. and Houssa, C.,** unpublished data, 1992.
54. **Besnard-Wibaut, C.,** Effectiveness of gibberellins and 6-benzyladenine on flowering of *Arabidopsis thaliana*, *Physiol. Plant.*, 53, 205, 1981.
55. **Seidlovà, F. and Krekule, J.,** Effects of kinetin on growth of the apical meristem and floral differentiation in *Chenopodium rubrum* L., *Ann. Bot.*, 41, 755, 1977.
56. **Bernier, G., Lejeune, P., Jacqmard, A., and Kinet, J.-M.,** Cytokinins in flower initiation, in *Plant Growth Substances 1988*, Pharis, R. P. and Rood, S. B., Eds., Springer-Verlag, Berlin, 1990, 486.
57. **Kinet, J.-M., Zune, V., Linotte, C., Jacqmard, A., and Bernier, G.,** Resumption of cellular activity induced by cytokinin and gibberellin treatments in tomato flowers targeted for abortion in unfavorable light conditions, *Physiol. Plant.*, 64, 67, 1985.
58. **Reyes, J., Jiménez-Garcia, Gonzalez, M. A., and Vazquez-Ramos, J. M.,** Benzyladenine-stimulation of nuclear DNA synthesis and cell division in germinating maize, *Seed Sci. Res.*, 1, 113, 1991.
59. **Tamas, I. A.,** Hormonal regulation of apical dominance, in *Plant Hormones and Their Role in Plant Growth and Development*, Davies, P. J., Ed., Martinus Nijhoff, Dordrecht, 1987, 393.
60. **Sachs, T. and Thimann, K. V.,** The role of auxins and cytokinins in the release of buds from dominance, *Am. J. Bot.*, 54, 136, 1967.
61. **Ali, A. and Fletcher, R. A.,** Xylem differentiation in inhibited cotyledonary buds of soybeans, *Can. J. Bot.*, 48, 1139, 1970.
62. **Usciati, M., Codaccioni, M., and Guern, J.,** Early cytological and biochemical events induced by a 6-benzylaminopurine application on inhibited axillary buds of *Cicer arietinum* plants, *J. Exp. Bot.*, 23, 1009, 1972.
63. **Mauseth, J. D.,** Cytokinin- and gibberellic acid-induced effects on the structure and metabolism of shoot apical meristems in *Opuntia polyacantha* (Cactaceae), *Am. J. Bot.*, 63, 1295, 1976.
64. **Michaux-Ferrière, N.,** Le contenu en ADN nucléaire des bourgeons latéraux actifs, inhibés et en cours de réactivation chez le *Polypodium vulgare*, *Can. J. Bot.*, 62, 1250, 1984.
65. **Turbull, C. G. N. and Hanke, D. E.,** The control of bud dormancy in potato tubers, *Planta*, 165, 359, 1985.
66. **Fisher, J. B.,** Control of the internodal intercalary meristem of *Cyperus alternifolius*, *Am. J. Bot*, 57, 1017, 1970.
67. **Kuang, A., Peterson, C. M., and Dute, R. R.,** Changes in soybean raceme and petiole anatomy induced by 6-benzylaminopurine, *Ann. Bot.*, 67, 23, 1991.
68. **Patau, K., Das, N. K., and Skoog, F.,** Induction of DNA synthesis by kinetin and indoleacetic acid in excised tobacco pith tissue, *Physiol. Plant.*, 10, 949, 1957.
69. **Fosket, D. E. and Short, K. C.,** The role of cytokinin in the regulation of growth, DNA synthesis and cell proliferation in cultured soybean tissues (*Glycine max* var. *Biloxi*), *Physiol. Plant.*, 28, 14, 1973.

70. **Fosket, D. E.,** The regulation of the plant cell cycle by cytokinins, in *Mechanisms and Control of Cell Division*, Rost, T. L. and Gifford, E. M., Jr., Eds., Dowden, Hutchinson, and Ross, Stroudsburg, PA, 1977, 62.
71. **Cooke, R. and Meyer, Y.,** Hormonal control of tobacco protoplast nucleic acid metabolism during *in vitro* culture, *Planta*, 152, 1, 1981.
72. **Libbenga, K. R. and Torrey, J. G.,** Hormone-induced endoreduplication prior to mitosis in cultured pea root cortex cells, *Am. J. Bot.*, 60, 293, 1973.
73. **Momotani, E., Kinoshita, I., Yokomura, E. I., and Tsuji, H.,** Rapid induction of synthesis and doubling of nuclear DNA by benzyladenine in intact bean leaves, *Plant Cell Physiol.*, 31, 621, 1990.
74. **Soeda, Y., Kinoshita, I., and Tsuji, H.,** Benzyladenine-induced increase in DNA content in the nucleus of bean leaf cells, *Plant Growth Regul.*, 9, 165, 1990.
75. **Kinoshita, I., Sanbe, A., and Yokomura, E. I.,** Increases in nuclear DNA content without mitosis in benzyladenine-treated primary leaves of intact and decapitated bean plants, *J. Exp. Bot.*, 42, 667, 1991.
76. **MacLeod, R. D.,** Changes in the mitotic cycle in lateral root meristems of *Vicia faba* following kinetin treatment, *Chromosoma*, 24, 177, 1968.
77. **Van't Hof, J.,** The action of IAA and kinetin on the mitotic cycle of proliferative and stationary phase excised root meristems, *Exp. Cell Res.*, 51, 167, 1968.
78. **Houssa, C., Jacqmard, A., and Bernier, G.,** Activation of replicon origins as a possible target for cytokinins in shoot meristems of *Sinapis*, *Planta*, 181, 324, 1990.
79. **Houssa, C., Jacqmard, A., and Bernier, G.,** unpublished data, 1991.
80. **Cairnie, A. B., Lamerton, L. F., and Steel, G. G.,** Cell proliferation studies in the intestinal epithelium of the rat, *Exp. Cell Res.*, 39, 539, 1965.
81. **Gonthier, R., Jacqmard, A., and Bernier, G.,** Changes in the cell-cycle duration and growth fraction in the shoot meristem of *Sinapis* during the floral transition, *Planta*, 170, 55, 1987.
82. **Van't Hof, J.,** Control points within the cell cycle, in *The Cell Division Cycle in Plants*, Bryant, J. A. and Francis, D., Eds., SEB Seminar Series, Cambridge University Press, Cambridge, 1985, 133.
83. **Campbell, J. L.,** Eukaryotic DNA replication, *Ann. Rev. Biochem.*, 55, 733, 1986.
84. **Van't Hof, J., Kuniyuki, A., and Bjerknes, C. A.,** The size and number of replicon families of chromosomal DNA of *Arabidopsis thaliana*, *Chromosoma*, 68, 269, 1978.
85. **Hand, R.,** Eucaryotic DNA: organization of the genome for replication. *Cell*, 15, 317, 1978.
86. **Van't Hof, J.,** Replicon size and rate of fork movement in early S of higher plant cells (*Pisum sativum*). *Exp. Cell Res.*, 103, 395, 1976.
87. **Laskey, R. A., Fairman, M. P., and Blow, J. J.,** S phase of the cell cycle, *Science*, 246, 609, 1989.
88. **Diffley, J. F. X. and Stillman, B.,** The initiation of chromosomal DNA replication in eukaryotes, *Trends Genet.*, 6, 427, 1990.
89. **Held, P. G. and Heintz, N. H.,** Eukaryotic replication origins, *Biochem. Biophys. Acta*, 1130, 235, 1992.
90. **Dunham, V. L. and Bryant, J. A.,** Enzymic controls of DNA replication, in *The Cell Division Cycle in Plants*, Bryant, J. A. and Francis, D., Eds., SEB Seminar Series, Cambridge University Press, Cambridge, 1985, 37.
91. **Aves, S. J. and Bryant, J. A.,** The molecular genetics and biochemistry of DNA replication, in *Molecular and Cell Biology of the Plant Cell Cycle*, Ormrod, J. and Francis, D., Eds., Kluwer, Dordrecht, 1993, 45.
92. **Davies, N. D., Bryant, J. A., Fitchett, P. N., and Van't Hof, J.,** Initiation of DNA replication in pea: isolation and characterisation of DNA replication origins, *Plant Mol. Biol.*, submitted.
93. **Callan, H. G.,** DNA replication in the chromosomes of eukaryotes, *Cold Spring Harbor Symp. Quant. Biol.*, 38, 195, 1974.
94. **Blumenthal, A. B., Kriegstein, H. J., and Hogness, D. S.,** The units of DNA replication in *Drosophila melanogaster* chromosomes, *Cold Spring Harbor Symp. Quant. Biol.*, 38, 205, 1974.
95. **Spradling, A. and Orr-Weaver, T.,** Regulation of DNA replication during *Drosophila* development, *Ann. Rev. Genet.*, 21, 373, 1987.
96. **Francis, D., Kidd, A. D., and Bennett, M. D.,** DNA replication in relation to DNA C values, in *The Cell Division Cycle in Plants*, Bryant, J. A. and Francis, D., Eds., SEB Seminar Series, Cambridge University Press, Cambridge, 1985, 61.
97. **Havelange, A., Bodson, M., and Bernier, G.,** Partial floral evocation by exogenous cytokinin in the long-day plant *Sinapis alba*, *Physiol. Plant.*, 67, 695, 1986.

98. **Piesco, N. P. and Alvarez, M. R.,** Nuclear cytochemical changes in onion roots stimulated by kinetin, *Exp. Cell Res.*, 73, 129, 1972.

99. **Webster, P. L. and Van't Hof, J.,** DNA synthesis and mitosis in meristems: requirements for RNA and protein synthesis, *Am. J. Bot.*, 57, 130, 1970.

100. **De la Torre, C., Gonzalez-Fernandez, H., and Gimenéz-Martin, G.,** Stringency at four regions of the plant cell cycle where proteins regulating its progression are synthesized, *J. Cell Sci.*, 94, 259, 1989.

101. **Kodama, H., Kawakami, N., Watanabe, A., and Komamine, A.,** Phase-specific polypeptides and poly(A)$^+$ RNAs during the cell cycle in synchronous culture of *Catharanthus roseus* cells, *Plant Physiol.*, 89, 910, 1989.

102. **Gonzalez-Fernandez, A., Giménez-Martin, G., Fernández-Gómez, M. E., and De la Torre, C.,** Protein synthesis requirements at specific points in the interphase of meristematic cells, *Exp. Cell Res.*, 88, 163, 1974.

103. **Gonzalez-Fernandez, A., Selman, A. M., Giménez-Martin, G., and De la Torre, C.,** Protein synthesis in late mitosis and in G1 required for cell cycle progression in synchronous onion root meristem cells, *Eur. J. Cell Biol.*, 49, 87, 1989.

104. **Ohnishi, N., Kodama, H., Ando, S., and Komamine, A.,** Synthesis of protein and mRNA is necessary for transition of suspension-cultured *Catharanthus roseus* cells from the G1 to the S phase of the cell cycle, *Physiol. Plant.*, 80, 95, 1990.

105. **Webster, P.L.,** Effects of cycloheximide on mitosis in *Vicia faba* root-meristem cells, *J. Exp. Bot.*, 24, 239, 1973.

106. **Garcia-Herdugo, G., Fernández-Gómez, M. E., Hidalgo, J., and Lopez-Saez, J. F.,** Effects of protein synthesis inhibition during plant mitosis, *Exp. Cell Res.*, 89, 336, 1974.

107. **Jouanneau, J. P.,** Renouvellement des protéines et effet spécifique de la kinétine sur des cultures de cellules de tabac, *Physiol. Plant.*, 23, 232, 1970.

108. **Simpson, S. F. and Torrey, J. G.,** Hormonal control of deoxyribonucleic acid and protein syntheses in pea root cortical explants, *Plant Physiol.*, 59, 4, 1977.

109. **Meyer, Y. and Chartier, Y.,** Hormonal control of mitotic development in tobacco protoplasts — two-dimensional distribution of newly-synthesized proteins, *Plant Physiol.*, 68, 1273, 1981.

110. **Teyssendier de la Serre, B., Jouanneau, J. P., and Peáud-Lenoël, C.,** Incorporation of N^6-benzyladenine into messenger poly(A)-RNA of tobacco cells incubated at stimulatory or cytostatic cytokinin concentration, *Plant Physiol.*, 74, 669, 1984.

111. **Bevan, M. and Northcote, D. H.,** Subculture-induced protein synthesis in tissue cultures of *Glycine max* and *Phaseolus vulgaris*, *Planta*, 152, 24, 1981.

112. **Pines, J. and Hunt, T.,** Molecular cloning and characterization of the mRNA for cyclin from sea urchin eggs, *EMBO J.*, 6, 2987, 1987.

113. **Draetta, G.,** Cell cycle control in eukaryotes: molecular mechanisms of cdc2 activation, *Trends Biochem. Sci.*, 15, 378, 1990.

114. **Lehner, C. F. and O'Farrell, P. H.,** The roles of *Drosophila* cyclins A and B in mitotic control, *Cell*, 61, 535, 1990.

115. **Nurse, P.,** Universal control mechanism regulating onset of M-phase, *Nature*, 344, 503, 1990.

116. **Pines, J. and Hunter, T.,** $p34^{cdc2}$: the S and M kinase? *New Biol.*, 2, 389, 1990.

117. **Solomon, M. J., Glotzer, M., Lee, T. H., Philippe, M., and Kirschner, M. W.,** Cyclin activation of $p34^{cdc2}$, *Cell*, 63, 1013, 1990.

118. **Whitfield, W. G. F., Gonzalez, C., Maldonado-Codina, G., and Glover, D. M.,** The A- and B-type cyclins of *Drosophila* are accumulated and destroyed in temporally distinct events that define separable phases of the G2-M transition, *EMBO J.*, 9, 2563, 1990.

119. **Fang, F. and Newport, J. W.,** Evidence that the G1-S and G2-M transitions are controlled by different cdc2 proteins in higher eukaryotes, *Cell*, 66, 731, 1991.

120. **Huberman, J. A.,** Cell cycle control of initiation of eukaryotic DNA replication, *Chromosoma*, 100, 419, 1991.

121. **Luca, F. C., Shibuya, E. K., Dohrmann, C. E., and Ruderman, J. V.,** Both cyclin A 97 and B 97 are stable and arrest cells in M-phase, but only cyclin B 97 turns on cyclin destruction, *EMBO J.*, 10, 4311, 1991.

122. **Reed., S. I.,** G_1-specific cyclins: in search of an S-phase-promoting factor, *Trends Genet.*, 7, 95, 1991.

123. **Reed., S. I.,** The role of p34 kinases in the G1 to S-phase transition, *Ann. Rev. Cell Biol.*, 8, 529, 1992.

124. **Hollingsworth, R. E. and Sclafani, R. A.,** DNA metabolism gene CDC7 from yeast encodes a serine (threonine) protein kinase, *Proc. Natl. Acad. Sci. U.S.A.*, 87, 6272, 1990.

125. **Din, S.-U., Brill, S. J., Fairman, M.-P., and Stillman, B.**, Cell-cycle-regulated phosphorylation of DNA replication factor A from human and yeast cells, *Genes Dev.*, 4, 968, 1990.
126. **Dutta, A. and Stillman, B.**, cdc2 family kinases phosphorylate a human cell DNA replication factor, RPA, and activate DNA replication, *EMBO J.*, 11, 2189, 1992.
127. **Moreno, S. and Nurse, P.**, Substrates for $p34^{cdc2}$: in vivo veritas? *Cell*, 61, 549, 1990.
128. **John, P. C. L., Sek, F. J., and Lee, M. G.**, A homolog of the cell cycle control protein $p34^{cdc2}$ participates in the division cycle of *Chlamydomonas*, and a similar protein is detectable in higher plants and remote taxa, *Plant Cell*, 1, 1185, 1989.
129. **Feiler, H. S. and Jacobs, T. W.**, Cell division in higher plants: a cdc2 gene, its 34-kDa product, and histone H1 kinase activity in pea, *Proc. Natl. Acad. Sci. U.S.A.*, 87, 5397, 1990.
130. **Colasanti, J., Tyers, M., and Sundaresan, V.**, Isolation and characterization of cDNA clones encoding a functional $p34^{cdc2}$ homologue from *Zea mays*, *Proc. Natl. Acad. Sci. U.S.A.*, 88, 3377, 1991.
131. **Ferreira, P. C. G., Hemerly, A. S., Villarroel, R., Van Montagu, M., and Inzé, D.**, The *Arabidopsis* functional homolog of the $p34^{cdc2}$ protein kinase, *Plant Cell*, 3, 531, 1991.
132. **Gorst, J. R., John, P. C. L., and Sek, F. J.**, Levels of $p34^{cdc2}$-like protein in dividing, differentiating and dedifferentiating cells of carrot, *Planta*, 185, 304, 1991.
133. **Hata, S., Kouchi, H., Suzuka, I., and Ishii, T.**, Isolation and characterization of cDNA clones for plant cyclins, *EMBO J.*, 10, 2681, 1991.
134. **Hirayama, T., Imajuku, Y., Anai, T., Matsui, M., and Oka, A.**, Identification of two cell-cycle-controlling cdc2 gene homology in *Arabidopsis thaliana*, *Gene*, 105, 159, 1991.
135. **Hirt, H., Pay, A., Gyorgyey, J., Bako, L., Németh, K., Bögre, L., Schweyen, R. J., Heberle-Bors, E., and Dudits, D.**, Complementation of a yeast cell cycle mutant by an alfalfa cDNA encoding a protein kinase homologous to $p34^{cdc2}$, *Proc. Natl. Acad. Sci. U.S.A.*, 88, 1636, 1991.
136. **Hemerly, A., Bergounioux, C., Van Montagu, M., and Inzé, D.**, Genes regulating the plant cell cycle: isolation of a mitotic-like cyclin from *Arabidopsis thaliana*, *Proc. Natl. Acad. Sci. U.S.A.*, 89, 3295, 1992.
137. **Martinez, M. C., Jorgensen, J. E., Lawton, M. A., Lamb, C. J., and Doerner, P. W.**, Spatial pattern of cdc2 expression in relation to meristem activity and cell proliferation during plant development, *Proc. Natl. Acad. Sci. U.S.A.*, 89, 7360, 1992.
138. **Francis, D.**, The cell cycle in plant development, *New Phytol.*, 122, 1, 1992.
139. **Jacobs, T.**, Control of the cell cycle, *Dev. Biol.*, 153, 1, 1992.
140. **Feiler, H. S. and Jacobs, T. W.**, Cloning of the pea cdc2 homologue by efficient immunological screening of PCR products, *Plant Mol. Biol.*, 17, 321, 1991.
141. **Mineyuki, Y., Yamashita, M., and Nagahama, Y.**, $p34^{cdc2}$ kinase homologue in the preprophase band, *Protoplasma*, 162, 182, 1991.
142. **Zavala, M. E. and Brandon, D. L.**, Localization of a phytohormone using immunocytochemistry, *J. Cell Biol.*, 97, 1235, 1983.
143. **Chen, C.-M., Erth, J. R., Leisner, S. M., and Chang, C.-C.**, Localization of cytokinin biosynthetic sites in pea plants and carrot roots, *Plant Physiol.*, 78, 510, 1985.
144. **Sossountzov, L., Maldiney, R., Sotta, B., Sabbagh, I., Habricot, Y., Bonnet, M., and Miginiac, E.**, Immunocytochemical localization of cytokinins in Craigella tomato and sideshootless mutant, *Planta*, 175, 291, 1988.
145. **Stroobants, C.**, Etude immunocytochimique de la compartimentation des cytokinines au cours de la rhizogenèse chez trois génotypes de tabac (*Nicotiana tabacum* L.). Relations avec le cycle cellulaire, Thèse de Doctorat, Université P. et M. Curie Paris VI, 1991.
146. **Stroobants, C., Sossountzov, L., and Miginiac, E.**, Immunocytolocalisation du riboside de la zéatine dans des feuilles de tabac isolées et bouturées, au cours des phases initiales de la rhizogenèse, *C.R. Acad. Sci.*, Paris, 312, 261, 1991.
147. **Lejeune, P., Kinet, J.-M., and Bernier, G.**, Cytokinin fluxes during floral induction in the long day plant *Sinapis alba*, *Plant Physiol.*, 86, 1095, 1988.
148. **Sotta, B., Lejeune, P., Maldiney, R., Kinet, J.-M., Miginiac, E., and Bernier, G.**, Cytokinin and auxin levels in apical buds of *Sinapis alba* following floral transition, in *Physiology and Biochemistry of Cytokinins in Plants*, Kamínek, M., Mok, D. W. S., and Zazímalová, E., Eds, SPB Academic Publishers, The Hague, 1992, 377.

149. **Machácková, I., Krekule, J., Edar, J., Seidlová, F., and Strnad, M.,** Cytokinins in photoperiodic of flowering induction in *Chenopodium* species, *Physiol. Plant.*, 87, 160, 1993.
150. **Grayling, A. and Hanke, D. E.,** Cytokinins in exudates from leaves and roots of red *Perilla*, *Phytochemistry*, 31, 1863, 1992.
151. **Kinet, J.-M., Houssa, P., and Bernier, G.,** unpublished data, 1990.
152. **Léonard, M. and Kinet, J.-M.,** The role of cytokinins and gibberellins in controlling inflorescence development in tomato, *Acta Hort.*, 134, 117, 1983.
153. **Phillips, I. D. J.,** Apical dominance, *Ann. Rev. Plant Physiol.*, 26, 341, 1975.
154. **Nishinari, N. and Syōno, K.,** Induction of cell division synchrony and variation of cytokinin contents through the cell cycle in tobacco cultured cells, *Plant Cell Physiol.*, 27, 147, 1986.
155. **Nishinari, N. and Syōno, K.,** Identification of cytokinins associated with mitosis in synchronously cultured tobacco cells, *Plant Cell Physiol.*, 21, 383, 1980.
156. **Trewavas, A. J.,** Growth substances, calcium and the regulation of cell division, in *The Cell Division Cycle in Plants*, Bryant, J. A. and Francis, D., Eds., SEB Seminar Series, Cambridge University Press, Cambridge, 1985, 133.
157. **Van't Hof, J. and Kovacs, C. J.,** Mitotic cycle regulation in the meristem of cultured roots: the principal control point hypothesis, in *The Dynamics of Meristem Cell Populations*, Miller, M. W. and Kuehnert, C. G., Eds., Plenum Press, New York, 1972, 1.
158. **Simpson, S. F. and Torrey, J. G.,** Hormonal control of deoxyribonucleic acid and protein syntheses in pea root cortical explants, *Plant Physiol.*, 59, 4, 1977.
159. **Simard, A.,** Initiation of DNA synthesis by kinetin and experimental factors in tobacco pith tissues *in vitro*, *Can. J. Bot.*, 49, 1541, 1971.
160. **Palmer, M. V. and Gunning, B. E. S.,** Cytokinin-induced mitotis in cultured explants of *Helianthus tuberosus* L. tuber tissue, *Aust. J. Plant Physiol.*, 11, 1, 1984.
161. **Matthysse, A. G. and Torrey, J. G.,** DNA synthesis in relation to polyploid mitoses in excised pea root segments cultured *in vitro*, *Exp. Cell Res.*, 48, 484, 1967.
162. **Hansen, C. E., Meins, F., Jr., and Aebi, R.,** Hormonal regulation of zeatin-riboside accumulation by cultured tobacco cells, *Planta*, 42, 520, 1987.
163. **Palni, L. M. S., Burch, L., and Horgan, R.,** The effect of auxin concentration on cytokinin stability and metabolism, *Planta*, 171, 231, 1988.
164. **Sachs, R. M.,** Stem elongation, *Ann. Rev. Plant Physiol.*, 16, 73, 1965.
165. **Banerjee, S. N.,** DNA synthesis in the root meristem cells of *Zea mays* dwarf mutants (d1 and d5), *Plant Cell Physiol.*, 9, 577, 1968.
166. **Loy, J. B.,** Hormonal regulation of cell division in the primary elongating meristems of shoots, in *Mechanisms and Control of Cell Division*, Rost, T. L. and Gifford, E. M., Jr., Eds., Dowden, Hutchinson, and Ross, Stroudsburg, PA, 1977, 92.
167. **Barlow, P. W. and Pilet, P. E.,** The effect of abscisic acid on cell growth, cell division and DNA synthesis in the maize root meristem, *Physiol. Plant.*, 62, 125, 1984.
168. **Nougarède, A., Rondet, P., Landré, P., and Rembur, J.,** Effet d'un traitement par l'acide abscisique sur la division cellulaire, les teneurs en ADN et l'élongation du bourgeon cotylédonaire de plants de pois décapités, *Can. J. Bot.*, 65, 907, 1987.
169. **Apelbaum, A. and Burg, S. P.,** Effect of ethylene on cell division and deoxyribonucleic acid synthesis in *Pisum sativum*, *Plant Physiol.*, 50, 117, 1972.
170. **Lieberman, M.,** Biosynthesis and action of ethylene, *Ann. Rev. Plant Physiol.*, 30, 533, 1979.
171. **Tabor, C. W. and Tabor, H.,** Polyamines, *Ann. Rev. Biochem.*, 53, 749, 1984.
172. **Evans, P. T. and Malmberg, R. L.,** Do polyamines have roles in plant development? *Ann. Rev. Plant Phys. Plant Mol. Biol.*, 40, 235, 1989.
173. **Pfosser, M., Königshofer, H., and Kandeler, R.,** Free, conjugated and bound polyamines during the cell cycle of synchronized cell suspension cultures of *Nicotiana tabacum*, *J. Plant Physiol.*, 136, 574, 1990.
174. **Maki, H., Ando, S., Kodama, H., and Komamine, A.,** Polyamines and the cell cycle of *Catharanthus roseus*, *Plant Physiol.*, 96, 1008, 1991.
175. **Tramontano, W. A., Evans, L. S., and McGinley, P. A.,** Effects of cytokinins on promotion of cell arrest in G2 by trigonelline and trigonelline concentrations in cultured roots of *Pisum sativum* and *Glycine max*, *Environ. Exp. Bot.*, 25, 83, 1985.
176. **Tramontano, W. A., Evans, L. S., McGinley, P. A., and Ciancaglini, E.,** Effects of cytokinins on the metabolism of nicotinic acid in cultured roots of *Pisum sativum* and *Glycine max* in relation to cell arrest in G2, *Environ. Exp. Bot.*, 25, 393, 1985.

177. **Klee, H., Horsch, R., and Rogers, S.,** *Agrobacterium*-mediated plant transformation and its further applications to plant biology, *Ann. Rev. Plant Physiol.*, 38, 467, 1987.

178. **Medford, J. I., Horgan, R., El-Sawi, Z., and Klee, H. J.,** Alterations of endogenous cytokinins in transgenic plants using a chimeric isopentenyl transferase gene, *Plant Cell*, 1, 403, 1989.

179. **Medford, J. I.,** Vegetative apical meristems, *Plant Cell*, 4, 1029, 1992.

180. **Jullien, M., Lesueur, D., Laloue, M., and Caboche, M.,** Isolation and preliminary characterization of cytokinin-resistant mutants of *Nicotiana plumbaginifolia*, in *Physiology and Biochemistry of Cytokinins in Plants*, Kamínek, M., Mok, D. W. S., and Zazímalova, E., Eds., SPB Academic Publishers, The Hague, 1992, 157.

181. **Blonstein, A. D., Parry, A. D., Horgan, R., and King, P. J.,** A cytokinin-resistant mutant of *Nicotiana plumbaginifolia* is wilty, *Planta*, 183, 244, 1991.

182. **Su, W. and Howell, S. H.,** A single genetic locus, ckr1, defines *Arabidopsis* mutants in which root growth is resistant to low concentrations of cytokinin, *Plant Physiol.*, 99, 1569, 1992.

183. **John, P. C. L., Zhang, K., Dong, C., Diederich, L., and Whightman, F.,** $p34^{cdc2}$ related proteins in control of cell cycle progression, the switch between division and differentiation in tissue development, and stimulation of division by auxin and cytokinin, *Aust. J. Plant Physiol.*, 20, 503, 1993.

184. **Hemerly, A. S., Ferreira, P., Engler, J. de A., Van Montagu, M., Engler, G., and Inzé, D.,** *cdc2a* expression in *Arabidopsis* is linked with competence for cell division, *Plant Cell*, in press.

Chapter 16

Cytokinin Binding Proteins and Receptors

Chris Brinegar

TABLE OF CONTENTS

I. INTRODUCTION

The quest for cytokinin receptors began a little over 20 years ago with the discovery that plant ribosomes possessed cytokinin-binding abilities.[1] Searching for cytokinin-binding proteins (CBPs) on ribosomes was an obvious choice, since the stimulatory effects of cytokinins on protein synthesis had been well established by then.[2] In fact, the whole mechanism of translation had been suspected as mediating the signals of cytokinins, ever since it was discovered that certain cytokinin derivatives were incorporated into some transfer RNA molecules (see Chapters 7 and 8). As it turns out, no direct evidence was ever found indicating that CBPs on ribosomes were involved in the mechanism of cytokinin action.

In 1986, at the NATO Advanced Science Institute Symposium on Plant Hormone Receptors, Firn[3] stated openly in the title of his paper what plant hormone (especially cytokinin) biologists had often grumbled under their breath: "Too many binding proteins, not enough receptors." The reason for this frustration was that no one had provided definitive proof that any of the reported CBPs had true "receptor-like" qualities. It was obvious that new approaches were needed for further progress in this area.

Fortunately, several investigations were just then beginning to study how cytokinins affected gene expression at the transcriptional and posttranscriptional levels (see Chapter 17). The advent of DNA hybridization methods allowed the study of numerous genes whose expression was regulated by cytokinin treatment. Many of these genes were involved in photosynthesis.[4,5] This spurred a number of laboratories to renew the search for the elusive cytokinin receptor(s), using leaves as starting material. As we shall see, this research produced a number of exciting discoveries (along with the usual number of puzzling reports) that are beginning to shed some light on the cytokinin receptor problem.

Why did it take so long? There are probably several reasons, but foremost on my list is that much of the early work on CBPs was of the "grind and find" school of science where one of many cytokinin binding assays were used to screen for binding proteins from whole extracts or particulate fractions from any plant material that happened to be in the greenhouse or on the shelf. Many investigators, including the author, were guilty of designing these simple experiments, in the hope that a cytokinin receptor with an exceptionally low dissociation constant (K_d) would magically reveal itself. Once all of the background nonspecific binding was subtracted, cytokinin binding was indeed detected, but K_d values were higher than we intuitively felt they should be for receptors.

Gradually, research techniques and experimental designs have become more sophisticated. Affinity chromatography using immobilized cytokinins is used routinely to purify CBPs from extracts. More attention is being devoted to the nature of the

0-8493-6252-0/94/$0.00+$.50

binding assays used, in order to avoid the dreaded accusation at scientific meetings that the detected binding is "nonspecific." And more importantly, researchers are beginning to select their starting tissues with a little more forethought; i. e., they are using tissues where cytokinin effects are well documented at the cellular and molecular levels.

In this presentation I will review the data on CBPs and putative cytokinin receptors, that have accumulated since the mid-1970s. In addition, there will be commentary on the nature of cytokinin binding at the molecular level and some concluding thoughts about the biological and semantic differences between receptors and binding proteins. The reader is also directed to earlier reviews that mention, or are dedicated to, cytokinin-binding proteins.[6-12]

II. THE NATURE OF CYTOKININ BINDING

Before reviewing the many CBPs that have appeared in the literature, I would like to ponder the nature of cytokinin binding itself. First, I would like to limit my comments to the purine cytokinins before mentioning the urea cytokinins.

It has been demonstrated many times with many CBPs (see later sections) that the ribosides of purine cytokinins have much lower affinities for the binding sites than their corresponding bases (and also have less activity in bioassays). Therefore, a ribosyl group at the purine's N^9 position must inhibit binding by either

1. Preventing hydrogen bonding at the N^9 position
2. Increasing the hydrophilic nature of the cytokinin, which may block its interaction with a hydrophobic binding "pocket"
3. Simple steric hindrance

The purine ring has an aromatic character that, along with the predominantly hydrophobic sidechain, contributes to the limited water solubility of cytokinins, but the unshared electron pairs or the hydrogen atoms on the ring nitrogens are still available for hydrogen bonding. The purine ring is also capable of tautomerization, with the predominant tautomeric form dependent on pH.[13] Therefore, the arrangement of double bonds within the nearly planar purine ring, and the subsequent protonation or deprotonation of the ring nitrogens, could have an effect on the hydrogen-bonding characteristics of a purine cytokinin. Most CBPs, however, demonstrate binding that is relatively constant in the physiological pH range (6 to 8) where one purine tautomer dominates.

The chemical nature of the purine ring in cytokinins is both a blessing and a bane to the study of cytokinin binding. On the one hand, all biologically active purine-based cytokinins have the same purine ring in common, so we know that the ultimate specificity of binding must be due to the cytokinin side chain (usually at the adenine N^6 position). But on the other hand, the ubiquitous nature of purine metabolites (especially adenine) makes it extremely difficult to distinguish exactly how much of the binding is due to the cytokinin side chain vs. the purine ring; i. e., are we detecting proteins that truly bind cytokinins *in vivo* or just proteins that normally bind adenine derivatives? A case in point is the ability of some lectins to bind cytokinins at their adenine-binding sites.[14,15] Many CBPs can also bind adenine, but usually with much less tenacity.

As mentioned above, the true specificity of cytokinin action and cytokinin binding is due to the side chain. The side chains of the most biologically active cytokinins, e. g., zeatin, kinetin, N^6-isopentenyladenine (i^{6}Ade), and N^6-benzyladenine (bzl^{6}Ade) are mainly hydrophobic. This implies that a cytokinin binding site must also have a certain degree of hydrophobic nature. Since benzyl groups or isopentenyl groups alone do not have any cytokinin activity, the purine ring of a cytokinin must play an indirect, but indispensible, role in mediating the binding. This can be accomplished by one (or both) of the following mechanisms:

1. The purine gives the cytokinin molecule a specific three-dimensional shape required for binding.
2. Before (or after) the high-affinity binding occurs at the side chain, hydrogen bonding via the purine ring is essential for the maintenance of a stable protein–cytokinin complex.

Evidence to support the first mechanism comes from the studies of Fox,[16] who used energy minimization calculations to obtain models of the most stable conformers of several cytokinins (Figure 1), and from Korszun et al.,[17] who analyzed X-ray crystal structures of cytokinins. The results of both studies indicate that N^6 side chains in cytokinins point away from the plane of the purine ring, at nearly right angles, producing L-shaped conformations. An exciting finding is that some stable conformers of a urea cytokinin (*N,N'*-diphenylurea) have a geometry similar to the purine cytokinins.[16] It has always been a curiosity as to why these urea derivatives, with a chemistry distinctly different from that of the purine cytokinins, exhibit cytokinin activity. This conformation study may help resolve that controversy.

In the late 1970s, several potential cytokinin derivatives designed for photoaffinity labeling of

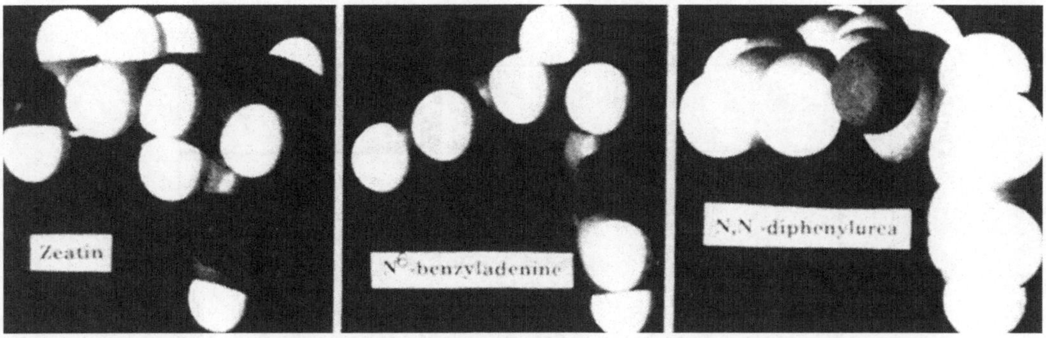

Figure 16-1 Space-filling models of energetically favorable conformers of zeatin, *N*^{6}-benzyladenine, and *N,N'*-diphenylurea. (From Fox, J.E., in *Physiology and Biochemistry of Cytokinins in Plants*, SPB Academic Publishers, The Hague, 1992, 127. With permission.)

CBPs were reported.[18,19] Keim and Fox used [^{14}C]-2-azido-*N*^{6}-benzyladenine (N$_3$bzl^{6}Ade) to label the single high-affinity binding site in the wheat embryo cytokinin binding factor (CBF-1).[20] The subsequent photoaffinity labeling study by Brinegar et al.[21] of the cytokinin binding site of CBF-1 gave some insight into how the purine moiety of bzl^{6}Ade may be involved in the binding mechanism. [^{14}C]-N$_3$bzl^{6}Ade was covalently attached to the CBF-1 cytokinin binding site by UV activation of the azido group. After proteolytic digestion, a ^{14}C-labeled 12-residue peptide was isolated by high-pressure liquid chromatography (HPLC), sequenced by classic Edman analysis, and also analyzed by a mass spectral sequencing method. The results (Figure 2) showed that the photoaffinity ligand attached specifically to one of the two histidine residues in the peptide, even though the other histidine was adjacent to the labeled one.

Since the reactive azido group was on the purine ring of the reagent, it was concluded that the labeled histidine, and possibly the adjacent residues, were required to stabilize the ligand via hydrogen bonding. Further sequence analysis showed that in some CBF-1 polypeptides, a tyrosine was substituted for the target histidine.[22] This observation confirmed the previous conclusion on the importance of the histidine residue for binding, since a tyrosine residue was never labeled by the photoaffinity reagent. Also, the fact that 3-deaza derivatives of cytokinins lose much of their biological activity,[23] in spite of having nearly identical surface conformational energies,[24] suggests the importance of purine hydrogen bonding at receptor sites.

Although the 2-azido derivative of bzl^{6}Ade has been the only photoaffinity reagent used to identify a cytokinin binding site, Mornet et al. synthesized *N*^{6}-(2-azidobenzyl)adenine, which has the reactive group on the side chain instead of on the purine.[25] The use of this reagent, which shows very high growth-promoting activity in the tobacco callus assay, would allow the hydrophobic part of the binding site of CBPs to be probed. The same investigators also synthesized 8-azido-*N*^{6}-benzyl-adenine, which showed activity at extremely low concentrations (rivaling that of zeatin). There are no reports, however, of the use of these latter two derivatives for the detection of binding sites on CBPs. If all three of these azido derivatives could be used to characterize just one cytokinin binding site, our understanding of the site's topological and amino acid requirements would greatly increase.

A prediction of a secondary structure based on a 59-residue sequence surrounding the binding site of CBF-1 (Figure 3) showed that a proline residue near the labeled histidine could induce a turn in the polypeptide backbone.[16] This might bring the histidine close enough to a phenylalanine side chain to allow these residues to hold bzl^{6}Ade in place — hydrophobic interactions could occur between phenylalanine and the benzyl group, while the purine ring could hydrogen bond with the histidine. Only an X-ray crystal-structure determination will prove or disprove this theory. Unfortunately, problems with CBF-1's limited solubility (<2 mg/ml) may preclude good crystallization.

Returning to the urea cytokinins, *N*-phenyl-*N'*(4-pyridyl)urea is reported to have much greater cytokinin activity than *N,N'*-diphenylurea.[26-29] Based on the previously mentioned cytokinin conformational studies, and the binding site data from CBF-1, perhaps the pyridyl nitrogen is involved in hydrogen binding, which could enhance a higher-affinity binding of the phenyl group to the hydrophobic component of the receptor's binding site. This type of hydrogen bonding could not easily occur with *N,N'*-diphenylurea and may account for the 100-fold-higher activity of the pyridyl

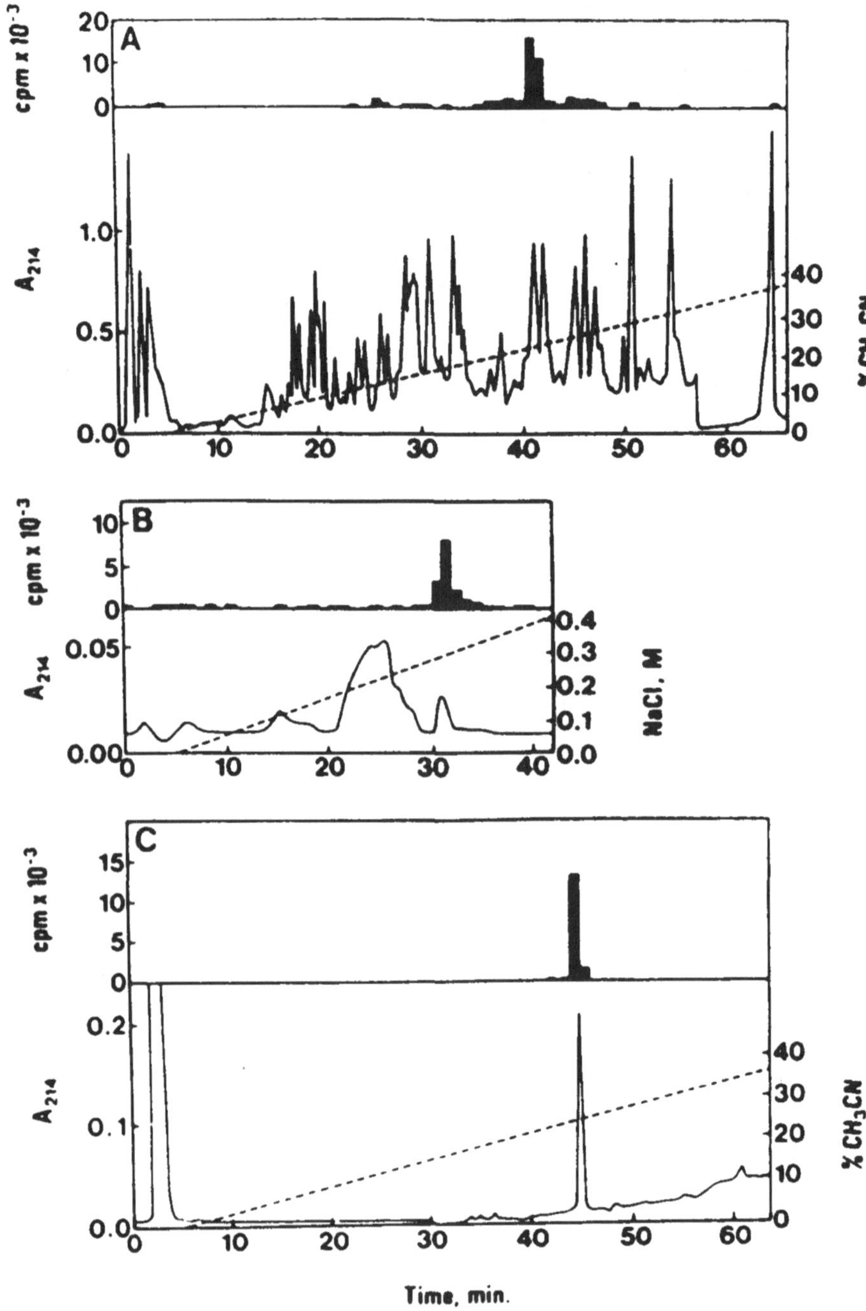

Ala-Phe-Leu-Gln-Pro-Ser-His-His-Asp-Ala-Asp-Glu

$[^{14}C]$-2-azido-bzl^{6}Ade

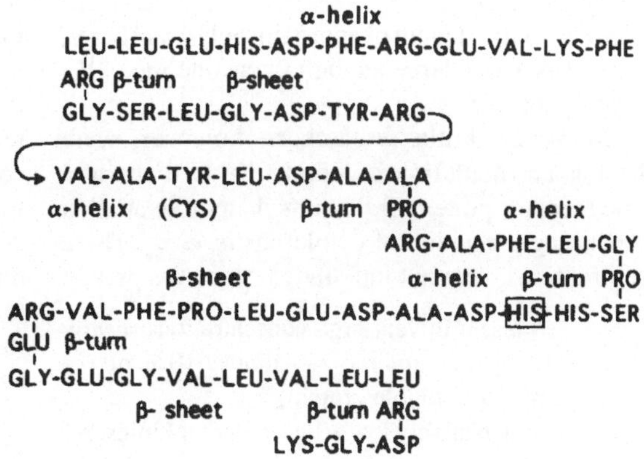

Figure 16-3 Predicted secondary structure of a 59-residue sequence from wheat embryo CBF-1 containing the 12-residue photoaffinity-labeled sequence from Figure 2. The histidine residue labeled by [^{14}C]-2-azido-N^6-benzyladenine is framed. (From Fox, J.E., in *Physiology and Biochemistry of Cytokinins in Plants*, SPB Academic Publishers, The Hague, 1992, 127. With permission.)

derivative in the tobacco callus assay. For a more detailed analysis of the phenylurea cytokinins, see Chapter 3.

Another promising method of studying receptor binding sites is by comparing the structure–activity relationships between cytokinin agonists and antagonists.[30-33] From these types of studies, Iwamura et al.[31,33] proposed that in addition to a hydrophobic receptor site, there is a basic residue that interacts with the imino protons of both the urea and purine cytokinins, via an electrostatic effect. This model has features in common with the CBF-1 model where the purine ring is thought to hydrogen bond to a histidine residue. For a description of cytokinin antagonists, see Chapter 4.

III. CYTOKININ BINDING PROTEINS AND RECEPTORS

A. CEREAL EMBRYO PROTEINS

Soon after the detection of cytokinin binding in ribosome preparations, Fox and Erion showed that a binding factor having the properties of a protein could be removed from wheat germ ribosomes by washing with high concentrations of potassium chloride.[34] Three cytokinin binding factors were subsequently detected in wheat germ extracts and called CBF-1, CBF-2, and CBF-3.[8] CBF-1 had the highest affinity for cytokinins, especially bzl^{6}Ade. Within a few years three separate laboratories had isolated CBF-1 and characterized its subunit structure and binding specificity.[35-38] Affinity chromatography using immobilized Concanavalin A,[35] kinetin riboside,[37] and N^6-benzyladenosine (bzl^{6}Ado)[38] proved to be effective steps in the purification of CBF-1. Preliminary studies showed that CBF-1 had K_d values for the synthetic cytokinins bzl^{6}Ade and kinetin, in the range of 10^{-6} to 10^{-7} *M* and could bind 1.0 to 1.7 mol of these cytokinins per mole of protein. Zeatin binding, however, was much less ($K_d = 1.6 \times 10^{-4}$ *M*).

Native molecular-weight estimates from gel filtration ranged from 122 to 180 kDa. Analysis by sodium dodecyl sulfate polyacrylamide gel electrophoresis (SDS-PAGE) yielded three heterogeneous sets of polypeptides in the size ranges of 12 to 15, 34 to 46, and 53 to 60 kDa, with the latter two sets being the most prominent. This apparent complex subunit structure of CBF-1 was later resolved when it was shown that CBF-1 in milled wheat germ, the starting material for all previous studies, was already partially degraded.[39,40] When isolated from unmilled seeds or dissected embryos, only one polypeptide, with a molecular weight of 54 kDa, was observed by SDS-PAGE. This observation led to a working model of a trimeric structure for CBF-1, composed of identical 54-kDa polypeptides arranged in such a manner that allowed for only one high-affinity cytokinin binding site. A substitution of tyrosine for histidine at the binding site may explain why all three subunits of the CBF-1 trimer are incapable of binding cytokinins.[22]

In vitro translation of mRNA from developing wheat embryos produced a similar-sized polypeptide that was immunoprecipitable with anti-CBF-1 antibodies.[40] A cDNA clone, isolated by immunochemical screening of a phage expression library, hybridized to a single mRNA species upon Northern

Figure 16-2 HPLC isolation of a trypsin/endoproteinase Glu-C peptide from [^{14}C]-2-azido-N^6 benzyladenine-labeled wheat embryo CBF-1. The labeled fractions from the reversed-phase (C_{18}) separation (top figure) were pooled and separated by anion-exchange HPLC (middle figure). The pooled labeled fractions from this separation were shown to contain one homogeneous peptide, by a final reversed-phase (C_8) analysis (bottom figure). The sequence of the isolated peptide and the site of attachment of the cytokinin are shown. (From Brinegar, A.C., et al., *Proc. Natl. Acad. Sci. U.S.A.*, 85, 5927, 1988. With permission.)

blotting. With a length of approximately 1600 bases, this mRNA was large enough to encode the CBF-1 polypeptide.[41]

Many cytokinin researchers, however, were having a difficult time accepting that CBF-1 was a true receptor protein or that it had any role at all in mediating the action of cytokinins *in vivo* . These reservations were not unfounded, since the protein

1. Is present in very high concentrations (nearly 50 mg per embryo — almost 10% of the embryo's soluble protein)[40]
2. Has a relatively low affinity for cytokinins,[35,38] compared to animal receptor-hormone interactions ($K_d < 10^{-10}$ *M*)
3. Binds naturally occurring cytokinins with significantly less affinity than the synthetic ones[42]
4. Cannot be detected in the embryos of all seeds, especially legumes[43,44]

Immunochemical and cytokinin-binding comparisons between wheat embryo CBF-1 and extracts from several cereal embryos and legume cotyledons are given in Table 1. Although no immunochemical cross-reactivity was evident between wheat CBF-1 and a cytokinin factor in oat embryos, the binding activity of the latter was two times greater. A later report indicated that the oat embryo CBP had binding properties similar to the wheat protein.[45] Cytokinin binding in barley embryos is similar to that in wheat and is likely due to a CBP, described by Reddy et al.,[46] that has a K_d for bzl^{6}Ade of 6.6×10^{-7} *M* and a native molecular weight of 155 kDa — both values nearly identical to those of wheat CBF-1. From the table it is obvious that strong immunochemical cross-reactivity between embryo CBPs does not always correlate with significant binding activity (in maize, for example).

The data that convinced everyone CBF-1 and related embryo CBPs could not be "classic" receptors came from immunocytochemical studies that localized CBF-1 in the protein bodies of the tissues surrounding the embryonic axis: the scutellum, coleoptile, coleorhiza, and epiblast (Figure 4).[42,44] In fact, the properties of CBF-1 were noted to be very similar to the vicilin (or 7S) class of seed storage proteins, in terms of its molecular weight, subunit structure, amino acid composition, and developmental expression.[40] The reported association of CBF-1 with ribosomes is probably a nonspecific ionic attraction that occurs when CBF-1 is released from protein bodies, upon homogenization of the embryo tissues.

So, are the cereal embryo CBPs just seed storage proteins that, in some cases, coincidentally bind cytokinins? It is very tempting to say yes, but there is one more question that needs to be answered before we should give in to that temptation. That question is, Are cytokinins bound to embryo CBPs inside the protein bodies? One must consider the possibility that some embryo CBPs sequester cytokinins within protein bodies in order to control their effective concentrations during embryogenesis or to store them for use in germination. Seed

Table 16-1 **bzl^{6}Ade Binding of Extracts from Several Cereal Embryos and Legume Cotyledons and their Immunochemical Cross-Reactivity with Anti-Wheat Embryo CBF-1 IgG[a]**

Species	Relative bzl^{6}Ade Binding[b]	Immunochemical Cross-reactivity[c]	
		Ouchterlony	Western
Wheat (*Triticum durum*)	100	+	+
Oats (*Avena sativa*)	199	–	–
Barley (*Hordeum vulgare*)	93	+	–
Triticale (*Triticosecale* sp.)	85	+	+
Rice (*Oryza sativa*)	48	–	+
Rye (*Secale cereale*)	39	+	+
Maize (*Zea mays*)	7	–	+
Soybean (*Glycine max*)	4	–	–
Pea (*Pisum sativum*)	3	–	–

[a] Data summarized from Brinegar, A.C. and Fox, J.E., in *Plant Hormone Receptors*, Springer-Verlag, Berlin, 1987, 177, with permission; [b] Binding was determined by equilibrium dialysis using [^{3}H]-bzl^{6}Ade. Binding data (cpm/mg protein) are expressed as a percentage of the wheat extract binding; [c] Extracts were tested for their ability to produce a precipitin line with anti-wheat embryo CBF-1 IgG in the Ouchterlony double diffusion assay. Blots of SDS-acrylamide gels of the extracts were treated with the IgG in the Western analysis to detect cross-reactive polypeptides.

storage tissues store proteins, oils, starch, phosphate, and vitamins, so why not hormones as well? The obvious experimental approach is to isolate protein bodies from wheat embryos and assay them for cytokinin content. Until this experiment has been performed, it is premature to rule out this role for cereal embryo CBPs.

B. LECTINS

Seed lectins, proteins with carbohydrate binding sites, have also been shown to bind cytokinins. While studying an adenine binding site on the 124-kDa lima bean (*Phaseolus lunatus*) lectin, investigators noted that a number of active cytokinins could compete with adenine for the binding site, with bzl^{6}Ade (K_d = 2.4 × 10^{-5} *M*) being the most effective.[14] Although the cytokinin affinity was low, the authors speculated that the abundance of the lectin " . . . would result in a major portion of the cytokinin present being bound to the lectin sites." Lectins from *Dolichos biflorus*, *Phaseolus vulgaris*, and *Glycine max* were also shown to bind adenine, and they presumably bind cytokinins at the same site. Cytokinin binding has also been reported at a hydrophobic site of a peanut agglutinin.[15] Again, the binding is very weak (K_d = 4.8 × 10^{-3} *M*), but bzl^{6}Ade was shown to enhance the lectin carbohydrate binding ability and promote the dissociation of the normally tetrameric protein to the dimeric state.

C. TOBACCO TISSUE PROTEINS

A small (4 kDa) CBP was isolated, by Takegami and Yoshida, from tobacco leaves.[47] As with wheat embryo CBF-1, the highest cytokinin affinities were for bzl^{6}Ade and kinetin, although the K_d for bzl^{6}Ade was only 4 × 10^{-5} *M* (approximately 100 times higher than the K_d of CBF-1). The protein could, however, still be isolated by affinity chromatography using immobilized bzl^{6}Ado.[48] The same investigators demonstrated that this protein could bind to tobacco 40S ribosomal subunits in the presence of cytokinins and suggested that it could be a subunit of a translation initiation factor.[49] Zhang et al. also reported an affinity-purified, low-molecular-weight CBP (4.4 kDa) from tobacco callus.[50] It, too, was associated with ribosomes, but was found in the cytoplasm, mitochondria, chloroplasts, and nuclei and on the cell surface as well.

A novel approach to the detection of cytokinin binding sites in tobacco callus was taken by Hamaguchi et al.[51] By using the fluorescent anticytokinin 4-*n*-butylamino-2-methylthiopyrido [2,3-*d*]-pyrimidine in the binding assay, they were able to detect two soluble CBPs (9.4 and 32 kDa). The larger protein had K_d values of 2 × 10^{-8} *M* and 1.9 × 10^{-7} *M* for zeatin and kinetin, respectively. Chen et al.[52] used an N^6-(Δ^2-isopentenyl)adenosine (i^{6}Ado)-derivatized Sepharose affinity column and gel filtration to partially purify a low-molecular-weight

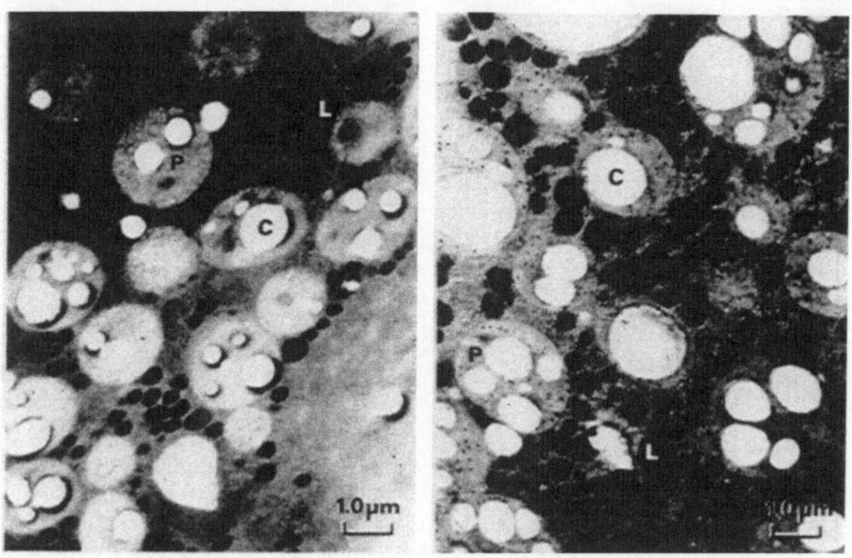

Figure 16-4 Immunocytochemical localization of wheat embryo CBF-1. Developing scutellar tissue was treated with preimmune IgG (left) or anti-CBF-1 IgG (right), followed by a colloidal gold-labeled second antibody (black dots). (P, protein body; C, crystalloid inclusion (phytic acid) within protein body; L, lipid body.) (From Brinegar, A.C. and Fox, J.E., in *Plant Hormone Receptors*, Springer-Verlag, Berlin, 1988, 177. With permission.)

(8.5 kDa) protein fraction from tobacco callus containing a high-affinity binding site for i^6Ade (K_d = 8.8×10^{-7} *M*). This protein could be the same as the previously mentioned 9.4-kDa CBP.

In a heroic effort, Momotani and Tsuji recently isolated 17 μg of a CBP from 9 kg of tobacco leaves.[53] Using both bzl^6Ado- and *t*-zeatin-affinity chromatography in their purification scheme, an apparently monomeric CBP of 31 kDa with a K_d for bzl^6Ade of 1.1×10^{-7} *M* and a single cytokinin binding site was purified to homogeneity. The authors imply that the very low yield of this protein may be due to low chromatographic recoveries rather than being a reflection of its true physiological abundance. This CBP is probably the same as the 32-kDa protein of Hamaguchi et al., mentioned previously.

Sussman and Kende detected a high-affinity cytokinin binding site in an 80,000x*g* particulate fraction of tobacco callus,[54] which was specific for biologically active halogenated derivatives of bzl^6Ade. Binding activity was labile to heat, implying that the binding site was associated with a protein. Gladun et al.[55] reported a cytokinin binding fraction from tobacco callus (having major polypeptides of 20 and 60 kDa) that stimulated RNA polymerase activity *in vitro* in the presence of cytokinins.

D. MAIZE SHOOT PROTEIN

A soluble 46-kDa protein isolated from 5-day-old etiolated maize shoots[56,57] possessed two high-affinity binding sites for zeatin, with K_d values of approximately 2.4×10^{-7} *M* and 1.2×10^{-8} *M*. The latter value is notable for two reasons: (1) it is the lowest K_d measured for a cytokinin binding site, and (2) zeatin, a naturally occurring cytokinin, was the ligand used in the binding assay. Only active cytokinins could displace zeatin, with the exception that ATP was also an effective competitor. The protein is present in very low concentrations, requiring four chromatographic steps and a 500-fold purification, to achieve a purity of approximately 90% (Figure 5). Attempts at sequencing have failed, presumably due to a blocked N terminus. Its low abundance, coupled with a very high affinity for zeatin, gives this protein the characteristics of a classic receptor.

E. BARLEY LEAF PROTEINS

Cytokinin binding in barley (*Hordeum vulgare*) leaves was detected, by Kharchenko et al.,[58] from extracts purified by affinity chromatography on bzl^6Ado-Sepharose. Elution of the bound proteins with 10^{-3} *M* bzl^6Ade was found to more specifically displace the cytokinin binding factor(s). Extensions of this work[59,60] resulted in the purification of a 40 to 45-kDa CBP having a K_d for zeatin of 1.5×10^{-8} *M*. Competition for the zeatin binding site by other cytokinins was in proportion to their biological cytokinin activities. The size and zeatin-binding affinity of this protein are nearly identical to the previously described maize shoot CBP. Romanov[61] pointed out the similarities between this class of cytokinin binding protein in barley, maize, wheat, and rye and proposed to call them the "major cytokinin binding proteins (mCBPs),"

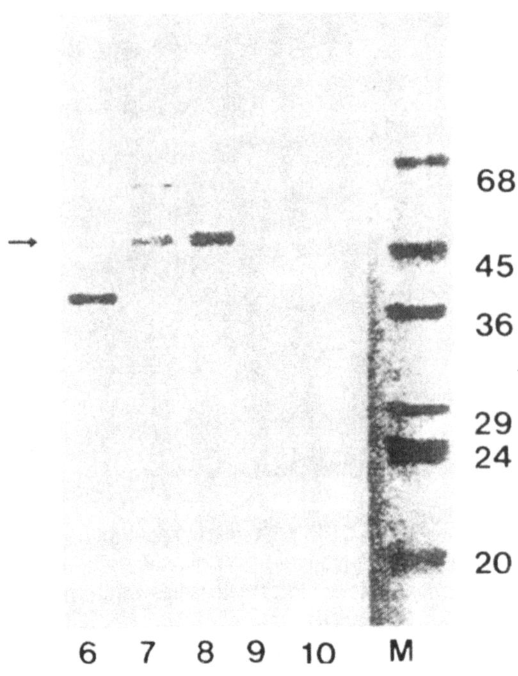

Figure 16-5 Electrophoretic analysis of fractions from hydrophobic interaction (Alkyl Superose) chromatography of maize shoot CBP (after previous chromatography on DEAE Bio-Gel, Sephacryl S200, and Mono Q columns). Proteins were eluted with a 40 to 0%-saturated ammonium sulfate gradient, and fractions 6–10 were run on a sodium dodecyl sulfate-polyacrylamide gel. The arrow indicates the 46-kDa polypeptide present in fractions containing zeatin binding activity. (From Romanov, G.A., Taran, V.Y., and Venis, M.A., in *Plant Physiol.*, 136, 208, 1990. With permission.)

since they have now been detected in many vegetative parts of cereal plants.[62,63] These proteins do, indeed, show many properties associated with receptors, but the term "major" must not be confused with "most important," until there is more information on their biological function.

There are a number of reports describing the effects of barley leaf CBPs on transcriptional regulation. Stimulation of RNA synthesis by cytokinins was observed several years ago,[64] but researchers led by Kulaeva in Moscow discovered that barley leaf CBPs in the presence of cytokinins could activate RNA synthesis even further.[58,65] Romanko et al. affinity-purified a cytokinin binding fraction that stimulated RNA synthesis in isolated nuclei and chloroplasts, but only in the presence of cytokinins.[66] Later they reported that a cytokinin binding fraction (containing major polypeptides of 57 and 64 kDa) from isolated chloroplasts stimulated RNA synthesis in chloroplasts, but not in nuclei.[67] Selivankina et al. showed that the combination of barley leaf CBPs and cytokinins increased the RNA polymerase activity in isolated barley leaf chromatin[68] and also increased the template activity of the chromatin itself.[69] Nonplant sources of RNA polymerase or chromatin were unaffected by such treatment. A wheat embryo CBP preparation (presumably CBF-1) could not substitute for the barley leaf CBP.[70]

The effects of CBPs on transcription were studied further by Fedina et al.[71] They were able to fractionate CBPs from the nucleoplasm and chromatin of barley leaf protoplasts and test their effects on the RNA polymerase activities in isolated nuclei, alone or in combination with cytokinins. Nucleoplasmic CBPs stimulated RNA polymerases I and II, but only in the presence of bzl^{6}Ade (Table 2). The chromatin proteins, however, could only enhance the activity of RNA polymerase I. This regulation was not cytokinin dependent. It is unclear whether any of these nuclear CBPs were in the earlier whole barley leaf cytokinin binding fractions. (For a detailed review of the effects of cytokinins on plant gene expression, see Chapter 17.)

Cytokinins are known to stimulate the activity of a protein kinase in barley leaf chromatin.[72-75] This protein kinase copurifies with, and can phosphorylate, some of the subunits of RNA polymerase I.[76] Cytokinins can increase this RNA polymerase-associated protein kinase activity nearly 400%, but if a 28 to 30-kDa cytosolic barley leaf cytokinin binding protein[73,77] is added to the *in vitro* assay, the cytokinins can also stimulate the RNA polymerase I activity. This CBP is apparently different than the 45-kDa barley leaf zeatin binding protein described earlier.[59,60]

An interesting feature of the 28 to 30-kDa cytosolic protein is that it either copurifies with a protein kinase or it has its own kinase activity.[74] And to make matters more interesting (or more frustrating), this kinase activity can be increased by treatment with abscisic acid (ABA). Bzl^{6}Ade cannot stimulate the kinase, but it can prevent its activation by ABA. The model for a cytokinin-induced RNA polymerase I activation that these data imply is quite complex and rather speculative at the moment, but involves (1) the cytokinin activation of a nuclear protein kinase that phosphorylates (but does not activate) RNA polymerase I, and (2) the activation of the RNA polymerase I by a cytoplasmic cytokinin-dependent factor (CBP?) that can presumably reach the nucleus via nuclear pores. (Signal transduction by cytokinins and the role of protein kinases is reviewed in more detail in Chapter 18.)

F. MITOCHONDRIAL PROTEINS

The effects of exogenously applied cytokinins on plant respiration have been studied in great detail (for a review, see Chapter 13). The inhibitory effect of bzl^{6}Ade on the alternative (or cyanide-resistant) respiratory pathway in plant mitochondria is especially well documented.[78-83] Through inhibitor studies, many sites of cytokinin action have been proposed within the mitochondrial electron transport system,[78,84,85] but there has been no direct evidence that implicates any particular mitochondrial protein. The immediate inhibitory effects

Table 16-2 **Effect of Barley Leaf Protoplast Nucleoplasmic CBP on the Activity of Nuclear RNA Polymerases**[a]

Treatment	Relative Activity (% of Control ATP Incorporation)	
	RNA Polymerase I	RNA Polymerase II
Nuclei (control)	100	100
Nuclei + bzl^{6}Ade[b]	120	125
Nuclei + CBP	114	124
Nuclei + CBP + bzl^{6}Ade	201	157

[a] Data from Fedina, A.B, Berkhanova, E.A., and Kharchenko, V.I., *Sov. Plant Physiol.*, 34, 263, 1987, with permission; [b] Concentration of bzl^{6}Ade was 5×10^{-7} *M*.

of bzl^{6}Ade on oxygen consumption by isolated mitochondria rule out the involvement of CBPs or receptors that regulate transcription or translation. The most likely targets of cytokinin action are proteins of the mitochondrial inner membrane, that directly or indirectly affect electron transport.

Brinegar et al. used a photoaffinity labeling approach in an attempt to identify CBPs in isolated mitochondria from etiolated mung bean hypocotyls where 15 to 20% of the total respiration was due to electron flow through the alternative pathway.[86] The photoaffinity reagent, [^{3}H]-2-azido-N^6-benzyladenine,[87] attached to several mitochondrial proteins. The most prominently labeled polypeptides were the α and β (catalytic) subunits of F_1ATPase and an unidentified 33 kDa polypeptide. These polypeptides are also the most abundant proteins in the mitochondria, so the specificity of the labeling was questionable. Pretreatment of mitochondria with ADP, to protect ADP binding sites on the F_1 subunits, or with bzl^{6}Ade, to out-compete labeling of true cytokinin sites, both resulted in the complete inhibition of labeling. These confusing results can be explained as follows: (1) the labeling was broadly specific for adenine derivatives, or (2) nonspecific labeling was enhanced by concentration of the lipophilic reagent into the protein-rich inner membrane.

More recent refinements of the labeling technique have shown that the 33 kDa polypeptide can be more specifically labeled than previously thought (Figure 6).[88] Electrophoretic comparison with partially purified F_1ATPase indicates that this protein is not the gamma subunit of F_1 (which has a molecular weight near 35 kDa). The 33 kDa protein appears to be an integral membrane protein and can be recovered by detergent solubilization after washing mitochondrial membranes with 6 *M* urea to remove F_1ATPase and other peripheral proteins. Attempts to identify this protein by N-terminal sequencing are underway.

G. FUNGAL AND MOSS PROTEINS

LéJohn and Cameron isolated a 6-kDa calcium binding glycoprotein from the cytoplasmic surface of the plasma membrane from the water molds *Bastocladiella emersonii* and *Achlya* spp.[89] The binding of calcium ions was found to be allosterically regulated by cytokinins; i. e., cytokinins caused the release of bound calcium into the cytoplasm. In a subsequent report, it was shown that auxins and cytokinins may act synergistically to control the release of calcium from the protein.[90] The authors suggested that one action of plant hormones may be to regulate cellular calcium levels by this mechanism.

Binding of cytokinin bases, but not ribosides, was demonstrated in particulate (13,000 to 18,000x*g*) fractions from extracts of the moss *Funaria hygrometrica*.[91] As in the case of wheat embryo CBF-1, zeatin binding was significantly less than bzl^{6}Ade binding. Heat or detergent treatment greatly reduced the binding capacity, suggesting that the binding moiety was a protein. Binding to ribosomes did not seem to occur to any great extent.

The moss system has not been pursued. The reason for this may be that in the same paper, the authors demonstrated that cytokinins can also bind competitively with nonbiological material such as talc (hydrous magnesium silicate). To many, this observation cast doubts on whether cytokinin binding data were biologically meaningful. Could it be that most of the reported binding data was just due to nonspecific physical adsorption? In my opinion, this single observation received more attention than it deserved. If one adds a finely ground mineral (especially a silicate) to a biological extract, one should expect to see some batch chromatography take place. The authors' warning, however, about being wary of nonbiologically significant hormone binding is still sound advice. A good example of such nonspecific binding was provided by the same laboratory in the study of cytokinin binding sites in tobacco callus tissue.[54] Both high- and low-affinity binding sites were detected, but the latter site appeared not to be associated with a protein and had very low specificity toward active cytokinin derivatives.

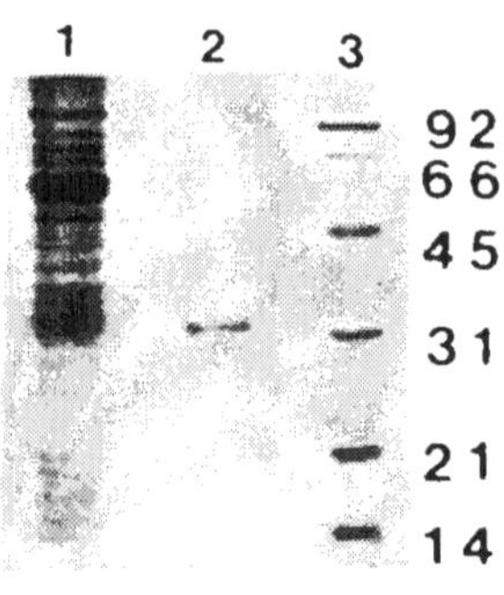

Figure 16-6 Photoaffinity labeling of mung bean mitochondrial proteins, using [^{3}H]-2-azido-N^6-benzyladenine. [Lane 1, Coomassie Blue stained mitochondrial proteins after labeling; Lane 2, fluorogram of Lane 1; Lane 3, molecular weight markers (kDa).] (From Shah, G. and Brinegar, A.C., Unpublished data, 1992.)

While on the topic of nonspecific binding, it is convenient here to note some studies on the validity of several commonly used hormone binding assays. Venis demonstrated that the commonly used ammonium sulfate precipitation method often yields higher apparent binding because of its enhancement of hydrophobic interactions between the ligand and binding proteins or between the ligand and the glass or plastic tubes used in the assay.[92] This was confirmed by Kamínek and Fox[93] in their comparison of equilibrium dialysis, ultrafiltration, and ammonium sulfate precipitation assays. The recommendation from both studies is that any binding detected by the ammonium sulfate precipitation assay must be confirmed by at least one other assay. A good case in point is a study on auxin binding sites where the investigators showed that three different assays gave similar results.[94]

H. MISCELLANEOUS PROTEINS

Kobayashi et al. described a high-affinity cytokinin binding site in an 80,000x*g* particulate fraction from carrot (*Daucus carota*) suspension cells.[95] The K_d for bzl^{6}Ade was very low (3.3×10^{-8} *M*). The specificity of cytokinin binding was in the order bzl^{6}Ade > kinetin > zeatin > i^{6}Ade > Ade, with the ribosides showing less binding affinity than their corresponding bases. Two CBPs have been detected in cucumber (*Cucumis sativus*) cotyledons. A peptide of 4 kDa, capable of dimerizing to an 8-kDa form, was partially purified by affinity chromatography and detected by its binding to i^{6}Ado trialcohol.[96] CBPs of these molecular weights are reminiscent of those detected in tobacco tissues described previously. This raises the question of whether such a class of binding protein is restricted to dicots. A larger zeatin binding protein was partially purified from the soluble fraction of cucumber cotyledons.[97] The K_d for zeatin was measured at 4×10^{-7} *M*, and its molecular weight of 43 kDa is similar to that of the "major CBPs" from the vegetative parts of cereals.

The isolation of CBPs from bean hypocotyl membranes[98] and cotton seedlings[99] were reported, but the data appear to be preliminary.

IV. SUMMARY AND CONCLUSIONS

It has been proposed that a certain set of criteria must be met by a CBP before it is called a receptor.[60] The protein in question must:

1. Recognize cytokinins and no other plant hormone
2. Demonstrate reversible binding and have low binding capacity for cytokinins
3. Have affinities for cytokinin derivatives that correlate positively with their biological (bioassay) activity
4. Be present in cytokinin-sensitive tissues
5. Have K_d values for cytokinins (especially zeatin) similar to, or lower than, the *in vivo* cytokinin concentrations of their parent tissues

Many of the CBPs mentioned in this review satisfy the first three criteria and sometimes the fourth, but very few satisfy the last requirement. The 40 to 45-kDa "major CBPs" in the vegetative parts of maize, barley, rye, and wheat seem to fulfill all five of the above requirements.

There is, however, one last criterion: there must be a demonstrated biological function for the protein. Evidence has been presented that cytoplasmic CBPs in barley leaves can stimulate transcription in the presence of cytokinins. If these cytoplasmic CBPs are, indeed, receptors, then their mode of action would be somewhat similar to animal steroid receptors — the proteins would carry cytokinins into the nucleus through nuclear pores where they exert their effects on gene expression. Such a cytoplasmic–nuclear exchange could also explain why CBPs have been detected in the nucleoplasm and chromatin of barley leaf protoplasts. All of the evidence to date on the biological function of these CBPs has been in the form of *in vitro* nucleotide incorporation experiments (e. g., Table 2). Another method, such as subcellular immunocytochemical localization, would go a long way in confirming the function of these proteins and their status as cytokinin receptors. Having cytokinin binding mutant lines where CBPs are inactive or reduced in abundance would be a gift from heaven, but there are few known mutants involving cytokinin function.[100]

The detection of membrane-bound cytokinin receptors has not been very successful. Cytokinin binding has been detected in "particulate" or crude membrane fractions, but the membrane and organellar composition of those fractions has never been determined. Interestingly, there has never been a report of cytokinin binding by isolated plasma membranes. It is certainly premature to discount the existence of plasma membrane cytokinin receptors or transport proteins, but are not most cytokinin bases hydrophobic enough to diffuse through membranes? Perhaps there is no need for membrane receptors.

If it appears that the cytoplasmic proteins are the true receptors, should all of the other reported CBPs be dismissed as anomalies? Definitely not! The premise that a hormone binding protein must have the characteristics of a receptor in order to be biologically relevant is nonsense. The possible role of seed proteins (embryo CBPs or lectins) in sequestering cytokinins during embryogenesis has already been discussed. What about all of the reports of cytokinin binding to ribosomes? In spite of numerous studies, no one has ever provided any convincing evidence that cytokinins or CBPs act directly on ribosomes. Until that time, ribosome binding should be considered an artifact. But that is not to say that all the CBPs that bind to ribosomes are not biologically functional. Their function just does not appear to be related to ribosomes. Finally, it is prudent to consider the possibilty that cytokinins can act directly on metabolic processes, without the aid of a receptor. A case in point is the immediate effect of cytokinins on mitochondrial respiration, as mentioned previously where a "binding protein" may actually be an electron transport protein. Miller made an impressive argument for the role of cytokinins in regulating oxidative processes and predicted that flavoproteins may be targets of cytokinins.[101]

It is easy for us to fall into a collective pattern of thinking that promotes certain biases, e. g., that the "cut-off" K_d that differentiates a receptor from just another binding protein is 10^{-8} *M*, that nonreceptor binding proteins do not count, that plant receptors must act like animal receptors, or that there is only one ubiquitous receptor that mediates all cytokinin action. As scientists, we should all know by now that nearly anything is possible. All of the confusion and controversy regarding cytokinin receptors will eventually be worked out if we continue to design good experiments, select the proper cytokinin-dependent systems to study, take advantage of new technical developments, and, above all, continue to keep our minds open.

REFERENCES

1. **Berridge, M. V., Ralph, R. K., and Letham, D. S.,** The binding of kinetin to plant ribosomes, *Biochem. J.*, 119, 75, 1970.
2. **Davies, J. W. and Cocking, E. C.,** Protein synthesis in tomato-fruit locule tissue. Incorporation of amino acids into protein by aseptic cell-free systems, *Biochem. J.*, 104, 23, 1967.
3. **Firn, R. D.,** Too many binding proteins, not enough receptors, in *Plant Hormone Receptors*, NATO ASI Series H, Cell Biology, Vol. 10, Kläembt, D., Ed., Springer-Verlag, Berlin, 1987, 1.
4. **Flores, S. and Tobin, E. M.,** Benzyladenine modulation of the expression of two genes for nuclear encoded chloroplast proteins in *Lemna gibba*. Apparent post-transcriptional regulation, *Planta*, 168, 340, 1986.
5. **Flores, S. and Tobin, E. M.,** Cytokinin modulation of LHCP mRNA levels: the involvement of post-transcriptional regulation, *Plant Mol. Biol.*, 11, 409, 1988.
6. **Venis, M. A.,** Hormone receptor proteins, *Curr. Adv. Plant Sci.*, 2, 21, 1973.
7. **Kende, H. and Gardner, G.,** Hormone binding in plants, *Ann. Rev. Plant Physiol.*, 27, 267, 1976.
8. **Fox, J. E. and Erion, J. L.,** Cytokinin-binding proteins in higher plants, in *Plant Growth Regulation*, Proc. 9th Int. Conf. on Plant Growth Substances, Pilet, P. E., Ed., Springer-Verlag, Berlin, 1977, 139.
9. **Keim, P., Erion J. L., and Fox, J. E.,** The current status of cytokinin-binding moieties, in *Metabolism and Molecular Activities of Cytokinins*, Guern, J. and Péaud-Lenoël, C., Eds., Springer-Verlag, Berlin, 1981, 179.
10. **Venis, M. A.,** *Hormone Binding Sites in Plants*, Longman, New York, 1985, 115.
11. **Romanov, G. A.,** Hormone-binding proteins of plants and the problem of reception of phytohormones, *Sov. Plant Physiol.*, 36, 136, 1989.
12. **Napier, R. M. and Venis, M. A.,** Receptors for plant growth regulators: recent advances, *J. Plant Growth Regul.*, 9, 113, 1990.
13. **Lehninger, A. L.,** *Biochemistry*, 2nd ed., Worth, New York, 1975, 312.
14. **Roberts, D. D. and Goldstein, I. J.,** Adenine binding sites of the lectin from lima beans (*Phaseolus lunatus*), *J. Biol. Chem.*, 258, 13820, 1983.
15. **Olsen, K. W., Zaluzec, E. J., Zaluzec, M. M., Fernandez, E. J., and Pavkovic, S. F.,** Crystallographic and binding studies on a cytokinin peanut agglutinin complex, *Biophys. J.*, 59, 296A, 1991.
16. **Fox, J. E.,** Molecular modeling of cytokinins and the CBF-1 receptor, in *Physiology and Biochemistry of Cytokinins in Plants*, Kamínek, M., Mok, D. W. S., and Zazímalová, E., Eds., SPB Academic Publishers, The Hague, 1992, 127.
17. **Korszun, Z., Knight, C., and Chen, C.-M.,** A stereochemical model for cytokinin activity, *FEBS Lett.*, 293, 1989.
18. **Theiler, J. B., Leonard, N. J., Schmitz, R. Y., and Skoog, F.,** Photoaffinity-labeled cytokinins: synthesis and biological activity, *Plant Physiol.*, 58, 803, 1976.
19. **Sussman, M. and Kende, H.,** The synthesis and biological properties of 8-azido-N^6-benzyladenine, a potential photoaffinity reagent for cytokinins, *Planta*, 137, 91, 1977.

20. **Keim, P. and Fox, J. E.,** Interaction of a radiolabeled cytokinin photoaffinity probe with a receptor protein, *Biochem. Biophys. Res. Commun.*, 96, 1325, 1980.
21. **Brinegar, A. C., Shabanowitz, J., Hauer, C., Cooper, G., Stevens, A., and Fox, J. E.,** Characterization of a benzyladenine binding site peptide isolated from a wheat cytokinin-binding protein: sequence analysis and identification of a single affinity-labeled histidine residue by mass spectrometry, *Proc. Natl. Acad. Sci. U.S.A.*, 85, 5927, 1988.
22. **Brinegar, A. C. and Fox, J. E.,** unpublished data, 1988.
23. **Rogozinska, J. H., Kroon, C., and Salemink, C. A.,** Deazapurine derivatives. IX. Influence of alterations in the purine ring on biological activity of cytokinins, *Phytochemistry*, 12, 2087, 1973.
24. **Kaneti, J. and Karanov, E.,** Conformational analysis of cytokinins and analogs, *Chem-Biol. Interactions*, 43, 73, 1983.
25. **Mornet, R., Theiler, J. B., Leonard, N. J., Schmitz, R. Y., Moore, F. H., III, and Skoog, F.,** Active cytokinins. Photoaffinty labeling agents to detect binding, *Plant Physiol.*, 64, 600, 1979.
26. **Bruce, M. I. and Zwar, J. A.,** Cytokinin activity of some substituted ureas and thioureas, *Proc. R. Soc. Lond., [Biol.]* 76, 245, 1976.
27. **Takahashi, S., Shudo, K., Okamoto, T., Yamada, K., and Isogai, Y.,** Cytokinin activity of N-phenyl-N′-(4-pyridyl)urea derivatives, *Phytochemistry*, 17, 1201, 1978.
28. **Okamoto, T., Shudo, K., Takahashi, S., and Isogai, Y.,** 4-Pyridylureas are surprisingly potent cytokinins. The structure–activity relationship., *Chem. Pharm. Bull.*, 29, 3748, 1981.
29. **Shudo, K., Okamoto, T., Isogai, Y., and Ohshima, K.,** Structure–activity relationships and agricultural applications of synthetic cytokinins, in *Physiology and Biochemistry of Cytokinins in Plants*, Kamínek, M., Mok, D. W. S., and Zazímalová, E., Eds., SPB Academic Publishers, The Hague, 1992, 189.
30. **Shimizu, R., Iwamura, H., Matsubara, S., and Fujita, T.,** Development of S-triazine anticytokinins and their quantitative structure–activity relationship, *J. Agric. Food Chem.*, 37, 236, 1989.
31. **Iwamura, H.,** Structure and biological activity of non-adenylic anticytokinins, in *Physiology and Biochemistry of Cytokinins in Plants*, Kamínek, M., Mok, D. W. S., and Zazímalová, E., Eds., SPB Academic Publishers, The Hague, 1992, 195.
32. **Iwamura, H., Masuda, N., Koshimizu, K., and Matsubara, S.,** Quantitative aspects of receptor binding of cytokinin agonists and antagonists, *J. Med. Chem.*, 26, 838, 1983.
33. **Iwamura, H., Murakami, S., Koshimizu, K., and Matsubara, S.,** Quantitative structure–activity relationships in cytokinin agonistic and antagonistic pyrido[2,3-*d*] pyrimidine derivatives: insights into receptor topology, *J. Med. Chem.*, 28, 577, 1985.
34. **Fox, J. E. and Erion, J. L.,** A cytokinin binding protein from higher plant ribosomes, *Biochem. Biophys. Res. Commun.*, 64, 694, 1975.
35. **Polya, G. M. and Davis, A. W.,** Properties of a high affinity cytokinin-binding protein from wheat germ, *Planta*, 139, 139, 1978.
36. **Polya, G. M. and Bowman, J. A.,** Ligand specificity of a high affinity cytokinin-binding protein, *Plant Physiol.*, 64, 387, 1979.
37. **Moore, F. H., III,** A cytokinin-binding protein from wheat germ. Isolation by affinity chromatography and properties, *Plant Physiol.*, 64, 594, 1979.
38. **Erion, J. L. and Fox, J. E.,** Purification and properties of a protein which binds cytokinin-active 6-substituted purines, *Plant Physiol.*, 67, 156, 1981.
39. **Brinegar, A. C. and Fox, J. E.,** Resolution of the subunit composition of a cytokinin-binding protein from wheat embryos, *Biol. Plant.*, 27, 100, 1985.
40. **Brinegar, A. C., Stevens, A., and Fox, J. E.,** Biosynthesis and degradation of a wheat embryo cytokinin-binding protein during embryogenesis and germination, *Plant Physiol.*, 79, 706, 1985.
41. **Brinegar, A. C. and Fox, J. E.,** The developmental expression and molecular cloning of the gene for a wheat embryo cytokinin-binding protein, in *Plant Genetics*, UCLA Symp. on Molecular and Cellular Biology, New Series, Vol. 35, Freeling, M., Ed., Alan R. Liss, New York, 1985, 147.
42. **Brinegar, A. C. and Fox, J. E.,** Studies on a wheat embryo cytokinin-binding protein, in *Current Topics in Plant Biochemistry and Physiology 1985*, Proc. 3rd Ann. Plant Biochemistry and Physiology Symp., Vol. 4, Randall, D. D., Ed., Interdisciplinary Plant Biochemistry and Physiology Program, Univ. of Missouri, 1985, 91.
43. **Fox, J. E. and Gregerson, E.,** Variation in a cytokinin binding protein among several cereal crop plants, in *Plant Growth Substances 1982*, Wareing, P. F., Ed., Academic Press, London, 1982, 207.

44. **Brinegar, A. C. and Fox, J. E.,** Immunocytological localization of a wheat embryo cytokinin-binding protein and its homology with proteins in other cereals, in *Plant Hormone Receptors*, NATO ASI Series H, Cell Biology, Vol. 10, Klämbt, D., Ed., Springer-Verlag, Berlin, 1987, 177.
45. **Kamínek, M. and Fox, J. E.,** Cytokinin binding protein from oat seeds. 14th Int. Congr. on Biochemistry, Prague, Abst. FR-C31-7, 1988.
46. **Reddy, A. S. N., Sopory, S. K., and Datta, A.,** Purification and characterization of a cytokinin binding protein from barley embryos, *Biochem. Int.*, 6, 181, 1983.
47. **Takegami, T. and Yoshida, K.,** Isolation and purification of cytokinin binding protein from tobacco leaves by affinity column chromatography, *Biochem. Biophys. Res. Commun.*, 67, 782, 1975.
48. **Yoshida, K. and Takegami, T.,** Isolation of cytokinin binding protein from tobacco leaves by bioaffinity chromatography and its partial characterization, *J. Biochem.*, 81, 791, 1977.
49. **Takegami, T. and Yoshida, K.,** Specific interaction of cytokinin binding protein with 40S ribosomal subunits in the presence of cytokinin *in vitro*, *Plant Cell Physiol.*, 18, 337, 1977.
50. **Zhang, K., Yao, Y., Lu, M., Gao, G., Ni, D., and Sun, H.,** Study on cytokinin binding protein from cell of Baibaili tobacco callus, *Acta Bot. Yunnanica*, 12, 80, 1990.
51. **Hamaguchi, N., Iwamura, H., and Fujita, T.,** Fluorescent anticytokinins as a probe for binding. Isolation of cytokinin-binding proteins from the soluble fraction and identification of a cytokinin-binding site on ribosomes of tobacco (*Nicotiana tabacum* cultivar Wisconsin No. 38) callus cells, *Eur. J. Biochem.*, 153, 565, 1985.
52. **Chen, C.M., Melitz, D. K., Petschow, B., and Ecker, R. L.,** Isolation of cytokinin binding protein from plant tissues by affinity chromatography, *Eur. J. Biochem.*, 108, 379, 1980.
53. **Momotani, E. and Tsuji, H.,** Isolation and characterization of a cytokinin-binding protein from the water-soluble fraction of tobacco leaves, *Plant Cell Physiol.*, 33, 407, 1992.
54. **Sussman, M. R. and Kende, H.,** *In vitro* cytokinin binding to a particulate fraction of tobacco cells, *Planta*, 140, 251, 1978.
55. **Gladun, A. A., Kalinin, R. L., and Sarnatskaya, V. V.,** Isolation and study of the cytokinin-binding protein function in tobacco tissue culture, *Fiziol. Biokhim. Kul't. Rast.*, 18, 333, 1986.
56. **Romanov, G. A., Taran, V. Y., and Venis, M. A.,** Cytokinin-binding protein from maize shoots, *J. Plant Physiol.*, 136, 208, 1990.
57. **Taran, V. Y., Romanov, G. A., and Venis, M. A.,** Purification of soluble zeatin-binding proteins from maize shoots, in *Physiology and Biochemistry of Cytokinins in Plants*, Kamínek, M., Mok, D. W. S., and Zazímalová, E., Eds., SPB Academic Publishers, The Hague, 1992, 165.
58. **Kharchenko, V. I., Romanko, E. G., Selivankina, S. Y., and Kulaeva, O. N.,** Purification of cytokinin binding proteins from barley leaves by affinity chromatography, *Sov. Plant Physiol.*, 30, 932, 1983.
59. **Romanov, G. A., Taran, V. Y., Chovjka, L., and Kulaeva, O. N.,** Specific binding of zeatin to a protein fraction from barley leaves and purification of cytokinin-binding proteins, *Sov. Plant Physiol.*, 33, 75, 1986.
60. **Romanov, G. A., Taran, V. Y., Chvojka, L., and Kulaeva, O. N.,** Receptor-like cytokinin-binding protein(s) from barley leaves, *J. Plant Growth Regul.*, 7, 1, 1988.
61. **Romanov, G. A.,** The current status of major cytokinin-binding proteins from cereal plants, in *Physiology and Biochemistry of Cytokinins in Plants*, Kamínek, M., Mok, D. W. S., and Zazímalová, E., Eds., SPB Academic Publishers, The Hague, 1992, 133.
62. **Romanov, G. A., Taran, V. Y., Kuleva, O. N., Chovjka, L., and Venis, M. A.,** Cytokinin-binding proteins from some cereal plants, in *Molecular Aspects of Hormonal Regulation of Plant Development*, Proc. 14th Biochemical Congr., Kutacek, M., Elliott, M. C., and Machácková, I., Eds., SPB Academic Pubishers, The Hague, 1990, 213.
63. **Romanov, G. A., Taran, V. Y., and Orudzhev, E. M.,** Cereal zeatin-binding proteins: age, organ, tissue and subcellular aspects, *Sov. Plant Physiol.*, 38, 808, 1991.
64. **Matthysse, A. G. and Abrams, M.,** A factor mediating interaction of kinins with the genetic material, *Biochim. Biophys. Acta*, 199, 511, 1970.
65. **Romanko, E. G., Selivankina, S. Y., Ovcharov, A. K., and Kulaeva, O. N.,** Activation of RNA synthesis in vitro by cytokinin-receptor complex from barley leaves, *Dokl. Akad. Nauk SSSR*, 255, 1009, 1981.
66. **Romanko, E. G., Selivankina, S. Y., and Ovcharov, A. K.,** Involvement of cytokinin binding proteins from barley leaves in the activation by cytokinin of RNA synthesis in isolated nuclei and choroplasts, *Sov. Plant Physiol.*, 29, 408, 1982.

67. **Romanko, E. G., Selivankina, S. Y., Moshkov, I. E., and Novikova, G. V.,** Effect of cytokinin-binding proteins isolated from chloroplasts on transcription, *Sov. Plant Physiol.*, 33, 823, 1986.
68. **Selivankina, S. Y., Romanko, E. G., Ovcharov, A. K., and Kharchenko, E. I.,** Participation of cytokinin-binding proteins from barley leaves in cytokinin-activation of chromatin-bound RNA-polymerase, *Sov. Plant Physiol.*, 29, 201, 1982.
69. **Selivankina, S. Y., Romanko, E. G., Burkhanova, E. A., Karavaiko, N. N., Zayakin, V. V., and Kulaeva, O. N.,** A comparison of effects of the cytokinin receptor complex on the template activity of chromatin and RNA-polymerase activity in barley leaves, *Biokhimiya*, 50, 47, 1985.
70. **Selivankina, S. Y., Romanko, E. G., Moshkov, I. E., Kharchenko, V. I., and Kulaeva, O. N.,** Complexes of cytokinin with cytokinin-binding proteins from barley leaves and with cytokinin-binding proteins from wheat embryos: comparison of their action on RNA synthesis in vitro, *Sov. Plant Physiol.*, 32, 402, 1985.
71. **Fedina, A. B., Burkhanova, E. A., and Kharchenko, V. I.,** Functional activity of cytokinin-binding nuclear proteins from barley leaf protoplasts, *Sov. Plant Physiol.*, 34, 263, 1987.
72. **Selivankina, S. Y., Romanko, E. G., Novikova, G. V., and Kulaeva, O. N.,** Cytokinin regulation of protein kinase bound to chromatin and RNA-polymerase I from barley leaves, *Dokl. Akad. Nauk SSSR*, 292, 766, 1987.
73. **Selivankina, S. Y., Romanko, E. G., Karavaiko, N. N., Moshkov, I. E., Novikova, G. V., and Kulaeva, O. N.,** Hormonal regulation of barley leaf cytosol protein kinase, *Dokl. Akad. Nauk SSSR*, 299, 254, 1988.
74. **Selivankina, S. Y., Romanko, E. G., Novikova, G. V., Muromtseva, D. G., and Kulaeva, O. N.,** Effect of cytokinin and other phytohormones on protein kinases associated with chromatin and RNA-polymerase I in barley leaves, *Sov. Plant Physiol.*, 35, 205, 1988.
75. **Selivankina, S. Y., Novikova, G. V., Pereverzeva, I. N., Moshkov, I. E., and Kulaeva, O. N.,** Occurrence of cytokinin-controlled C-type protein kinase in barley leaves, *Sov. Plant Physiol.*, 37, 641, 1990.
76. **Kulaeva, O. N., Novikova, G. V., Selivankina, S. Y., Karavaiko, N. N., and Moshkov, I. E.,** Protein kinases and cytokinin-binding proteins in plant cell response to cytokinin, in *Physiology and Biochemistry of Cytokinins in Plants*, Kamínek, M., Mok, D. W. S., and Zazímalová, E., Eds., SPB Academic Publishers, The Hague, 1992, 121.
77. **Kulaeva, O. N., Karavaiko, N. N., Moshkov, I. E., Selivankina, S. Y., and Novikova, G. V.,** Isolation of a protein with cytokinin-receptor properties by means of anti-idiotype antibodies, *FEBS Lett.*, 261, 410, 1990.
78. **Miller, C. O.,** Cytokinin inhibition of respiration in mitochondria from six plant species, *Proc. Natl. Acad. Sci. U.S.A.*, 77, 4731, 1980.
79. **Miller, C. O.,** Cytokinin modification of mitochondrial function, *Plant Physiol.*, 69, 1274, 1982.
80. **Miller, C. O.,** Cytokinin inhibition of respiration by cells and mitochondria of soybean, *Glycine max* (L.) Merrill, *Planta*, 146, 503, 1989.
81. **Dizengremel, P., Chauveau, M., and Roussaux, J.,** Inhibition by adenine derivatives of the cyanide-insensitive electron transport pathway of plant mitochondria, *Plant Physiol.*, 70, 585, 1982.
82. **Musgrave, M. E. and Siedow, J. N.,** A relationship between cyanide-resistant respiration and plant responses to cytokinins, *Plant Physiol.*, 64, 161, 1985.
83. **Musgrave, M. E., Miller, C. O., and Siedow, J. N.,** Do some plant responses to cytokinins involve the cyanide-resistant pathway? *Planta* 172, 330, 1987.
84. **Ralph, R. K. and Wojcik, S. J.,** Cytokinin effects on mitochondria and calcium metabolism, *Plant Sci. Lett.*, 25, 321, 1982.
85. **Chauveau, M., Dizengremel, P., and Roussaux, J.,** Interaction of benzylaminopurine with electron transport in plant mitochondria during malate oxidation, *Plant Physiol.*, 73, 945, 1983.
86. **Brinegar, A. C., Blumenthal, S., and Cooper, G.,** Photoaffinity labeling of mung bean mitochondrial proteins using [^{3}H]-2-azido-6-benzylaminopurine, in *Physiology and Biochemistry of Cytokinins in Plants*, Kamínek, M., Mok, D. W. S., and Zazímalová, E., Eds., SPB Academic Publishers, The Hague, 1992, 301.
87. **Cooper, G., Bourell, J., Kamínek, M., and Fox, J. E.,** Method for synthesis of 2-azido-N^6-*m*-tritobenzylaminopurine, a photoaffinity label for cytokinin-binding proteins in plants, *J. Labelled Comp. Radiopharm.*, 25, 957, 1988.

88. **Shah, G. and Brinegar, A. C.,** unpublished data, 1992.

89. **LéJohn, H. B. and Cameron, L. E.,** Cytokinins regulate calcium binding to a glycoprotein from fungal cells, *Biochem. Biophys. Res. Commun.*, 54, 1053, 1973.

90. **LéJohn, H. B.,** A rapid and sensitive auxin binding system for detecting N^6-substituted adenines and some urea and thiourea derivatives that show cytokinin activity in cell division tests, *Can. J. Biochem.*, 53, 768, 1975.

91. **Gardner, G., Sussman, M. R., and Kende, H.,** *In vitro* cytokinin binding to a particulate cell fraction from protonemata of *Funaria hygrometrica*, *Planta*, 143, 67, 1978.

92. **Venis, M. A.,** Hormone-binding studies and the misuse of precipitation assays, *Planta*, 162, 502, 1984.

93. **Kamínek, M. and Fox, J. E.,** Comparison of the sensitivity and reliability of cytokinin-binding assays using highly purified soluble binding protein, in *Physiology and Biochemistry of Cytokinins in Plants*, Kamínek, M., Mok, D. W. S., and Zazímalová, E., Eds., SPB Academic Publishers, The Hague, 1992, 461.

94. **Kaur, S. and Kapoor, H. C.,** Validity of different binding assays for detection of soluble auxin-binding sites in plants, *Plant Sci.*, 72, 151, 1990.

95. **Kobayashi, K., Zbell, B., and Reinert, J.,** High affinity binding site for cytokinin to a particulate fraction in carrot (*Daucus carota*) suspension cells, *Protoplasma*, 106, 145, 1981.

96. **Jayabaskaran, C.,** Isolation of a cytokinin-binding protein from *Cucumis sativus* cotyledons, *Proc. Indian Acad. Sci. Plant Sci.*, 98, 269, 1988.

97. **Jayabaskaran, C.,** Isolation and characterization of a cytokinin-binding protein from cucumber cotyledons, *Plant Growth Regul.*, 9, 9, 1990.

98. **Kharchenko, V. I. and Tereshchenko, A. F.,** Cytokinin-binding proteins of bean hypocotyl membranes, *Dokl. Akad. Nauk Ukr. SSR. Ser. B. Geol. Khim. Biol. Nauki.*, 0, 73, 1986.

99. **Avlyanov, Kh. K., Sidorov, G. V., Veshkurova, O. N., Takanaev, A. A., and Salikhov, Sh. I.,** Isolation of cytokinin-binding proteins from cotton-plant seedlings, *Chem. Nat. Compd.*, 24, 582, 1988.

100. **Reid, J. B.,** Phytohormone mutants in plant research, *J. Plant Growth Regul.*, 9, 97, 1990.

101. **Miller, C.,** Possible regulatory roles of cytokinins; NADH oxidation by peroxidase and a copper interaction, *Plant Physiol.*, 79, 908, 1985.

Chapter 17

Cytokinins and Plant Gene Regulation

Dring N. Crowell and Richard M. Amasino

TABLE OF CONTENTS

I. INTRODUCTION

Cytokinin, discovered in 1955, was shown to stimulate division and sustained growth of tobacco pith cells when provided in combination with auxin and was thus described as a "cell division factor".[1] The name "cytokinin" was later adopted to describe this effect on cytokinesis.[2] However, since its discovery, cytokinin has been shown to exert other effects on plant growth and development as well. For example, cytokinin delays tissue senescence[3,4] and promotes shoot organogenesis,[5] lateral bud release,[6] chloroplast development,[7,8] leaf expansion,[9] and tumor formation.[10-12] The dramatic nature of these effects suggests that changes in gene expression may mediate the response of plant cells to cytokinin. The goal of this article is to review the progress that has been made on the regulation of plant gene expression by cytokinin.

One of the challenges in writing a review on cytokinins and plant gene regulation is choosing a logical way to organize it. Research in this area has been done over many years and in many different systems (whole plants, excised organs, cell suspension cultures, etc.). Since a chronological order of presentation did not seem satisfactory, we attempted to discuss this vast amount of work in the context of physiological responses to cytokinin. We should emphasize, however, that this order of presentation may be arbitrary, since cytokinin-induced changes in gene expression measured in the context of one physiological response may, in some cases, actually be involved in another response (or in several responses).

II. CYTOKININ-REGULATED GENES AND CELL DIVISION

Several researchers studied the effects of cytokinin on plant gene expression in systems where cell growth and division require cytokinin (see Chapter 15). For the most part, this work was done in cytokinin-dependent tobacco or soybean cell suspension cultures. These cultures were first depleted of cytokinin, either by culturing the cells in cytokinin-free medium for several days or by allowing the cells to reach stationary phase, and then divided into two aliquots, one of which was transferred to a cytokinin-free medium and the other to a cytokinin-containing medium. Changes in the abundance or rate of synthesis of specific mRNAs or polypeptides were then measured as a function of time.

In an early study on cytokinin regulation of gene expression in cytokinin-dependent soybean suspension cultures, cytokinin-induced polyribosome formation and synthesis of new proteins was demonstrated, but no increase in the rate of synthesis of total RNA or poly(A) RNA was observed, leading Fosket and Tepfer to propose that "...in the absence of hormone, certain genes are transcribed,

0-8493-6252-0/94/$0.00+$.50

but their messengers are not translated."[13] Hence, a major effect of cytokinin on gene expression was argued to be at the level of increased translation of pre-existing mRNAs. (However, as discussed below, cytokinin clearly has effects on the abundance of specific mRNAs in plant cells.) These observations were made before any cytokinin-induced cell growth was detected (≤24 h), suggesting that some of these newly synthesized proteins might play a role in cell growth and division.

A later study on cytokinin-induced changes in gene expression revealed the existence of 20 mRNAs that rapidly accumulated following cytokinin treatment of cytokinin-starved soybean suspension cultures (as above, these cells required cytokinin for growth).[14] These changes in mRNA levels preceded the greening and cell growth induced by cytokinin in this system. One cytokinin-induced mRNA, called cim1, accumulated 21-fold within 2 h of cytokinin addition (although accumulation was evident within 1 h), suggesting that induction of the cim1 gene is an early response to cytokinin. This mRNA strongly accumulated in response to various biologically active cytokinins, but only weakly accumulated in response to less-active cytokinin analogs. Interestingly, auxin also had a mild stimulatory effect on the accumulation of this mRNA. Several of these cytokinin-induced mRNAs were found to accumulate specifically in mature, expanded leaves and cotyledons of intact soybean plants, consistent with a role in chloroplast development or in the differentiation of green tissues, while others accumulated in tissues undergoing cell division or expansion (i.e., young, unexpanded leaves, shoot tips, stems, root tips, and whole roots, but not mature leaves and cotyledons), consistent with a role in cell growth. Furthermore, two mRNAs from the latter group were shown, by nucleotide sequence analysis of the corresponding cDNAs, to correspond to ribosomal protein genes.

In another study, Bauw et al. examined changes in the abundance of *Nicotiana plumbaginifolia* cell proteins following cytokinin treatment of cytokinin-depleted cell cultures.[15] Within 72 h of cytokinin treatment, changes in the pattern of protein accumulation were detected by two-dimensional polyacrylamide gel electrophoresis (PAGE), and some of these proteins were identified by amino-terminal sequence analysis. None of the proteins that accumulated upon cytokinin addition was related to sequences contained in the National Biomedical Research Foundation Protein Sequence Data Bank (release 9), but β-1,3-glucanase was found to be an abundant, cytokinin-repressed protein (see below for discussion of other genes that are down-regulated by cytokinin). In fact, other researchers reported decreases in the abundance of β-1,3-glucanase and its corresponding mRNA,[16-18] as well as other defense-related proteins and mRNAs[19-21] (e.g., chitinase, pathogenesis-related proteins, etc.) in response to cytokinin. Paradoxically, several reports of cytokinin-induced accumulation of defense-related mRNAs also appeared.[22-25] These observations suggest that defense-related genes respond to the stress of cytokinin depletion or overabundance.

The studies described above clearly demonstrate that cytokinin influences various aspects of gene expression (i.e., mRNA and protein accumulation) in plant cell sytems where cytokinin is required for growth. However, the roles of these proteins in cytokinin-mediated cell growth and division remain elusive.

III. CYTOKININ-REGULATED GENES AND DIFFERENTIATION

A. LATERAL BUD INDUCTION AND SHOOT ORGANOGENESIS

Numerous researchers have studied cytokinin-regulated gene expression in the context of tissue differentiation. As stated above, some mRNAs that accumulate in response to cytokinin-induced growth of cultured soybean cells may actually be involved in chloroplast development or differentiation of green tissues.[14] Other researchers, however, have studied cytokinin regulation of gene expression in systems where organogenesis or tissue differentiation is a physiological consequence of cytokinin treatment. For example, changes in the pattern of newly synthesized proteins upon cytokinin-mediated adventitious bud induction in Norway Spruce were found to closely resemble the changes observed during *in vitro* germination of spruce embryos.[26] These changes occurred after 24 h of cytokinin treatment (no changes were apparent after 5 h). Similarly, cytokinin-induced proliferation of axillary shoots in apple plantlets was shown, over a period of 30 days, to cause an increase in the abundance of a 1.8 kilobase transcript and a decrease in the abundance of a 0.7 kilobase transcript.[25] The former was shown to correspond to a 31-kDa proline-rich protein that is related to the stress-inducible hydroxyproline-rich glycoproteins (HRGPs) of the plant cell wall, and the latter was shown to correspond to a protein that is highly homologous to the precursor of the 10-kDa polypeptide of photosystem II from spinach.

Many reports have appeared in the literature that describe the effects of introducing into plants a cytokinin biosynthetic gene from the T-DNA of *Agrobacterium tumefaciens* (*ipt*). Expression of this gene in transgenic plants causes cytokinin overproduction and the formation of shoots that exhibit various degrees of abnormal morphology. Thus, differences in gene expression between transformed and normal shoots may not be directly caused by cytokinin overproduction, but by indirect mechanisms resulting from the establishment of a transformed phenotype. In one such report, increases and decreases in the abundance of specific translatable poly(A) RNAs, as detected by *in vitro* translation and two-dimensional PAGE of translation products, were reported in potato.[27] Similarly, Memelink et al. reported that tobacco shoots transformed with the *ipt* gene from *Agrobacterium tumefaciens* accumulate transcripts corresponding to defense-related, stress-induced proteins, including extensin, chitinase, and pathogenesis-related protein I.[22,23] These mRNAs, which normally accumulate in roots, but not in leaves, of normal tobacco plants, were shown to accumulate in leaves exposed to elevated levels of cytokinin. In another report Memelink et al. demonstrated that tobacco shoots transformed with the *ipt* gene accumulate root-specific mRNAs, as judged by two-dimensional PAGE of *in vitro* translation products.[28] (The significance of root-specific expression of stress-induced genes is discussed below.) In addition, relatively normal tobacco plants transformed with the *ipt* gene were shown to contain threefold more cytokinin and *rbcL* mRNA than untransformed controls (these transformants were selected for attenuated-expression of the transgene).[29] This mRNA, which encodes the large subunit of ribulose-1,5-bisphosphate carboxylase, is transcribed in the chloroplast. The mechanism responsible for this increase has not been addressed, but it is possible that cytokinins affect chloroplast gene expression directly or that leaves of transformed plants simply contain greater numbers of chloroplasts.

B. COTYLEDON, LEAF, AND CHLOROPLAST DEVELOPMENT

1. Cotyledon and Leaf Development

Excised cotyledons have been widely used to study cytokinin-induced changes in gene expression in the context of developmental changes. Cytokinin promotes water uptake, expansion, turnover of storage materials, and chloroplast development in excised cotyledons.[30,31] These changes were found to be associated, over a period of 4 days, with changes (increases and decreases) in the accumulation of various proteins and with an increase in hydroxypyruvate reductase activity in excised cotyledons.[30,31] This enzyme, which is involved in the glycolate pathway, may participate in the osmotic expansion of cotyledons through carbohydrate metabolism.[32] Consistent with this hypothesis, Howard and Witham found that two isoforms of invertase are induced by cytokinin in detached radish cotyledons.[33] Cytokinin treatment of excised cotyledons has also been shown to increase the rate of elongation of ribosomal RNA nascent chains by RNA polymerase I[34] and to cause accumulation of ribosomal RNAs.[35] In a later study Chen et al. detected changes in the population of translatable mRNAs in excised pumpkin cotyledons 1 h after cytokinin treatment by two-dimensional PAGE of *in vitro* translation products.[36] These studies indicate that cytokinin-induced development of excised cotyledons is associated with changes in the accumulation of various gene products.

Cytokinin-induced changes in gene expression have also been studied in leaves.[22-25,28,37] In an early study, nitrate reductase activity was shown to be induced by cytokinin in tobacco leaves.[37] Similarly, nitrate reductase activity has been shown to increase in excised *Agrostemma githago* embryos within an hour of cytokinin treatment.[38] Thus, cytokinin appears to regulate nitrate assimilation in various plant systems.

2. Chloroplast Development

The cytokinin-induced accumulation of chloroplast proteins and their corresponding transcripts has been studied in many plant systems, including cell suspension cultures, excised cotyledons, and whole plants. Most of these gene products are associated with the photosynthetic apparatus and have been used as molecular markers for cytokinin- and/or light-induced chloroplast development (see Chapter 14). For example, cytokinin-induced accumulation of the light-harvesting chlorophyll a/b binding protein and/or mRNA has been reported in cultured tobacco cells,[39-42] watermelon cotyledons,[43] cucumber cotyledons,[44] and whole *Lemna gibba* plants.[45,46] Similarly, increases in ribulose-1,5-bisphosphate carboxylase activity, subunit polypeptides (i.e., the nuclear-encoded small subunit and/or the chloroplast-encoded large subunit), and/or mRNAs following cytokinin treatment have been reported in tobacco cell cultures,[39] watermelon cotyledons,[31] cucumber cotyledons,[47] pumpkin cotyledons,[48-50] etiolated rye leaves,[51] whole *Lemna gibba* plants,[45] and phenotypically normal tobacco plants transformed with the *ipt* gene.[29] NADP-glyceraldehyde 3-phosphate dehydrogenase and other chloroplast enzymes have also been shown to increase in the presence of cytokinin.[47,51] Furthermore, Dominov et al. described a novel mRNA from *Nicotiana plumbaginifolia* suspension cultures that accumulates during cytokinin-induced greening.[52] In contrast, cytokinin causes the synthesis of the 55-kDa and 57-kDa subunits of CF_1 to decrease.[53] Most of these changes were observed over a period of days,[31,40-43,45,48] though some occurred over several hours.[44,47,49,52]

C. FLORAL DEVELOPMENT

Several cytokinin-responsive tobacco genes have been isolated that are expressed during *in vitro* floral initiation.[54] In this tobacco "thin-cell-layer" system, zeatin promotes vegetative shoot formation and kinetin promotes floral bud formation, phenomena that may reflect quantitative differences in the activities of the two cytokinins. This system was used to identify mRNAs that accumulate over a period of days in thin-cell-layer explants initiating floral meristems (i.e., in response to kinetin), but not in thin-cell-layer explants initiating vegetative meristems (i.e., in response to zeatin). Two of the mRNAs specific to thin cell layers undergoing floral initiation were shown to accumulate in floral apices during normal plant development and were also found, along with a third mRNA, to occur at high levels in the roots of intact plants. These observations suggest that cytokinin-responsive genes may play a role in the switch from vegetative to floral development.

D. NODULATION

Another system in which changes in development have been associated with cytokinin-induced changes in gene expression is nodulation of the tropical legume *Sesbania rostrata*.[55] The early nodulin gene *SrEnod2*, which encodes a (presumably stress-inducible) proline-rich protein, was shown to be expressed in *S. rostrata* seedlings treated with exogenous cytokinin, in nodules induced by *Azorhizobium caulinodans* ORS571, and in crown gall tumors induced by *Agrobacterium tumefaciens*. The *SrEnod2* mRNA was only modestly induced by auxin and accumulated in response to cytokinin over the course of 10 h (although some *SrEnod2* mRNA was detectable after 2 h of cytokinin stimulation). Thus, cytokinin may serve as a signal in the establishment of symbiotic relationships between legumes and nitrogen-fixing bacteria.

E. SENESCENCE

Cytokinin delays the onset of senescence in plants. Consistent with this effect being related to changes in gene expression, Shinshi et al. demonstrated that β-1,3-glucanase and chitinase mRNA levels, which increase in response to auxin or cytokinin depletion of cultured tobacco tissues, are higher in the lower (older) leaves than in the upper (younger) leaves of tobacco plants.[21] Hence, these mRNA levels correlated with the auxin content, cytokinin content, developmental age, and senescence of the leaves. Similarly, Crowell et al. demonstrated that an mRNA corresponding to a "disease-resistance-response protein" accumulated in cultured soybean cells during cytokinin or auxin starvation and during the maturation and senescence of leaves.[19,20] Like β-1,3-glucanase and chitinase mRNAs, this mRNA also accumulated to high levels in roots.[19-21,24] The role of these gene products in senescence is unclear, but all three were associated with stress and pathogenesis, suggesting a putative role in generalized stress responses.

IV. REGULATION OF CYTOKININ-RESPONSIVE GENES BY OTHER FACTORS

Many cytokinin-responsive genes were found to respond to environmental factors such as light, other phytohormones, or stress. Light, specifically red light, was shown to induce the accumulation of the light-harvesting chlorophyll a/b binding protein (Cab), the small subunit of ribulose-1,5-bisphosphate carboxylase (SSU) and the corresponding mRNAs, implicating phytochrome in the regulation of expression of photosynthetic genes.[44-51] In addition, numerous researchers observed that cytokinin induces the accumulation of these proteins and mRNAs when applied in the dark[43-51] or in the light.[39-42,49] Since cytokinin mimics many of the effects of red light on plant development,[9,56] it is perhaps not surprising that cytokinin and light have some similar effects on gene expression. In fact, cytokinin stimulated the accumulation of the Cab protein in excised cucumber cotyledons when added to dark-grown tissue (prior to illumination), but not when given to tissue grown in continuous light.[44] In other systems, however, the effects of cytokinin and light were found to be cumulative.[48,49] Interestingly, Abdelghani et al. found that cytokinin-induced accumulation of Cab and SSU mRNAs occurred in tobacco cell suspensions only if the cells were maintained in continuous light.[39] These differences in the cytokinin response under various light conditions probably reflect differences in the physiological state of the tissues. Nitrate reductase activity was also induced by light or, in the dark, by cytokinin and/or gibberellic acid (GA_3) in tobacco leaves.[37] Cytokinin was later shown to enhance light induction of nitrate reductase transcript levels in etiolated barley leaves.[57] Furthermore, the abundance of phytochrome was shown to be responsive to both light and cytokinin (red light or cytokinin treatment of etiolated cucumber cotyledons decreased phytochrome mRNA levels by 55 to 60%

within 4 h).[58] This observation suggests a molecular interaction between the effects of light and cytokinin (light- and cytokinin-induced desensitization of plant cells to light).

Several cytokinin-responsive genes respond to other phytohormones, including auxin,[14,16-21,52,55,59-61] gibberellic acid,[37] abscisic acid,[38,62] and ethylene.[24,63] In *Agrostemma githago*, for example, the rapid induction of nitrate reductase by cytokinin was inhibited by abscisic acid,[38] but required ethylene.[63] These observations are consistent with the hypothesis that relative concentrations of many phytohormones, rather than absolute levels of individual phytohormones, determine developmental processes.[5] Organic compounds and inorganic ions were also shown to influence the accumulation of various cytokinin-inducible gene products. For example, sucrose induced nitrate reductase gene transcription in *Arabidopsis thaliana,*[64] and nitrate induced the accumulation of nitrate reductase in *Agrostemma githago* by a mechanism independent of cytokinin (perhaps due to the existence of distinct nitrate reductase genes that respond to either cytokinin or nitrate).[38,63,65,66] In addition, KCl enhanced the cytokinin-induced accumulation of the chlorophyll a/b binding protein in cucumber cotyledons.[44]

Numerous cytokinin-inducible or cytokinin-repressible gene products were shown to accumulate in response to stress or are related to gene products that respond to stress.[15-25,55] In one case a transcript that accumulated in response to cytokinin or auxin starvation of cultured soybean cells was also shown to accumulate in roots and senescent tissues of whole soybean plants.[19,20] In fact, many defense-related, cytokinin-responsive genes were shown to be strongly expressed in roots.[19-25,28] The accumulation of stress-inducible gene products in roots and senescent tissues is not surprising, since senescent tissues undergo cell death, and roots undergo wounding during the penetration of lateral roots through the cortex and epidermis of the main root.[20] These findings suggest that manipulation of endogenous or exogenous cytokinin levels can induce stress-responsive genes that may not be directly involved in cytokinin-mediated effects.

V. MECHANISM OF REGULATION OF PLANT GENES BY CYTOKININ

The cytokinin-induced accumulation of photosynthetic proteins and their corresponding mRNAs suggests cytokinin control of RNA synthesis or turnover.[29,39,42,43,45,46,49,50] Accordingly, establishment of the photosynthetic apparatus during cytokinin-induced chloroplast development has been proposed to be bipartite, involving increased transcription of nuclear genes that specifically code for plastid proteins, as well as increased transcription of chloroplast genes in general.[49] However, the careful work of Flores and Tobin indicates that the cytokinin-induced accumulation of the mRNAs for the light-harvesting chlorophyll a/b binding protein and the small subunit of ribulose-1,5-bisphosphate carboxylase occurs primarily by a posttranscriptional mechanism in *Lemna gibba.*[45,46]

Inhibitors of transcription (e.g., cordycepin or actinomycin D) and translation (e.g., cycloheximide) were used to study the mechanism of induction of various plant genes by cytokinin.[30,49,65,67] For example, Chatfield and Armstrong showed that cytokinin-induced accumulation of cytokinin oxidase activity in *Phaseolus vulgaris* is sensitive to both cordycepin and cycloheximide.[67a] In addition, Lerbs et al. showed that cordycepin inhibits the accumulation of the small subunit of ribulose-1,5-bisphosphate carboxylase in response to cytokinin, suggesting a role for transcription in this response.[49] However, while these observations demonstrate that transcription is necessary for the synthesis of cytokinin-responsive mRNAs, they do not rule out the possibility that cytokinin induces mRNA accumulation by decreasing mRNA turnover. On the other hand, Lu et al. obtained direct evidence in etiolated barley leaves for the role of transcription in cytokinin enhancement of light induction of nitrate reductase mRNA accumulation, and its inhibition by abscisic acid, by measuring changes in the rate of transcription of the nitrate reductase gene.[57,62]

A role for transcription in cytokinin regulation of specific plant genes has been confirmed by the isolation and characterization of plant gene promoters that are activated by cytokinin.[59-61] Interestingly, one of these promoters, isolated from *Oryza sativa*, is induced by both cytokinin and auxin.[60] Similarly, the mannopine synthase gene promoters of *Agrobacterium tumefaciens* are induced by cytokinin, auxin, and wounding.[61] The gene-5 promoter from the T-DNA of the *Agrobacterium tumefaciens* plasmid pTiAch5 is also activated by both cytokinin (at concentrations as low as 0.05 n*M*) and auxin (at concentrations as low as 50 n*M*) and, consequently, has been fused to the coding region of the *E. coli uidA* (GUS) gene, introduced into tobacco, and used as a sensitive indicator of cytokinin and auxin in culture medium.[59] These observations lend support to the hypothesis that cytokinin and auxin signal transduction pathways have some common effects on the pattern of gene expression in plants. Also consistent with this hypothesis, Crowell et al. demonstrated that both cytokinin and auxin induce the accumulation of various mRNAs in cultured soybean cells,[14] and Dominov et al. showed that both hormones stimulate the accumulation of a novel mRNA in *Nicotiana plumbaginifolia* suspension cultures.[52] Nuclear run-off transcription experiments demonstrated that synthesis of this *N. plumbaginifolia* "multiple-stimulus-response" mRNA first increased and then decreased, in response to these hormones, resulting in a transient accumulation of the mRNA. This "feedback" regulation of transcription appeared to cause cells previously exposed to hormones to be desensitized to subsequent treatments with the same hormones. However, cells desensitized to auxin were hypersensitive to cytokinin, and vice versa. Furthermore, a protein kinase inhibitor (staurosporine) blocked hormone-induced mRNA accumulation, whereas a protein phosphatase inhibitor (okadaic acid) stimulated accumulation, suggesting that hormone activation involves phosphorylation of signal transduction proteins, and "feedback inhibition" involves dephosphorylation of these proteins. In contrast, we have recently found that cytokinin-induced accumulation of the cim1 mRNA in cultured soybean cells is stimulated by staurosporine and blocked by okadaic acid.[67b] Together, these observations suggest that different cytokinin signal transduction pathways may regulate the expression of different plant genes.

VI. FUTURE DIRECTIONS

The mechanism by which cytokinin regulates the expression of plant genes remains largel unexplored. Though cytokinin has been shown to influence molecular processes that are known to participate in signal transduction, such as phosphatidylinositol turnover,[68,69] changes in cytosolic Ca^{2+} levels,[70] and changes in protein phosphorylation,[71] the role of these processes in cytokinin action is unclear (see Chpaters 16 and 18). The work described in this review provides a solid foundation for future studies on cytokinin signal transduction. For example, the characterization of cytokinin-responsive gene promoters will undoubtedly lead to the identification of cytokinin-responsive DNA sequences and transcription factors. Furthermore, cytokinin-regulated genes will be used as markers for studying the role of G-proteins, phosphatidylinositol, Ca^{2+}, and protein phosphorylation in cytokinin action. In the next few years, these studies are likely to reveal many of the molecular events that mediate the effects of cytokinin on gene expression and, thus, on the growth and development of plants.

While this manuscript was in press, several cytokinin-responsive genes were reported. These include the phosphoenolpyruvate carboxylase gene from *Mesembryanthemum crystallinum*,[72,73] the phosphoenolpyruvate carboxylase and carbonic anhydrase genes from maize,[74] the nitrate reductase gene from *Triticum aestivum* L.,[75] the 3-hydroxy-3-methylglutaryl-coenzyme A reductase gene from *Camptotheca acuminata*,[76] the *zfpA* gene from *Physcomitrella patens*,[77] three homeotic floral genes from *Antirrhinum majus*,[78] two photosynthetic genes from *Pinus ponderosa* Dougl.,[79] and several senescence-associated genes from *Avena sativa* L.[80] Benzyladenine-induced changes in translatable mRNAs were also observed in excised cucumber cotyledons.[81] Finally, cytokinin-modulated polyadenylation of nitrate reductase mRNA in tobacco suspension cultures was reported[82] as well as transcriptional regulation of the hydroxypyruvate reductase gene by cytokinin in etiolated pumpkin cotyledons.[83,84]

REFERENCES

1. **Miller, C. O., Skoog, F., Von Saltza, M. H., and Strong, F. M.,** Kinetin, a cell division factor from deoxyribonucleic acid, *J. Am. Chem. Soc.*, 78, 1392, 1955.
2. **Skoog, F., Strong, F. M., and Miller, C. O.,** Cytokinins, *Science*, 148, 532, 1965.
3. **Noodén, L. D., Guiamét, J. J., Singh, S., Letham, D. S., Tsuji, J., and Schneider, M. J.,** Hormonal control of senescence, in *Plant Growth Substances*, Pharis, R. P. and Rood, S. B., Eds., Springer-Verlag, Berlin, 1990, 537.
4. **Richmond, A. E. and Lang, A.,** Effect of kinetin on protein content and survival of detached *Xanthium* leaves, *Science*, 125, 650, 1957.
5. **Skoog, F. and Miller, C. O.,** Chemical regulation of growth and organ formation in plant tissues cultured in vitro, *Symp. Soc. Exp. Biol.*, 11, 118, 1957.
6. **Sachs, T. and Thimann, K. V.,** Release of lateral buds from apical dominance, *Nature*, 201, 939, 1964.
7. **Parthier, B.,** The role of phytohormones (cytokinins) in chloroplast development, *Biochem. Physiol. Pflanz.*, 174, 173, 1979.
8. **Stetler, D. A. and Laetsch, W. M.,** Kinetin-induced chloroplast maturation in cultures of tobacco tissue, *Science*, 149, 1387, 1965.
9. **Miller, C. O.,** Similarity of some kinetin and red light effects, *Plant Physiol.*, 31, 318, 1956.

10. **Akiyoshi, D. E., Klee, H., Amasino, R. M., Nester, E. W., and Gordon, M. P.,** T-DNA of *Agrobacterium tumefaciens* encodes an enzyme of cytokinin biosynthesis, *Proc. Natl. Acad. Sci. U.S.A.*, 81, 5994, 1984.
11. **Akiyoshi, D. E., Morris, R. O., Hinz, R., Mischke, B. S., Kosuge, T., Garfinkel, D. J., Gordon, M. P., and Nester, E. W.,** Cytokinin/auxin balance in crown gall tumors is regulated by specific loci in the T-DNA, *Proc. Natl. Acad. Sci. U.S.A.*, 80, 407, 1983.
12. **Inzé, D., Follin, A., Van Lijsebettens, M., Simoens, C., Genetello, C., Van Montagu, M., and Schell, J.,** Genetic analysis of the individual T-DNA genes of *Agrobacterium tumefaciens*; further evidence that two genes are involved in indole-3-acetic acid synthesis, *Mol. Gen. Genet.*, 194, 265, 1984.
13. **Fosket, D. E. and Tepfer, D. A.,** Hormonal regulation of growth in cultured plant cells, *In Vitro*, 14, 63, 1978.
14. **Crowell, D. N., Kadlecek, A. T., John, M. C., and Amasino, R. M.,** Cytokinin-induced mRNAs in cultured soybean cells, *Proc. Natl. Acad. Sci. U.S.A.*, 87, 8815, 1990.
15. **Bauw, G., De Loose, M., Inzé, D., Van Montagu, M., and Vandekerckhove, J.,** Alterations in the phenotype of plant cells studied by NH_2-terminal amino acid-sequence analysis of proteins electroblotted from two-dimensional gel-separated total extracts, *Proc. Natl. Acad. Sci. U.S.A.*, 84, 4806, 1987.
16. **Eichholz, R., Harper, J., Felix, G., and Meins, F., Jr.,** Evidence for an abundant 33,000-dalton polypeptide regulated by cytokinins in cultured tobacco tissues, *Planta*, 158, 410, 1983.
17. **Felix, G. and Meins, F., Jr.,** Purification, immunoassay and characterization of an abundant, cytokinin-regulated polypeptide in cultured tobacco tissues, *Planta*, 164, 423, 1985.
18. **Mohnen, D., Shinshi, H., Felix, G., and Meins, F., Jr.,** Hormonal regulation of β-1,3-glucanase messenger RNA levels in cultured tobacco tissues, *EMBO J.*, 4, 1631, 1985.
19. **Crowell, D. N. and Amasino, R. M.,** Induction of specific mRNAs in cultured soybean cells during cytokinin or auxin starvation, *Plant Physiol.*, 95, 711, 1991.
20. **Crowell, D. N., John, M. E., Russell, D., and Amasino, R. M.,** Characterization of a stress-induced, developmentally regulated gene family from soybean, *Plant Mol. Biol.*, 18, 459, 1992.
21. **Shinshi, H., Mohnen, D., and Meins, F., Jr.,** Regulation of a plant pathogenesis-related enzyme: inhibition of chitinase and chitinase mRNA accumulation in cultured tobacco tissues by auxin and cytokinin, *Proc. Natl. Acad. Sci. U.S.A.*, 84, 89, 1987.
22. **Memelink, J., Hoge, J. H. C., and Schilperoort, R. A.,** Cytokinin stress changes the developmental regulation of several defence-related genes in tobacco, *EMBO J.*, 6, 3579, 1987.
23. **Memelink, J., de Pater, B. S., Hoge, J. H. C., and Schilperoort, R. A.,** T-DNA hormone biosynthetic genes: phytohormones and gene expression in plants, *Dev. Genet.*, 8, 321, 1987.
24. **Simmons, C. R., Litts, J. C., Huang, N., and Rodriguez, R. L.,** Structure of a rice β-glucanase gene regulated by ethylene, cytokinin, wounding, salicylic acid and fungal elicitors, *Plant Mol. Biol.*, 18, 33, 1992.
25. **Watillon, B., Kettmann, R., Boxus, P., and Burny, A.,** Characterization of two gene transcripts modulated by cytokinins in micropropagated apple (*Malus domestica* [L.] Borkh) plantlets, *Plant Physiol.*, 96, 479, 1991.
26. **Stabel, P., Eriksson, T., and Engström, P.,** Changes in protein synthesis upon cytokinin-mediated adventitious bud induction and during seedling development in Norway Spruce, *Picea abies*, *Plant Physiol.*, 92, 1174, 1990.
27. **Burrell, M. M., Temple, S., and Ooms, G.,** Changes in translatable poly(A) RNA from differentiated potato tissues transformed with shoot-inducing Ti T_L-DNA of *Agrobacterium tumefaciens*, *Plant Mol. Biol.*, 6, 213, 1986.
28. **Memelink, J., de Vries, S. C., Schilperoort, R. A., and Hoge, J. H. C.,** Changes in the tissue-specific prevalence of translatable mRNAs in transgenic tobacco shoots containing the T-DNA cytokinin gene, *Plant Mol. Biol.*, 11, 625, 1988.
29. **Yusibov, V. M., Il, P. C., Andrianov, V. M., and Piruzian, E. S.,** Phenotypically normal transgenic T-*cyt* tobacco plants as a model for the investigation of plant gene expression in response to phytohormonal stress, *Plant Mol. Biol.*, 17, 825, 1991.
30. **Chen, C.-M. and Leisner, S. M.,** Cytokinin-modulated gene expression in excised pumpkin cotyledons, *Plant Physiol.*, 77, 99, 1985.
31. **Longo, G. P., Pedretti, M., Rossi, G., and Longo, C. P.,** Effect of benzyladenine on the development of plastids and microbodies in excised watermelon cotyledons, *Planta*, 145, 209, 1979.

32. **Huff, A. K. and Ross, C. W.,** Promotion of radish cotyledon enlargement and reducing sugar content by zeatin and red light, *Plant Physiol.*, 56, 429, 1975.

33. **Howard, H. F. and Witham, F. H.,** Invertase activity and the kinetin-stimulated enlargement of detached radish cotyledons, *Plant Physiol.*, 73, 304, 1983.

34. **Ananiev, E. D., Karagyozov, L. K., and Karanov, E. N.,** Effect of cytokinins on ribosomal RNA gene expression in excised cotyledons of *Cucurbita pepo* L., *Planta*, 170, 370, 1987.

35. **Mikulovich, T. P., Wollgiehn, R., Khokhlova, W. A., Neumann, D., and Kulaeva, O. N.,** Synthesis of plastid and cytoplasmic ribosomal RNAs in isolated pumpkin cotyledons. II. Effect of cytokinin and light, *Biochem. Physiol. Pflanz.*, 172, 101, 1978.

36. **Chen, C.-m., Ertl, J., Yang, M.-S., and Chang, C.-C.,** Cytokinin-induced changes in the population of translatable mRNA in excised pumpkin cotyledons, *Plant Sci.*, 52, 169, 1987.

37. **Lips, S. H. and Roth-Bejerano, N.,** Light and hormones: interchangeability in the induction of nitrate reductase, *Science*, 166, 109, 1969.

38. **Kende, H., Hahn, H., and Kays, S. E.,** Enhancement of nitrate reductase activity by benzyladenine in *Agrostemma githago*, *Plant Physiol.*, 48, 702, 1971.

39. **Abdelghani, M. O., Suty, L., Chen, J. N., Renaudin, J.-P., and Teyssendier de la Serve, B.,** Cytokinins modulate the steady-state levels of light-dependent and light-independent proteins and mRNAs in tobacco cell suspensions, *Plant Sci.*, 77, 29, 1991.

40. **Axelos, M., Barbet, J., and Péaud-Lenoël, C.,** Influence of cytokinins on the biosynthesis of light-harvesting chlorophyll *a/b* proteins in tobacco cell suspensions: detection by radioimmunological assay, *Plant Sci. Lett.*, 33, 201, 1984.

41. **Axelos, M. and Péaud-Lenoël, C.,** The apoprotein of the light-harvesting chlorophyll a/b complex of tobacco cells as a molecular marker of cytokinin activity, *Plant Sci. Lett.*, 19, 33, 1980.

42. **Teyssendier de la Serve, B., Axelos, M., and Péaud-Lenoël, C.,** Cytokinins modulate the expression of genes encoding the protein of the light-harvesting chlorophyll a/b complex, *Plant Mol. Biol.*, 5, 155, 1985.

43. **Longo, G. P. M., Bracale, M., Rossi, G., and Longo, C. P.,** Benzyladenine induces the appearance of LHCP-mRNA and of the relevant protein in dark-grown excised watermelon cotyledons, *Plant Mol. Biol.*, 14, 569, 1990.

44. **Ohya, T., Naito, K., and Suzuki, H.,** Combined effect of benzyladenine and potassium on the level of light-harvesting chlorophyll a/b protein in detached cucumber cotyledons, *Z. Pflanzenphysiol.*, 108, 39, 1982.

45. **Flores, S. and Tobin, E. M.,** Benzyladenine modulation of the expression of two genes for nuclear-encoded chloroplast proteins in *Lemna gibba*: apparent post-transcriptional regulation, *Planta*, 168, 340, 1986.

46. **Flores, S. and Tobin E. M.,** Cytokinin modulation of LHCP mRNA levels: the involvement of post-transcriptional regulation, *Plant Mol. Biol.*, 11, 409, 1988.

47. **Harvey, B. M. R., Lu, B. C., and Fletcher, R. A.,** Benzyladenine accelerates chloroplast differentiation and stimulates photosynthetic enzyme activity in cucumber cotyledons, *Can. J. Bot.*, 52, 2581, 1974.

48. **Klyachko, N. L., Parthier, B., Chayanova, S. S., Volodarsky, A. D., and Kulaeva, O. N.,** Ribulose-1,5-bisphosphate carboxylase synthesis in detached cytokinin-treated pumpkin cotyledons, *Biochem. Physiol. Pflanz.*, 175, 712, 1980.

49. **Lerbs, S., Lerbs, W., Klyachko, N. L., Romanko, E. G., Kulaeva, O. N., Wollgiehn, R., and Parthier, B.,** Gene expression in cytokinin- and light-mediated plastogenesis of *Cucurbita* cotyledons: ribulose-1,5-bisphosphate carboxylase/oxygenase, *Planta*, 162, 289, 1984.

50. **Parthier, B., Lehmann, J., Lerbs, S., Lerbs, W., and Wollgiehn, R.,** Cytokinin control in the synthesis of plastid proteins: ribulose-1,5-bisphosphate carboxylase, in *Plant Growth Substances*, Wareing, P. F., Ed., Academic Press, London, 1982, 225.

51. **Feierabend, J. and de Boer, J.,** Comparative analysis of the action of cytokinin and light on the formation of ribulose-bisphosphate carboxylase and plastid biogenesis, *Planta*, 142, 75, 1978.

52. **Dominov, [illegible] A., Stenzler, L., Lee, S., Schwarz, [illegible] J., Leisner, S., and Howell, S. H.,** C[illegible]okinins and auxins control the expres[illegible] of a gene in *Nicotiana plumbaginif[illegible]* cells by feedback regulation, *Plant Cell*, 4, 451, 1992.

53. **Colijn, C. M., Sijmons, P., Mol, J. N. M., Kool, A. J., and Nijkamp, H. J. J.,** Light and benzylaminopurine induce changes in ultrastructure and gene expression in plastids of *Petunia hybrida* cell cultures, *Curr. Genet.*, 6, 129, 1982.
54. **Meeks-Wagner, D. R., Dennis, E. S., Tran Thanh Van, K., and Peacock, W. J.,** Tobacco genes expressed during in vitro floral initiation and their expression during normal plant development, *Plant Cell*, 1, 25, 1989.
55. **Dehio, C. and de Bruijn, F. J.,** The early nodulin gene *SrEnod2* from *Sesbania rostrata* is inducible by cytokinin, *Plant J.*, 2, 117, 1992.
56. **Miller, C. O.,** The relationship of the kinetin and red-light promotions of lettuce seed germination, *Plant Physiol.*, 33, 115, 1958.
57. **Lu, J.-I., Ertl, J. R., and Chen, C.-m.,** Cytokinin enhancement of the light induction of nitrate reductase transcript levels in etiolated barley leaves, *Plant Mol. Biol.*, 14, 585, 1990.
58. **Cotton, J. L. S., Ross, C. W., Byrne, D. H., and Colbert, J. T.,** Down-regulation of phytochrome mRNA abundance by red light and benzyladenine in etiolated cucumber cotyledons, *Plant Mol. Biol.*, 14, 707, 1990.
59. **Boerjan, W., Genetello, C., Van Montagu, M., and Inzé, D.,** A new bioassay for auxins and cytokinins, *Plant Physiol.*, 99, 1090, 1992.
60. **Claes, B., Smalle, J., Dekeyser, R., Van Montagu, M., and Caplan, A.,** Organ-dependent regulation of a plant promoter isolated from rice by 'promoter-trapping' in tobacco, *Plant J.*, 1, 15, 1991.
61. **Langridge, W. H. R., Fitzgerald, K. J., Koncz, C., Schell, J., and Szalay, A. A.,** Dual promoter of *Agrobacterium tumefaciens* mannopine synthase genes is regulated by plant growth hormones, *Proc. Natl. Acad. Sci. U.S.A.*, 86, 3219, 1989.
62. **Lu, J.-I., Ertl, J. R., and Chen, C.-m.,** Transcriptional regulation of nitrate reductase mRNA levels by cytokinin-abscisic acid interactions in etiolated barley leaves, *Plant Physiol.*, 98, 1255, 1992.
63. **Schmerder, B. and Borriss, H.,** Induction of nitrate reductase by cytokinin and ethylene in *Agrostemma githago* L. embryos, *Planta*, 169, 589, 1986.
64. **Cheng, C.-L., Acedo, G. N., Cristinsin, M., and Conkling, M. A.,** Sucrose mimics the light induction of *Arabidopsis* nitrate reductase gene transcription, *Proc. Natl. Acad. Sci. U.S.A.*, 89, 1861, 1992.
65. **Dilworth, M. F. and Kende, H.,** Comparative studies on nitrate reductase in *Agrostemma githago* induced by nitrate and benzyladenine, *Plant Physiol.*, 54, 821, 1974.
66. **Dilworth, M. F. and Kende, H.,** Control of nitrite reductase activity in excised embryos. of *Agrostemma githago*, *Plant Physiol.*, 54, 826, 1974.
67a. **Chatfield, J. M. and Armstrong, D. J.,** Regulation of cytokinin oxidase activity in callus tissues of *Phaseolus vulgaris* L. cv. Great Northern, *Plant Physiol.*, 80, 493, 1986.
67b. **Crowell, D. N.,** Unpublished observations.
68. **Connett, R. J. A. and Hanke, D. E.,** Changes in the pattern of phospholipid synthesis during the induction by cytokinin of cell division in soybean suspension cultures, *Planta*, 170, 161, 1987.
69. **Falkenau, C., Heim, S., and Wagner, K. G.,** Effect of cytokinins on the phospholipid phosphorylation of the suspension cultured *Catharanthus roseus* cells, *Plant Sci.*, 50, 173, 1987.
70. **Hepler, P. K. and Wayne, R. O.,** Calcium and plant development, *Ann. Rev. Plant Physiol.*, 36, 397, 1985.
71. **Ralph, R. K., Bullivant, S., and Wojcik, S. J.,** Effects of kinetin on phosphorylation of leaf membrane proteins, *Biochim. Biophys. Acta*, 421, 319, 1976.
72. **Schmitt, J. M. and Piepenbrock, M.,** Regulation of phosphoenolpyruvate carboxylase and crassulacean acid metabolism induction in *Mesembryanthemum crystallinum* L. by cytokinin. Modulation of leaf gene expression by roots?, *Plant Physiol.*, 99, 1664, 1992.
73. **Thomas, J. C., McElwain, E. F., and Bohnert, H. J.,** Convergent induction of osmotic stress-responses. Abscisic acid, cytokinin, and the effects of NaCl, *Plant Physiol.*, 100, 416, 1992.
74. **Sugiharto, B., Burnell, J. N., and Sugiyama, T.,** Cytokinin is required to induce the nitrogen-dependent accumulation of mRNAs for phosphenolpyruvate carboxylase and carbonic anhydrase in detached maize leaves, *Plant Physiol.*, 100, 153, 1992.
75. **Banowetz, G. M.,** The effects of endogenous cytokinin content on benzyladenine-enhanced nitrate reductase induction, *Physiol. Plant.*, 86, 341, 1992.
76. **Maldonado-Mendoza, I. E., Burnett, R. J., and Nessler, C. L.,** Phytohormone regulation of a 3-hydroxy-3-methylglutaryl-coenzyme A reductase (HMGR) gene from *Camptotheca acuminata*, *Suppl. Plant Physiol.*, 102, 24, 1993 (*abstract*).

77. **Kasten, B., Wehe, M., Kruse, S., Reutter, K., Abel, W. O., and Reski, R.,** The plastome-encoded *zfpA* gene of a moss contains procaryotic as well as eucaryotic promoter consensus sequences and its RNA abundance is modulated by cytokinin, *Curr. Genet.*, 22, 327, 1992.
78. **Estruch, J. J., Granell, A., Hansen, G., Prinsen, E., Redig, P., Van Onckelen, H., Schwarz-Sommer, Z., Sommer, H., and Spena, A.,** Floral development and expression of floral homeotic genes are influenced by cytokinins, *Plant J.*, 4, 379, 1993.
79. **Mazari, A. and Camm, E. L.,** Effect of cytokinins on plastid development and photosynthetic polypeptides during organogenesis of *Pinus ponderosa* Dougl. cotyledons cultured in vitro, *Plant Cell Tissue Org. Culture,* 33, 81, 1993.
80. **Klerk, H., Rebers, M., and Van Loon, L. C.,** Effects of light and regulators on senescence-related changes in soluble proteins in detached oat (*Avena sativa* L.) leaves, *Plant Growth Regul.*, 13, 137, 1993.
81. **Teramoto, H., Momotani, E., and Tsuji, H.,** Benzyladenine-induced changes in the translatable mRNA population in excised cucumber cotyledons, *Physiol. Plant.*, 87, 584, 1993.
82. **Suty, L., Moureaux, T., Leydecker, M.-T., and Teyssendier de la Serve, B.,** Cytokinin affects nitrate reductase expression through the modulation of polyadenylation of the nitrate reductase mRNA transcript, *Plant Sci.*, 90, 11, 1993.
83. **Jin, G., Andersen, B. R., Chen, R., Ertl, J. R., and Chen, C.-m.,** Transcriptional regulation of hydroxypyruvate reductase transcript by cytokinin in etiolated pumpkin cotyledons, *Suppl. Plant Physiol.*, 102, 129, 1993 (*abstract*).
84. **Jin, G., Chen, R., Daniel, S., Becker, W. M., and Chen, C.-m.,** Cytokinin-modulated binding of *trans*-acting regulatory factors to the promoter of a cucumber hydroxypyruvate reductase gene, *Suppl. Plant Physiol.*, 102, 129, 1993 (*abstract*).

Chapter 18

Cytokinin and Signal Transduction

Mary Jane Saunders

TABLE OF CONTENTS

I. SUMMARY

Plant cells contain all the elements for a calcium-based signal transduction chain that can couple the stimulus of cytokinin to its physiological responses (including cell division). These include a highly regulated low cytoplasmic calcium level, a plasma membrane, endomembrane calcium pumps and channels, and spatially controlled calcium-dependent regulatory proteins and kinases. In addition, recent research has led to the description of plant homologs to important cell cycle regulatory proteins found in animals. These include the cdc2 gene product $p34^{cdc2}$, cyclins, and nuclear lamins. These proteins are potential substrates for Ca^{2+}-dependent kinases and phosphatases, linking a cytokinin signal to progress through the cell cycle. Although much of the evidence for calcium as a second messenger in plants is still in its initial stages, evidence has accumulated that correlates cytokinin stimulation of several different responses to increases in intracellular calcium by activation of plasma membrane ion channels. The best documented system is the cytokinin-stimulation of asymmetrical cell division in the moss *Funaria*, which is characterized by spatial and temporal changes in intracellular [Ca^{2+}] from 250 to 750 n*M*. Future research may relate these changes in [Ca^{2+}] to specific regulation of the cell cycle. However, it is unlikely that all cytokinin effects will be explained by invoking calcium as a second messenger, and we must recognize that the mechanism of action of plant hormones may be quite complex and diverse.

II. INTRODUCTION

The role of calcium ions (Ca^{2+}) as a second messenger in plant cells has moved from being merely hypothetical to being increasingly well documented over the past 10 years. There are several excellent reviews about its cellular regulation and its role in plant signal transduction.[1-8] Current models for signal transduction (the molecular mechanism that couples cellular response to external stimuli) in both plants and animals share four main components:

1. The stimulus binds to a receptor.
2. A second messenger is activated (usually Ca^{2+} or cyclic AMP).
3. There is a cascade or amplifying effect within the cell by stimulation of second messenger-regulated kinases, phosphatases, or binding proteins.
4. The physiological response.

0-8493-6252-0/94/$0.00+$.50

In this chapter, I will attempt to link

1. Cytokinin as a signal
2. Ca^{2+} as a second messenger
3. Activation of cell cycle regulatory kinases and phosphatases
4. Entry into M phase (mitosis) from G_2 of the cell cycle

Plants have a low cytosolic [Ca^{2+}] that is regulated by Ca^{2+} pumps and channels, Ca^{2+}-modulated regulatory proteins such as calmodulin (CaM), Ca^{2+}- and CaM-regulated protein kinases and phosphatases, and a potential phosphoinositol signal pathway including inositol trisphosphate and diacylglycerol.[3,5,7,9-14] In addition, recent evidence suggests that the plant cell cycle may be regulated by kinases and phosphatases that are similar in structure and function to animal cell cycle control proteins.[15-17] One of the substrates for these enzymes, the nuclear lamins, were recently described in plants.[18,19] The phosphorylation of the lamins, which regulates nuclear envelope breakdown, may be a critical control point in the stimulation of mitosis. This chapter will concentrate on evidence for Ca^{2+} as a second messenger and speculate on the events downstream from this message, that could link cytokinin to cell division.

III. CALCIUM AS A SECOND MESSENGER IN PLANTS

A. INTRACELLULAR [CA^{2+}]

The intracellular [Ca^{2+}] of a wide variety of plant cells was measured using different techniques, i.e., microinjected or transgenically expressed aequorin, quin 2, indo-1, fura-2, fluo-3, and microelectrodes. In all cases the free [Ca^{2+}] of resting or quiescent cells was estimated to be between 50 and 250 n*M* and rose to 300 to 750 n*M* in stimulated cells.[20-36] A wide variety of stimuli that can modulate the intracellular [Ca^{2+}] were shown, including cytokinin, touch, inositol trisphosphate, gibberellin, auxin, phytochrome, light, and abscisic acid. Precise control of the cytoplasmic [Ca^{2+}] by regulating movement across both the plasma membrane and the endomembrane system is essential for signal transduction.[7,13,32,37] Chloroplasts, mitochondria, tonoplast fractions, endoplasmic reticulum vesicles, and plasma membrane fractions were all shown to actively transport Ca^{2+} by a Ca^{2+} uniport, Ca^{2+}-ATPases, a Ca^{2+}/H^+ exchange mechanism that is dependent on a proton gradient, or nonspecific Ca^{2+} channels.[7,13,38-44] To increase cytoplasmic Ca^{2+}, there is evidence for both a receptor-mediated inositol 1,4,5, trisphosphate Ca^{2+} release system,[41,45-56] and evidence for at least three classes of voltage-gated Ca^{2+} channels.[13,57-69] Evidence is accumulating that vacuoles may have two classes of Ca^{2+} channels: one gated by inositol 1,4,5 trisphosphate (similar to animal endoplasmic reticulum channels), and the other voltage gated.[7,67,69] The voltage-dependent tonoplast channels found in plants may play a large role in maintaining the cytosolic [Ca^{2+}].[69] Other ligand-regulated channels have not been described in plants, but have been hypothesized to exist for cytokinin regulated Ca^{2+} transport across the plasma membrane.[34,64,70]

B. CA^{2+}-DEPENDENT REGULATORY PROTEINS

Calmodulin was purified and characterized from fungi, algae, and higher plants, and has a primarily cytosolic distribution that can change during mitosis to being highly concentrated within the spindle apparatus.[71-73] There is an ever increasing list of Ca^{2+}- or CaM-regulated protein kinases in plants, which includes

1. A Ca^{2+}-dependent protein kinase that phosphorylates serine and threonine residues of plasma membrane proteins[74]
2. A putative phosphatidylserine-activated protein kinase C[55,75,76]
3. A plasma membrane-associated Ca^{2+}- and CaM-activated protein kinase with phosphatidylinositol kinase activity[77,78]
4. A Ca^{2+} stimulated protein kinase that phosphorylates the plasma membrane H^+-ATPase[79]
5. A nuclear casein type II kinase[80]
6. A plasma membrane phosphatidylinositol kinase[81]
7. Ca^{2+}-dependent kinases.[82-85] Some of the Ca^{2+}-dependent kinases have a subunit that is homologous to calmodulin.[82,84]

There is an increasing awareness of the importance of the potential substrates of these Ca^{2+}-modulated kinases, for cell regulation by Ca^{2+}-based messenger systems in plants.[5,10,11,86,87] The role of phosphatases in mitosis and the cell cycle in plants is just starting to be investigated since the initial reports of homology between plant and animal phosphatases and the finding that inhibitors of these phosphatases can be effective in plants. Plants have homologs of animal type-1, type-2A, and type-2C protein phosphatases.[88-92] Phosphatase inhibitors have effects on the timing of anaphase separation of chromosomes and can result in asynchronous separation of individual chromatids in plants. They are also implicated in the regulation of cell-cycle-specific proteins in plants.[1,92]

C. PLANT CELL CYCLE

Since cytokinin can act as a mitotic stimulus, it must affect the cell cycle and the entry of cells from interphase into mitosis. Therefore, recent advances in understanding the control of the cell cycle in animal and yeast cells (especially the transition from G_2 to M) can have important implications for plant cells. These control points may be influenced by a signal transduction chain that culminates in activation of specific kinases that phosphorylate proteins essential for mitosis, such as histones and lamins. Plant cell division and the plant cell cycle were recently reviewed, and many similarities to the better studied animal model are emerging.[15,93-96] Mitosis promoting factor (MPF), with its associated $p34^{cdc2}$ kinase activity and the corresponding *cdc2* gene homolog, were described in pea, carrot, corn, and *Arabidopsis*.[15,97-100] This protein was immunologically detected in both higher and lower plants, including *Arabidopsis*, oats, *Chlamydomonas*, and brown and red algae.[101] MPF was shown to phosphorylate lamins,[102] and purified $p34^{cdc2}$ induces lamina disassembly in isolated nuclei.[103] Lamins were recently described in pea and carrot cells and may be a substrate for these cell-cycle-regulated kinases in plants as they are in animal cells.[18,19] Phosphorylation of a putative lamin A by a casein kinase-II-like protein kinase was also reported recently.[104] Cytokinin-induced phosphorylation of proteins with molecular weights between 55 and 65 kDa (similar in size to lamins) was described in the moss *Funaria*.[105] If lamins are a substrate for calcium-stimulated plant kinases and phosphatases, they may be a key intermediary protein in the cytokinin-triggered signal transduction pathway leading to cell division.

IV. CYTOKININ AND SIGNAL TRANSDUCTION

A. CYTOKININ AND CA^{2+}

Logical candidates for external stimuli that stimulate the production of second messengers are the plant hormones. It is only recently that the receptors for these important molecules are being characterized, and we are a long way from understanding their diverse mechanisms of action and interaction in both the isolated plant cell and the whole plant. Although Ca^{2+} has been proposed more than once as a second messenger for auxin, cytokinin, gibberellin, and abscisic acid, there is strong evidence for a cytokinin-Ca^{2+} interaction that fits the model of a signal transduction chain. However, all the hormones may, in some cells, interact with, regulate, or be regulated by some aspect of Ca^{2+} metabolism.

There is increasing evidence that cytokinins may exert their effects on plant cells by modulating the intracellular [Ca^{2+}], in both a temporal and spatial manner, within a given cell. The early work was summarized by Hepler and Wayne and included a review of cytokinin-induced calcium uptake in *Achlya*, and synergistic interactions between Ca^{2+} and cytokinin in ethylene production, delay of senescence in leaf disks, increased cotyledon dry weight, membrane protein phosphorylation, and betacyanin synthesis.[2] Hepler and Wayne also described the evidence to date that cytokinin stimulates cell division in the moss *Funaria*, by increasing the intracellular [Ca^{2+}]. There has been additional research in several of these areas, which supports the basic hypothesis that cytokinin can modulate intracellular [Ca^{2+}].

Yip and Yang found that thidiazuron, a cytokinin-active urea derivative, mimicked the effect of benzyladenine (BA) in stimulating Ca^{2+}-dependent ethylene production in mung bean hypocotyls.[106] The relationship between the structure and activity of thidiazuron and other thiadiazolylurea derivatives in inducing synergistic ethylene production with Ca^{2+} is similar to that needed for the growth of callus tissue. It is of interest that phosphate is required to see an effect. The authors concluded that both N^6-substituted adenine cytokinins and substituted urea derivatives with cytokinin activity have a similar mode of action and, in this system, may activate calmodulin, inducing the synthesis of aminocyclopropane-1-carboxylic acid.

Pan et al. showed that aluminum inhibits lateral branch formation on soybean shoots, which can be overcome by exogenous cytokinin applications to the shoot.[107] The authors proposed that aluminum acts by a three-pronged mechanism in this system. First, aluminum inhibits Ca^{2+} uptake essential for both cell division in the meristematic region of the root and continued cytokinin biosynthesis in this region. Second, there is reduced Ca^{2+} transport to lateral buds. Therefore, both the cytokinin and Ca^{2+} supply to the developing lateral buds is lowered, inhibiting cell division and growth. They also proposed that the inhibition of CaM by aluminum binding may also play a role.

Pooviah and co-workers have continued their investigation into the relationship between Ca^{2+} and senescence.[3] During early stages of senescence in apples, phosphorylation of proteins with molecular weights of 95, 91, 53, and 50 kDa was promoted by calcium and calmodulin. This was greatly decreased during the later stages of senescence.[108] There was also a decrease in ATP-dependent Ca^{2+} uptake by a microsomal fraction of carnation petals that occurred just before the onset of senescence.[109]

Elliott has continued to investigate the role of calcium in cytokinin-dependent betacyanin production in *Amaranthus,* which is blocked by relatively high levels of CaM inhibitors.[110-113] A23187 plus exogenous Ca^{2+} slightly stimulated the response in the absence of added hormone. Adding 10 μg/ml dioctanoylglycerol increased the effect by 52%, but the absolute amount of betacyanin produced was still far below the levels induced by benzyladenine.[111] Elliott et al. purified a putative protein kinase C from both the cytosolic and membrane fractions of this tissue.[75,76] Both kinase preparations cross-react with antiserum raised against rat brain protein kinase C. The cytosol-derived enzyme could be activated by added phosphatidylserine, but little or no stimulation of the membrane-derived enzyme was seen. The authors suggested that this was because the prior calcium-generated membrane association would result in a completely activated enzyme state. Recently, Elliott and Yuguang showed that cytokinin increases Ca^{2+} accumulation in isolated protoplasts of *Amaranthus* protoplasts derived from the cotyledon, 20 to 40% over controls.[70] The authors concluded that the initial rapid response indicates that cytokinin may affect a Ca^{2+} channel, pump, or antiport, directly or indirectly, through protein phosphorylation, rather than work via long-term effects on protein or membrane synthesis. Secondary effects may arise from destruction of the proton gradient and new membrane synthesis. There is indirect evidence for cytokinin effects on Ca^{2+} in *Plantago,* since addition of benzyladenine retarded growth reduction in limiting [Ca^{2+}], indicating that Ca^{2+} uptake into cells may be increased.[114]

B. CYTOKININ AND CA^{2+} IN *FUNARIA*

1. Evidence for Ca^{2+} Involvement

Previous research on cytokinin-induced asymmetric plant cell division in target caulonema cells of the moss *Funaria hygrometrica* implicated a spatially controlled rise in intracellular Ca^{2+} as a second messenger.[115] Both the spatial and temporal changes in [Ca^{2+}] were proposed to regulate the profound cytoplasmic rearrangements associated with plant cell growth and differentiation noted after treatment with exogenous hormone. To date, this is the best characterized system that links cytokinin action to a rise in cytoplasmic [Ca^{2+}].

Funaria grows as a filamentous mat of cells composed of two cell types: (1) chloronema cells with large chloroplasts and transverse crosswalls, and (2) target caulonema cells with small, spindle-shaped chloroplasts and oblique cross walls. The elongated target cells respond to cytokinin induction by a localized swelling at the distal end of target cells, then nuclear migration to that region 20 to 22 h after treatment, followed by initial cell formation. This initial cell continues to divide, producing a mass of cells termed a bud.

Previous research produced ample evidence that substantiates the hypothesis that cytokinin-induced cell division in *Funaria* is Ca^{2+}-mediated. Utilizing the fluorescent calcium chelator probe chlorotetracycline, increases in membrane-associated Ca^{2+} after benzyladenine (BA) treatment have been detected and correlated with the migrating nucleus, the site of initial asymmetric division, and the rapidly dividing cells of the bud.[116] In addition, the artificial generation of a rise in intracellular Ca^{2+}, by A23187, induced bud initiation in the absence of externally applied BA.[117] Prevention of a natural rise in intracellular Ca^{2+} after hormone treatment, utilizing low levels of Ca^{2+} transport inhibitors, such as La^{3+}, D 600, and verapamil, as well as incubation in Ca^{2+}-free medium, was shown to block both nuclear migration and subsequent cell division.[64]

Conrad and Hepler examined the role of protein kinase C in BA-induced bud formation, by analyzing the effects of the phorbol ester 12-*O*-tetradecanolphorbol 13-acetate (TPA).[60] When TPA or A23187 are used separately, there is a slight stimulation of budding. However, when applied together, there is a dramatic increase in the number and development of buds, indicating that buds can be initiated by activation of the putative protein kinase C and Ca^{2+}-CaM portions of the Ca^{2+} messenger system. Conrad and Hepler also examined the involvement of voltage-dependent Ca^{2+} channels, using dihydropyridine compounds (DHP) that can alter Ca^{2+} movement through potentially sensitive channels. The DHP agonists (+)202-791 and CGP 28392 (100 μ*M*) induce bud initials on every target cell including the tip cell. The DHP antagonist (−)202-791 can block budding induced by optimal levels of BA. Nifedipine also blocks cytokinin-induced budding, which can be reversed by a pulse of UV light. The authors concluded that cytokinin activation of Ca^{2+} channels is the first mode of action of the hormone. They also suggested there are additional effects of cytokinin that may be unrelated to Ca^{2+} because the initial cells rarely complete development.

2. Plasma Membrane Ca^{2+} Channels

Ionic currents around BA-treated cells of *Funaria* were explored using a nonintrusive vibrating probe.[118] The results indicated that untreated caulonema cells have maximum inward current at the nuclear region; the addition of BA induces an increase in inward current along the length of the target cell, which subsequently shifts to the presumptive bud site, and after establishment of the growth zone at the presumptive bud site, the inward current falls.

Recently, calcium fluxes around quiescent and cytokinin-stimulated dividing caulonema cells of *Funaria* were measured using a Ca^{2+}-selective vibrating microelectrode. These results were compared with previous measurements of current density, using the nonspecific vibrating probe, and the portion of the current carried by Ca^{2+} ions were determined. Electrodes were constructed as described in Kühtreiber and Jaffe and contained a 20-μm long column of a Ca^{2+}-sensitive FLUKA cocktail.[119] The electrodes were vibrated at 0.5 Hz (with a vibrational amplitude of 10 μm) close to the cell surface, and the difference in voltage at the vibrational extremes were measured (sample frequency 1 KHz), averaged, and stored on a computer. This is referred to as the phase-sensitive voltage difference. The results were superimposed on the microscope video image of the cells in a vector form indicating the position of the electrode, the direction of vibration, and the signal magnitude.

A phase-sensitive voltage difference of less than 15 μV was measured along the length of nonstimulated caulonema cells. This is equivalent to a steady influx of Ca^{2+} of less than 0.2 pmol/cm²/sec along the length of the cell. In mature cells (six to seven cells back from the growing tip) the Ca^{2+} flux was almost undetectable. This was in contrast to the current density pattern of quiescent caulonema cells, which exhibited a small inward current at distal and proximal ends of ~0.25 μA/cm², with a zone of higher current at the nuclear region (0.5 μA/cm²). Since the Ca^{2+} flux measured is equivalent to a current density of less than 0.04 μA/cm², ions other than Ca^{2+} must have been responsible for the larger current both along the length of the cell and at the nuclear zone in the quiescent cells.

After cytokinin treatment an increase in Ca^{2+} flux was detected at the presumptive bud site at the distal end of caulonema cells. The inward Ca^{2+} flux increased to 1.38 to 2.76 pmol/cm²/sec within 5 h after treatment (a measured difference of 100 to 200 μV). This is equivalent to a current density of 0.26 to 0.52 μA/cm². The greatest current density previously reported was 1.2 μA/cm² (measured 1 h after cytokinin treatment), which fell to 0.25 μA/cm² after 24 h, concurrent with the establishment of the new growth zone. Although there was a 14-fold increase in the amount of Ca^{2+} entering the cell at this point, there must have been ions other than Ca^{2+} involved in the establishment of the new growth zone and the determination of the initial bud site, since the Ca^{2+} component was only 40% of the total current density.

However, in contrast to the current changes that dropped after the establishment of the bud, a high inward Ca^{2+} flux remained after the growth zone was established. The inward Ca^{2+} flux at the surface of the initial bud cell was equivalent to that at the growing tip and appeared to be the major component of the current. The detected increase in Ca^{2+} influx after cytokinin treatment at the presumptive bud site was of interest because it preceded morphological change. After the new growth zone was established, the total current fell while the Ca^{2+} component remained high and was similar in magnitude to tip cells. Again, Ca^{2+} influx is essential for polarized growth. In this case there was additional evidence that Ca^{2+} was acting as a second messenger both in the establishment of the new growth zone and in triggering cell division.

3. Intracellular [Ca^{2+}]

Changes in intracellular [Ca^{2+}] after cytokinin treatment in protonema cells of *Funaria*, were measured using the pentapotassium salt of indo-1.[34] The extent of dye loading strongly depended on lowering the pH of the incubation medium to 5.0. Exposing dye-loaded cells briefly with Mn^{2+} did not quench fluorescence, suggesting that the source of fluorescence was from the cytoplasm and not from the cell wall. Indo-1 remained responsive to changes in $[Ca^{2+}]_i$ in *Funaria* cells. The $[Ca^{2+}]_i$ in quiescent cells (with and without extracellular Ca^{2+}) was 250 n*M*, which is within the range of reported $[Ca^{2+}]_i$ of other plant cells.

Treatment of cells with extracellular cytokinin in 4 m*M* Ca^{2+} induced a threefold increase in $[Ca^{2+}]_i$ to 750 n*M* in target caulonema cells. This increase was not observed in a Ca^{2+}-free medium. These target cells responded to cytokinin treatment by an asymmetric division, while nontarget chloronema cells did not divide. Cytokinin appears to increase $[Ca^{2+}]_i$ by extracellular Ca^{2+} uptake. However, nontarget chloronema cells and tip cells also responded to cytokinin treatment by increasing $[Ca^{2+}]_i$. The differential physiological response of these cell types to hormonal stimulation must lie further down the signal transduction chain. These observations confirm previous research using Ca^{2+} channel agonists and antagonists. Additional research is necessary to determine the time course for induction by cytokinin as related to Ca^{2+} influx in individual caulonema cells, and to directly measure the effect of Ca^{2+} channel agonists and antagonists in this system.

C. CYTOKININ AND CELL CYCLE REGULATION

Since cytokinin can act as a mitotic stimulus, it must affect the cell cycle and the entry of cells from interphase into mitosis. Recent advances in understanding the control of cell cycles in animal and yeast cells (especially the transition from G_2 to M) may have important implications for plant cells. These control points may be influenced by a signal transduction chain that culminates in activation of specific kinases that phosphorylate proteins essential for mitosis, such as histones and lamins. Mitosis promoting factor has been shown to phosphorylate lamins, and purified $p34^{cdc2}$ induces lamina disassembly in isolated nuclei.

The nuclear envelope of eukaryotic cells is responsible for segregating the nucleoplasm from the cytoplasm, regulating transport through nuclear pores, maintaining chromatin organization in interphase, and disassembling and reassembling during mitosis. A major structural component of the inner face of the nuclear envelope in vertebrates and invertebrates is the nuclear lamina, an array of one to three principal extrinsic membrane proteins, lamins A, B, and C (M_r 70,000–60,000), which together form a fibrous network.[120,121]

Lamins are highly homologous in sequence and structure to intermediate filament (IF) proteins and were recently classified as type-V IF.[122,123] Biochemical and molecular analyses have indicated that lamins A and C are more closely related to each other than to lamin B. Structurally, all IF proteins share a common central α-helical rod domain with long heptad repeat patterns of hydrophobic residues. At the NH_2 and COOH terminal ends of the rod domains, all IF classes share two highly conserved sequences. The rod portion is flanked by variable, non-α-helical head and tail domains.

During interphase, lamin B binds to a receptor on the inner nuclear envelope, and lamins A and C bind to both lamin B and decondensed chromatin. It is hypothesized that during prophase of mitosis, a complex cascade of phosphorylation events culminates in phosphorylation of the lamins, releasing lamins A and C, while lamin B remains attached to the membrane.[124,125] The nuclear envelope fragments into small vesicles as the chromosomes condense and the mitotic spindle forms. During telophase, the lamins become dephosphorylated, associate with chromatin, and polymerize into a fibrous network, and the nuclear envelope reforms around the decondensed chromatin.[125,126] Thus, the nuclear lamins are hypothesized to be responsible for both the structural integrity and the organization of the nucleus during interphase, and function as key regulators of the cell cycle.

One to five lamins have been described from a wide variety of vertebrates and invertebrates, including humans, rats, mice, birds, frogs, clams, and fruit flies.[127,128] There has been some speculation about the presence of intermediate filaments, lamin-type proteins, and cell-cycle-regulated phosphorylation events, in plants. Antigenically determined intermediate filaments that are related to type-III IF[129-131] and type-II IF were reported in the cytoplasm of higher plant cells.[132] Monoclonal antibodies to plant nuclear matrix proteins recognize several nuclear proteins that are immunologically related to intermediate filaments.[133] Monoclonal antibodies to a high-salt Triton X-100 insoluble fraction (which is characteristic of IF) from *Chlamydomonas* produces both a diffuse cytoplasmic fluorescence as well as a perinuclear fluorescence, in onion root tips.[134]

There is biochemical and immunological evidence for the presence and localization of four lamin proteins (A, B_1, B_2, and C) in *Pisum sativum* (pea) nuclei and, by extension, nuclear type-V intermediate filaments.[18] These plant proteins were purified by procedures that were developed for the isolation of animal lamins, share a common epitope with both animal intermediate filaments and animal lamins, and are localized to the nuclear matrix. In addition, the putative plant lamins A, B_1, and C can be solubilized from the nuclear matrix by NaOH treatment, while B_2 remains with the pelletable fraction. Both the putative plant lamins B_1 and B_2 bind antihuman lamin B, although they have lower molecular weights than rat and HeLa cell lamin B used as controls. Two functionally different forms of lamin B were also identified in both avian and mammalian cells.[135] The two forms of lamin B identified in plants may also have different functions, since B_1 may be more abundant and more easily solubilized than B_2.

A major difference between plant and animal lamins is their localization. While animal lamins are found exclusively around the inner periphery of the nuclear envelope, the proteins that have been identified as nuclear intermediate filaments and lamin B extend throughout the nuclear matrix.[18,19] Since anti-lamin B does not discriminate between the two forms of plant lamin B, it cannot be determined if they are differentially distributed, as might be indicated by the solubilization data. It may be that there are additional functions for lamin-like proteins in plants, which involve matrix organization that is supplied by other filamentous proteins in animal cells. Since nuclear lamins are proposed to represent the ancestral class of intermediate filaments, it is logical that of all classes of intermediate filaments, lamins may be conserved in plants. Previous reports of immunological detection of intermediate filaments or nuclear phosphoproteins in plants may, in fact, have described some nuclear lamin proteins.[129-134,136]

One would imagine that plant nuclei have a structure similar to that of other eukaryotes, and similar regulation pathways as the nuclear envelope breaks down in mitosis. A proposed model for cytokinin stimulation of cell division includes the following features. Cytokinin binds to a receptor that may be either on the plasma membrane or internal. Calcium channels are opened either on the plasma membrane (as in *Funaria*) or internally on the endomembrane system. The increase in intracellular Ca^{2+} is temporally and spatially regulated within the cytoplasm, and Ca^{2+} binds to Ca^{2+}-dependent regulatory proteins such as calmodulin. The subsequent kinase and phosphatase activity leads to the synthesis and/or activation of mitosis promoting factor ($p34^{cdc2}$ plus cyclin). The mitotic apparatus is assembled, lamins are phosphorylated, chromatin condenses, and the nuclear envelope breaks down as cells enter mitosis.

V. CONCLUSIONS

There is much excitement in plant science today about the recent advances in our understanding of signal transduction in plants. Because this field is so new, much of our reasoning is based on parallels with animal signal transduction systems. As plant scientists accumulate more information about the regulation of Ca^{2+}, protein kinases and phosphatases, the cell cycle, and plant hormones, a picture may emerge that deviates from the animal model. We are still a long way from understanding the interactions between cytokinin, Ca^{2+} signaling, activation of cell-cycle-regulated proteins, and mitosis. If plant scientists attack this problem from many directions, a unified model may emerge.

ACKNOWLEDGMENTS

The author acknowledges the research support of the U.S. Department of Agriculture and the National Science Foundation, and the assistance of Crystal Gambino and Amy McNulty in the preparation of this chapter.

REFERENCES

1. **Roux, S. J. and Slocum, R. D.,** Role of calcium in mediating cellular functions important for growth and development in higher plants, *Calcium Cell Funct.*, 3, 409, 1982.
2. **Hepler, P. K. and Wayne, R. O.,** Calcium and plant development, *Ann. Rev. Plant Physiol.* 36, 397, 1985.
3. **Pooviah, B. W. and Reddy, A. S. N.,** Calcium messenger system in plants, *Crit. Rev. Plant Sci.*, 6, 47, 1987.
4. **Boss, W. F. and Morre, D. J.,** Eds., *Second Messengers in Plant Growth and Development*, Alan R. Liss, New York, 1989.
5. **Saunders, M. J.,** Calcium and plant hormone action, in *Hormone Perception and Signal Transduction in Animals and Plants*, Roberts, J., Kirk, C., and Venis, M., Eds., The Company of Biologists Limited, Cambridge, 1990, 271.
6. **Trewavas, A. J. and Gilroy, S.,** Signal transduction in plant cells, *Trends Genet.*, 7, 356, 1991.
7. **Johannes, E., Brosnan, J. M., and Sanders, D.,** Calcium channels and signal transduction in plant cells, *BioEssays*, 13, 331, 1991.
8. **Williamson, J. R. and Monck, J. R.,** Hormone effects on cellular Ca^{2+} fluxes, *Ann. Rev. Physiol.*, 51, 107, 1989.
9. **Rincon, M. and Boss, W. F.** Second-messenger role of phosphoinositides, in *Inositol Metabolism in Plants*, Morre, D. J., Boss, W. F., and Loewus, F. A., Eds., Alan R. Liss, New York, 1990, 173.
10. **Ranjeva, R. and Boudet A. M.,** Phosphorylation of proteins in plants: regulatory effects and potential involvement in stimulus/response coupling, *Ann. Rev. Plant Physiol.*, 38, 73, 1987.
11. **Blowers, D. P. and Trewavas, A. J.,** Second messengers: their existence and relationship to protein kinases, in *Second Messengers in Plant Growth and Development*, Boss, W. F. and Morre, D. J., Eds., Alan R. Liss, New York, 1989, 1.
12. **Cohen, P.,** Classification of protein-serine/threonine phosphatases: identification and quantitation in cell extracts, *Methods Enzymol.*, 201, 389, 1991.
13. **Schroeder, J. I. and Thuleau, P.,** Ca^{2+} channels in higher plant cells, *Plant Cell*, 3, 555, 1991.
14. **Drbak, B. K.,** Plant signal perception and transduction: the role of the phosphoinositide system, *Essays Biochem.*, 26, 27, 1991.
15. **Jacobs, T.,** Control of the cell cycle, *Develop. Biol.*, 153, 1, 1992.
16. **Zhang, D. H., Callaham, D. A., and Hepler, P. K.,** Regulation of anaphase chromosome motion in *Tradescantia* stamen hair cells by calcium and related signaling agents, *J. Cell Biol.*, 111, 171, 1990.
17. **Wolniak, S. M. and Larsen, P. M.,** Changes in the metaphase transit times and the pattern of sister chromatid separation in stamen hair cells of *Tradescantia* after treatment with protein phosphatase inhibitors, *J. Cell Sci.*, 102, 691, 1992.

18. **McNulty, A. K. and Saunders, M. J.,** Purification and immunological detection of pea nuclear intermediate filaments: evidence for plant nuclear lamins, *J. Cell Sci.*, 103, 407, 1992.
19. **Frederick, S. E., Mangan, M. E., Carey, J. B., and Gruber, P. J.,** Intermediate filament antigens of 60 and 65 kDa in the nuclear matrix of plants: their detection and localization, *Exp. Cell Res.*, 199, 213, 1992.
20. **Williamson, R. E. and Ashley C.C.,** Free Ca^{2+} and cytoplasmic streaming in the alga *Chara*, *Nature*, 296, 647, 1982.
21. **Gilroy, S., Hughes, W. A., and Trewavas, A. J.,** The measurement of intracellular calcium levels in protoplasts from higher plant cells, *FEBS Lett.*, 192, 217, 1986.
22. **Brownlee, C., Wood, J. W., and Briton, D.,** Cytoplasmic free calcium in single cells of centric diatoms. The use of fura-2, *Protoplasma*, 140, 118, 1987.
23. **Bush, D. S. and Jones, R. L.,** Measurement of cytoplasmic calcium in aleurone protoplasts using indo-1 and fura-2, *Cell Calcium*, 8, 455, 1987.
24. **Felle, H.,** Auxin causes oscillations of cytosolic free calcium and pH in *Zea mays* coleoptiles, *Planta*, 174, 495, 1988.
25. **Felle, H.,** Ca^{2+}-selective microelectrodes and their application to plant cells and tissues, *Plant Physiol.*, 91, 1239, 1989.
26. **Felle, H.,** Cytoplasmic free calcium in *Riccia fluitans* L. and *Zea mays* L.: interaction of Ca^{2+} and pH, *Planta*, 176, 248, 1988.
27. **Larsen, P. M., Chen, T. L., and Wolniak, S. M.,** Quin2-induced metaphase arrest in stamen hair cells can be reversed by 1,2-dioctanoylglycerol but not by 1,3-dioctanoylglycerol, *Eur. J. Cell Biol.*, 48, 212, 1989.
28. **Chae, Q., Park, H. J., and Hong, S. D.,** Loading of quin2 into the oat protoplast and measurement of cytosolic calcium ion concentration changes by phytochrome action, *Biochim. Biophys. Acta*, 1051, 115, 1990.
29. **Gilroy, S., Read, N. D., and Trewavas, A. J.,** Elevation of cytoplasmic calcium by caged calcium or caged inositol trisphosphate initiates stomatal closure, *Nature*, 346, 769, 1990.
30. **McAnish, M. R., Brownlee, C., and Hetherington, A. M.,** Abscisic acid-induced elevation of guard cell cytosolic Ca^{2+} precedes stomatal closure, *Nature*, 343, 186, 1990.
31. **Williams, D. A., Cody, S. H., Gehring, C. A., Parish, R. W., and Harris, P. J.,** Confocal imaging of ionised calcium in living plant cells, *Cell Calcium*, 11, 291, 1990.
32. **Schroeder J. I. and Hagiwara, S.,** Repetitive increases in cytosolic Ca^{2+} of guard cells by abscisic acid activation of nonselective Ca^{2+} permeable channels, *Proc. Natl. Acad. Sci. U.S.A.*, 87, 9305, 1990.
33. **Gilroy, S., Fricker, M., Read, N. D., and Trewavas, A. J.,** Role of calcium in signal transduction of *Commelina* guard cells, *Plant Cell*, 3, 333, 1991.
34. **Hahm, S. H. and Saunders, M. J.,** Cytokinin increases intracellular Ca^{2+} in Funaria: detection with Indo-1, *Cell Calcium*, 12, 675, 1991.
35. **Knight, M. R., Campbell, A. K., Smith, S. M., and Trewavas, A. J.,** Transgenic plant aequorin reports the effects of touch and cold-shock and elicitors on cytoplasmic calcium, *Nature*, 352, 524, 1991.
36. **Timmers, A. C., Reiss H. D., and Schell, J. H.,** Digitonin-aided loading of Fluo-3 into embryogenic plant cells, *Cell Calcium*, 12, 515, 1991.
37. **Evans, D. E.,** Regulation of cytoplasmic free calcium by plant cell membranes, *Cell Biol. Inter. Reports*, 12, 383, 1988.
38. **Beilby, M. J.** Calcium and plant action potentials, *Plant Cell Environ.*, 7, 415, 1984.
39. **Kreimer, G., Melkonian, M., and Latzko, E.,** An electrogenic uniport mediates light-dependent Ca^{2+} fluxes across the envelope of intact spinach chloroplasts, *Planta*, 166, 515, 1985.
40. **Kreimer, G., Melkonian, M., and Latzko, E.,** An electrogenic uniport mediates light-dependent Ca^{2+} influx into intact spinach chloroplasts, *FEBS Lett.*, 180, 253, 1985.
41. **Schumaker, K. S. and Sze, H.,** Inositol 1,4,5-trisphosphate releases Ca^{2+} from vacuolar membrane vesicles of oat roots, *J. Biol. Chem.*, 262, 3944, 1987.
42. **Graf, P. and Weiler, E. W.,** ATP-driven Ca^{2+} transport in sealed plasma membrane vesicles prepared by aqueous two-phase partitioning from leaves of *Commelina communis*, *Physiol. Plant.*, 75, 469, 1989.
43. **Hedrich, R., Busch, H., and Raschke, K.,** Ca^{2+} and nucleotide dependent regulation of voltage dependent anion channels in the plasma membrane of guard cells, *EMBO J.*, 9, 3889, 1990.
44. **White, P. J., Banfield, J., and Diaz, M.,** Unidirectional Ca^{2+} fluxes in roots of rye (*Secale cereale* L.). Comparison of excised roots with roots of intact plants, *J. Exp. Bot.*, 43, 1061, 1992.

45. **Boss, W. F. and Massel, M. O.,** Polyphosphoinositides are present in plant tissue culture cells, *Biochem. Biophys. Res. Commun.*, 132, 1018, 1985.
46. **Ranjeva, R., Carrasco, A., and Boudet, A. M.,** Inositoltrisphosphate stimulates the release of calcium from intact vacuoles from *Acer* cells, *FEBS Lett.*, 230, 137, 1988.
47. **Brosman, J. M. and Sanders, D.,** Inositol trisphosphate-mediated Ca^{2+} release in beet microsomes is inhibited by heparin, *FEBS Lett.*, 260, 70, 1990.
48. **Boss, W. F.,** Phosphoinositide metabolism: its relation to signal transduction in plants, in *Second Messengers in Plant Growth and Development*, Boss, W. F. and Morre, D. J., Eds., Alan R. Liss, New York, 1989, 29.
49. **Morre, J. D.,** Stimulus-response coupling in auxin regulation of plant cell elongation, in *Second Messengers in Plant Growth and Development*, Boss, W. F. and Morre, D.J., Eds, Alan R. Liss, New York, 1989, 81.
50. **Alexandre, J., Lassalles, J. P., and Kado, R. T.,** Opening of Ca^{2+} channels in isolated red beet root vacuole membrane by inositol-1,4,5-trisphosphate, *Nature*, 343, 567, 1990.
51. **Lehle, L.,** Phosphatidylinositol metabolism and its role in signal transduction in growing plants, *Plant Mol. Biol.*, 15, 647, 1990.
52. **Thiel, G., MacRobbie, E. A., and Hanke, D. E.,** Raising the intracellular level of inositol 1,4,5-trisphosphate changes plasma membrane ion transport in characean algae, *EMBO J.*, 9, 1737, 1990.
53. **Kamada, Y. and Muto, S.,** Ca^{2+} regulation of phosphatidylinositol turnover in the plasma membrane of tobacco suspension culture cells, *Biochim. Biophys. Acta*, 1093, 72, 1991.
54. **Allan, E., Dawson, A., Drbak, B., and Roberts, K.,** GTP causes calcium release from a plant microsomal fraction, *Cell Signal*, 1, 23, 1989.
55. **Zherelova, O. M.,** Protein kinase C is involved in regulation of Ca^{2+} channels in plasmalemma of *Nitella syncarpa*, *FEBS Lett.*, 242, 330, 1989.
56. **Gilroy, S., Read, N. D., and Trewavas, A. J.,** Elevation of cytoplasmic calcium by caged calcium or caged inositol triphosphate initiates stomatal closure, *Nature*, 346, 769, 1990.
57. **Andrejauskas, E., Hertel, R., and Marme, D.,** Triiodobenzoic acid affects [^{3}H] verapamil binding to plant and animal membrane fractions and smooth muscle contraction, *Biochem. Biophys. Res. Commun.*, 138, 1269, 1988.
58. **Shina, T. and Tazawa, M.,** Ca^{2+}-activated Cl-channel in plasmalemma of *Nitellopsis obtusa*, *J. Mem. Biol.*, 99, 137, 1987.
59. **Shina, T. and Tazawa, M.,** Demonstration and characterization of a Ca^{2+} channel in tonoplast-free cells of *Nitellopis obtusa*, *J. Memb. Biol.*, 96, 263, 1987.
60. **Conrad, P. A. and Hepler, P. K.,** The effect of 1,4-dihydropyridines on the initiation and development of gametophore buds in the moss *Funaria*, *Plant Physiol.*, 86, 684, 1988.
61. **Graziana, A., Fosset, M.M., Ranjeva, R., Hetherington, A.M., and Lazdunski, M.,** Ca^{2+} channel inhibitors that bind to plant cell membranes block Ca^{2+} entry into protoplasts, *Biochemistry*, 27, 764, 1988.
62. **Graziana, A., Ranjeva, R., and Tessie, J.,** External electric fields stimulate the electrogenic calcium/sodium exchange in plant protoplasts, *J. Biochem.*, 29, 8313, 1990.
63. **Harvey, H. J., Venis, M. A., and Trewavas, A. J.,** Partial purification of a protein from maize (*Zea mays*) coleoptile membranes binding the Ca^{2+} channel antagonist verapamil, *Biochem. J.*, 257, 95, 1989.
64. **Saunders, M. J. and Hepler P. K.,** Calcium antagonists and calmodulin inhibitors block cytokinin-induced bud formation in *Funaria*, *Dev. Biol.*, 99, 41, 1983.
65. **Tester, M. and Harvey, H. J. J.,** Verapamil-binding fraction forms Ca^{2+} channels in planar lipid bilayers, in *Plant Mem. Transport*, Dainty, J., de Michelis, M. I., Marre, E., and Rasi-Caldogno, F., Eds., Elsevier, New York, 1989, 277.
66. **DuPont F. M. and Morrissey P. J.,** Subunit composition and Ca^{2+}-ATPase activity of the vacuolar ATPase from barley roots, *Arch. Biochem. Biophys.*, 294, 341, 1992.
67. **Blackford, S., Rea, P. A., and Sanders, D.,** Voltage sensitivity of H^+/Ca^{2+} antiport in higher plant tonoplast suggests a role in vacuolar calcium accumulation, *J. Biol. Chem.*, 265, 9617, 1990.
68. **Fairley-Grenot, K. A. and Assmann, S. M.,** Permeation of Ca^{2+} through K^+ channels in the plasma membrane of *Vicia faba* guard cells, *J. Membr. Biol.*, 128, 103, 1992.
69. **Pantoja, O., Gelli, A., and Blumwald, E.,** Voltage-dependent calcium channels in plant vacuoles, *Science*, 255, 1567, 1992.
70. **Elliott, D. C. and Yuguang, Y.,** Cytokinin and fusicoccin effects on calcium transport in *Amaranthus* protoplast, *Plant Sci.*, 65, 243, 1989.

71. **Marme, D.,** The role of calcium and calmodulin in signal transduction, in *Second Messengers in Plant Growth and Development*, Boss, W. F. and Morre, D. J., Eds., Alan R. Liss, New York, 1989, 57.
72. **Braam, J.,** Regulation of expression of calmodulin and calmodulin-related genes by environmental stimuli in plants, *Cell Calcium*, 13, 457, 1992.
73. **Braam, J.,** Regulated expression of the calmodulin-related TCH genes in cultured *Arabidopsis* cells: induction by calcium and heat shock, *Proc. Natl. Acad. Sci. U.S.A.*, 89, 3213, 1992.
74. **Sakamotto, H. and Shibata, S.,** Calcium-dependent protein phosphorylation in morning glory hypocotyls, *Phytochemistry*, 31, 2251, 1992.
75. **Elliott, D. C. and Kokke, Y. S.,** Partial purification and properties of protein kinase C type enzyme in plants, *Phytochemistry*, 25, 39, 1986.
76. **Elliott, D. C., Fournier, A., and Kokke, Y. S.,** Phosphatidylserine activation of plant protein kinase C, *Phytochemistry*, 27, 3725, 1988.
77. **Blowers, D. P. and Trewavas, A. J.,** Phosphatidylinositol kinase activity of a plasma membrane-associated calcium-activated protein kinase from pea, *FEBS Lett.*, 238, 87, 1988.
78. **Collinge, M. and Trewavas, A. J.,** The location of calmodulin in the pea plasma membrane, *J. Biol. Chem.*, 264, 8865, 1989.
79. **Schaller, G. E. and Sussman, M. R.,** Phosphorylation of the plasma-membrane H^+-ATPase of oat roots by a calcium-stimulated protein kinase, *Planta*, 173, 509, 1988.
80. **Grasser, K. D., Maier, U. G., and Felix, G.,** A nuclear casein type II kinase from maize endosperm phosphorylation HMG proteins, *Biochem. Biophys. Res. Commun.*, 162, 456, 1989.
81. **Gross, W., Yang, W., and Boss, W. F.,** Release of carrot plasma membrane-associated phosphatidylinositol kinase by phospholipase A2 and activation by a 70 kDa protein, *Biochim. Biophys. Acta*, 1134, 73, 1992.
82. **Harmon, A. C., Putnam-Evans, C., and Cormier, M. J.,** A calcium-dependent but calmodulin independent protein kinase from soybean, *Plant Physiol.*, 83, 830, 1987.
83. **Polya, G. M., Morrice, N., and Wettenhall, R. E. H.,** Substrate specificity of wheat embryo calcium-dependent protein kinase, *FEBS Lett.*, 253, 137, 1989.
84. **Suen, K. L. and Choi, J. H.,** Isolation and sequence analysis of a cDNA clone for a carrot calcium-dependent protein kinase: homology to calcium/calmodulin-dependent protein kinases and to calmodulin, *Plant Mol. Biol.*, 17, 581, 1991.
85. **Polya, G. M., Chandra, S., Chung, R., Neumann, G. M., and Hoj, P. B.,** Purification and characterization of wheat and pine small basic protein substrates for plant calcium-dependent protein kinase, *Biochim. Biophys. Acta*, 1120, 273, 1992.
86. **Saunders, M. J.,** Calcium mediation of cytokinin-induced cell division, in *Molecular and Cellular Aspects of Calcium in Plant Development*, Trewavas, A. J., Ed., Plenum Press, New York, 1986, 185.
87. **Park, M. H. and Chae, Q.,** Intracellular protein phosphorylation in oat (*Avena sativa* L.) protoplasts by phytochrome action. 1. Measurement of action spectra for the protein phosphorylation, *Biochem. Biophys. Res. Commun.*, 162, 9, 1989.
88. **Cheng, H. F. and Tao, M.,** Purification and characterization of phosphotyrosyl-protein phosphatase from wheat seedlings, *Biochim. Biophys. Acta*, 998, 271, 1989.
89. **Huber, S. C. and Huber, J. L.,** Activation of sucrose-phosphate synthase from darkened spinach leaves by an endogenous protein phosphatase, *Arch. Biochem. Biophys.*, 282, 421, 1990.
90. **MacKintosh, C. and Cohen, P.,** Identification of high levels of type 1 and type 2A protein phosphatases in higher plants, *Biochem. J.*, 262, 335, 1989.
91. **MacKintosh, C., Coggins, J., and Cohen, P.,** Plant protein phosphatases. Subcellular distribution, detection of phosphatase 2C and identification of protein phosphatase 2A as the major quinate dehydrogenase phosphatase, *Biochem. J.*, 273, 733, 1991.
92. **Nitschke, K., Fleig, U., Schell, J., and Palme, K.,** Complementation of the cdc dis2-11 cell cycle mutant of *Schizosaccharomyces pombe* by a protein phosphatase from *Arabidopsis thaliana*, *EMBO J.*, 11, 1327, 1992.
93. **Bryant, J. and Francis, D.,** *The Cell Division Cycle in Plants*, Cambridge University Press, Cambridge, 1985.
94. **Baskin, T. and Cande, W.,** The structure and function of the mitotic spindle in flowering plants, *Ann. Rev. Plant Physiol. Plant Mol. Biol.*, 41, 277, 1990.
95. **Davidson, D.,** Cell division, in *Plant Physiology: A Treatise*, Steward, F., Ed., Academic Press, New York, Vol. 10, 1991, 341.

96. **Wick, S.,** Spatial aspects of cytokinesis in plant cells, *Curr. Opin. Cell Biol.*, 3, 253, 1991.
97. **Feiler, H. M. and Jacobs, T.,** Cloning of the pea cdc2 homologue by efficient immunological screening of PCR products, *Plant Mol. Biol.*, 17, 321, 1990.
98. **Colasanti, J., Tyers, M., and Sundresan, V.,** Isolation and characterization of cDNA clones encoding a functional $p34^{cdc2}$ homologue, in *Zea mays*, *Proc. Natl. Acad. Sci. U.S.A.*, 88, 3377, 1991.
99. **Ferreira, P., Hemerly, A., Villarroel, R., Van Montagu, M., and Inze, D.,** The *Arabidopsis* functional homolog of the $p34^{cdc2}$ protein kinase, *Plant Cell*, 3, 531, 1991.
100. **Gorst, J. R., John, P. C. L., and Sek, F. J.,** Levels of $p34^{cdc2}$ like protein in dividing, differentiating and dedifferentiating cells of carrot, *Planta*, 185, 304, 1991.
101. **John, P. C. L., Sek, F. J., and Lee, M. G.,** A homologue of the cell division cycle control protein $p34^{cdc2}$ participates in the cell division cycle of *Chlamydomonas* and a similar protein is detectable in higher plants and remote taxa, *Plant Cell*, 1, 1185, 1989.
102. **Dessev, G., Iovcheva-Dessev, C., Bischoff, J. R., Beach, D., and Goldman, R.,** A complex containing $p34^{cdc2}$ and cyclin B phosphorylates the nuclear lamin and disassembles nuclei of clam oocytes in vitro, *J. Cell Biol.*, 112, 523, 1991.
103. **Peter, M., Nakagawa, J., Doree, M., Labbe, J. C., and Nigg, E. A.,** In vitro disassembly of the nuclear lamina and M-phase-specific phosphorylation of lamins by cdc2 kinase, *Cell*, 61, 591, 1990.
104. **Li, H. and Roux, S. J.,** Casein kinase II protein kinase is bound to lamina-matrix and phosphorylates lamin-like protein in isolated pea nuclei, *Proc. Natl. Acad. Sci. U.S.A.*, 89, 8434, 1992.
105. **Saunders, M. J.,** Calcium and cytokinin in mosses, in *Bryophytes: Physiology and Biochemistry*, Chopra, R. N. and Bhathla S. C., Eds., CRC Press, Boca Raton, FL, 1090, 103.
106. **Yip, W. and Yang, S. F.,** Effect of thidiazuron, a cytokinin-active urea derivative, in cytokinin-dependent ethylene production systems, *Plant Physiol.*, 80, 515, 1986.
107. **Pan, W. L., Hopkins, A. G., and Jackson, W. A.,** Aluminum-inhibited shoot in soybean: a possible consequence of impaired cytokinin supply, *Commun. Soil Sci. Plant Anal.*, 19, 1143, 1988.
108. **Paliyath, G. and Poovaiah, B. W.,** Calcium- and calmodulin-promoted phosphorylation of membrane proteins during senescence in apples, *Plant Cell Physiol.*, 26, 977, 1985.
109. **Paliyath, G. and Thompson, J. E.,** Senescence-related changes in ATP-dependent uptake of calcium into microsomal vesicles from carnation petals, *Plant Physiol.*, 88, 295, 1988.
110. **Elliott, D. C.,** Calmodulin inhibitor prevents plant hormone response, *Biochem. Int.*, 1, 290, 1980.
111. **Elliott, D. C.,** Calcium involvement in plant hormone action, in *Molecular and Cellular Aspects of Calcium in Plant Development*, Trewavas, A. J., Ed., Plenum Press, New York, 1986, 285.
112. **Elliott, D. C.,** Inhibition of cytokinin-regulated responses by calmodulin-binding compounds, *Plant Physiol.*, 72, 215, 1983.
113. **Elliott, D. C., Batchelor, S. M., Cassar, R. A., and Marinos, N. G.,** Calmodulin-binding drugs affect responses to cytokinin, auxin, and gibberellic acid, *Plant Physiol.*, 72, 219, 1983.
114. **Kuiper, D.,** Growth responses of *Plantago major* L. ssp. *pleiosperma* (Pilger) to changes in mineral supply, *Plant Physiol.*, 87, 555, 1988.
115. **Saunders, M. J.,** Cytokinin: Localized perception and response, in *Plant Growth Substances 1988*, Pharis, R. P. and Rood, S. B., Eds., Springer-Verlag, Berlin, 511, 1990.
116. **Saunders, M. J. and Hepler, P. K.,** Localization of membrane-associated calcium following cytokinin treatment in *Funaria* using chlorotetracycline, *Planta*, 152, 272, 1981.
117. **Saunders, M. J. and Hepler, P. K.,** Calcium ionophore A23187 stimulates cytokinin-like mitosis in *Funaria*, *Science*, 217, 943, 1982.
118. **Saunders, M. J.,** Cytokinin activation and redistribution of plasma-membrane ion channels in *Funaria*, *Planta*, 167, 402, 1986.
119. **Kühtreiber, W. M. and Jaffe, L. F.,** Detection of extracellular calcium gradients with a calcium specific vibrating electrode, *J. Cell Biol.*, 110, 1565, 1990.
120. **Aebi, U., Cohn, J., Buhle, L., and Gerace, L.,** The nuclear lamina is a meshwork of intermediate-type filaments, *Nature*, 323, 560, 1986.
121. **Gerace, L.,** Nuclear lamina and organization of nuclear architecture, *TIBS*, 11, 443, 1986.
122. **Fisher, D., Chaudhary, N., and Blobel, G.,** cDNA sequencing of nuclear lamins A and C reveals primary and secondary homology to intermediate filament proteins, *Proc. Natl. Acad. Sci. U.S.A.*, 83, 6450, 1986.

123. **Osborn, M. and Weber, K.,** Intermediate filament proteins: a multigene family distinguishing major cell lineages, *TIBS*, 11, 469, 1986.
124. **Gerace, L. and Blobel, G.,** The nuclear envelope lamina is reversibly depolymerized during mitosis, *Cell*, 19, 277, 1980.
125. **Ward, G. E. and Kirshner, M. W.,** Identification of cell cycle-regulated phosphorylation sites on nuclear lamin C, *Cell*, 61, 561, 1990.
126. **Glass, J. R. and Gerace, L.,** Lamins A + C bind and assemble at the surface of mitotic chromosomes, *J. Cell Biol.*, 111, 1047, 1990.
127. **Krohne, G. and Benavente, R.,** The nuclear lamins: multigene family of proteins in evolution and differentiation, *Exp. Cell Res.*, 162, 1, 1986.
128. **Osman, M., Paz, M., Landesman, Y., Fainsod, A., and Gruenbaum, Y.,** Molecular analysis of the *Drosophila* nuclear lamin gene, *Genomics*, 8, 217, 1990.
129. **Dawson, P. J., Hulme, J. S., and Lloyd, C. W.,** Monoclonal antibody to intermediate filament antigen cross-reacts with higher plant cells, *J. Cell Biol.*, 100, 1793, 1985.
130. **Goodbody, K. C., Hargreaves, A. J., and Lloyd, C. W.,** On the distribution of microtubule-associated intermediate filament antigens in plant cells, *J. Cell Sci.*, 93, 427, 1989.
131. **Hargreaves, A. L., Dawson, P. J., Butcher, G. W., Larkins, A., Goodbody, K. C., and Lloyd, C. W.,** A monoclonal antibody raised against cytoplasmic fibrillar bundles from carrot cells, and its cross-reaction with animal intermediate filaments, *J. Cell Sci.*, 92, 371, 1989.
132. **Ross, J.H., Hutchings, A., Butcher, G.W., Lane, E.B., and Lloyd, C.W.,** The intermediate filament-related system of high plant cells shares an epitope with cytokeratin 8, *J. Cell Sci.*, 99, 91, 1991.
133. **Beven, A., Guan, Y., Pert, J., Cooper, C., and Shaw, P.,** Monoclonal antibodies to plant nuclear matrix reveal intermediate filament-related components within the nucleus, *J. Cell Sci.*, 98, 293, 1991.
134. **Parke, J. M., Miller, C. C. J., Cowell, I., Dodson, A., Dowding, A., Downes, M., Duckett, J. G., and Anderton, B. J.,** Monoclonal antibodies against plant proteins recognize animal intermediate filaments, *Cell Motil. Cytoskeleton*, 8, 312, 1987.
135. **Lehner, C. F., Kurer, V., Eppenberger, H. M., and Nigg, E. A.,** The nuclear lamin protein family in higher vertebrates, *J. Biol. Chem.*, 261, 13293, 1986.
136. **Harper, J. D. I., Rao, P. N., and John, P. C.,** The mitosis-specific monoclonal antibody MPM-2 recognizes phosphoproteins associated with the nuclear envelope in *Chlamydomonas reinhardtii* cells, *Eur. J. Cell Biol.*, 51, 272, 1990.

Chapter 19

Cytokinin Mutants

Trevor L. Wang

TABLE OF CONTENTS

I. RETROSPECTIVE

Studies of mutants have served to underline the important contribution that genetic analysis has made over the last 15 years to advancing our understanding of hormone biosynthesis and action.[1-5] It is perhaps disappointing, therefore, that cytokinin mutants form a very small part of this contribution. In the last review of this subject,[6] it was difficult to find any evidence of genetic variation used for the analysis of cytokinins, and there was only one group of plant mutants available — those of the moss *Physcomitrella patens.* In the intervening 6 years, however, efforts have been made to isolate mutants affecting cytokinin metabolism in both *Nicotiana* and *Arabidopsis*. In some instances this has not amounted to much effort on the cytokinin front, as the mutants were selected for other reasons and subsequently turned out to have phenotypes that might result from a disturbance to cytokinin relations in the plant.

It is our lack of knowledge of the physiology and biochemistry of cytokinins that has made it so difficult to isolate mutants. For example, what would mutants that had a decreased production of cytokinin look like? Would they be pale green and have a high degree of apical dominance or would they sprout roots from their axils? Would they be stunted due to a lack of cell division? Recent successes in some species, including results from transgenics, have provided a few answers, but they may have helped more, in an ancillary way, by promoting new interest.

In this chapter I shall discuss cytokinin mutants that have been induced, usually by chemical means, since no spontaneous mutants are known. Plants transformed with foreign DNA, most frequently with bacterial genes, e.g., the cytokinin biosynthetic gene from *Agrobacterium tumefaciens*, to produce individuals with altered hormone levels will not be covered, since they are included elsewhere in this volume (see Chapter 21). Furthermore, the use of the term "mutant" to describe such plants[7] should be avoided, as it is inappropriate and could therefore be misleading, especially when the term "transgenic" is in common use.

II. TARGETS FOR SCREENING

Identifying screening targets is probably the most difficult component in the process of isolating

0-8493-6252-0/94/$0.00+$.50

cytokinin mutants. There are two reasons for this, which become apparent when one investigates studies of other hormone mutants. First, there is a lack of spontaneous mutants that can be repaired by application of the hormone to provide an easily recognized target and, second, reproducible effects of the hormone on physiological and biochemical processes have not been observed in many instances. For the gibberellins, examples would be the numerous internode mutants and the stimulation of germination and amylase activity. Another driving force behind the isolation of such mutants, of course, is the economic value to the food and agriculture industries of dwarfism and the regulation of germination.

In contrast, there are few processes in which a clear role for the cytokinins has been established. This can be illustrated by several examples. It was demonstrated most graphically in a review of cytokinins in 1987 where it was self evident in the title: "Cytokinins: plant hormones in a search of a role."[8] It was apparent with respect to plant developmental processes in the most recent of reviews on apical dominance,[9] where their influence was recognized but considered secondary. Moreover, to a certain extent, this volume also attests to the problem. Indeed, in a way this can become a circular argument, since the obtaining of cytokinin mutants for apical dominance, for example, would provide strong evidence of their role in the process, while we would not look at this process as a screening target if there was not prior physiological evidence that cytokinins had an effect. Hence, designing biologically based, rational (biorational) screens for cytokinin mutants is most difficult.

We are left, therefore, with attempting to identify processes where there are well-known and frequently observed effects. The best of these will also have a quantitative element. For any plant hormone, not just the cytokinins, these will be processes that have been used as biological assays for the hormone. Again, for the gibberellins these would be stem length (e.g., lettuce hypocotyl assay) and germination (e.g., α-amylase induction), exactly the processes for which most mutants have been isolated. Thus, one is led to the bioassays (reviewed by Letham[10]). Table 1 illustrates those bioassays that would provide suitable biorational targets for the isolation of mutants. The mutants that have been isolated so far have fallen into the categories of greening phenomena, cell growth *in vitro*, and morphogenesis. Of the processes, phenomena such as cell division may well lead to lethality in a mutant, and thus their isolation is unlikely. To circumvent this problem, conditional mutants would need to be sought, e.g., temperature sensitives, but as has been pointed out,[27] this is not always practicable and requires a considerable amount of effort. It may be possible to isolate mutants unable to synthesize cytokinins, as cytokinin-requiring mutants, although exogenous supply will not necessarily repair a mutant unable to make a hormone, since temporal and spatial regulation of its synthesis may also be required. Attempts to isolate cytokinin-requiring mutants of *Physcomitrella patens* have so far yielded no such mutant, although only about 10,000 mutagen-treated plants have been screened.[28]

In addition to these processes, there is an obvious one in resistance to cytokinin. In many cases the chemicals used will be toxic and inhibitory. Therein lies a danger, since resistance to a chemical can be nonphysiological and merely represent an artifact. Resistance mutants can be of two forms: either direct or indirect. In direct mutants the mutation will lead to a change in the metabolism, uptake, or transport of the hormone, or the perception of the hormone (sensitivity to the hormone), whereas indirect mutants may represent an "avoidance" of the effect of the hormone, e.g., the target for the hormone is no longer present or it is

Table 19-1 **Processes Influenced by Cytokinin, their Bioassays, and Relevant Mutants**

Process	Assay	Mutant[a]
Greening phenomena	Delay of senescence[11,12]	–
	Leaf/cotyledon greening[13]	+[14]
Cell division	Tobacco callus[15]	+[16,17]
	Soybean callus[18]	–
Cell expansion	Radish cotyledon[11,19]	–
	Spirodela frond[11]	–
Morphogenesis	Callus shoot formation[20]	–
	Moss gametophore induction[21,22]	+[23,24]
Pigment formation	*Amaranthus* betacyanin[25,26]	–

[a] A + sign indicates mutants have been isolated and gives the relevant references.

circumvented, thus the mutant can survive and is selected. This latter point will be covered later under the individual mutants, e.g., wilty mutants in *Nicotiana*. With resistance mutants in which screening is essentially untargeted and based on toxicity, the chance of isolating the latter type of mutant is much greater.

A further problem that has been identified in mutants isolated as hormone-resistant is that of cross-resistance. In both *Arabidopsis* and *Nicotiana* (see below), mutants have been isolated that show a high degree of cross-resistance. It is said that the occurrence of this phenomenon lends support to the hypothesis that the plant hormones in question have common or integrated pathways during development.[1,29,30] The simplest form of cross-resistance would be associated with an effect on the level of the hormone where one hormone acts on an enzyme involved with the metabolism of another or, if one prefers to consider sensitivities, where one hormone regulates the level of another's receptor. Alternatively, it can be postulated that cross-resistance is simply an illustration that the mutant lesion is not linked directly to the action of either hormone (see Section III.A).

One can see, therefore, that targeting a highly cytokinin-specific response should yield mutants more appropriate to an understanding of cytokinin action. There is only one example where a class of mutants has been isolated in a totally untargeted fashion, and that is the OVE mutants of *P. patens*. In this instance these mutants bore such a distinct morphological phenotype and occurred at such a frequency as to be readily retrieved by total isolation.[24]

III. THE MUTANTS

A. *NICOTIANA*

1. The *aba1* Locus (*ckr, iba,* and *esg*)

Several groups have produced auxin and cytokinin-resistant mutants of *Nicotiana plumbaginifolia* in the last few years. Interestingly, several of these mutants provide evidence that screening for hormone mutants in this way can be fraught with problems. All the groups used hormone concentrations (NAA, IBA, or BA) at levels greater than 10^{-5} *M* in the culture medium. These could be considered toxic levels, therefore, rather than physiological concentrations, and were inhibitory to the growth of the wild type. Hence, plants were scored for reduced inhibition of growth in the presence of the hormone. Ethane methyl sulfonate (EMS) was the mutagen in all cases.

Blonstein et al. reported a mutant, ckr1, that showed considerably more seedling growth than the wild type, on a medium supplemented with 20 μM BA.[31] The resistance to the hormone was inherited as a monogenic recessive character. This mutant also showed differences from the wild type in its sensitivity to auxin, its requirement for GA to germinate, and its susceptibility to wilt. The latter provided a clue to the real basis of this mutant; it was later shown that the mutant was defective in ABA biosynthesis[32] and was a wilty mutant of the *flacca* and *sitiens* type in tomato. Mutants at three other loci in *Nicotiana*, *iba1*, *iba2*, and *iba3*, have been isolated on the basis of their resistance to high concentrations of auxin.[33] Mutants at the *iba1* locus also had altered responses to ABA and the GA biosynthesis inhibitor paclobutrazol and displayed differences to the wild type in their germination and sensitivity to water stress.

Subsequently, it has been shown that mutants of the IBA and ckr type fall into the same complementation group together with a third mutant, Esg152.[29] This last mutant was isolated on the basis of its *e*arly and *s*ynchronous *g*ermination (precocious germination), which would be consistent with a defect in ABA biosynthesis. All these mutants are wilty, resistant to auxin, and, by inference, to cytokinin, and show the Esg phenotype. Since the lesion in one of them, CKR1, has been shown to be in the last step of ABA biosynthesis, the authors renamed this locus *aba1*, in accordance with the terminology for *Arabidopsis thaliana*.[34] Explanations for the pleiotropic effects of this locus have been proposed based on the known interactions of the different hormones in stomatal physiology and the alterations in endogenous levels of the hormones measured in tomato ABA mutants.[29] A much simpler explanation, however, for the selection of ABA mutants in a cytokinin resistance screen was proposed by Horgan.[27] He postulated that the presence of high levels of cytokinin imposed a selection pressure on germinability. Mutant seeds were selected, therefore, that had low ABA levels and thus could germinate more readily, circumventing the toxic levels of cytokinin. This is a clear example of the avoidance phenomenon mentioned above and shows that mutants arising from resistance to a compound may not have a direct connection with physiological responses of the plant to the compound in question. This view is reinforced by the fact that Rousselin et al. also found a strict correlation between tolerance to auxin and the precocious germination of the mutants at the *ckr*, *iba*, and *esg* loci;[29] auxin acting in precisely the same way as cytokinin, by maintaining a selection pressure in favor of precocious germination. Hence, the lesion in the CKR mutant of *Nicotiana* is not founded in cytokinin relations.

2. The *zea* Loci

Recently, six new mutants were isolated from *N. plumbaginifolia*.[35] Again, these mutants are based on resistance to a high level (10^{-4} *M*) of cytokinin, but in this instance zeatin, rather than N^6-benzyladenine, was used. Complementation analysis showed that mutants at three loci, *zea1*, *zea2*, and *zea3*, were isolated; all showed resistance to both purine and nonpurine (diphenylurea-type) cytokinins. The *zea1* mutant showed increased resistance to zeatin, compared with the wild type, at concentrations as low as 0.1 μ*M*. The mutants grew well at this concentration, exhibiting increased root formation, while the wild type was severely inhibited (see Figure 1). At higher concentrations the mutant showed callus formation and excessive shoot formation, while the wild type showed almost no development at all. The resistance to cytokinin, however, was not exhibited in tissue culture per se, though in the absence of hormone, microcalli showed an increased capacity for root formation, demonstrating that this trait was not related to the high hormone level originally included in the medium for screening. These mutants did not show the same ABA-related phenomena as those at the *aba1* locus and may represent, therefore, mutants related to the action of cytokinin, the catabolism of cytokinin not being affected according to preliminary data.[35]

3. The *hl* Loci

In addition to the mutants affecting the physiology and development of the plant, two loci have been identified that influence the growth of *N. tabacum* L. leaf cells in culture. These have been termed the habituated leaf loci *hl-1* and *Hl-2*, the latter being a dominant mutant.[16,17] The loci determine whether the leaf cells will require cytokinin for growth in culture. These loci and others in *Phaseolus* that affect growth in culture will be discussed in more detail in Chapters 12 and 20. Neither of the tobacco loci has any effect on the phenotype of the plant, though it has been suggested from complementation studies between *hl-1* plants and wild-type transgenics bearing defective Ti plasmids, that genes at the *hl-1* locus have an oncogenic function similar to gene 4 on the plasmid.[36]

B. *ARABIDOPSIS*

The use of *Arabidopsis* for molecular genetic studies on plants makes it an obvious species for the isolation of hormone mutants, including cytokinin mutants. However, the latter have been very slow to emerge. Mutants affecting ABA, GA, ethylene, and auxin relations have been isolated (see the review by Klee and Estelle[1]), but until recently none has been isolated for the cytokinins. Of those that are now in the literature, only one group was isolated directly in a search for cytokinin mutants; the other two were isolated for other reasons and have been included here because they may prove to have a basis in cytokinin relations. The one class isolated with cytokinin mutants in mind, *ckr*, is based on a decreased inhibition of root growth.

1. The *ckr* Locus

Su and Howell isolated a series of mutants resistant to BA, based on the inhibition of root growth by the hormone.[37] Mutagenized seeds, again using EMS, were plated on agar in slits, and the plates were placed on edge so that root growth was

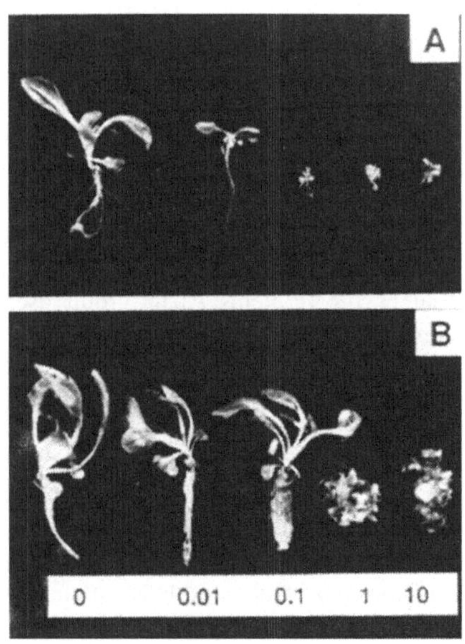

Figure 19-1 Response of wild type (A) and *zea1* mutant (B) of *Nicotiana plumbaginifolia*, to exogenous cytokinin. Numbers indicate concentration (μ*M*) of zeatin in the culture medium of seedlings. (From Jullien, M., et al., in *Physiology and Biochemistry of Cytokinins in Plants*, Kamínek, M., Mok, D. W. S., and Zazímalová, E., Eds., SPB Academic Publishing, The Hague, 1992, 159. With permission.)

encouraged to proceed through the agar and facilitate the isolation of mutants with altered root responses. The wild type was particularly sensitive to cytokinin, showing a substantial inhibition of primary elongation at only 5 n*M* BA. Interestingly, at the same time, root-hair elongation was stimulated. This stimulation was progressive up to a maximum at 50 μ*M*; thereafter, BA became inhibitory. The inhibition of primary elongation was related to the logarithm of the BA concentration up to 50 μ*M*. To avoid toxic effects of the hormone, the mutants were selected on 2.5 μ*M*. At this concentration, *Arabidopsis* is described as having a characteristic "cytokinin root syndrome" with "shorter, fuzzier roots." All the mutants isolated were found to be alleles at a single locus termed *ckr*. The mutants were single-gene recessives with resistance to BA inhibition of primary root growth and to BA-stimulated root-hair elongation, as indicated in Figure 2. The phenotype of these mutants on cytokinin, therefore, is significantly different from those of the *zea1* locus in *N. plumbaginifolia* where large numbers of roots per se were formed.

In the absence of cytokinin, the authors described the *ckr* mutants as having only subtle differences from the wild type. Shoots of the *ckr* mutants were very similar to those of the wild type, in terms of stature, branching pattern, and fertility, with only minor differences in color and shape at the seedling stage. The roots showed extra elongation and a slightly weaker gravitropic response and had shorter root hairs.

As mentioned above, cross-resistance often occurs when mutants are selected on the basis of resistance to inhibition by high levels of a hormone. *Arabidopsis* mutants are no exception to this. The auxin mutants at the *aux* and *axr* loci all show cross-resistance to other hormones,[30,38,39] often several, which has been attributed to interactions between the hormones.[1] Resistance to high levels of hormone, however, may invoke avoidance selection. To test this, Su and Howell examined the sensitivity of the *ckr* mutants to auxin, but found no cross-resistance.[37] They suggested, therefore, that the mutants were true cytokinin-response mutants and not just root morphology mutants

Figure 19-2 Response of wild type (left) and *ckr1*-109 mutant (right) of *Arabidopsis*, to exogenous cytokinin. Plants were photographed after 8 days of growth in the vertical position and in the presence of 2.5 μ*M* benzyladenine. (Reproduced with permission of the copyright holder.)

(a possible avoidance selection). The mild nature of the lesion in the *ckr* mutants, however, led the authors to suggest further that the mutants may be leaky or have organ-specific perception defects.

2. The *det* Loci

There are other candidates in *Arabidopsis* that may yet prove to be cytokinin mutants. Chory and co-workers sought EMS-induced mutants that can undergo light-regulated developmental responses in the dark. Mutants were isolated that show characteristics of *de-et*iolation in the absence of light.[40,41] These mutants fall into three complementation groups, *det1*, *det2*, and *det3*. Mutants at the *det1* locus undergo leaf and chloroplast development, anthocyanin accumulation, and the accumulation of mRNAs for light-regulated chloroplast- and nuclear-encoded genes, when held in the dark[40] (Figure 3). In contrast, mutants at the *det2* locus show an aberrant phenotype in the light, including a lack of leaf senescence following seed set, and although exhibiting most of the characteristics of the *det1* mutants in the dark, they do not initiate chloroplast development.[41] The phenotypes of the two mutants are additive in the double recessive mutant, indicating independent pathways. Mutants at the *det1* locus are small and pale green and lack apical dominance when grown in the light, indicating that *Det1* has a function in light-grown plants. This information together with data from tissue culture greening experiments[14] indicates that the *det1* locus may be involved with cytokinin relations.

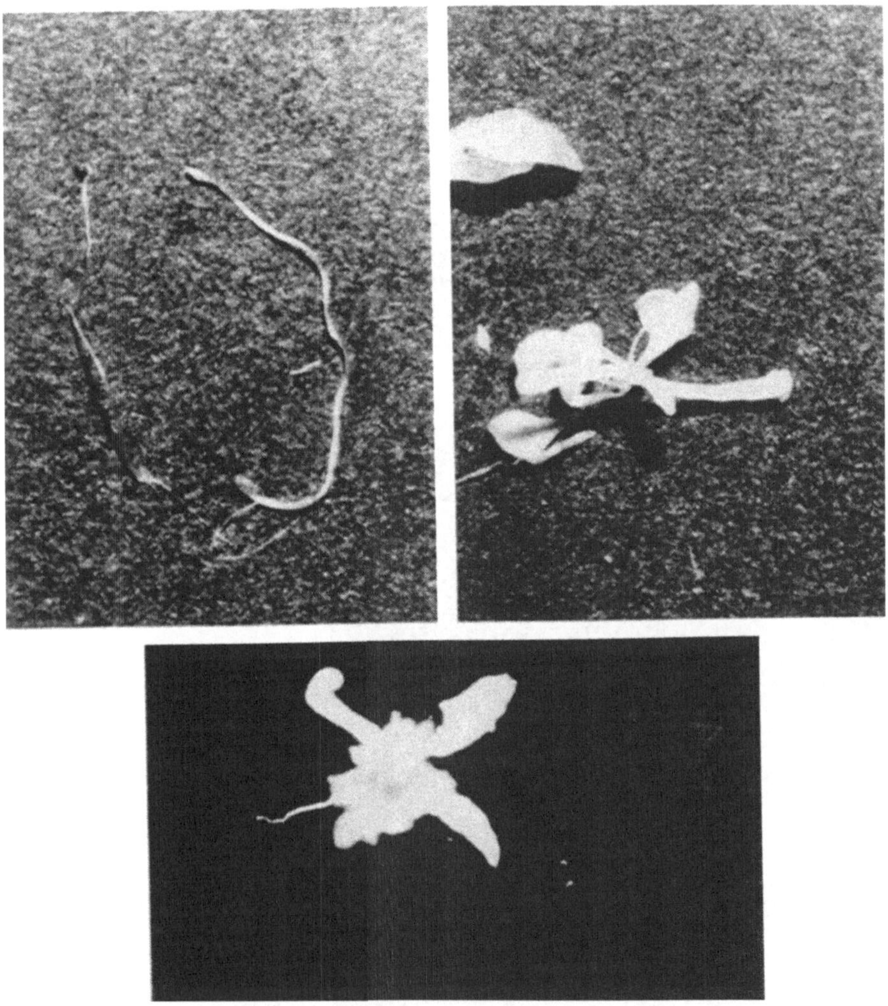

Figure 19-3 Growth response of wild type (top left) and *det1* mutant (top right) of *Arabidopsis* in the dark, and the corresponding effect of cytokinin on the wild type (bottom). Plants were grown for 7 days in the dark or for 20 days in the presence of isopentenyladenine (20 mg/l) in the dark.[14]

Recently, Chory et al. showed that much of the phenotype of the *det1* mutants can be mimicked by the addition of cytokinin to the medium of the wild type in the dark (Figure 3),[14] including the inhibition of hypocotyl elongation, chloroplast development, and expression of light-regulated genes. One of the most significant differences, however, was the lack of root growth in the treated wild-type plants compared to *det1* mutants. A range of cytokinins were tested, and all were found to be active, though significant effects were only observed in the micromolar-or-greater range. The *det1* plants themselves appeared to be more sensitive to the hormone, since an inhibition of germination was observed at only one third the level of cytokinin required by the wild type. The results indicate that abnormally high levels of cytokinin, or an increased sensitivity to the hormone, may be the basis of the *det1* phenotype. Currently, the latter appears to be more likely, since measurements of the cytokinin level in the wild type and mutants have shown that they are the same.[42]

3. Fasciation Loci

One further group of *Arabidopsis* mutants deserves mention here, and that is the one containing the fasciated mutants (*fas, clv*, and *fuf*).[43,44] Many bacteria induce abnormalities in growth, which are often termed fasciation.[45,46] In such bacteria there are genes on their plasmids that confer virulence to the organism. Some of these genes encode enzymes with isopentenyltransferase (IPT) activity. It is possible, therefore, that fasciated mutants have a basis in cytokinin relations. A word of caution should also be added here; although the word describing the disease and the *Arabidopsis* mutants (and phenotypically similar mutants in other species[47]) may be the same, the phenotypes are not. The disease is characterized by, for example,[45] "a loss of apical dominance and the development of multiple shoots, giving rise to leafy galls," whereas the mutants are characterized by having abnormalities in phyllotaxis and the apex, leading to a broadening and flattening, with bifurcation of the stem in extreme cases.[43] The main feature lacking in the *Arabidopsis* mutants is the decrease in apical dominance.[48] The two phenotypes, therefore, may have quite distinct bases: the bacterially induced based on cytokinin overproduction; the mutants on another perhaps noncytokinin, but meristem-based, phenomenon.

C. *PHYSCOMITRELLA PATENS*

Shortly after the isolation of the first cytokinin, kinetin, by Miller et al.,[49] Gorton and Eakin demonstrated that this substance can increase the formation of leafy shoots on moss protonemata.[50] By using the moss, therefore, the cytokinins were shown to influence the morphogenesis not only of tissue cultures, but also of intact plants. This response of moss tissue is specific to cytokinins, including diphenylurea types,[51] and is both quantitative and qualitative. The quantitative nature of the response has allowed moss tissue to be used as a bioassay for the cytokinins.[22] Its qualitative nature shows that cytokinins influence not only the morphogenesis of the moss, but also the morphology of the leafy shoots. At low concentrations (nanomolar) the number of leafy shoots is increased, and these have a normal morphology; at high concentrations (micromolar) the leafy shoots become "twisted and distorted" or remain as "callus-like tissue," in the words of the original experimenters[22] (see Figure 4). In *P. patens* the response is similar, but depends on the length of exposure and the support medium;[23,52,53] the greatest sensitivity (10^{-9} *M*) is obtained in liquid culture. In this medium both cytokinin ribosides and bases are active, though this was not found in earlier work on *Funaria*.[54,55]

Several hundred developmentally[23,24] and physiologically[56] abnormal mutants of the moss, *P. patens* have been isolated, many of which show abnormal responses to plant hormones. There are also plastid mutants of *P. patens*, whose morphology is influenced by exogenous cytokinin.[57] The latter is covered in detail in Chapter 14 and hence will not be outlined here. In order to follow the different mutants and their present classification, it is necessary to provide a brief outline of *P. patens* development; a more detailed description can be found elsewhere.[58]

Protonemal filaments develop directly from the germinating spore, by the continued growth and division of the filament apical cell. The first cells produced following spore germination are rich in chloroplasts and have perpendicular cross-walls (chloronemata). Later, cells with fewer chloroplasts, oblique cross-walls, and greater vacuolation (caulonemata) are produced. The subapical cells of chloronemal filaments can also divide and give rise to side branches of chloronemata. Subapical cells of caulonemal filaments divide to give rise to an initial cell that has one of four fates: it can give rise to a chloronemal filament, a caulonemal filament, or a leafy shoot (gametophore; see Figure 4), or it can quiesce. The cytokinins interact with light at a number of developmental stages, including the production of the gametophores. The latter is phytochrome mediated, will not occur in darkness, and represents a change in growth form from filamentous to one with radial symmetry. Protonemal filaments and leafy shoots together form

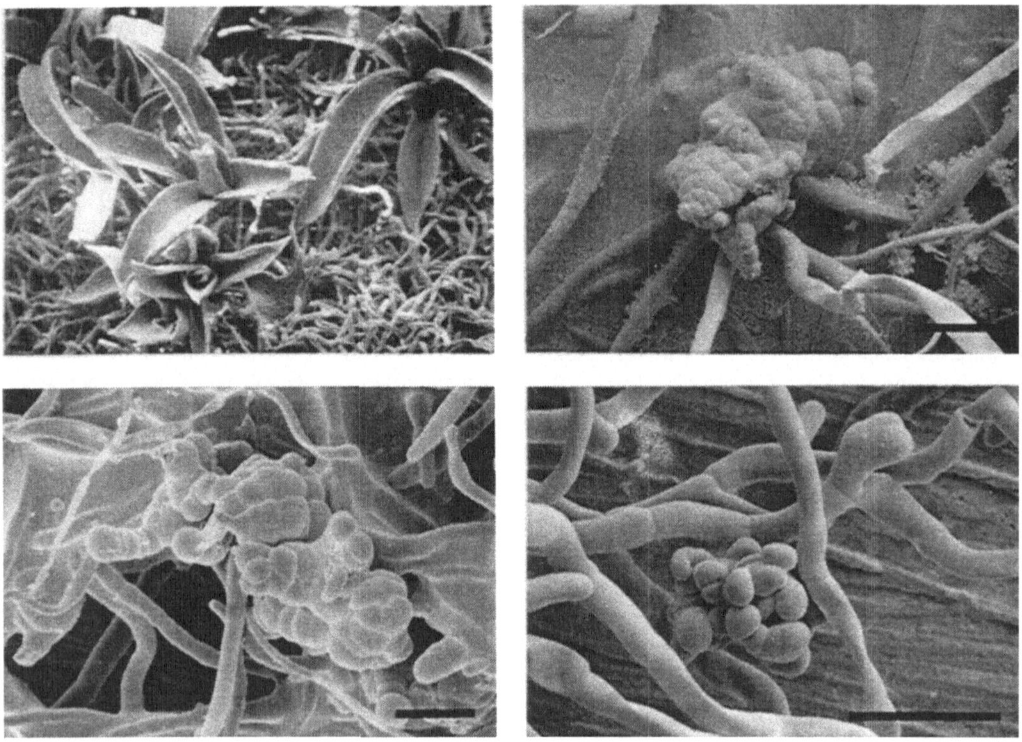

Figure 19-4 Morphology of moss gametophores. Wild type (top left), wild type treated with cytokinin (top right), OVE (bottom left), and cytokinin-repaired *gam*⁻ (bottom right) strains were observed by scanning electron microscopy of gold sputter-coated frozen gametophytic tissue. Cytokinin treatment consisted of the addition of BA (final concentration 1 μ*M*) to a 3-week old culture on cellophane-overlaid agar. (Bar = 100 μm.)

the gametophyte, which is the dominant generation of the life cycle and is haploid. The leafy shoots bear the sex organs and gametes. Following fertilization, the zygote develops to give rise to the diploid sporophyte. The moss gametophyte, therefore, represents a relatively simple type of development where each step in cell differentiation can be readily recognized and observed directly in living tissue. Furthermore, the promotory effect of cytokinin on leafy-shoot formation is akin to the effect this hormone has on shoot formation in tobacco tissue cultures.[20]

1. Mutant Designations

Ashton et al. described the isolation of a number of auxin- and cytokinin-resistant mutants in auxotrophic backgrounds.[23] The spores were treated with EMS or methyl-nitroso-nitrosoguanidine and screened on either NAA or BA, at various concentrations from 5 to 500 μ*M*. Eight categories of mutants were recognized, one resistant to both NAA and BA (1), two (2 and 3) resistant to NAA, and five (4–8) resistant to BA. Each category was designated by its morphology, resistance, and repair. A further two categories of mutants (9 and 10) were also isolated. These overproduced gametophores and were designated OVE mutants. Subsequently, some category-1 mutants were found to be repaired by BA under red light and were redesignated category-2A mutants, which necessitated an alteration to 2B of the category-2 mutants.[59] These categories are complicated and may even contain members that represent alleles at the same locus. Hence, they have now been reassigned gene symbols relating them more closely to their phenotypes rather than their mode of selection:[58] those that are blocked in the formation of caulonemata (*cal*⁻) and consist almost entirely of chloronemata; those that cannot form gametophores (*gam*⁻); those that overproduce gametophores (*ove*).

2. The *ove* Mutants

The *ove* mutants are the only cytokinin mutants retrieved by total isolation without any selection pressure. They were recovered simply on the basis of their morphologies following mutagenesis.[24] All overproduce gametophores, which fall into two types: either slightly or grossly abnormal. Those with grossly abnormal gametophores, when grown in the presence of wild-type plants, induce those plants to produce abnormal gametophores themselves. The phenotypes of the OVE mutants, therefore, are reminiscent of the wild type treated with cytokinin at various levels. Furthermore, when cultured under continuous medium feeding, OVE mutants revert to a wild-type phenotype, under these conditions — a colony consisting of chloronemata only — indicating leachable substances are required to achieve both the OVE and wild-type phenotypes.[58,60]

Genetic analysis of the mutants with an extreme phenotype (category 9) indicated three complementation groups (A–C), whereas those with less-severe phenotypes (category 10) fell into a fourth group (D; Table 2).[6,52,61] Several other mutants could not be assigned clearly to groups, for a number of reasons; in the case of one, because it was a dominant phenotype.[61] This analysis indicated that at least three genes can be mutated to produce the extreme phenotype and one can be mutated to bring about the less-severe type. Since most of these mutants are recessive and they occur at a relatively high frequency, a loss of function in the gene product is the simplest explanation. In addition, the wild type treated with cytokinin (1 μM or above) is a phenocopy of the OVE mutant, which indicates that it is due to an elevation in the level of cytokinin produced by the moss. This has indeed proved to be the case for some of these mutants.

Two cytokinins were isolated from the culture medium of OVE mutants. These were identified by gas chromatography-mass spectrometry (GC-MS) as isopentenyladenine and zeatin, the former being, by far, the main one.[62] A representative from each of the complementation groups A–C was analyzed and shown to secrete a high level of cytokinin into the culture medium: a hundredfold that of the wild type in the most extreme instance (Table 3). All three groups were also shown to incorporate significant amounts (Table 3) of radiolabeled adenine into isopentenyladenine, though incorporation was almost undetectable in the wild type.[63] Cytokinin was only detected in OVE tissue, using single-ion monitoring (SIM) GC-MS or radioimmunoassay,[64] since by calculation even the best cytokinin overproducer would yield very little cytokinin from its tissue compared to its culture medium.[53] The concentration of cytokinin based on labeling experiments, however, indicates that it is at much higher values in the tissue than in the culture medium.[63] Using OVE mutants, however, it was possible to immunolocalize cytokinin[65] in moss tissue where it exists at a high concentration in the cell wall. In contrast, *Agrobacterium*-induced tumor tissue in the same study showed cytokinin mainly in the cytoplasm.

An interesting feature of cytokinin production in moss was revealed by the isolation of an OVE mutant that was assigned to complementation group A. This mutant was isolated conventionally as an extreme OVE, but was thought to have reverted when stored at low temperature (15°C). Subsequently, the mutant (409) was found to be a temperature-sensitive overproducer.[66] A full analysis revealed that both wild-type and OVE mutants are temperature sensitive for cytokinin production (Table 3) and that production in OVE409 at low temperature was within the wild-type range, while at a higher temperature (24°C), it fell within the OVE range. Furthermore, its production indicated that only about a fivefold increase (as reflected in secreted cytokinin) is required for the OVE phenotype to be expressed. This finding is supported by data for OVE200,[53] which at lower temperatures (20°C) produces levels similar to OVE409 at 24°C, but retains the OVE phenotype (Table 3).

Table 19-2 **Complementation Groups of Gametophore-Overproducing Mutants of *Physcomitrella patens*[a]**

Group	Category[24]	Mutant Number[b]
A	9	OVEA78, OVEA102, OVEA131, OVEA201, OVEA409
B	9	*ove*B100, OVEB300, OVEB302
C	9	*ove*C200
D	10	*ove*D90, *ove*D97, *ove*D99, *ove*D107A
Unassigned[c]	9	OVE79, OVE134, OVE202
Unassigned[d]	9	OVE130
Unassigned[e]	10	OVE701

[a] Data from Wang, T. L., in *Cytokinins — Plant Hormones in Search of a Role*, Horgan, R. and Jeffcoat, B., Eds., British Plant Growth Regulator Group, Bristol, U.K., 1987, 55; Cove, D. J., Wang, T. L., and Laloue, M., unpublished data, 1992, and Cove, D. J., et al., in *The Plant Genome. Proc. IV John Innes Symp.*, Davies, D. R. and Hopwood, D. A., Eds., John Innes Charity, Norwich, U.K., 1980, 231; [b] Lower-case italics indicate single Mendelian recessive gene; [c] Not in Group A; [d] Not in Group B; [e] Dominant mutation.

Table 19-3 **Cytokinin Production and Incorporation of Labeled Adenine into Isopentenyladenine by Gametophore-Overproducing Mutants of *Physcomitrella patens***

Strain	Growth Temperature °C	Cytokinin Content (nM)[a]	% Incorporation of ^{14}C-adenine[b]
Wild type	25	<1[c]	nd
	24	0.42[d]	–
	20	0.08[e]	–
	15	0.07[d]	–
OVEA78	25	100[c]	0.47
	24	62.3[d]	–
	20	35.8[e]	–
	15	15.5[d]	–
OVEA409	24	2.3[d]	–
	15	0.19[d]	–
OVEA201	25	60[c]	0.12
*ove*B100	25	25[c]	–
*ove*C200	25	23[c]	0.16
	20	1.3[e]	–

[a] Cytokinin content of the culture medium accumulated over a 3–4 week period; [b] Percent incorporation of ^{14}C-adenine into ^{14}C-isopentenyladenine, measured following TLC and HPLC with crystallization to constant specific activity. (A dash indicates not assessed; nd, none detected.) (Data from Wang, T. L., Horgan, R., and Cove, D., *Plant Physiol.*, 68, 735, 1981.); [c] Quantified by UV absorption of purified compounds prior to identification by GCMS chemical ionisation analysis using direct probe insertion.[61]; [d] Quantified by multiple-ion-monitoring ACE GC-MS of permethyl derivatives, using ^{15}N-labeled internal standards.[65]; [e] Quantified by multiple-ion-monitoring CI GC-MS of permethyl derivatives, using ^{15}N-labeled internal standards.[52]

The data above clearly indicate that there is an increased production of cytokinin in OVE mutants. The bases of the lesions, however, are not known, but the possibilities have been discussed in detail elsewhere.[24,53,62] For reasons given earlier, a loss of function is likely. For this to be the case and for there to be a direct effect, one would expect either loss of catabolic activity in the OVE or loss of a repression of biosynthesis. From preliminary data,[53] differences in the metabolism of isopentenyladenine in wild type and OVE cannot account for the overproduction of cytokinin. Furthermore, differences in tRNA turnover also cannot account for this overproduction.[67] A simple but indirect route that could be effected in several ways because of the complexity of the particular biosynthetic pathway (and hence accommodate the several complementation groups identified) would be a general increase in purine anabolism, which is manifest only in the cytokinins through the fact that these N^6-substituted purines possess unique biological activity.[62]

3. The *cal*⁻ Mutants

The *cal*⁻ mutants produce little or no caulonemata and, hence, cannot produce gametophores. Very little work has been carried out on these mutants, but at least three subclasses are recognized.[23,58,59] All of them show a certain degree of resistance to either NAA or BA, since none can be restored to the wild-type phenotype though caulonemal development, and some gametophore production can be achieved. One group is totally resistant to either hormone and must represent, therefore, defects in genes coding for the perception of the hormones or in the signal transduction chain to caulonemata formation.

4. The *gam*⁻ Mutants

Again, this class of mutant has been little studied. It contains at least two subclasses, one of which can be restored to the wild-type phenotype with the addition of cytokinin (Figure 4). Since this class produces both chloronemata and caulonemata, it has been proposed that it represents cytokinin nonsynthesizing mutants.[23] It has proved to be very difficult to measure the hormone in these mutants,[64,68] but bearing in mind that there is little enough produced by the wild type, this is hardly unexpected and, in part, corroborates the original hypothesis.

The other class of *gam*⁻ mutants is totally resistant to cytokinin application. They produce both filamentous cell types, but will not produce gametophores. The level of cytokinin in one representative of this class, BAR1, is similar to the wild type[53] and so, too, is its capacity to bind cytokinin.[6] This class should contain mutants unable to perceive cytokinin and those in which gene(s) involved in gametophore formation are mutated.

IV. PERSPECTIVE

It is obvious in this review that there are only a handful of mutants that can be counted as "cytokinin mutants" *sensu strictu* in clear contrast to the other groups of hormones. Some of the reasons have been touched on, but primarily it appears to be our lack of knowledge about the cytokinins that impairs the isolation of mutants. Those mutants that have been isolated remain in the preliminary phase of investigation; to date, most analysis has been minimal and primarily descriptive. The only exception to this is the amount of information that has been obtained through the unique OVE mutants of *P. patens*.

The majority of the mutants described above have been single-gene recessives, though the cytokinin mutants, like those of the other groups of hormones,[69] have representatives within each of the three moss classes that have dominant phenotypes.[23,61] Single-gene recessives often represent loss of function, but it is interesting that in the mutants analyzed so far,[35,53,64] loss of catabolic activity was not detected. Most of the mutants, therefore, are likely to be in the regulation of biosynthesis or the mode of action chain. Bleeker pointed out that many of the dominant mutants isolated for plant hormones show pleiotropic effects and are likely to be mutants early in the perception and signal transduction chain.[69] Nevertheless, this cannot be the rule, since the *aba1* (CKR1) cytokinin-resistant mutant is a single-gene recessive and exhibits pleiotropy (see above). A much simpler explanation has been put forward to explain the phenotype of this particular mutant.[27]

A priority must be, therefore, the isolation of further mutants on a biorational basis. Conditional mutants have proved to be very difficult to obtain, as mentioned in Section II, and thus, if cytokinin is required for normal cell division and growth, mutants that would otherwise be lethal will be difficult to isolate. Obvious and relatively straightforward targets, however, from our current knowledge, would be greening phenomena (including senescence) and apical dominance. The involvement of cytokinins in the latter has been reinforced by results on transgenics where overexpression of the *IPT* gene leads to a loss of apical dominance.[1,70] Although a screen for apical dominance mutants is just as likely to turn up a mutant defective in auxin biosynthesis or one with a defective apical meristem, mutants which overproduce cytokinin, and thus lack dominance, should be obtainable, judging by the precedence set by the moss OVE mutants.

Further, *Arabidopsis* greening mutants are being sought at present by Chory and co-workers:[14] in this instance, mutants that remain etiolated even in the presence of cytokinin and are thus "cytokinin-insensitives." The use of *Arabidopsis* has obvious molecular biological advantages and opens up the possibility of direct gene isolation by chromosome walking, genomic subtraction, or gene tagging,[71,72] though in the case of the mutants isolated so far, none have come from tagged germplasm. In physiological terms, however, *Arabidopsis* has many disadvantages mainly because we know almost nothing of its cytokinin relations. Targeted screening for mutants, therefore, is much impeded. If one were to choose a higher plant for cytokinin mutant isolation per se, radish might be more appropriate, since there is considerable knowledge of its cytokinin physiology and metabolism,[73] and it is used as a bioassay for both greening and cell expansion, so that biorational screening can be undertaken. Furthermore, it is diploid and has a rapid cycling time.

The moss *P. patens*, although a lower plant, produces both auxin and cytokinin, and its development is regulated by auxin and cytokinin. Recent advances[74,75] mean that molecular techniques can be applied readily to this organism; both transformation and antisense technology are currently being applied.[76,77] Furthermore, predicted improvements in transformation rates, by the use of biolistics, may mean that direct gene cloning by complementation may become feasible, especially with *gam*⁻ mutants that, when repaired, will be readily recognizable at high plating densities against a protonemal background. If this is the case, then coupled with the ease of isolation of mutants, these techniques will provide *P. patens* with distinct molecular and cellular advantages over several other systems.

ACKNOWLEDGMENTS

I should like to thank Joanne Chory, Stephen Howell, Marc Jullien, and Ottoline Leyser for supplying me with unpublished data to include in this article and for allowing me to reproduce their photographs. I should like to thank David Cove for his kind advice on the manuscript, also Roy Davies for reading the text, and SPB Academic Publishing and Plant Physiology for allowing me to reproduce Figures 1 and 2.

REFERENCES

1. **Klee, H. and Estelle, M.,** Molecular genetic approaches to plant hormone biology, *Ann. Rev. Plant Physiol. Plant Mol. Biol.*, 42, 529, 1991.
2. **King, P. J.,** Plant hormone mutants, *Trends Genet.*, 4, 157, 1988.

3. **Reid, J. B.,** The genetic control of growth via hormones, in *Plant Hormones and Their Role in Plant Growth and Development*, Davies, P. J., Ed., Kluwer, Dordrecht, 1988, 318.
4. **Reid, J. B.,** Phytohormone mutants in plant research, *J. Plant Growth Regul.*, 9, 97, 1990.
5. **Scott, I. M.,** Plant hormone response mutants, *Physiol. Plant.*, 78, 147, 1990.
6. **Wang, T. L.,** Cytokinin mutants, in *Cytokinins — Plant Hormones in Search of a Role, Monograph 14*, Horgan, R. and Jeffcoat, B., Eds., British Plant Growth Regulator Group, Bristol, 1987, 55.
7. **Schmulling, T., Schell, J., and Spena, A.,** Construction of cytokinin overproducing mutants in higher plants, in *Physiology and Biochemistry of Cytokinins in Plants*, Kamínek, M., Mok, D. W. S., and Zazímalová, E., Eds., SPB Academic Publishing, The Hague, 1992, 87.
8. **Horgan, R. and Jeffcoat, B.,** Eds., *Cytokinins — Plant Hormones in Search of A Role*, British Plant Growth Regulator Group, Bristol, 1987.
9. **Cline, M. G.,** Apical dominance, *Bot. Rev.*, 57, 318, 1991.
10. **Letham, D. S.,** Cytokinins, in *Phytohormones and Related Compounds: A Comprehensive Treatise*, Letham, D. S., Goodwin, P. B., and Higgins, T. G. V., Eds., Elsevier/North Holland, Amsterdam, 1978, 205.
11. **Letham, D. S.,** A comparison of the activities of zeatin and other cytokinins in five bioassays, *Planta*, 74, 228, 1967.
12. **Osborne, D. J. and McCalla, D. R.,** Rapid bioassay for kinetin and kinins using senescing leaf tissue, *Plant. Physiol.*, 36, 219, 1961.
13. **Fletcher, R. A. and McCullagh, D.,** Cytokinin-induced chlorophyll formation in cucumber cotyledons, *Planta*, 101, 88, 1971.
14. **Chory, J., Aguilar, N., and Peto, C. A.,** The phenotype of *Arabidopsis thaliana det1* mutants suggests a role for cytokinins in greening, in *Molecular Biology of Plant Development*, Jenkins, G. I. and Schuch, W., Eds., The Company of Biologists, Cambridge, 1991, 21.
15. **Linsmaier, E. M. and Skoog, F.,** Organic growth factor requirements of tobacco tissue cultures, *Physiol. Plant.*, 18, 100, 1965.
16. **Meins, F., Jr., Foster, R., and Lutz, J.D.,** Evidence for a Mendelian factor controlling the cytokinin requirement of cultured tobacco cells, *Dev. Genet.*, 4, 129, 1983.
17. **Meins, F., Jr. and Foster, R.,** A cytokinin mutant derived from cultured tobacco cells, *Dev. Genet.*, 7, 159, 1986.
18. **Miller, C. O.,** Naturally occurring cytokinins, in *Biochemistry and Physiology of Plant Growth Substances*, Wightman, F. and Setterfield, G., Eds., Runge Press, Ottawa, 1968, 33.
19. **Letham, D. S.,** A cytokinin bioassay using excised radish cotyledons, *Physiol. Plant.*, 25, 391, 1971.
20. **Skoog, F. and Miller, C. O.,** Chemical regulation of growth and organ formation in plant tissue cultures, in *The Biological Action of Growth Substances*, Porter, H. K., Ed., Cambridge University Press, Cambridge, 1957, 118.
21. **Brandes, H. and Kende, H.,** Studies on the cytokinin-controlled bud formation in moss protonemata, *Plant Physiol.*, 43, 827, 1968.
22. **Hahn, H. and Bopp, M.,** A cytokinin test with high specificity, *Planta*, 83, 115, 1968.
23. **Ashton, N. W., Grimsley, N. H., and Cove, D. J.,** Analysis of gametophytic development in the moss, *Physcomitrella patens*, using auxin and cytokinin resistant mutants, *Planta*, 144, 427, 1979.
24. **Ashton, N. W., Cove, D. J., and Featherstone, D. R.,** The isolation and physiological analysis of mutants of the moss, *Physcomitrella patens*, which over-produce gametophores, *Planta*, 144, 437, 1979.
25. **Conrad, K.,** A sensitized *Amaranthus* cytokinin bioassay, *Biochem. Biophy. Pflanzen.*, 165, 531, 1974.
26. **Elliott, D. C.,** Ionic regulation for cytokinin dependent betacyanin synthesis in *Amaranthus tricolor* seedlings, *Plant Physiol.*, 63, 264, 1979.
27. **Horgan, R.,** Present and future prospects for cytokinin research, in *Physiology and Biochemistry of Cytokinins in Plants*, Kamínek, M., Mok, D. W. S., and Zazímalová, E., Eds., SPB Academic Publishing, The Hague, 1992, 3.
28. **Cove, D. J.,** personal communication, 1992.
29. **Rousselin, P., Kraepiel, Y., Maldiney, R., Miginiac, E., and Caboche, M.,** Characterization of three hormone mutants of *Nicotiana plumbaginifolia*: evidence for a common ABA deficiency, *Theor. Appl. Genet.*, 85, 213, 1992.
30. **Wilson, A. K., Pickett, F. B., Turner, J. C., and Estelle, M.,** A dominant mutation in *Arabidopsis* confers resistance to auxin, ethylene and abscisic acid, *Mol. Gen. Genet.*, 222, 377, 1990.
31. **Blonstein, A. D., Parry, A. D., Horgan, R., and King, P. J.,** A cytokinin-resistant mutant of *Nicotiana plumbaginifolia* is wilty, *Planta*, 183, 244, 1991.

32. **Parry, A. D., Blonstein, A. D., Babiano, M. J., King, P. J., and Horgan, R.,** Abscisic-acid metabolism in a wilty mutant of *Nicotiana plumbaginifolia*, *Planta*, 183, 237, 1991.
33. **Bitoun, R., Rousselin, P., and Caboche, M.,** A pleiotropic mutation results in cross-resistance to auxin, abscisic acid and paclobutrazol, *Mol. Gen. Genet.*, 220, 234, 1990.
34. **Koornneef, M., Jorna, M. L., Brinkhorst-van der Swan, D. L. C., and Karssen, C. M.,** The isolation of abscisic acid (ABA) mutants by selection of induced revertants in nongerminating gibberellin sensitive lines of *Arabidopsis thaliana* (L.) Heynh., *Theor. Appl. Genet.*, 61, 385, 1982.
35. **Jullien, M., Lesueur, D., Laloue, M., and Caboche, M.,** Isolation and preliminary characterisation of cytokinin-resistant mutants of *Nicotiana plumbaginifolia*, in *Physiology and Biochemistry of Cytokinins in Plants*, Kamínek, M., Mok, D. W. S., and Zazímalová, E., Eds., SPB Academic Publishing, The Hague, 1992, 157.
36. **Hansen, C. E. and Meins, F., Jr.,** Evidence for a cellular gene with potential oncogenic activity in plants, *Proc. Natl. Acad. Sci. U.S.A.*, 83, 2492, 1986.
37. **Su, W. and Howell, S. H.,** Single genetic locus, *Ckr1*, defines *Arabidopsis* mutants in which root growth is resistant to low concentrations of cytokinin, *Plant Physiol.*, 99, 1569, 1992.
38. **Pickett, F. B., Wilson, A. K., and Estelle, M.,** The aux1 mutation of *Arabidopsis* confers both auxin and ethylene resistance, *Plant Physiol.*, 94, 1462, 1990.
39. **Lincoln, C., Britton, J. H., and Estelle, M.,** Growth and development of the *axr1* mutants of *Arabidopsis*, *Plant Cell*, 2, 1071, 1990.
40. **Chory, J., Peto, C., Pratt, L., and Ausubel, F.,** *Arabidipsis thaliana* mutant that develops as a light-grown plant in the absence of light, *Cell*, 58, 991, 1989.
41. **Chory, J., Nagpal, P., and Peto, C. A.,** Phenotypic and genetic analysis of *det2*, a new mutant that affects light-regulated seedling development in *Arabidopsis*, *Plant Cell*, 3, 445, 1991.
42. **Chory, J.,** personal communication, 1992.
43. **Leyser, H. M. O. and Furner, I. J.,** Characterisation of three shoot apical meristem mutants of *Arabidopsis thaliana*, *Development*, 116, 397, 1992.
44. **Medford, J. I., Behringer, F. J., Callos, J. D., and Feldman, K. A.,** Normal and abnormal development in the *Arabidopsis* vegetative shoot apex, *Plant Cell*, 4, 631, 1992.
45. **Crespi, M., Messens, E., Caplan, A. B., van Montagu, M., and Desomer, J.,** Fasciation induction by the pathogen *Rhodococcus fascians* depends upon the linear plasmid encoding a cytokinin synthase gene, *EMBO J.*, 11, 795, 1992.
46. **Murai, N., Skoog, F., Doyle, M. E., and Hanson, R. S.,** Relationships between cytokinin production, presence of plasmids, and fasciation caused by strains of *Corynebacterium fascians*, *Proc. Natl. Acad. Sci. U.S.A.*, 77, 619, 1980.
47. **Blixt, S.,** Mutation genetics in *Pisum*, *Agri. Hort. Genet.*, 30, 1, 1972.
48. **Leyser, H. M. O.,** personal communication, 1992.
49. **Miller, C. O., Skoog, F., Okumura, F. S., Von Saltza, M. H., and Strong, F. M.,** Isolation, structure and synthesis of kinetin, a substance promoting cell division, *J. Am. Chem. Soc.*, 78, 1373, 1956.
50. **Gorton, B. S. and Eakin, R. E.,** Development of the gametophyte in the moss, *Tortella caespitosa*, *Bot. Gaz.*, 119, 31, 1957.
51. **Cove, D. J., Wang, T. L., and Laloue, M.,** Unpublished data, 1992.
52. **Featherstone, D. R.,** Studies of mutants of the moss which over-produce gametophores, Ph.D. thesis, Cambridge University, Cambridge, 1980.
53. **Wang, T. L., Futers, T. S., McGeary, F., and Cove, D. J.,** Moss mutants and the analysis of cytokinin metabolism, in *The Biosynthesis and Metabolism of Plant Hormones*, Crozier, A. and Hillman, J. R., Eds., Cambridge University Press, Cambridge, 1984, 135.
54. **Speiss, L.,** Comparative activity of isomers of zeatin and ribosylzeatin on *Funaria hygrometrica*, *Plant Physiol.*, 55, 583, 1975.
55. **Whitaker, B. D. and Kende, H.,** Bud formation in *Funaria hygrometrica*: a comparison of the activities of three cytokinins and their ribosides, *Planta*, 121, 93, 1974.
56. **Knight, C. D. and Cove, D. J.,** Genetic analysis of tropic responses, *Environ. Exp. Bot.*, 29, 57, 1989.
57. **Abel, W. O., Knebel, W., Koop, H.-U., Marienfeld, J. R., Quader, H., Reski, R., Schnepf, B., and Spörlein, B.,** A cytokinin sensitive mutant of the moss, *Physcomitrella patens*, defective in chloroplast division, *Protoplasma*, 152, 1, 1989.
58. **Cove, D. J.,** Regulation of development in the moss, *Physcomitrella patens*, in *Development — The Molecular Genetic Approach*, Russo, V. E. A., Brody, S., Cove, D. J., and Ottolenghi, S., Eds., Springer-Verlag, Berlin, 1992, 179.

59. **Ashton, N. W., Cove, D. J., Wang, T. L., and Saunders, M. J.,** Developmental studies of *Physcomitrella patens* using auxin and cytokinin sensitivity mutants, in *Plant Growth Substances 1988*, Pharis, R. P. and Rood, S. B., Eds., Springer-Verlag, Berlin, 1990, 57.
60. **Cove, D. J., Ashton, N. W., Featherstone, D. R., and Wang, T. L.,** The use of mutant strains in the study of hormone action and metabolism in the moss *Physcomitrella patens*, in *The Plant Genome. Proc. IV John Innes Symp.*, Davies, D. R. and Hopwood, D. A., Eds., John Innes Charity, Norwich, UK, 1980, 231.
61. **Featherstone, D. R., Cove, D. J., and Ashton, N. W.,** Genetic analysis by somatic hybridization of cytokinin overproducing developmental mutants of the moss, *Physcomitrella patens, Mol. Gen. Genet.*, 222, 217, 1990.
62. **Wang, T. L., Horgan, R., and Cove, D.,** Cytokinins from the moss *Physcomitrella patens*, *Plant Physiol.*, 68, 735, 1981.
63. **Wang, T. L., Beutelmann, P., and Cove, D. J.,** Cytokinin biosynthesis in mutants of the moss *Physcomitrella patens*, *Plant Physiol.*, 68, 739, 1981.
64. **Wang, T. L.,** unpublished data, 1984.
65. **Eberle, J., Wang, T. L., Cook, S., Wells, B., and Weiler, E. W.,** Immunoassay and ultrastructural localization of isopentenyladenine and related cytokinins using monoclonal antibodies, *Planta*, 172, 289, 1987.
66. **Futers, T. S., Wang, T. L., and Cove, D. J.,** Characterisation of a temperature-sensitive gametophore over-producing mutant of the moss, *Physcomitrella patens*, *Mol. Gen. Genet.*, 203, 529, 1986.
67. **Perry, K. and Cove, D. J.,** Transfer RNA pool sizes and half lives in wild-type and cytokinin over-producing strains of the moss, *Physcomitrella patens*, *Physiol. Plant.*, 67, 680, 1986.
68. **Futers, T. S.,** The mode of action of cytokinin in the moss, *Physcomitrella patens*, Ph.D. thesis, University of Leeds, Leeds, UK, 1984.
69. **Bleeker, A. B.,** The use of dominant mutations in the study of hormonal responses in plants, in *Progress in Plant Growth Regulation*, Karssen, C. M., van Loon, L. C., and Vreugdenhil, D., Eds., Kluwer, Dordrecht, 1992, 28.
70. **Medford, J. I., Horgan, R., El-Sawi, Z., and Klee, H. J.,** Alterations of endogenous cytokinins in transgenic plants using a chimeric isopentenyl transferase gene, *Plant Cell*, 1, 403, 1989.
71. **Schmidt, R. and Dean, C.,** Physical mapping of the *Arabidopsis thaliana* genome, in *Genome Analysis, Vol IV: Strategies for Physical Mapping*, Davies, K. E. and Tilghman, S. M., Eds., Cold Spring Harbor Laboratory Press, Cold Spring Harbor, NY, 1992, 71.
72. **Coupland, G.,** Transposon tagging in *Arabidopsis*, in *Methods in Arabidopsis Research*, Koncz, C., Chua, N.-H., and Schell, J., Eds., *World Sci. Press*, Singapore, 1992, 290.
73. **Letham, D. S. and Palni, L. M. S.,** The biosynthesis and metabolism of cytokinins, *Ann. Rev. Plant Physiol.*, 34, 163, 1983.
74. **Cove, D.J., Kammerer, W., Knight, C. D., Leech, M. J., Martin, C. R., and Wang, T. L.,** Developmental studies of the moss, *Physcomitrella patens*, in *Molecular Biology of Plant Development*, Jenkins, G. I. and Schuch, W., The Company of Biologists, Cambridge, 1991, 21.
75. **Wang, T. L., Leech, M. J., Martin, C. R., Knight, C. D., Kammerer, W., and Cove, D. J.,** Molecular genetic approaches to studying cytokinin action in moss, in *Physiology and Biochemistry of Cytokinins in Plants*, Kamínek, M., Mok, D. W. S., and Zazímalová, E., Eds., SPB Academic Publishing, The Hague, 1992, 149.
76. **Schaefer, D., Zryd, J.-P., Knight, C. D., and Cove, D. J.,** Stable transformation of the moss, *Physcomitrella patens*, *Mol. Gen. Genet.*, 226, 418, 1991.
77. **Leech, M. J., Kammerer, W., Cove, D. J., Wang, T. L., and Martin, C. R.,** Expression of *myb*-related genes in the moss, *Physcomitrella patens*, *Plant J.*, 3, 51, 1993.

Chapter 20

Habituation of Cultured Cells for Cytokinins

Frederick Meins, Jr.

TABLE OF CONTENTS

I. INTRODUCTION

Cultured plant tissues sometimes lose their exogenous requirement for specific growth factors and thereafter can be serially subcultured in the absence of the added factor. This phenomenon, discovered by Gautheret in 1942,[1] is usually called *habituation*.[2] Habituation has attracted the interest of students of genetics and developmental biology, for several reasons. First, a key event in plant tumor transformation is the loss by cells of their requirement for growth hormones, suggesting that habituation might provide insight into how the capacity for neoplastic growth is established.[3] Second, stable changes are also an important feature of normal plant development, and habituation provides a model system for studying cell determination under precisely controlled conditions in culture.[4] Finally, comparative studies of habituated and hormone-requiring cell lines have helped to unravel the complex relationship between the response of cells to hormones and hormone metabolism.

Based on studies of cultured carrot tissues, Gautheret proposed that auxin habituation results from some form of stable physiological modification, i.e., a developmental or *epigenetic* modification.[5] Other early workers took the alternative view and speculated that habituation has as its basis classic genetic mutations.[2,6] Insight into the contribution of these two types of heritable modification has come from detailed studies of habituation for cytokinins.

This chapter deals with cytokinin habituation and related phenomena that occur during normal and neoplastic development. The discussion begins with the nature of the heritable change, then turns to genetic, developmental, and physiological factors influencing habituation, and ends with the proposal of molecular mechanisms that can account for both the stability and reversibility of the habituation process.

II. THE NATURE OF THE STABLE CHANGE

A. ALTERATIONS IN CELLULAR HEREDITY

Explants of tobacco pith parenchyma tissue require exogenous auxin and cytokinin for sustained

0-8493-6252-0/94/$0.00+$.50

growth in culture on an otherwise complete medium containing inorganic salts, sucrose, and vitamins; i.e., the tissues exhibit an auxin- and cytokinin-requiring phenotype (A^-C^-).[7,8] Fox[9] identified two types of stable hormone variants from cultured tobacco pith tissue:[9] (1) cytokinin habituated lines that exhibit a cytokinin-independent, auxin-dependent phenotype (A^-C^+); and (2) fully habituated lines that exhibit a cytokinin- and auxin-independent phenotype (A^+C^+).

In principle, the stable changes that gave rise to these variants could reflect the selection of variant cell types present at low frequencies in the original tissue explant, patterns of cell–cell interactions that lead to stable alternative states, or modifications in cellular heredity that are transmitted to daughter cells during mitosis.[10] By cloning cells from habituated tissues and establishing that the clones express the same phenotype as the parent tissue, it was shown that cytokinin-habituated[11,12] and fully habituated[13] variants of tobacco result from an alteration in cellular heredity.

One striking result of these cloning experiments was the finding that individual clones varied widely in their relative growth rate on media with and without the hormones for which they were habituated.[11,12] Therefore, habituated tissues consist of a heterogeneous population of cells differing in the degree of habituation. Meins and Binns[14] monitored the degree of habituation of numerous subclones derived from a cloned line of cytokinin-habituated tobacco pith tissue. The average degree of habituation of the subclones was approximately equal to the value obtained with the parent clone, indicating that there is no cell selection during cloning and, hence, that the frequency of clones reflects the cellular constitution of the tissues. The distribution of subclones gradually increased in degree of habituation with prolonged culture of the parent cloned line. These results show that cytokinin habituation is a gradual, progressive process at the cellular level.

B. EPIGENETIC MODIFICATION VS. GENETIC MUTATION

Two types of mechanisms have been proposed for the heritability of the habituated state: *genetic mutation* and *epigenetic change*. Classic genetic mutations result from rare, nondirected alterations in genetic constitution, such as point mutations, deletions, and gross chromosomal changes. Mutations usually occur at very low rates per cell generation, are not readily reversible, and are transmitted meiotically.[15]

The concept of epigenetic change has its origins in developmental biology and has come to denote cellular heritable alterations associated with developmental events that do not result in permanent genetic changes. Epigenetic changes differ from genetic mutations, in several important ways:[10]

1. The heritable cellular change is directed; i.e., it occurs regularly in response to specific inducers at rates greater than 10^{-3} per cell generation.
2. The variant phenotype, although mitotically transmissible, is potentially reversible at high rates.
3. The range of phenotypes generated by epigenetic changes is limited by the genetic potential of the cell.
4. By definition, epigenetic changes are not transmitted meiotically.

Convincing evidence that epigenetic changes can account for cytokinin habituation has come from studies of pith cells of Havana 425 tobacco, which habituate rapidly under inductive conditions, can be cloned readily, and remain totipotent for many cell generations in culture. Primary explants of Havana 425 pith habituate for cytokinins when incubated on auxin-containing medium at 35°C, approximately 10°C above the standard culture temperature, or in the presence of trace amounts of the cytokinin kinetin.[16,17] Under these conditions some of the explants form dense, white, hyperplastic nodules clearly discernible from the surrounding translucent tissue. Using the nodules as a marker, it was possible to estimate a minimum rate of habituation.[16] The rate found, $>5 \times 10^{-3}$ per cell generation, is 100 to 1000 times faster than the rates reported for somatic mutation in *Nicotiana* species.[18,19]

The tendency of tobacco pith to habituate is affected by the physiological and developmental state of the tissue. The incidence of habituation in primary explants after 35°C treatment depends on the season of the year.[16] The incidence is high in the spring and greatly reduced in autumn and winter, for pith from greenhouse-grown plants. There are also striking gradients in the incidence of induced and spontaneous habituation in pith tissue sampled from different parts of the plant.[17,20,21] The incidence is generally very high in young, growing regions near the top of the plants and drops toward the bottom of the plant. Tissue near the periphery of mature pith adjacent to the internal phloem habituates more readily than does tissue at the center of the pith.

Normal, fertile plants can be regularly regenerated from clones of cytokinin-habituated pith cells.[12] Pith tissues from these regenerates, like untreated seed-grown plants, once again exhibit a non-habituated phenotype, indicating that cytokinin habituation is a potentially reversible process. These results have been independently confirmed using cloned lines derived from Wisconsin 38 tobacco.[22] Reversion to the nonhabituated state, like habituation, occurred at very high rates, $\approx 4 \times 10^{-3}$ per cell generation, during the regeneration process.[23] This observation and the fact that habituation is extremely rapid and depends on both the physiological and developmental state of pith tissue provide strong evidence that cytokinin habituation results from epigenetic changes rather than from mutation.

C. VARIATION IN THE COMPETENCE OF CELLS TO HABITUATE

Cells from different parts of the tobacco plant vary in their requirement for cytokinin and in their competence to habituate under inductive conditions. Cloning experiments revealed that primary pith tissue consists of two types of cells: *inducible* C^- cells that habituate rapidly when incubated at 35°C, and *noninducible* C^- cells that do not habituate under these conditions.[24] Primary explants of stem cortex, on the other hand, grow rapidly on medium without added cytokinin, indicating that this tissue exhibits a *constitutive* C^+ phenotype.[25] Finally, primary explants of leaf lamella grow very slowly, or not at all, on cytokinin-free medium. In contrast to pith explants, these tissues do not habituate under conditions that induce the rapid habituation of competent pith cells.[25] Therefore, three distinct states of cytokinin requirement can be discerned in tobacco: an "off" state, i.e., the noninducible C^- phenotype that is represented by leaf tissues and some pith cells; an "on" state, i.e., the constitutive C^+ phenotype, represented by stem-cortex tissue; and a "regulated" state, the inducible C^- phenotype, that is represented by some pith cells.

Cell cloning and plant regeneration tests show that these different states have an epigenetic basis. Of clones isolated from noninducible C^- pith and leaf tissue and C^+ tissues of stem-cortex, $\approx$90 to 95% exhibit the cytokinin requirement of the tissue from which they were derived.[24,25] Although a few of the clones appear to lose their tissue-specific phenotype, the great majority do not, indicating that these states of cytokinin requirement are inherited by individual cells. These results show that alterations in cellular heredity can occur during development of the tobacco plants, to give cells exhibiting different cytokinin requirements in culture. To find out whether these changes are potentially reversible, plants were regenerated from cloned lines of C^- pith and leaf cells and C^+ cortex cells, and the cytokinin requirement of the tissues from the regenerates was assayed. Regardless of the clone used as a source for the regenerated plant, the pith, cortex, and leaf tissues exhibited the phenotype expected for comparable tissues from untreated, seed-grown plants.[26,27]

At some point during the plant regeneration process, different epigenetic states of cytokinin requirement are "reset" to give tissue types that express the "correct" phenotype. Plant regeneration is accomplished by subculturing tissues on a medium containing the cytokinin kinetin, but no exogenous auxin. Under these conditions a small fraction of cells in the tissue give rise to initials that eventually form buds.[28] In the case of C^+ clones, these newly formed buds, 2 to 4 mm in length, show a C^- phenotype, even though cells cloned from the surrounding unorganized tissue still exhibit a C^+ phenotype.[23] These results suggest that the resetting process occurs at some stage during organogenesis, rather than as a direct response to cytokinin treatment.

Early studies of habituation emphasized the difference between "normal" cells, which require specific growth factors, and hormone-autotrophic habituated cells.[29] We now recognize that the habituated phenotype can be expressed by callus tissues derived from primary explants; i.e., the C^+ phenotype is "normal" for certain tissue types. Thus, rapid habituation in culture appears to reflect types of stable changes that also occur during normal development in the plant. Cells derived from tissues in different parts of the plant can differ both in their cytokinin requirement and their tendency for habituation (Figure 1). The picture that emerges is that these cells express alternative developmental states of determination that are stable, but potentially reversible, and leave the cell totipotent. The "regulated" state characteristic of some pith cells appears to be rather unstable, and cells of this type habituate readily under inductive conditions. On the other hand, the "on" and "off" states are extremely stable, at least on the tissue level. They persist for long periods when tissues are serially subcultured, and changes in state occur primarily during organogenesis.

III. GENETIC FACTORS IN CYTOKININ HABITUATION

A. SPECIFIC GENE LOCI

Epigenetic changes involve alteration in the expression of preexistent genetic potentialities of the cell.[30] Therefore, in principle, karyotypic alterations that commonly occur when plant cells are serially propagated in culture could also give rise to variants altered in their cytokinin requirement. Moreover, it may be possible to obtain mutations in

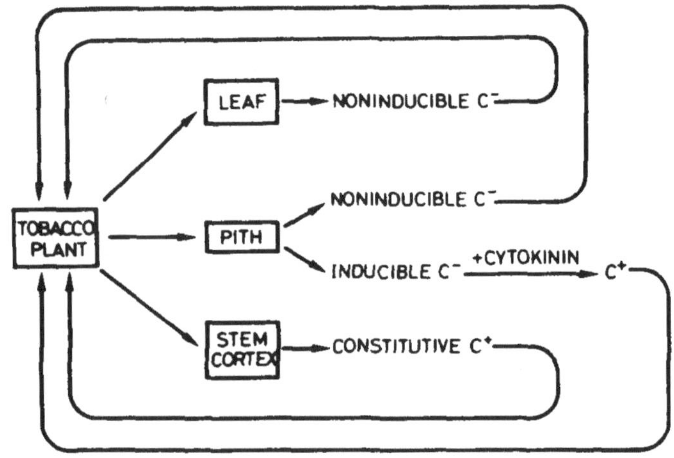

Figure 20-1 Summary of epigenetic changes in the state of the cytokinin requirement of tobacco cells established by cloning and plant regeneration. (From Meins, F., Jr., in *Molecular Basis of Plant Development*, Goldberg, R., Ed., Alan R. Liss, New York, 1989, 13. With permission.)

specific genes that affect the cytokinin requirement or tendency for habituation. One well-documented case is the rapid habituation for cytokinin and auxin, observed for tissues of tumor-prone, interspecific hybrids of *Nicotiana*,[31,32] described in Section IV.C, below. The important point is that cytokinin habituation can have different proximal genetic and epigenetic causes.

Early studies aimed at investigating the correlation between habituation and alterations in karyotype were largely inconclusive.[33] No correlation was found between ploidy level and the degree of habituation or capacity for reversion of cytokinin-habituated pith cells of tobacco.[34] On the other hand, a detailed study of cloned, fully habituated tobacco lines showed that a centric chromosome fragment, m, is correlated with the tumorigenicity of tissues in grafting tests,[35,36] which presumably reflects the degree of hormone autotrophic growth.

Of greater interest are specific genetic loci implicated in cytokinin habituation. The first of these loci was identified in a survey of numerous genotypes of *Phaseolus vulgaris* and *P. lunatus*.[37] In these studies cultures were established from the hypocotyls of seedlings and scored for cytokinin requirement. Of the ten *P. vulgaris* genotypes screened, nine showed a cytokinin autotrophic phenotype in culture. In contrast, of the ten *P. lunatus* phenotypes screened, six showed a strictly cytokinin-dependent phenotype. These results clearly show that there are varietal differences in cytokinin requirement and, hence, that this phenotype is genetically regulated. The results of sexual crosses between strictly C^+ and C^- lines of *P. vulgaris* were consistent with the hypothesis that the cytokinin requirement is regulated primarily at a single nuclear locus, but that additional loci may affect the degree of habituation and the growth rate of the tissues in culture.[37]

The tendency of bean tissues to habituate for cytokinin is also under genetic regulation.[38] Mok et al.[39] showed that *P. lunatus* tissues could be induced to habituate for cytokinin by treatment with the cell-division factor diphenylurea (DPU), but not with the cytokinin kinetin at the concentrations tested. Five of 16 genotypes screened were not responsive to DPU, nor did they habituate spontaneously in culture.

Two genes regulating cytokinin habituation have been identified in tobacco. The first of these, *Habituated leaf-1 (Hl-1)*, is a partially dominant nuclear locus.[40] In contrast to wild-type plants, which exhibit tissue-specific patterns of cytokinin requirement, pith, leaf, and stem-cortex tissues of *Hl-1/Hl-1* plants exhibit a C^+ phenotype in culture. Thus, the *Hl-1* allele appears to convert the cytokinin phenotype of leaf and pith cells to that of stem-cortex cells. In spite of this dramatic effect in culture, homozygous plants grow and develop normally under greenhouse conditions. The only effects detectable are a slightly reduced yield of seed and a delay in flowering.

Mutations at a second *Habituated leaf* locus, designated *Hl-2*, were recovered by subculturing cloned lines of noninducible C^- leaf cells, on media with reduced concentrations of cytokinin.[41] After two or three transfers, C^+ lines were obtained, and leaf tissues from plants derived from the cloned C^+ lines exhibited a C^+ phenotype when returned to culture. This trait segregated, in breeding tests, in the manner expected for a single, dominant Mendelian factor and was inherited at a locus different from that for the Hl-1 trait.[41] In terms of gross morphology, seed set, and growth rate, homozygous *Hl-2/Hl-2* plants are indistinguishable from wild-type plants.

A number of mutants have been described that could provide tools for studying the physiological and genetic basis for habituation. Included in this group are hormone auxotrophic mutants, mutants showing altered sensitivity of tissues to exogenous hormones,[42-44] and hormone-autotrophic mutants generated by transforming cells with chimeric genes encoding cytokinin- and auxin-biosynthetic genes of Ti-plasmid origin.[44] Of more direct bearing on habituation are mutants generated by introducing host genes tagged with T-DNA. Hayashi et al. transformed auxin-requiring tobacco protoplasts with a T-DNA vector containing a plant-expressed drug resistance marker and multiple transcriptional enhancers derived from the Cauliflower Mosaic Virus 35S RNA promoter.[45] Insertion of these enhancers into the plant genome would be expected to drive the high-level expression of adjacent host genes. By selecting for drug-resistant auxin autotrophic transformants, it was possible to identify a specific gene that is overexpressed in the autotrophic mutant and promotes auxin autotrophic growth when introduced into wild-type protoplasts.

B. PSEUDODIRECTED VARIATION

The discovery of the Hl-2 trait in tobacco was of particular interest because it demonstrated that cytokinin habituation can result from genetic changes at a specific locus as well as from epigenetic changes in the same plant species. Studies of this trait also provided insight into variation that occurs in a "grey zone" between mutation and epigenetic change. One puzzling aspect of Hl-2 was the ease with which mutants could be recovered from cultured materials, suggesting that this mutation might result from a directed process. The important criterion for a directed process is the rate at which the event occurs on a per-cell-generation basis. Unfortunately, in the case of cytokinin habituation, these rates cannot be measured directly because proliferation of tissues is required to distinguish between the C^+ and C^- phenotypes.

One approach to this problem, which is not entirely satisfactory, is to estimate rates of variation from considerations of the dynamics of the cell population. For an exponentially dividing population, the frequency of C^+ cells (q) as a function of time depends on (1) the doubling constants of C^- and C^+ cells (k_1 and k_2, respectively), (2) the rate of conversion of C^- cells to C^+ cells (μ), (3) the rate of reversion of C^+ cells to C^- cells (ν), and (3) the frequency of C^+ cells at the start of the experiment. By growing tissues on media that favor the relative growth of C^- or C^+ cells, the population should eventually reach an equilibrium in which the rates of interconversion and rates of proliferation are balanced. When there is strong selection for C^- cells,

$$\mu \approx \hat{q}S$$

where $\hat{q}$ is the equilibrium frequency and S is the selection coefficient defined as $S = 1 - (k_2/k_1)$. Similarly, when there is strong selection for C^+ cells,

$$\nu \approx -S(1 - \hat{q})/(1 - S)$$

In preliminary measurements, a value for the forward rate μ of about 5×10^{-3} conversions per cell generation was obtained for three cloned lines of C^- leaf cells.[46a] More recent experiments with cloned lines of C^- and C^+ leaf cells led to the surprising conclusion that both the forward (μ) and backward (ν) rates of variation are on the order of 10^{-2} per cell generation.[46b]

The finding that different C^+ and C^- states of leaf cells can persist in regenerated plants and are then inherited in crosses according to well-established laws of heredity leads to the suggestion that this rapid form of variation has as its basis some form of unconventional genetic modification. Genetic variation most commonly results from rare random events rather than from directed changes.[47,48] Variation in the cytokinin requirement of tobacco leaf cells occurs so rapidly that the distinction between random and induced events is blurred. Seemingly directed changes induced by cytokinins at the tissue level appear to result from selection for the different cell types acting on an interconverting population of the two cell types. Possible mechanisms for this form of *pseudodirected* variation will be discussed later.

IV. CYTOKININ HABITUATION AND NEOPLASTIC GROWTH

A. CROWN GALL DISEASE

Several neoplastic diseases are known in plants: crown gall and hairy root, which are caused by plasmids of bacterial origin;[49,50] Black's wound tumor, caused by a double-stranded RNA virus;[51] and Kostoff genetic tumors, which result in certain interspecific sexual hybrids.[52] Regardless of the proximal cause, the common feature of these plant neoplasias is that acquisition of the capacity for growth autonomy is associated with a persistent loss in the requirement of the cultured tumor cells for auxin and cytokinin.[53] Thus, in a broad sense, habituation can be viewed as a type of neoplastic transformation. Indeed, some, but not all, fully habituated tissues that arise as spontaneous variants are tumorigenic in grafting tests.[33]

In the case of crown gall, which has been studied in greatest detail, there is considerable evidence that the key evident in neoplastic transformation is the acquisition of the capacity of cells for cytokinin autotrophic growth. Crown gall tumors result when a portion of the Ti plasmid carried by *Agrobacterium tumefaciens*, the T-DNA, is integrated into the nuclear genome of the host cell.[54] Several genes in the T-DNA encode enzymes involved in the metabolism of auxins and cytokinins. Of particular importance are the loci *tms1* and *tms2*, which encode enzymes for indole-3-acetic acid synthesis, and the *ipt* locus, which encodes isopentenyl transferase, a key enzyme in cytokinin biosynthesis. Deletion analyses of T-DNA provide strong evidence that the expression of these genes has a major role in determining the pattern of neoplastic growth and development.[55,56]

Nutritional studies of a graded series of progressively transformed *Vinca rosea* tissues showed that the acquisition of cytokinin autotrophy is correlated with the onset of neoplastic growth.[57] Transformation of tobacco cells with the nopaline-plasmid pTiT37 results in teratomas that are fully hormonal independent in culture and tumorigenic in grafting tests.[58] In contrast, tobacco cells transformed with a mutant pTiT37 plasmid defective at the *ipt* locus require cytokinin, but not auxin, for growth in culture and are not tumorigenic.[58,59] These results lead to the conclusion that cytokinin autotrophy is regulated at the *ipt* locus and that the acquisition of cytokinin autotrophy is the key event in tumor initiation.

Tumor reversal experiments indicate that epigenetic mechanisms ultimately determine whether or not the genetic predisposition for neoplastic growth in crown gall transformants is realized.[60] When the cut stem of tobacco plants is inoculated with the T37 strain of *A. tumefaciens*, organoid teratomas form, consisting of numerous abnormal leaves and shoots. Cloning experiments have established that individual teratoma cells are multipotential and give rise to the tissue types and organs characteristic of the teratoma.[61] When the cloned tissues are grafted onto the cut-stem tip of tobacco plants from which the axillary buds have been removed, they give rise to shoots that become progressively more normal in function, gross morphology, and histological detail.[61,62] Nevertheless, these shoots still contain T-DNA,[63] do not form roots, and are immune to *A. tumefaciens* transformation. Unlike comparable normal tissues, explants cultured from diploid flower parts, pith, and leaves of the suppressed teratoma shoots are hormone independent in culture and revert to the typical teratomatous morphology. In more detailed experiments, individual epidermal hair cells and all but 3 of 760 clones of leaf mesophyll cells obtained from the suppressed tumor shoots were found to revert to the tumor phenotype, when cultured.[64,65] Therefore, crown gall transformants can differentiate, participate in organogenesis, and exhibit normal growth control while still retaining their neoplastic character in a covert state.

Early anatomic studies by Smith[66] established that crown gall tumors of the stem arise most frequently from cortex tissue at the site of bacterial inoculation. This tissue, which in tobacco exhibits the constitutive C^+ phenotype,[25] also forms spontaneous tumors in suppressed tumor shoots and is responsible for the abnormal structure of the stem early in tumor reversal.[67] These findings suggest that expression of T-DNA genes essential for tumorigenesis depends on epigenetic regulation of host genes concerned with cytokinin autotrophic growth.

More direct evidence for this hypothesis comes from studies of mutants of the nopaline-type pTiT37 plasmid defective at the *ipt* locus.[59] Tobacco cells transformed with these *ipt*$^-$ plasmids can form fertile, normal appearing plants. Nevertheless, cortex tissue explanted from the regenerated plants exhibit a hormone-independent and neoplastic phenotype in culture. Explants of pith tissue also retain the potential for neoplastic growth, but only express this potentiality after being cultured under conditions that induce the cytokinin-habituation normal pith tissue. Therefore, the epigenetic C^+ phenotype can complement the defective *ipt*$^-$ locus of the T-DNA, to restore the tumor phenotype.

Using a similar strategy, Hansen and Meins compared the properties of wild-type tobacco and *Hl-1/Hl-1* tobacco cells transformed with wild-type pTiT37 plasmid and an *ipt*$^-$ derivative.[58] Whereas *hl-1/hl-1* cells carrying the *ipt*$^-$ T-DNA were not tumorous, *Hl-1/Hl-1* cells carrying this plasmid showed the three characteristics typical of neoplastic crown gall cells: full hormone autotrophy in culture, inability to form plants, and tumorigenicity in grafting tests. These findings lead to the conclusion that the *Hl-1* allele has oncogenic functions similar to those of the *ipt* gene in T-DNA. Nevertheless, it is clear that Hl-1 is not a cellular homolog of *ipt*: no plant DNA sequences have been found to cross-hybridize with *ipt* sequences, *Hl-1* does not replace *ipt* functions in tumor inception, and the presence of the *Hl-1* allele does not result in the marked elevation in cytokinin levels typical of *ipt*-transformed cells.[68]

Further insight into the function of the *Hl-1* locus has come from transfection experiments[69a] with the *tms1* and *tms2* genes from a limited-host-range Ti plasmid.[70] Cloned lines of *Hl-1/Hl-1 tms1 tms2* transformants exhibit a fully hormone autotrophic phenotype in culture; however, unlike the *Hl-1/Hl-1 ipt⁻* transformants described above, these transformants are not tumorigenic in grafting tests and can be induced to form complete, normal-appearing, fertile plants. Control plants derived from transformants of wild-type cells occasionally formed small overgrowths at wound sites.[69b] In striking contrast, wounding *Hl-1/Hl-1 tms1 tms2* plants consistently induced the formation of large "rooty" tumors. Therefore, combinations of foreign auxin genes and the *Hl-1* allele result in a tumor-prone phenotype. This phenotype is characteristic of certain interspecific *Nicotiana* hybrids[52] and also arises in *Nicotiana rustica* plants overexpressing the *6b onc* gene[71] of the Ti plasmid.[72]

In summary, expression of the tumor phenotype in crown gall requires both a genetic predisposition for neoplastic growth, acquired by integration of foreign genes, and stable epigenetic regulation of these genes. Specific host genetic and epigenetic states affecting the cytokinin requirement of the untransformed cells appear to be functional equivalents of specific T-DNA genes such as the *ipt* and *6b onc* loci.

B. HAIRY ROOT

Genes with functions similar to those of the *tms* and *ipt* loci of the Ti plasmid are responsible for the neoplastic phenotype in hairy root plants. Hairy root tumors are highly organized root teratomas induced by Ri plasmids carried by virulent strains of *Agrobacterium rhizogenes*.[73] During tumor inception with agropine strains of the bacterium, two regions of the plasmid, T_L- and T_R-DNA, are integrated into the genome of the host cells.[50,74,75] Ri-plasmid transformed roots from many plant species form complete plants when cultured *in vitro*.[76,77] These aberrant plants, which are analogous to suppressed tumor shoots in crown gall, exhibit the hairy-root syndrome: reduced apical dominance of roots and stems; wrinkled, elongated leaves; high rates of root growth in culture; and a loss of root geotropism.[77] Unlike suppressed tumor shoots, however, the aberrant phenotype and T-DNA sequences are transmitted into the next seed generation. Hairy root plants can also revert to a completely normal phenotype.[78] This phenotype is stably transmitted meiotically and by vegetative propagation and results from the persistent inactivation of T-DNA gene expression, rather than from the loss or rearrangement of T-DNA sequences.

Mutational analysis of the Ri plasmid[79,80] and the introduction of Ri genes, alone or in combination, into plants[81-84] has shown that T_R-DNA carries two loci functionally and structurally homologous to the *tms1* and *tms2* loci of the Ti plasmid.[80,85,86] Although there is some variability depending on the host plant, the genes responsible for the hairy root syndrome — designated *rolA*, *rolB*, *rolC*, and *rolD*[80]— are located in the T_L-DNA.[81-83,87] There is considerable interaction between *rol* genes, and the most pronounced phenotype occurs in transformants containing the full set of genes, but individual genes introduced into plants can also give the aberrant phenotype, e.g., *rolA* in tomato[84] and *rolB* in tobacco.[81]

The expression of *rol* genes mimics certain physiological and morphogenetic effects of auxins and cytokinins.[82] These genes have cytokinin-like effects on tumor growth: they can complement a defective *ipt* locus of the Ti plasmid to restore tumoricity in *Kalanchoe diagremontiana*,[80] even though there is no sequence similarity between the genes and the *ipt* locus.[85] The *rol* genes also have auxin-like effects: they increase the tendency for root initiation, increase the rate of root growth, and increase the sensitivity of organs and cells to exogenous auxins.[83,88,89] There is a particularly striking effect on the response of the membrane resting potential to exogenous auxins. Protoplasts from tobacco plants transformed individually with *rolA* and *rolB* show an approximately 1000-fold increase in auxin sensitivity, whereas *rolC* has a somewhat smaller 10- to 100-fold effect.[90]

A more direct effect of *rol* genes on hormone metabolism is suggested by studies in which the genes were expressed in *E. coli* and the encoded proteins assayed for enzymatic activity. Using this approach, evidence was obtained that *rolC* encodes a cytokinin-β-glucosidase.[91] *RolB* encodes an activity that can release indoles from indoxyl-β-glucosides. It has been proposed that because these substrates are similar to IAA glucosides, they may be important in mobilizing stored forms of auxin.[92]

The available evidence leads to the conclusion that *rol* genes have some functional equivalent to the *ipt* locus of the Ti plasmid in tumorigenicity.[80] It is not known if this functional equivalence extends to host genes important in cytokinin habituation.

C. GENETIC HYBRID TUMORS

Studies of genetic tumors provide further evidence for the concept that tumorigenesis depends both on genetic predisposition and epigenetic changes involving hormone autotrophy. Genetic tumors, which have been described for many plant families, most commonly arise in interspecific hybrids.[31] The tumors of *Nicotiana* hybrids have been studied in detail.[31,32,52,93] In the case of *N. glauca* x *langsdorffii* hybrids (GGLL), early development of the plant is quite normal. Later, after the onset of flowering, teratomatous overgrowths arise on the plants, usually on regions of the stem subject to stress, such as wound sites and leaf scars. These teratomas have their origin in mature cortical cells just below the epidermis. Although tumor inception can be promoted by various chemicals and irradiation, there is no evidence that an infectious agent is involved.

With few exceptions, *Nicotiana* species can be classified into two groups called "plus" and "minus."[94] Hybrids formed between species within each group are nontumorous, whereas hybrids formed between species in different groups form tumors. The available evidence suggests that one parent in the cross contributes a small number of genetic factors, or a single genetic factor, controlling tumor initiation, and the other parent contributes a number of genetic factors that influence the degree of tumor expression.[95-97]

Cultured lines established from tumor-prone *Nicotiana* hybrids exhibit a fully habituated phenotype in culture.[98] The conditions required for establishing these lines differ for normal-appearing hybrid tissues and tissues explanted from spontaneous teratomas: teratoma tissues exhibit a constitutive habituated phenotype, whereas the normal-appearing tissues require an initial treatment with auxin to trigger habituation.[99] Nevertheless, in the case of GGLL, cultured tissues of both types have been shown to be tumorigenic when grafted onto the *N. glauca* parent.[100,101] Therefore, teratomatous and normal-appearing tissue from tumor-prone plants have the potential for neoplastic growth.

The relationship between hormone autotrophy in culture, hormone content, and expression of the tumor phenotype in the *Nicotiana* hybrids is complex.[31] Cultured tumor-prone hybrid tissues contain elevated levels of auxins and cytokinins[101-103] and exhibit a fully habituated phenotype. Bayer and Ahuja made crosses of the nontumorous hybrid *N. debneyi* x *N. tabacum* and the tumorous hybrid *N. debneyi-tabacum* x *N. langiflora*.[104] Indole-3-acetic acid content of tissues, and expression of the tumor phenotype, was correlated in the segregants, indicating that, at least in the case of auxin, elevated hormone levels and tumorigenicity have the same genetic basis. However, hormone autonomy of the hybrids cannot be accounted for by a simple model in which parental contributions of the capacity for auxin- and cytokinin-autotrophic growth are additive.[98] Similar conclusions may be drawn from experiments in which leaf disks from *N. langsdorffii*, *N. glauca*, and the tumor-prone GGLL hybrid were transformed with Ti plasmids defective at the *tms* and *ipt* loci.[105] A high incidence of fully habituated calli was obtained with *N. glauca* cells transformed with either the tms^-ipt^+ plasmid or the tms^+ipt^- plasmid. In contrast, only the wild-type tms^+ipt^+ plasmid gave rise to fully habituated calli in the *N. langsdorffii* host. Evidently, neither the contribution of parental genes nor complementation of these genes with T-DNA genes for auxin and cytokinin biosynthesis is additive.

Although predisposition for neoplastic development depends on combining genes of the appropriate parental species, tumor initiation appears to be a potentially reversible epigenetic phenomenon. Parasexual hybrids can be obtained by fusing protoplasts prepared from leaves of *N. glauca* and *N. langsdorffii*.[106,107] The resultant hybrid clones can grow on a hormone-free medium that does not support growth of cloned lines derived from either parent. The hybrid clones, which exhibit the typical teratoma phenotype, can be induced to form shoots, which when grafted onto *N. glauca* plants, develop normally, eventually flower and set seed, but spontaneously form teratomas. These results show that individual genetic-hybrid cells are totipotent and capable of both neoplastic and normal development.

There are several hints that epigenetic modification of genes concerned with cytokinin autotrophy are important in tumor initiation. A nontumorous mutant that is hormone requiring in culture has been obtained by X-irradiating populations of GGLL.[108] Although some of the data are conflicting, it appears that the nontumorous character is inherited as a monogenic, dominant trait.[109,110] Feng et al. reported that the tumor phenotype could be restored by transforming cells of this mutant with the Ti plasmid *ipt* gene.[103] These transformants were fully habituated in culture, had elevated levels of cytokinin, and exhibited the teratomatous phenotype typical of GGLL tissues. The teratomatous phenotype could also be restored physiologically by maintaining mutant tissues on cytokinin-containing medium, suggesting that the mutant has a lesion affecting cytokinin metabolism.

Of particular interest is the report that the neoplastic phenotype can be restored in a stable fashion by treating the mutant cells with azacytidine, which is known to block the methylation of cytosines in DNA.[111] Reversion was associated with a drop in the methylcytosine content of the bulk DNA, suggesting that the mutant arose as a result of "epimutation,"[112] i.e., an epigenetic, stable alteration in the pattern of DNA methylation.

A further link between genes important for cytokinin-autotrophic growth and the tumor phenotype has come from studies of *N. glauca* homologs of *rol* genes carried by Ri plasmids.

DNA from uninfected *N. glauca* plants was found to contain sequences about 80% identical to part of the T_L region of the Ri plasmid.[75,113] This region in the *N. glauca* genome contains complete, uninterrupted reading frames corresponding to *rolB* and *rolC* and a second copy of the *rolC* open reading frame interrupted by a stop codon. Ichikawa and Syono reported that this region of the *N. glauca* genome also contains sequences homologous to ORF 13 and ORF 14 of the T_L-DNA.[32] Based on RNA-blot analyses, Ichikawa et al. obtained evidence that although the *rolB* and *rolC* homologs are not expressed in *N. glauca*, they are expressed in cultured GGLL hybrid tissues.[114] Moreover, expression of these genes was blocked by exogenous auxin added to the culture medium. This leads to the proposal that the tumor-prone phenotype in GGLL results from the activation of normally silent *rol*-type genes resident in the *N. glauca* genome.

The studies of plant tumor diseases summarized above lead to the concept that neoplastic growth in plants is a multistep process. First, a competent or tumor-prone state is established by mutation, by plasmid-mediated transformation, by interspecific hybridization, or by a combination of these genetic events. Second, cells competent for neoplastic-development must be triggered to express these potentialities in a stable fashion. These changes are potentially reversible and depend on epigenetic mechanisms that appear to regulate hormone-autotrophic growth in normal development (Figure 2). Each tumor disease shows a distinctive pattern of "normal" and neoplastic phases, ranging from fully autonomous, unorganized crown gall tumors to tumor-prone genetic hybrid plants.

V. PHYSIOLOGICAL CHANGES ASSOCIATED WITH CYTOKININ HABITUATION

A. 5-BROMODEOXYURIDINE SENSITIVITY

The thymidine analog 5-bromodeoxyuridine (BUdR), which selectively inhibits differentiation of animal cells,[115] blocks the cytokinin-autotrophic growth of cultured tobacco cells.[116] In the presence of 10^{-4} *M* BUdR, cloned lines of cytokinin-habituated tobacco pith cells show the same cytokinin dose response as nonhabituated cells. Whereas cytokinin-habituated cells on a cytokinin-free medium are as sensitive to BUdR as cytokinin-requiring cells on a cytokinin-containing medium, in the presence of cytokinin the habituated cells are about 100-fold less sensitive than cytokinin-requiring cells.[116] Thus, habituated and nonhabituated cells differ in BUdR tolerance. The degree of tolerance depends on the state of habituation. Cells autotrophic for both auxin and cytokinin, such as fully habituated cells, crown gall cells, and tumor-prone *Nicotiana* hybrid cells, are BUdR tolerant even in

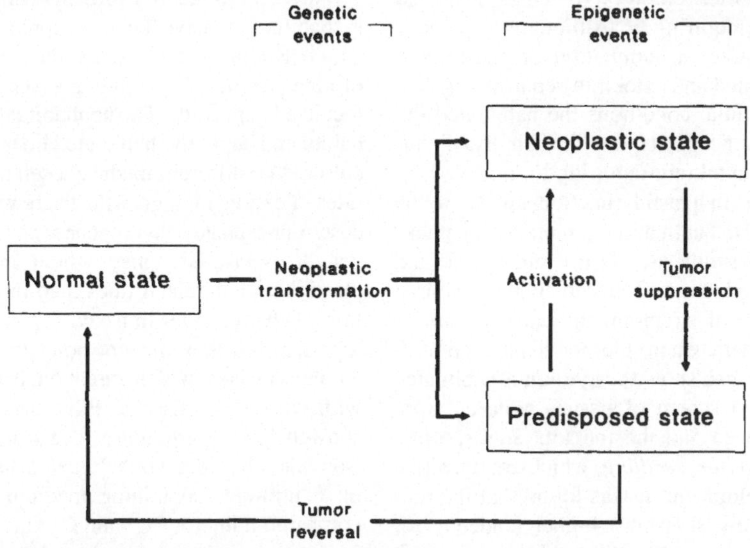

Figure 20-2 The key genetic and epigenetic events in plant neoplastic transformation associated with changes in cytokinin requirement. The predisposed state is the suppressed tumor shoot in crown gall disease, hairy root syndrome or normal appearing plants in hairy root disease, tumor-prone plants in *Nicotiana* genetic hybrids, and *Hl-1/Hl-1* plants transformed with *tms* genes. True tumor reversal with loss of oncogenes has only been demonstrated for crown gall transformants. (From Meins, F., Jr., in *Molecular Basis of Plant Development,* Goldberg, R., Ed., Alan R. Liss, New York, 1989, 13. With permission.)

the absence of added cytokinin.[117,118] Mutants of tobacco selected for BUdR tolerance are not cytokinin autotrophic, suggesting that the cellular changes resulting in BUdR tolerance are a consequence of rather than a cause of cytokinin-autotrophic growth.[119]

When proliferating cells are treated with BUdR, a portion of the thymines in replicating DNA are replaced by 5-bromouracil (BU). Studies with animal cells suggest that incorporation of BU into the cell genome is required for inhibition of differentiated function.[115] Cultured cells exhibiting different states of cytokinin habituation incorporate BU into DNA to roughly the same extent,[120] and treatment with cytokinins does not appear to affect incorporation.[116,120] Bezdek and Vyskot have evidence that BUdR tolerance results from debromination of BU in DNA and that this process is promoted by cytokinins.[120]

B. CHLOROPLAST DIFFERENTIATION AND FUNCTION

It has long been recognized that cytokinin physiology and chloroplast differentiation are linked. For example, cytokinins stimulate plastid differentiation[121] as well as the synthesis of specific chloroplast proteins such as the chlorophyll a/b binding protein.[122] Several studies established a correlation between cytokinin habituation and plastid function. Cultured lines of tobacco pith tissue that are cytokinin habituated or fully habituated accumulate significantly less chlorophyll than do lines that are cytokinin requiring.[123] Although exogenous cytokinin promotes accumulation of chlorophyll by both habituated and cytokinin-requiring tissues, even under optimal conditions the habituated tissues have a lower capacity for chlorophyll accumulation than do cytokinin-requiring tissues.

Deficiencies in plastid function appear to be causally related to habituation. Streptomycin blocks plastid protein synthesis, plastid replication, and chlorophyll production. Tobacco tissues cultured in the presence of streptomycin can give rise to stable variants deficient in chlorophyll production.[124] These variants also show the cytokinin-habituated phenotype. In a screen of tissues cultured from wild-type tobacco and the mutants *Sulfur*, *white seedling*, and *yellow seedling*, which are defective in plastid development, it was found that the mutants consistently showed a higher tendency for cytokinin habituation than did wild-type plants.[125]

Based on observations of this type, Kamínek and Lustinec proposed an explanation for habituation, based on the assumption that plastids are a sink for cytokinins.[123] According to this hypothesis, the primary event in cytokinin habituation is a defect in plastid function. Both cytokinin-requiring and cytokinin-habituated cells have the capacity to produce cytokinins. However, in cytokinin-requiring cells, cytokinins are metabolized (or compartmentalized) in plastids, and hence, steady-state levels of cytokinins are insufficient to support cell proliferation. In cytokinin-habituated cells, on the other hand, defects in plastid function result in steady-state levels of cytokinins, sufficient to support proliferation.

C. MARKER PROTEINS

Understanding the cellular mechanisms for habituation will probably depend on finding suitable molecular markers that show different patterns of expression in habituated and nonhabituated cells. The most commonly used approach, the identification of proteins correlated with habituation, is difficult for several reasons. First, habituation is a physiological trait and can have different proximal causes. For example, there is, at present, no compelling evidence that epigenetic and genetic forms of habituation have the same underlying mechanism. Second, cytokinin-habituated lines exhibit marked clonal differences, and clonal variation is a continuous process in culture. Finally, because the hormone requirements for growth of habituated and nonhabituated lines differ, it is difficult to distinguish between biochemical changes associated with responses to the hormones and intrinsic differences of the two cell types.

Attempts to identify proteins with a causal role in habituation have been disappointing. For example, Kamínek et al. studied the isozyme patterns of nine enzymes,[126] including several known to be localized in plastids. The nonhabituated, cytokinin-habituated, and fully habituated lines sampled were cultured on different media chosen to give optimal rates of exponential growth. There were fewer glucose-6-phosphate dehydrogenase and NADP-malate dehydrogenase isozymes, which are localized in plastids, in habituated lines than in nonhabituated lines. Other changes in isozyme patterns were also reported, but they did not point to specific alterations associated with habituation. In a similar study with sugar beet tissues, fully habituated tissues showed lower peroxidase levels and numbers of isozymes than did nonhabituated tissues cultured on a hormone-containing medium.[127,128] Carlson compared numerous C^+ and C^- lines of Wisconsin 38 tobacco pith tissue cultured on different media.[22] He reported that 4% of the 240 polypeptides analyzed by one- and two-dimensional gel electrophoresis differed in the two phenotypes.

The pitfalls of comparing tissues on different culture media are illustrated by studies of ^{35}S-methionine-labeled proteins extracted from C^+ and

C^- tobacco tissues cultured on combinations of auxin and cytokinin giving optimal growth rates.[129] Exponentially growing, cloned lines of C^-, C^+, fully habituated, and crown gall tissues were assayed. A ^{35}S-labeled polypeptide with an apparent M_r of 33,000 on SDS-polyacrylamide gel electrophoretograms was found in the habituated, but not the nonhabituated, lines. This difference reflects the hormonal constitution of the culture medium used to raise the tissues, rather than the tissue type. The 33-kDa polypeptide, which was later identified as β-1,3-glucanase,[130] is downregulated at the mRNA level by combinations of auxin and cytokinin present in the culture medium. Comparison of tissues incubated for short periods of time on the same, hormone-free medium indicated there is a negative, rather than positive, correlation between β-1,3-glucanase production and expression of a C^+ phenotype. β-1,3-Glucanase is downregulated relative to C^- lines in habituated and crown gall lines,[129] which contain elevated cytokinin levels.[131]

D. CYTOKININ CONTENT

In principle, the cytokinin-autotrophic growth of habituated cells could result from increased steady-state levels of physiologically relevant cytokinins, an increased sensitivity of the cells to endogenous cytokinins, or a combination of both mechanisms. The few comparative studies of cytokinin content in C^- and C^+ tissues are difficult to interpret. Plant cells produce a variety of metabolically interconvertable cytokinins, and it has not been established which of these are relevant for cell proliferation.[132] Lines of cytokinin-habituated, fully habituated, and tumor cells have been shown to exhibit dramatic clonal and physiological variation in cytokinin content.[102,131,133] Finally, comparison of C^+ and C^- tissues in comparable physiological states requires supplying the C^- tissues with cytokinins. Exogenous cytokinins induce cytokinin oxidase activity in cultured tissues, which could result in the degradation of endogenous cytokinins.[134]

Although the literature is conflicting, there is indirect evidence that the C^+ state is associated with an increased potential for cytokinin production.[131] Hansen et al.[131] compared the zeatin riboside content of C^+ lines on auxin medium, and C^- lines on auxin plus cytokinin medium growing at comparable rates. Under these conditions, no zeatin riboside, i.e., <0.5 pmol/g FW, was detected in the 14 cloned lines tested. In striking contrast, when incubated on a hormone-free medium, which does not sustain the growth of either cell type, the C^+ lines, but not the C^- lines, accumulated high levels of zeatin riboside. The media effects appear to be due to auxin, which markedly inhibits cytokinin accumulation, probably by promoting cytokinin degradation.[131,135,136] Thus, C^+ lines have a greater capacity to accumulate zeatin riboside than do the C^- lines. In mixed cultures, C^+ cells have been shown to produce diffusible factors that promote the division of neighboring C^- cells on an auxin-containing medium.[11,137] C^+ cells also have reduced levels of β-1,3-glucanase, which is downregulated by cytokinins.[129] Taken together, the data suggest that in the presence of auxin, C^+ cells have higher steady-state levels of zeatin riboside or functionally related cytokinins than do C^- cells, but these levels are below the limit of detection in the assay used.

The limited data available do not rule out the possibility that the primary event in cytokinin habituation is a change in responsiveness of the cells to cytokinins or the production of signaling molecules downstream of cytokinins in the regulation of cell proliferation. Possible candidates for the signaling molecules are the dehydroconiferyl glucosides (DHGs), which promote cell division in cultured tobacco pith and leaf tissues. DHGs are present in cytokinin-habituated tissues, tissues treated with cytokinins, and cytokinin-producing crown gall tissues, but are present at greatly reduced levels in nonproliferating tissues.[138,139]

VI. A BIOCHEMICAL SWITCH MODEL FOR CYTOKININ HABITUATION

The central developmental problem posed by cytokinin habituation is how different states of cytokinin requirement are propagated in populations of dividing cells. Several conservative genetic mechanisms can, in principle, account for both the stability and potential reversibility of the habituated state.[140] Of particular interest is the modification of cytosines in DNA by methylation. Cell-heritable regulation of gene expression by DNA methylation is well documented.[112] There is also direct evidence implicating DNA methylation in the reversion of "mutant" C^- genetic-hybrid tumor cells to the C^+ state[111] and in the reactivation of *ipt* genes in Ti-plasmid transformants.[141,142] What is yet to be established is a role for DNA methylation in forms of cytokinin habituation that do not involve the introduction of foreign genes into the plant cell.

The cellular inheritance of habituation can also be explained by self-perpetuating regulatory circuits.[140,143] The basic idea is that certain autocatalytic regulatory systems can exist in alternative steady states. The simplest example is one in which a metabolite promotes its own biosynthesis. Assuming there is a mechanism for degrading the metabolite, this positive-feedback loop can exist in an "off" state in which the steady-state level of the metabolite is low, or in an "on" state in which the steady-state level of the metabolite is high. Provided the system persists when cells divide, it is possible for two genetically identical cells to inherit the "on" and "off" states.

A biochemical switch model for cytokinin habituation has been proposed that specifies that the C^+ state is maintained by a positive-feedback loop in which cytokinins, or physiologically related cell-division-promoting substances, either induce their own biosynthesis or inhibit their own degradation.[143,144] This heuristic model is based on several premises:

1. Cells exhibit the C^+ phenotype because they have higher steady-state levels of cytokinins than do C^- cells.
2. The rate of cytokinin synthesis is a biphasic function of cellular cytokinin concentration.
3. Cytokinins are degraded by a first-order process, as has been established by experiments with radiolabeled cytokinins.[145]

Figure 3 shows graphic representations of the dependence of degradation and synthesis on the cytokinin concentration proposed by the model. The points of intersection of the synthesis and degradation curves correspond to steady-state concentrations of cytokinin ($\hat{c}$) where the net rate of cytokinin accumulation is zero. Where and how often the two curves intersect depend on the kinetic constants. Thus, by varying the slope of the degradation curve, three different types of behavior can be obtained (Figure 3A). For large slopes (curve 1), the curves intercept at one point corresponding to a low cytokinin concentration. Analysis of this steady state shows that it is stable; i.e., it is at the bottom of a valley in the stability diagram (Figure 3B). For small slopes (curve 3), the curves intersect at one point corresponding to a high cytokinin concentration; this steady state is also stable. Finally, for intermediate slopes (curve 2), the curves intersect three times. Stability analyses show that these intersections specify three steady states: high and low states, which are stable, separated by an unstable intermediate state.

The graphic representation shows that the model can account for the three cytokinin states observed experimentally:

1. A stable, high-cytokinin phenotype in which cells express a C^+ "on" state.
2. A stable, low-cytokinin phenotype in which cells express a C^- "off" state.
3. A bistable phenotype in which cells can be induced to shift between a C^+ and a C^- state.

The variety and stability of the different phenotypes would depend on the kinetic parameters and, hence, ultimately on the genotype of the cells.

The biochemical switch model makes specific predictions that have been experimentally verified with cultured tobacco tissues. First, inducible C^- cells found in pith tissues should shift to the C^+ phenotype, when treated with exogenous cytokinin. Kinetin induces cytokinin habituation of pith explants. The low concentrations required — 1000-fold less than required for cell division — and

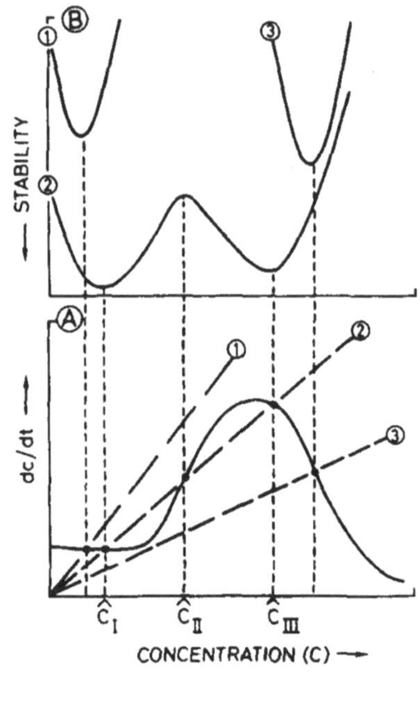

Figure 20-3 Graphic representation of the biochemical switch model for epigenetic changes in cytokinin requirement. (A) rates of cytokinin synthesis (———) and degradation (----) for three different first-order degradation constants (curves 1, 2, and 3) chosen to illustrate different classes of steady states. (B) Stability diagram showing the behavior expected for the steady states obtained for different degradation constants. The stability curves E(c) are defined by the relationship $E(c) = -\int(dc/dt)dc + \text{constant}$.[144] (From Meins, F., Jr., in *Molecular Basis of Plant Development*, Goldberg, R., Ed., Alan R. Liss, New York, 1989, 13. With permission.)

sharp concentration threshold are expected for an autocatalytic system.[16] Second, C^+ pith cells should shift back to the inducible C^- state, under conditions that block cytokinin accumulation. As judged by the cytokinin requirement and capacity to cross-feed neighboring cells, reducing the incubation temperature from 25 to 16°C or reducing the levels of auxin in the culture medium inhibits cytokinin accumulation.[137,146] When C^+ pith tissues are incubated under these nonpermissive conditions and then shifted to permissive conditions, the tissues exhibit an inducible C^- phenotype and revert to a C^+ phenotype when treated with cytokinin.

If cytokinins can act to induce their own production, then treatment with synthetic cytokinins should induce an accumulation by cells of natural cytokinins. This was the case: C^- pith cells and fully habituated cells of tobacco accumulate *trans*-zeatin riboside and isopentenyladenosine, respectively, when cultured on kinetin-containing medium in the absence of auxin.[135] Moreover, using tobacco cells immobilized in columns on a support matrix, it was shown that treatment with the synthetic cytokinin benzyladenine results in a pulse of *trans*-zeatin riboside and *trans*-zeatin, which is released into the bathing medium.[147]

The weakest point in the model is the uncertainty of how cytokinin content is related to the capacity for cytokinin-independent growth of cells, alluded to earlier. The available evidence strongly supports some type of autocatalytic mechanism involving cytokinins, but at present, it is not possible to rule out the possibility that cytokinins other than the ones measured or an entirely different class of cell-division-promoting substances are the key components of the feedback loop.

VII. CONCLUDING REMARKS

Cytokinin habituation involves gradual, progressive changes in cellular heredity. These changes can result from epigenetic changes that are reset during plant regeneration and from genetic alterations that persist through meiosis and are inherited at specific loci. The fact that certain mutations (and epimutations) can mimic epigenetic states of cytokinin requirement is consistent with the concept that habituation ultimately depends on the expression of specific genes. This is further illustrated by tumor transformation resulting from the incorporation of foreign genes or complete chromosome sets into the cell. There is compelling evidence that expression of different cytokinin states in these systems ultimately has an epigenetic basis as well.

The cellular mechanisms for cytokinin habituation are not known. The C^- and C^+ states are associated with differences in BUdR sensitivity, plastid differentiation, and the expression of specific proteins. It has yet to be demonstrated that these differences have a causal role in habituation. There is indirect evidence implicating two types of mechanisms — DNA methylation and biochemical switching — in maintaining states of cytokinin requirement. Both types of mechanisms, which are not mutually exclusive, could be involved. For example, mixed mechanisms involving autocatalytic switching, DNA methylation, and genetic transposition appear to generate the different classes of stability and developmental regulation observed for expression of the *Spm* transposable element in maize.[148] In the case of habituation, DNA methylation may maintain the more stable "on" and "off" states; whereas a purely autocatalytic mechanism may account for the far-less-stable alternation of states exhibited by inducible pith cells. The extreme plasticity predicted by biochemical switch models also provides an appealing explanation for the pseudodirected changes resulting from the rapid alternation of cells between the C^- and C^+ states.

The major limitation in testing these hypotheses is the availability of suitable markers for identifying genes directly involved in habituation. The future looks promising. Specific mutants affecting habituation have been characterized, technology for cloning mapped genes with physiological phenotypes is developing rapidly, and recent results show it is feasible to clone genes responsible for hormone-autotrophic growth.

REFERENCES

1. **Gautheret, R. J.,** Hétéro-auxines et cultures de tissus végétaux, *Bull. Soc. Chim. Biol.*, 24, 13, 1942.
2. **White, P. R.,** Neoplastic growth in plants, *Qt. Rev. Biol.*, 26, 1, 1951.
3. **Braun, A. C.,** The activation of two growth-substance systems accompanying the conversion of normal to tumor cells in crown-gall, *Cancer Res.*, 16, 53, 1956.
4. **Meins, F., Jr. and Binns, A. N.,** Cell determination in plant development, *Bioscience*, 29, 221, 1979.
5. **Gautheret, R. J.,** Sur la variabilité des proportiétés physiologiques des cultures de tissus végétaux, *Rev. Gen. Bot.*, 62, 1, 1955.
6. **de Ropp, R. S.,** The crown-gall problem, *Bot. Rev.*, 17, 629, 1951.
7. **Jablonski, J. R. and Skoog, F.,** Cell enlargement and cell division in excised tobacco pith tissue, *Physiol. Plant.*, 7, 16, 1954.

8. **Naylor, J., Sander, G., and Skoog, F.,** Mitosis and cell enlargement with cell division in excised tobacco pith tissue, *Physiol. Plant.*, 7, 25, 1954.
9. **Fox, J. E.,** Growth factor requirements and chromosome number in tobacco tissue cultures, *Physiol. Plant.*, 16, 793, 1963.
10. **Meins, F., Jr., and Lutz, J.,** Epigenetic changes in tobacco cell culture, in *Emergent Techniques for the Improvement of Crops*, Rubenstein, I., et al., Eds., University of Minnesota Press, Minneapolis, 1980, 220.
11. **Tandeau de Marsac, N. and Jouanneau, J.-P.,** Variation de l'exigence en cytokinine de lignées clonales de cellules de tabac, *Physiol. Veg.*, 10, 369, 1972.
12. **Binns, A. and Meins, F., Jr.,** Habituation of tobacco pith cells for factors promoting cell division is heritable and potentially reversible, *Proc. Natl. Acad. Sci. U.S.A.*, 70, 2660, 1973.
13. **Lutz, A.,** Obtention de plantes de tabac à partir de cultures unicellulaires provenant d'une souche anergée, *C. R. Acad. Sci. D*, 262, 1856, 1966.
14. **Meins, F., Jr. and Binns, A. N.,** Epigenetic variation of cultured somatic cells: evidence for gradual changes in the requirement for factors promoting cell division, *Proc. Natl. Acad. Sci. U.S.A.*, 74, 2928, 1977.
15. **Sparrow, A. H. and Sparrow, R. C.,** Spontaneous somatic mutation frequencies for flower color in several *Tradescantia* species and hybrids, *Environ. Exp. Bot.*, 16, 23, 1976.
16. **Meins, F., Jr. and Lutz, J.,** The induction of cytokinin habituation in primary pith explants of tobacco, *Planta*, 149, 402, 1980.
17. **Meins, F., Jr., Lutz, J., and Foster, R.,** Factors influencing the incidence of habituation for cytokinin of tobacco pith tissue in culture, *Planta*, 150, 264, 1980.
18. **Carlson, P. S.,** Mitotic crossing-over in a higher plant, *Genet. Res.*, 24, 109, 1974.
19. **Sand, S. A., Sparrow, A. H., and Smith, H. H.,** Chronic gamma irradiation effects on the mutable V and stable R loci in a clone of *Nicotiana*, *Genetics*, 45, 289, 1960.
20. **Turgeon, R.,** Cytokinesis, cell expansion, and the potential for cytokinin-autonomous growth in tobacco pith, *Plant Physiol.*, 70, 1071, 1982.
21. **Jackson, J. A. and Lyndon, R. F.,** Cytokinin habituation in juvenile and flowering tobacco, *Plant Physiol.*, 132, 575, 1988.
22. **Carlson, P. S.,** Peptides of normal and variant cells of tobacco, *Dev. Genet.*, 1, 3, 1979.
23. **Meins, F., Jr. and Binns, A. N.,** Rapid reversion of cell-division-factor habituated cells in culture, *Differentiation*, 23, 10, 1982.
24. **Meins, F., Jr., Lutz, J., and Binns, A. N.,** Variation in the competence of tobacco pith cells for cytokinin habituation in culture, *Differentiation*, 16, 71, 1980.
25. **Meins, F., Jr. and Lutz, J.,** Tissue specific variation in the cytokinin habituation of cultured tobacco cells, *Differentiation*, 15, 1, 1979.
26. **Meins, F., Jr. and Foster, R.,** Reversible, cell-heritable changes during the development of tobacco pith tissues, *Dev. Biol.*, 108, 1, 1985.
27. **Meins, F., Jr. and Foster, R.,** Transdetermination of plant cells, *Differentiation*, 30, 188, 1986.
28. **Meins, F., Jr., Foster, R., and Lutz, J.,** Quantitative studies of bud initiation in cultured tobacco tissues, *Planta*, 155, 473, 1982.
29. **Gautheret, R. J.,** The nutrition of plant tissue culture, *Annu. Rev. Plant Physiol.*, 6, 433, 1955.
30. **Nanney, D. L.,** Epigenetic control systems, *Proc. Natl. Acad. Sci. U.S.A.*, 44, 712, 1958.
31. **Bayer, M. H.,** Genetic tumors: physiological aspects of tumor formation in interspecific hybrids, in *Molecular Biology of Plant Tumors*, Kahl, G. and Schell, J.S., Eds., Academic Press, New York, 1983, 33.
32. **Ichikawa, T. and Syono, K.,** Tobacco genetic tumors, *Plant Cell Physiol.*, 32, 1123, 1991.
33. **Meins, F., Jr.,** Habituation of cultured plant cells, in *Molecular Biology of Plant Tumors*, Kahl, G. and Schell, J. S., Eds. Academic Press, New York, 1982, 3.
34. **Binns, A. N. and Meins, F., Jr.,** Chromosome number and the degree of cytokinin habituation of cultured tobacco pith cells, *Protoplasma*, 103, 179, 1980.
35. **Mouras, A. and Lutz, A.,** Denombrement et analyse morphologique des chromosomes chez quelques clones issus d'une souche anergiée de tabac: relation avec l'anergie, le pouvoir tumoral et l'organogenèse, *Genetica*, 50, 195, 1979.
36. **Mouras, A. and Lutz, A.,** Mise en évidence d'une relation entre pouvoir tumoral et présence d'un élément chromosomique marqueur chez des clones issus d'une aner-giée de tabac, *C. R. Acad. Sci. D*, 290, 975, 1980.
37. **Mok, M. C., Mok, D. W. S., Armstrong, D. J., Rabakoarihanta, A., and Kim, S.-G.,** Cytokinin autonomy in tissue cultures of *Phaseolus*: a genotype-specific and heritable trait, *Genetics*, 94, 675, 1980.

38. **Armstrong, D. J., Kim, S.-G., Mok, M. C., and Mok, D. W. S.,** Genetic regulation of cytokinin metabolism in *Phaseolus* tissue cultures, in *Metabolism and Molecular Activities of Cytokinins*, Guern, J. and Péaud-Lenoël, C., Eds., Springer-Verlag, Berlin, 1981, 97.
39. **Mok, M. C., Kim, S. -G., Armstrong, D. J., and Mok, D. W. S.,** Induction of cytokinin autonomy by N,N′-diphenylurea in tissue cultures of *Phaseolus lunatus* L., *Proc. Natl. Acad. Sci. U.S.A.*, 76, 3880, 1979.
40. **Meins, F., Jr., Foster, R., and Lutz, J. D.,** Evidence for a Mendelian factor controlling the cytokinin requirement of cultured tobacco cells, *Dev. Genet.*, 4, 129, 1983.
41. **Meins, F., Jr. and Foster, R.,** A cytokinin mutant derived from cultured tobacco cells, *Dev. Genet.*, 7, 159, 1986.
42. **King, P. J.,** Plant hormone mutants, *Trends Genet.*, 4, 157, 1988.
43. **Reid, J. B.,** Phytohormone mutants in plant research, *J. Plant Growth Regul.*, 9, 97, 1990.
44. **Klee, H. and Estelle, M.,** Molecular genetic approaches to plant hormone biology, *Ann. Rev. Plant Physiol. Plant Mol. Biol.*, 42, 529, 1991.
45. **Hayashi, H., Czaja, I., Lubenow, H., Schell, J., and Walden, R.,** Activation of a plant gene by T-DNA tagging: auxin-independent growth in vitro, *Science*, 258, 1350, 1992.
46a. **Meins, F., Jr.,** Cell heritable changes during development, in *Plant Genetics*, Freeling, M., Ed., Alan R. Liss, New York, 1985, 45.
46b. **Meins, F., Jr.,** Unpublished results.
47. **Hayes, W.,** *The Genetics of Bacteria and their Viruses*, 2nd ed., John Wiley & Sons, New York, 1968.
48. **Cairns, J., Overbaugh, J., and Miller, S.,** The origin of mutants, *Nature*, 335, 142, 1988.
49. **Chilton, M. -D., Drummond, M. H., Merlo, D. J., Sciaky, D., Montoya, A. L., Gordon, M. P., and Nester, E. W.,** Stable incorporation of plasmid DNA into higher plant cells: the molecular basis of crown-gall tumorigenesis, *Cell*, 11, 263, 1977.
50. **Willmitzer, L., Sanchez-Serrano, J., Buschfeld, E., and Schell, J.,** DNA from *Agrobacterium rhizogenes* is transferrred to and expressed in axenic hairy root plant tissues, *Mol. Gen. Genet.*, 186, 16, 1982.
51. **Black, L. M.,** Plant tumors of viral origin, *Prog. Exp. Tumor Res.*, 15, 110, 1972.
52. **Smith, H. H.,** Plant genetic tumors, *Prog. Exp. Tumor Res.*, 15, 138, 1972.
53. **Braun, A. C.,** Plant tumors, *Biochim. Biophys. Acta*, 516, 167, 1978.
54. **Binns, A. N. and Thomashow, M. F.,** Cell biology of *Agrobacterium* infection and transformation of plants, *Ann. Rev. Microbiol.*, 42, 575, 1988.
55. **Ooms, G., Hooykaas, P. J. J., Moolenaar, G., and Schilperoort, R. A.,** Crown gall plant tumors of abnormal morphology, induced by *Agrobacterium tumefaciens* carrying mutated octopine Ti plasmids; analysis of T-DNA functions, *Gene*, 14, 33, 1981.
56. **Garfinkel, D. J., Simpson, R. B., Ream, L. W., White, F. F., Gordon, M. P., and Nester, E. W.,** Genetic analysis of crown gall: fine structure map of the T-DNA by site-directed mutagenesis, *Cell*, 27, 143, 1981.
57. **Braun, A. C.,** A physiological basis for autonomous growth of the crown-gall tumor cell, *Proc. Natl. Acad. Sci. U.S.A.*, 44, 344, 1958.
58. **Hansen, C. E. and Meins, F., Jr.,** Evidence for a cellular gene with potential oncogenic activity in plants, *Proc. Natl. Acad. Sci. U.S.A.*, 83, 2492, 1986.
59. **Binns, A. N.,** Host and T-DNA determinants of cytokinin autonomy in tobacco cells transformed by *Agrobacterium tumefaciens*, *Planta*, 158, 272, 1983.
60. **Meins, F., Jr.,** Tumor reversal and tumor suppression in plants, in *Influence of the Host on Tumor Development*, Herberman, R.B., Ed., Kluwer, Dordrecht, 1989, 37.
61. **Braun, A. C. and Wood, H. N.,** Suppression of the neoplastic state with the acquisition of specialized function in cells, tissues and organs of crown gall teratomas of tobacco, *Proc. Natl. Acad. Sci. U.S.A.*, 73, 496, 1976.
62. **Turgeon, R., Wood, H. N., and Braun, A. C.,** Studies on the recovery of crown gall tumor cells, *Proc. Natl. Acad. Sci. U.S.A.*, 73, 3562, 1976.
63. **Yang, F. -M., Montoya, A., Merlo, D. J., Drummond, M. H., Chilton, M. -D., Nester, E. W., and Gordon, M. P.,** Foreign DNA sequences in crown gall teratomas and their fate during the loss of the tumorous trait, *Mol. Gen. Genet.*, 177, 707, 1980.
64. **Huff, M. and Turgeon, R.,** Neoplastic potential of trichomes isolated from tobacco crown gall teratomas, *Differentiation*, 19, 93, 1981.
65. **Binns, A. N., Wood, H. N., and Braun, A. C.,** Suppression of the tumorous state in crown-gall teratomas of tobacco: a clonal analysis, *Differentiation*, 19, 97, 1981.

66. **Smith, E. F.,** Embryomas in plants (produced by bacterial inoculations), *Johns Hopkins Hosp. Bull.*, 28, 1, 1917.
67. **Turgeon, R.,** Structure of grafted crown-gall teratoma shoots of tobacco: regulation of transformed cells, *Planta*, 153, 42, 1981.
68. **Meins, F., Jr. and Hansen, C. E.,** Elevated *trans*-zeatin riboside content is not required for the neoplastic growth of Ti plasmid-transformed tobacco cells, in *Physiology and Biochemistry of Cytokinins in Plants*, Kamínek, M., Mok, D. W. S., and Zazímalová, E., Eds. SPB Academic Publishing, The Hague, 1992, 83.
69a. **Klee, H. J., Yanofsky, M. F., and Nester, E. W.,** Vectors for transformation of higher plants, *Biotechnol.*, 3, 637, 1985.
69b. **Meins, F., Jr.,** Unpublished results.
70. **Aebi, R.,** Eigenschaften eines Pflanzenhormongenes mit potentiell onkogenen Funktionen: Komplementationsstudien mit einem klonierten Hormongen, Diploma Dissertation, University of Basel, Basel, 1988.
71. **Hooykaas, P. J. J., den Dulk-Ras, H., and Schilperoort, R. A.,** The *Agrobacterium tumefaciens* T-DNA gene 6^b is an *onc* gene, *Plant Mol. Biol.*, 11, 791, 1988.
72. **Tinland, B., Fournier, P., Heckel, T., and Otten, L.,** Expression of a chimeric heat-shock-inducible *Agrobacterium 6b* oncogene in *Nicotiana rustica*, *Plant Mol. Biol.*, 18, 921, 1992.
73. **White, F. F. and Sinkar, V. P.,** Molecular analysis of root induction by *Agrobacterium rhizogenes*, in *Plant DNA Infectious Agents*, Hohn, T. and Schell, J., Eds. Springer-Verlag, New York, 1987, 149.
74. **Chilton, M. -D., Tepfer, D. A., Petit, A., David, C., Casse-Delbart, F., and Tempé, J.,** *Agrobacterium rhizogenes* inserts T-DNA into the genomes of host plant root cells, *Nature*, 295, 432, 1982.
75. **White, F. F., Ghidossi, G., Gordon, M. P., and Nester, E. W.,** Tumor induction by *Agrobacterium rhizogenes* involves transfer of plasmid DNA to the plant genome, *Proc. Natl. Acad. Sci. U.S.A.*, 79, 3193, 1982.
76. **Ackermann, C.,** Pflanzen aus *Agrobacterium rhizogenes*-Tumoren an *Nicotiana tabacum*, *Pl. Sci. Lett.*, 8, 23, 1977.
77. **Tepfer, D.,** Transformation of several species of higher plants by *Agrobacterium rhizogenes*: sexual transmission of the transformed geneotype and phenotype, *Cell*, 37, 959, 1984.
78. **Sinkar, V. P., White, F. F., Furner, I. J., Abrahamsen, M., Pythoud, F., and Gordon, M. P.,** Reversion of aberrant plants transformed with *Agrobacterium rhizogenes* is associated with the transcriptional inactivation of the T_L-DNA genes, *Plant Physiol.*, 86, 584, 1988.
79. **Sinkar, V. P., Pythoud, F., White, F. F., Nester, E. W., and Gordon, M. P.,** *rolA* locus of the Ri plasmid directs developmental abnormalities in transgenic tobacco plants, *Gene Dev.*, 2, 688, 1988.
80. **White, F. F., Taylor, B. H., Huffman, G. A., Gordon, M. P., and Nester, E. W.,** Molecular and genetic analysis of the transferred DNA regions of the root-inducing plasmid of *Agrobacterium rhizogenes*, *J. Bacteriol.*, 164, 33, 1985.
81. **Cardarelli, M., Mariotti, D., Pomponi, M., Spano, L., Capone, I., and Costantino, P.,** *Agrobacterium rhizogenes* T-DNA genes capable of inducing hairy root phenotype, *Mol. Gen. Genet.*, 209, 475, 1987.
82. **Schmülling, T., Schell, J., and Spena, A.,** Single genes from *Agrobacterium rhizogenes* influence plant development, *EMBO J.*, 7, 2621, 1988.
83. **Spano, L., Mariotti, D., Cardarelli, M., Branca, C., and Costantino, P.,** Morphogenesis and auxin sensitivity of transgenic tobacco with different complements of Ri T-DNA, *Plant Physiol.*, 87, 479, 1988.
84. **van Altvorst, A. C., Bino, R. J., van Dijk, A. J., Lamers, A. M. J., Lindhout, W. H., van der Mark, F., and Dons, J. J. M.,** Effects of the introduction of *Agrobacterium rhizogenes rol* genes on tomato plant and flower development, *Plant Sci.*, 83, 77, 1992.
85. **Huffman, G. A., White, F. F., Gordon, M. P., and Nester, E. W.,** Hairy-root-inducing plasmid: physical map and homology to tumor-inducing plasmids, *J. Bacteriol.*, 157, 269, 1984.
86. **Offringa, I. A., Melchers, L. S., Regensberg-Tunik, A. J. G., Costantino, P., Schilperoort, R. A., and Hooykaas, P. J. J.,** Complementation of *Agrobacterium tumefaciens* tumor inducing *aux* mutants by genes from the T_R-region of the Ri plasmid of *Agrobacterium rhizogenes*, *Proc. Natl. Acad. Sci. U.S.A.*, 83, 6935, 1986.
87. **Spena, A., Schmülling, T., Koncz, C., and Schell, J. S.,** Independent and synergistic ac[illegible]ity of *rol A, B* and *C* loci in stimulating al[illegible]ormal growth in plants, *EMBO J.*, 6, 3891, [illegible]987.

88. **Shen, W. H., Petit, A., Guern, J., and Tempé, J.,** Hairy roots are more sensitive to auxin than normal roots, *Proc. Natl. Acad. Sci. U.S.A.*, 85, 3417, 1988.
89. **Shen, W. H., Davioud, E., David, C., Barbier-Brygoo, H., Tempé, J., and Guern, J.,** High sensitivity to auxin is a common feature of hairy root, *Plant Physiol.*, 94, 554, 1990.
90. **Maurel, C., Barbier-Brygoo, H., Spena, A., Tempé, J., and Guern, J.,** Single *rol* genes from the *Agrobacterium rhizogenes* TL-DNA alter some of the cellular responses to auxin in *Nicotiana tabacum*, *Plant Physiol.*, 97, 212, 1991.
91. **Estruch, J. J., Chriqui, D., Grossmann, K., Schell, J., and Spena, A.,** The plant oncogene *rolC* is responsible for the release of cytokinins from glucoside conjugates, *EMBO J.*, 10, 2880, 1991.
92. **Estruch, J. J., Schell, J., and Spena, A.,** The protein encoded by the *rolB* plant oncogene hydrolyses indole glucosides, *EMBO J.*, 10, 3125, 1991.
93. **Ahuja, M. R.,** Genetic control of tumor formation in higher plants, *Qt. Rev. Biol.*, 40, 329, 1965.
94. **Näf, U.,** Studies on tumor formation in *Nicotiana* hybrids. I. The classification of parents into two etiologically significant groups, *Growth*, 22, 167, 1958.
95. **Ahuja, M. R.,** A cytogenetic study of heritable tumors in *Nicotiana* species hybrids, *Genetics*, 47, 865, 1962.
96. **Ahuja, M. R.,** An hypothesis and evidence concerning the genetic components controlling tumor formation in *Nicotiana*, *Mol. Gen. Genet.*, 103, 176, 1968.
97. **Smith, H. H.,** The inheritance of genetic tumors in *Nicotiana* hybrids, *J. Hered.*, 79, 277, 1988.
98. **Schaeffer, G. W. and Smith, H. H.,** Auxin-kinetin interaction in tissue cultures of Nicotiana species and tumor-conditioned hybrids, *Plant Physiol.*, 38, 291, 1963.
99. **Cheng, T. -Y. and Smith, H. H.,** The influence of genomes on the autonomous growth of pith cultures of *Nicotiana glauca-langsdorffii* hybrids, *Planta*, 113, 29, 1973.
100. **White, P. R.,** Transplantation of plant tumors of genetic origin, *Cancer Res.*, 4, 791, 1944.
101. **Kehr, A. E. and Smith, H. H.,** Genetic tumors in Nicotiana hybrids, *Brookhaven Symp.*, 6, 55, 1954.
102. **Nandi, S. K., Palni, L. M. S., and Parker, C. W.,** Dynamics of endogenous cytokinins during the growth cycle of a hormone-autotrophic genetic tumor line of tobacco, *Plant Physiol.*, 94, 1084, 1990.
103. **Feng, X.-H., Dube, S. K., Bottino, P. J., and Kung, S.-D.,** Restoration of shooty morphology of a nontumorous mutant of *Nicotiana glauca* X *N. langsdorffii* by cytokinin and the isopentenyltransferase gene, *Plant Mol. Biol.*, 15, 407, 1990.
104. **Bayer, M. H. and Ahuja, M. R.,** Tumor formation in *Nicotiana*: auxin levels and auxin inhibitors in normal and tumor-prone genotypes, *Planta*, 79, 292, 1968.
105. **Nacmias, B., Ugolini, S., Ricci, M. D., Pellegrini, M. G., Bogani, P., Bettine, P., Inzé, D., and Buiatti, M.,** Tumor formation and morphogenesis on different *Nicotiana* sp. and hybrids induced by *Agrobacterium tumefaciens* T-DNA mutants, *Dev. Genet.*, 8, 61, 1987.
106. **Carlson, P. S., Smith, H. H., and Dearing, R. D.,** Parasexual interspecific plant hybridization, *Proc. Natl. Acad. Sci. U.S.A.*, 69, 2292, 1972.
107. **Smith, H. H., Kao, K. M., and Combatti, N. C.,** Interspecific hybridization by protoplast fusion in *Nicotiana*, *J. Hered.*, 67, 123, 1976.
108. **Izard, C.,** Obtention et fixation de lignées tumorales et non tumorales a partir de mutations expérimentales de l'hybride *N. glauca* X *N. langsdorffii*, *C. R. Acad. Agric.*, 43, 325, 1957.
109. **Smith, H. H. and Stevenson, H. Q.,** Genetic control and radiation effects in Nicotiana tumors, *Z. Vererbung.*, 92, 100, 1961.
110. **Durante, M., Geri, C., Buiatti, M., Baroncelli, S., Parenti, R., Ronchi, V. N., Martini, G., Grenci, F. C., Grisvard, J., and Guillé, E.,** DNA heterogeneity and genetic control of tumorigenesis in Nicotiana tumorous and nontumorous genotypes, *Dev. Genet.*, 3, 25, 1982.
111. **Durante, M., Cecchini, E., Natali, L., Citti, L., Geri, C., Parenti, R., and Ronchi, V. N.,** 5-Azacytidine induced tumorous transformation and DNA hypomethylation in *Nicotiana* tissue cultures, *Dev. Genet.*, 10, 298, 1989.
112. **Holliday, R.,** The inheritance of epigenetic defects, *Science*, 238, 163, 1987.
113. **Furner, I. J., Huffman, G. A., Amasino, R. M., Garfinkel, D. J., Gordon, M. P., and Nester, E. W.,** An *Agrobacterium* transformation in the evolution of the genus *Nicotiana*, *Nature*, 319, 422, 1986.
114. **Ichikawa, T., Ozeki, Y., and Syono, K.,** Evidence for the expression of the *rol* genes of *Nicotiana glauca* in genetic tumors of *N. glauca* X *N. langsdorffii*, *Mol. Gen. Genet.*, 220, 177, 1990.
115. **Rutter, W. J., Pictet, R. L., and Morris, P. W.,** Toward molecular mechanisms of developmental processes, *Ann. Rev. Biochem.*, 42, 601, 1973.

116. **Meins, F., Jr.**, 5-Bromodeoxyuridine: a specific inhibitor of cytokinin habituation in tobacco cell culture, *Planta*, 129, 239, 1976.

117. **Vyskot, B., Karpfel, Z., and Bezdek, M.**, Hormonal control of 5-bromodeoxyuridine (BUdR) tolerance in crown-gall tumors and habituated tissues, *Planta*, 137, 247, 1977.

118. **Bezdek, M. and Vyskot, B.**, Spontaneous 5-bromodeoxyuridine tolerance of *Nicotiana* genetic tumours, *Biochem. Physiol. Pflanzen*, 174, 479, 1979.

119. **Kandra, G. and Maliga, P.**, Is BUdR-resistance a consequence of cytokinin habituation in *Nicotiana tabacum*? *Planta*, 1992.

120. **Bezdek, M. and Vyskot, B.**, DNA synthesis in cytokinin-autotrophic tobacco cells, *Planta*, 152, 215, 1981.

121. **Stetler, D. A. and Laetsch, W. M.**, Kinetin-induced chloroplast maturation in cultures of tobacco tissue, *Science*, 149, 1387, 1965.

122. **Axelos, M. and Péaud-Lenoël, C.**, The apoprotein of the light-harvesting chlorophyll a/b complex of tobacco cells as a molecular marker of cytokinin activity, *Plant Sci. Lett.*, 19, 33, 1980.

123. **Kamínek, M. and Lustinec, J.**, Reduced chlorophyll synthesis in cytokinin-autonomous strains of tobacco callus, *Z. Pflanzenphysiol.*, 73, 65, 1974.

124. **Kamínek, M. and Lustinec, J.**, Induction of cytokinin-autonomy and chlorophyll-deficiency in tobacco callus tissue by streptomycin, *Z. Pflanzenphysiol.*, 74, 81, 1974.

125. **Vyskot, B. and Novak, F. J.**, Habituation and organogenesis in callus cultures of chlorophyll mutants of *Nicotiana tabacum* L., *Z. Pflanzenphysiol.*, 81, 34, 1977.

126. **Kamínek, M., Hadacova, V., and Lustinec, J.**, Origin of cytokinin- and auxin-autonomy and changes in specific proteins in tobacco callus tissue, *Biol. Plant.*, 23, 228, 1981.

127. **Kevers, C., Coumans, M., De Greef, W., Hofinger, M., and Gaspar, T.**, Habituation in sugarbeet callus: auxin content, auxin protectors, peroxidase pattern and inhibitors, *Physiol. Plant.*, 51, 281, 1981.

128. **Hagège, D., Kevers, C., Crèvecoeur, M., Tollier, M. T., Monties, B., and Gaspar, T.**, Peroxidases, growth and differentiation of habituated sugarbeet cells, in *Biochemical, Molecular, and Physiological Aspects of Plant Peroxidases*, Lobarzewski, J., Greppin, H., Penel, C., and Gaspar, T., Eds. University of Geneva, 1991, 281.

129. **Eichholz, R., Harper, J., Felix, G., and Meins, F., Jr.**, Evidence for an abundant 33,000-dalton polypeptide regulated by cytokinins in cultured tobacco tissues, *Planta*, 158, 410, 1983.

130. **Felix, G. and Meins, F., Jr.**, Purification, immunoassay and characterization of an abundant, cytokinin-regulated polypeptide in cultured tobacco tissues. Evidence the protein is a β-1,3-glucanase, *Planta*, 164, 423, 1985.

131. **Hansen, C. E., Meins, F., Jr., and Aebi, R.**, Hormonal regulation of zeatin-riboside accumulation by cultured tobacco cells, *Planta*, 172, 520, 1987.

132. **Kamínek, M.**, Progress in cytokinin research, *TIBTECH*, 10, 159, 1992.

133. **van Onckelen, H., Rudelsheim, P., Hermans, R., Horemans, S., Messens, E., Hernalsteens, J. -P., van Montagu, M., and De Greef, J.**, Kinetics of endogenous cytokinin, IAA and ABA levels in relation to the growth and morphology of tobacco crown gall tissue, *Plant Cell Physiol.*, 25, 1017, 1984.

134. **Motyka, V. and Kamínek, M.**, Regulation of cytokinin catabolism in tobacco callus cultures, in *Progress in Plant Cellular and Molecular Biology*, Nijkamp, H. J. J., van der Plas, L. H. W., and van Aartrijk, J., Eds., Kluwer, Dordrecht, 1990, 492.

135. **Hansen, C. E., Meins, F., Jr., and Milani, A.**, Clonal and physiological variation in the cytokinin content of tobacco-cell lines differing in cytokinin requirement and capacity for neoplastic growth, *Differentiation*, 29, 1, 1985.

136. **Palni, L. M. S., Burch, L., and Horgan, R.**, The effect of auxin concentration on cytokinin stability and metabolism, *Planta*, 174, 231, 1988.

137. **Binns, A. N. and Meins, F., Jr.**, Cold sensitive expression of cytokinin habituation by tobacco pith cells in culture, *Planta*, 145, 365, 1979.

138. **Binns, A. N., Chen, R. H., Wood, R. H., and Lynn, D. G.**, Cell division promoting activity of naturally occurring dehydrodiconiferyl glucosides: do cell wall components control cell division? *Proc. Natl. Acad. Sci. U.S.A.*, 84, 980, 1987.

139. **Teutonico, R. A., Dudley, M. W., Orr, J. D., Lynn, D. G., and Binns, A. N.**, Activity and accumulation of cell division-promoting phenolics in tobacco tissue cultures, *Plant Physiol.*, 97, 288, 1991.

140. **Meins, F., Jr.**, Habituation: heritable variation in the requirement of cultured plant cells for hormones, *Ann. Rev. Genet.*, 23, 395, 1989.

141. **Amasino, R. M. and John, M. C.,** Extensive changes in DNA methylation patterns accompany activation of a silent T-DNA ipt gene in *Agrobacterium tumefaciens*-transformed plant cells, *Mol. Cell Biol.*, 9, 4298, 1989.

142. **van Slogteren, G. M. S., Hooykaas, P. J. J., and Schilperoort, R. A.,** Silent T-DNA genes in plant lines transformed by *Agrobacterium tumefaciens* are activated by grafting and by 5-azacytidine treatment, *Plant Mol. Biol.*, 3, 333, 1984.

143. **Meins, F., Jr. and Binns, A.,** Epigenetic clonal variation in the requirement of plant cells for cytokinins, in *The Clonal Basis for Development*, Subtelny, S. and Sussex, I.M., Eds., Academic Press, New York, 1978, 185.

144. **Meins, F., Jr.,** A biochemical switch model for cell-heritable variation in cytokinin requirement, in *Molecular Basis of Plant Development*, Goldberg, R., Ed., Alan R. Liss, New York, 1989, 13.

145. **Laloue, M., Pethe-Terrine, C., and Guern, J.,** Uptake and metabolism of cytokinins in tobacco cells: studies in relation to the expression of their biological activities, in *Metabolism and Molecular Activities of Cytokinins*, Guern, J. and Péaud-Lenoël, C., Eds. Springer-Verlag, Berlin, 1981, 80.

146. **Binns, A. N.,** Habituation of tobacco pith cells for factors promoting cell division, Ph.D. thesis, Princeton University, Princeton, NJ, 1979.

147. **Vankova, R., Kamínek, M., Eder, J., and Vanek, T.,** Dynamics of production of *trans*-zeatin and *trans*-zeatin riboside by immobilized cytokinin-autonomous and cytokinin-dependent tobacco cells, *J. Plant Growth Regul.*, 6, 147, 1987.

148. **Fedoroff, N., Masson, P., and Banks, J.-A.,** Mutations, epimutations, and the developmental programming of the maize *suppressor-mutator* transposable element, *Bioessays*, 10, 139, 1989.

Chapter 21

Transgenic Plants and Cytokinin Biology

Harry J. Klee

TABLE OF CONTENTS

I. INTRODUCTION

Much of the credit for advances in our understanding of cytokinin biology can be attributed to the rapid pace of technology development. For example, the ability to detect ever-decreasing amounts of hormones and their metabolites has had a major impact on the types of experiments that can be done. Another significant advance in plant biology is the ability to create transgenic plants. It has been only ten years since the first reports of transgenic plants,[1,2] yet transgene technology has revolutionized approaches to plant physiology and biochemistry. The impact on hormone biology is particularly apparent. Already, plants with altered levels of auxins, cytokinins, and ethylene are widely available, and the first agricultural applications of this technology are approaching commercialization.[3-5] The goal of this review is to address the impact of transgenic plants on our understanding of cytokinins. Analysis of transgenic plants with altered cytokinin synthesis has provided some interesting and unexpected insights into cytokinin biology.

Cytokinins have clearly been shown to have a critical role in plant growth and differentiation. Most of what is known today about the role of cytokinins in development was determined by exogenous application experiments. One of the most important experiments involved the growth of tobacco callus on a medium containing different ratios of auxin to cytokinin.[6] The experiments of Skoog and Miller demonstrated that plant tissue could be made to differentiate into shoots or roots, by altering the ratio of auxin to cytokinin. These experiments were the foundation for understanding the mechanism of *Agrobacterium* crown gall tumorigenesis. However, it is widely accepted that exogenous hormone application experiments are subject to many complications that make interpretation of the results difficult. In particular, it is difficult to quantitate how much of an exogenously applied dose of a hormone is translocated in biologically active form to a target tissue. Transgenic plants offer an alternative that allows for reproducible hormone perturbations. Hormone levels can, in theory, be precisely manipulated both spatially and temporally. There are, however, limitations to the technology, and I will attempt to point these out in the course of this review.

II. ISOPENTENYL TRANSFERASE

Plants are believed to synthesize cytokinins from isopentenyl adenosine 5′-monophosphate (IPA). Although isopentenyl transferase (ipt) enzyme activity has been demonstrated in plants,[7] the enzyme has not been purified to homogeneity. Nor has the gene encoding ipt in plants been cloned. Fortunately, there is an alternate source for an ipt gene. This gene was first identified in the T-DNA of *Agrobacterium tumefaciens*.[8-10] Analysis of crown gall tumor tissues strongly suggested that a gene located in the T-DNA affected the level of cytokinins in transformed tissue.[11a] This led several groups to determine, almost simultaneously, that the tmr locus encodes an isopentenyl transferase activity. The *Agrobacterium* gene has been used to manipulate cytokinin levels in transgenic plants.

III. PRODUCTION OF TRANSGENIC PLANTS

Many transcriptional promoters having various temporal and tissue-specific patterns of expression are readily available, and construction of chimeric genes for plant expression is now routine. It is a

0-8493-6252-0/94/$0.00+$.50

relatively straightforward process to construct a chimeric gene and to introduce that gene into *Agrobacterium*-based plant transformation vectors. The nature of *Agrobacterium*-mediated plant transformation imposes certain limitations on our ability to manipulate cytokinin biosynthesis. As an example, the procedure for generating transgenic tobacco plants is typical of many dicot transformation systems. First, leaf pieces are exposed to a solution of *Agrobacterium* containing a plant transformation vector. The vector usually contains a selectable marker gene that permits the transformed cell to grow on a selective medium. The most commonly used selectable marker confers resistance to kanamycin. The medium usually contains a combination of auxins and cytokinins, which encourage growth and shoot formation. A cell that has been transformed will divide several times, frequently producing an undifferentiated callus. Groups of transformed cells will then become organized into a shoot. Once the shoot has elongated sufficiently, it can be excised from the callus and transferred to a root-inducing medium. It is at the stage of root formation that the most difficulty is encountered with *ipt*-containing shoots. Root induction in tobacco and most other dicots requires a high auxin-to-cytokinin ratio. Expression of an *ipt* transgene at this stage of regeneration can completely suppress root formation. Even low levels of *ipt* expression can suppress root initiation. What makes transformation particularly difficult is that most genes, even highly tissue-specific ones, are expressed at detectable levels in undifferentiated callus.[11b] For example, the promoter for the a′ subunit of B-conglycinin is very tissue specific, being expressed only in developing seeds. Nonetheless, the gene is expressed during regeneration, and only after the shoots are fully regenerated does it become properly regulated.[12] The problem of "leaky" expression of promoters during regeneration has been a significant problem in producing plants with altered cytokinin balance. It is for this reason that most of the work in the field has concentrated on heat-shock promoters. Generally, these promoters are among the most tightly regulated of transcriptional promoters, having almost undetectable expression at normal temperatures, and high expression at elevated temperatures.[13]

IV. TRANSGENIC PLANTS

Several groups independently obtained transgenic plants in which the *ipt* gene was under the control of various heat-shock promoter elements.[14-17] For the most part, all of these plants are very similar in their properties, and a number of important conclusions can be drawn from the work. The conclusions can be separated into biochemical and developmental effects.

Biochemical effects of *hsp*/*ipt* gene fusions are as follows. All of the results are consistent with formation of IPA being the rate-limiting step in synthesis of biologically active cytokinins. For example, in heat-shocked tobacco plants, the IPA increased threefold, while the *trans*-zeatin (Z) and *trans*-zeatin riboside (ZR) cytokinins increased by over 30-fold.[14] Following heat shock, very substantial increases in cytokinins occurred. Twenty to 200-fold increases in Z, ZR, and zeatin riboside 5′-monophosphate (ZMP) were observed, suggesting that the plant is capable of rapidly converting IPA to more active cytokinins.[14,15,17] Although it has not been thoroughly examined, transgenic plants also have very high levels of the conjugated zeatin 7-β-D-glucoside.

Developmental effects are as follows. There are several distinct morphological consequences of introducing an *hsp*/*ipt* gene into tobacco plants. In all cases plants are reduced in stature by up to 50%. This reduction is due to a decrease in the internode length rather than the node number.[14,15] Initially, it was surprising that the stature reduction occurred in plants whether or not they received a heat shock to induce the gene. Despite the obvious stature reduction observed by three independent groups, none was able to detect *ipt* RNA in the absence of heat shock. When cytokinin levels were examined in nonheat-shocked plants, there were significant increases in several types of cytokinin. These ranged from twofold for IPA to sixfold for ZMP.[14] Apparently, these minor changes are sufficient to cause the observed stature reduction.

In addition to the reduced stature, transgenic plants also have smaller leaves and a greatly reduced root system. Apparently, elevated levels of cytokinin are inhibitory not only to root initiation, but to root growth in general. Whether the reduced root mass is due to overall slower growth or an alteration in the pattern of branching has yet to be determined. The reduced root mass of transgenic plants may be at least partially responsible for the reduced stature and leaf size of the plants. Similar inhibitory effects on root growth were also observed in transgenic plants with either increased[12] or decreased[18] auxin levels. It would thus appear that roots respond negatively to any alteration in auxin and cytokinin.

Transgenic plants were also shown to have a less-developed vascular system, exhibiting quantitatively less xylem.[14] There are reports that cytokinin is promotive of vascular differentiation in cell culture systems (for a review see Roberts et al.[19a]). These seemingly contradictory observations are not mutually exclusive. Cytokinin is clearly necessary for the growth of most cultured cells. There may be a need for cytokinin in culture systems that is not necessarily related to the formation of differentiated vascular tissues. It is also possible that the reduced xylem content of transgenic plants is a secondary effect of a reduced root mass. The true cause of the reduced xylem growth in transgenic plants will require further analysis.

An increase in cytokinin levels in transgenic tobacco plants clearly leads to reduced apical dominance.[14,15,17] Transgenic plants have much more extensive growth of lateral shoots than do nontransgenic controls. This effect has also been observed in transgenic petunia plants containing the *ipt* gene.[19b] This observation is consistent with previous results with exogenous application of cytokinin to dormant lateral buds.[20] The effect of cytokinin on apical dominance, although clear, is only part of the story. Transgenic plants that have altered auxin levels also have greatly altered patterns of apical dominance. When the level of auxin (indole-3-acetic acid) is reduced by 10- to 20-fold, lateral buds are also released from apical dominance.[18] The effect that is achieved very much resembles that seen in the *ipt*-overexpressing plants. When the level of auxin is increased tenfold, the opposite effect is achieved, i.e., plants are almost completely apically dominant.[12] Further, the dormant lateral buds in these plants can be released from dormancy by exogenous application of cytokinin.[19b] The picture that emerges is that apical dominance is not controlled by auxin or cytokinin, but by the ratio of the two hormones. In fact, the plant seems to be relatively insensitive to the absolute concentration of each hormone. Rather, the plant appears to sense the ratio of the two hormones. A high auxin:cytokinin ratio will result in the suppression of growth of lateral shoots, while a low ratio results in growth. The results are consistent with the idea that the two hormones act antagonistically on the same signal transduction pathway and reinforces the observations of Skoog and Miller in which the ratio of auxin to cytokinin was the determinant of whether tobacco callus grew as shoots, roots, or callus.[6]

One interesting aspect of the phenotypic consequence of increased cytokinins is the lack of homeotic-type changes in development. Virtually all of the changes observed in plants containing *hsp/ipt* genes are quantitative in nature. What have been principally changed in all of the transgenic plants are the rate of cell division and the ultimate sizes of cells. Yet the experiments of Skoog and Miller clearly indicate the potential for greatly altering the developmental fate of cells.[6] However, one study involving a different type of *ipt* transgene suggests that the story may be more complex. Estruch et al.,[21] in order to circumvent the problems of *ipt* gene expression during regeneration, constructed a chimeric gene consisting of a strong constitutive promoter (the CaMV 35S), the *ipt* coding sequence, and the maize Ac transposon inserted between the promoter and the *ipt* sequence. The gene is inactive until the Ac element transposes and reconstitutes an active *35S/ipt* gene. Plants obtained from transformation with this gene become somatic mosaics as a consequence of transposition. In effect, what are created are islands of localized high cytokinin concentrations. The consequence of this alteration was the formation of adventitious shoots principally derived from vascular parenchyma cells. This result confirms that cytokinin can indeed reprogram the developmental fate of cells. It also reinforces the conclusion that cytokinins act to modulate cell division, since the parenchyma cells in these transgenic plants would not normally divide further. It is very important, however, to keep these experiments in perspective. That a continuous and extremely high dose of cytokinin causes certain cells to aberrantly initiate organogenesis does not mean that this is the normal role of cytokinin in development. Further, what may be most important here is the gradient of hormone and not the absolute level, since the level of ZR in the aberrant shoots is not any higher than that achieved in heat-shocked leaves containing the *hsp/ipt* genes. Nonetheless, the *35S/ipt* results indicate that further work will be needed to clarify the role, or lack of one, for cytokinin in tissue differentiation.

A role for cytokinin in the control of senescence has been postulated.[22] Effects on senescence of transgenic plants containing an *hsp/ipt* gene were reported. Smart et al. applied localized heat shock to leaves of transgenic plants and observed that the heat-shocked areas remained green for extended periods.[17] They also reported that nonheat-shocked transgenic plants remained green for longer than untransformed control plants. These results are consistent with results obtained with the application of cytokinin to detached leaves, where senescence is delayed.[23] Since no experiments were done to distinguish between delayed senescence and reduced chlorophyll degradation, these results should be considered preliminary. Nonetheless, they are suggestive of a possible agronomic application for transgenic plants (see below).

V. FUTURE OPPORTUNITIES

There is good evidence linking cytokinins to several important physiological processes. Of particular agronomic significance are the applications of cytokinin in senescence, floral initiation, and determination of sink strength. For example, abortion of flowers and fruits can be prevented by exogenous application of cytokinins in soybeans,[24] *Vicia faba*,[25] and tomato,[26] among other species. Using available molecular tools, it may be possible to elevate cytokinin levels in a specific tissue, to prevent abortion and ultimately lead to earlier fruit set or increased yield. It may also be possible to increase yield in some monocarpic plants by delaying the onset, or prolonging the period, of senescence. In this context, the results of Smart et al. are particularly intriguing.[17] There are significant uncertainties in the manipulation of endogenous hormone levels. Nonetheless, the experimental tools are, in many cases, available. The rate-limiting step in many of these systems is the ability to obtain transgenic plants with appropriate patterns of expression. For example, the transformation of leguminous plants, although possible, remains difficult. The addition of a gene that may interfere with normal regeneration further complicates the ability to obtain transgenic plants. Bombardment of intact tissues with a particle gun may circumvent much of the potential difficulty for some species such as soybean.[27] Hopefully, the critical experiments to test hypotheses will be done within the next several years.

Another application of cytokinin genes, which has been proposed for a number of years, is as an aid in regeneration of recalcitrant plant species. Generally, this approach has not been successful in obtaining transgenic plants. Transgenic shoots derived from teratomas were obtained.[28,29] Unfortunately, the shoots and plants derived from them were almost always morphologically abnormal and frequently sterile. Clearly the ease with which shoots can be recovered in *ipt*-overproducing tissues has merit. A method that would permit *ipt* gene shutoff following shoot formation could still be valuable for some plant species. In theory the reverse of the transposon approach of Estruch et al.[21] could lead to *ipt* gene shutoff. To date, this line of experimentation has not been reported, however.

Up to this point, researchers have been limited to the use of a gene that synthesizes IPA. This is a significant limitation for the types of experiments that can be undertaken. By contrast, auxin and ethylene can be either increased or decreased, in any of several different ways, in transgenic plants.[3-5,12,18] It is very likely that the situation for cytokinins will change in the very near future. Bacteria have been a rich source of genes that metabolize phytohormones and their precursors. One interesting potential bacterial alternative to *ipt* is the *rolC* gene of *Agrobacterium rhizogenes*. There is some evidence that it may be capable of hydrolyzing cytokinin conjugates, although this is far from clear.[30] If true, this enzyme would be especially useful for examining the roles and locations of certain cytokinin conjugates in development. Also, progress is being made in the purification of plant enzymes involved in cytokinin metabolism (see Chapters 10 and 11). Once the genes encoding the plant cytokinin biosynthetic enzymes do become available, it will be very straightforward to determine their spatial and temporal regulation, with transcriptional promoter fusions and antisense gene constructions in transgenic plants.

To put all of the transgenic plant work in perspective, it is clear that these plants are very useful tools for increasing our knowledge of both the biochemistry and developmental biology of cytokinins. We have both confirmed and learned much in the last 5 years. However, much remains to be done. The potential of practical applications of cytokinin biology to agriculture has yet to be achieved. All of the tools are available, and hopefully, significant progress can be achieved in the near future.

REFERENCES

1. **Barton, K., Binns, A., Matzke, A., and Chilton, M.- D.,** Regeneration of intact tobacco plants containing full length copies of genetically engineered T-DNA of R1 progeny, *Cell*, 32, 1033, 1983.
2. **Horsch, R., Fraley, R., Rogers, S., Sanders, P., Lloyd, A., and Hoffmann, N.,** Inheritance of functional foreign genes in plants, *Science*, 223, 496, 1984.
3. **Oeller, P. W., Min-Wong, L., Taylor, L. P., Pike, D. A., and Theologis, A.,** Reversible inhibition of tomato fruit senescence by antisense RNA, *Science*, 254, 437, 1991.
4. **Hamilton, A. J., Lycett, G. W., and Grierson, D.,** Antisense gene that inhibits synthesis of the hormone ethylene in transgenic plants, *Nature*, 346, 284, 1990.
5. **Klee, H. J., Hayford, M. B., Kretzmer, K. A., Barry, G. F., and Kishore, G.M.,** Control of ethylene synthesis by expression of a bacterial enzyme in transgenic tomato plants, *Plant Cell*, 3, 1187, 1991.
6. **Skoog, F. and Miller, C. O.,** Chemical regulation of growth and organ formation in plant tissues cultured *in vitro*, *Symp. Soc. Exp. Biol.*, 11, 188, 1957.

7. **Chen, C.,** Cytokinin biosynthesis in cell free systems, in *Plant Growth Substances 1982*, Wareing, P., Ed., Academic Press, New York, 1982, 155.
8. **Barry, G. F., Rogers, S. G., Fraley, R. T., and Brand, L.,** Identification of a cloned cytokinin biosynthetic gene, *Proc. Natl. Acad. Sci. U.S.A.*, 81, 4776, 1984.
9. **Akiyoshi, D., Klee, H., Amasino, R., Nester, E. W., and Gordon, M.,** T-DNA of Agrobacterium tumefaciens encodes an enzyme of cytokinin biosynthesis, *Proc. Natl. Acad. Sci. U.S.A.*, 81, 5994, 1984.
10. **Schröder, G., Waffenschmidt, S., Weiler, E. W., and Schröder, J.,** The T-region of Ti plasmids codes for an enzyme synthesizing indole-3-acetic acid, *Eur. J. Biochem.*, 138, 387, 1984.
11a. **Akiyoshi, D. E., Morris, R. O., Hinz, R., Mischke, B. S., Kosuge, T., Garfinkel, D. J., Gordon, M. P., and Nester, E. W.,** Cytokinin auxin balance in crown gall tumors is regulated by specific loci in the T-DNA, *Proc. Natl. Acad. Sci. U.S.A.*, 80, 407, 1983.
11b. **Klee, H. J.,** et al., unpublished results.
12. **Klee, H. J., Horsch, R. B., Hinchee, M. A., Hein, M. B., and Hoffmann, N.L.,** The effects of overproduction of two *Agrobacterium tumefaciens* T-DNA auxin biosynthetic gene products in transgenic petunia plants, *Genes Dev.*, 1, 86, 1987.
13. **Bond, U. and Schlesinger, M. J.,** Heat-shock proteins and development, *Adv. Gen.*, 24, 1, 1987.
14. **Medford, J. I., Horgan, R., El-Sawi, Z., and Klee, H. J.,** Alterations of endogenous cytokinins in transgenic plants using a chimeric isopentenyl transferase gene, *Plant Cell*, 4, 403, 1989.
15. **Smigocki, A.C.,** Cytokinin content and tissue distribution in plants transformed by a reconstructed isopentenyl transferase gene, *Plant Mol. Biol.*, 16, 105, 1991.
16. **Schmülling, T., Beinsberger, S., De Greef, J., Schell, J., Van Onckelen, H., and Spena, A.,** Construction of a heat inducible chimaeric gene to increase the cytokinin content in transgenic plant tissue, *FEBS Lett.*, 249, 401, 1989.
17. **Smart, C., Scofield, S., Bevan, M., and Dyer, T.,** Delayed leaf senescence in tobacco plants transformed with tmr, a gene for cytokinin production in *Agrobacterium*, *Plant Cell*, 3, 647, 1991.
18. **Romano, C. P., Hein, M. B., and Klee, H. J.,** Inactivation of auxin in tobacco transformed with the indoleacetic acid-lysine synthetase gene of Pseudomonas savastanoi, *Genes Dev.*, 5, 438, 1991.
19a. **Roberts, L. W., Gahan, P. B., and Aloni, R.,** *Vascular Differentiation and Plant Growth Regulators*, Springer-Verlag, New York, 1988.
19b. **Klee, H. J.,** Unpublished results.
20. **Cline, M. G.,** Apical Dominance, *Bot. Rev.*, 57, 318, 1991.
21. **Estruch, J., Prinsen, E., Van Onckelen, H., Schell, J., and Spena, A.,** Viviparous leaves produced by somatic activation of an inactive cytokinin-synthesizing gene, *Science*, 254, 1364, 1991.
22. **Goldthwaite, J.,** Hormones in plant senescence, in *Plant Hormones and Their Role in Plant Growth and Development*, Davies, P., Ed., Martinus Nijhoff, Boston, 1987, 553.
23. **Richmond, A. and Lang, A.,** Effect of kinetin on protein content and survival of detached *Xanthium* leaves, *Science*, 125, 650, 1957.
24. **Carlson, D., Dyer, D., Cotterman, C. D., and Durley, R.,** The physiological basis for cytokinin induced increases in pod set in IX93-100 soybeans, *Plant Physiol.*, 84, 233, 1987.
25. **Chapman, G. and Sadjadi, A.,** Exogenous growth substances and internal competition in *Vicia faba* L., *Z. Pflanzenphysiol.*, 104, 265, 1981.
26. **Kinet, J., Hurdebise, D., Parmentier, A., and Stainier, R.,** Promotion of inflorescence development by growth substance treatments in tomato plants grown in insufficient light conditions, *J. Am. Soc. Hort. Sci.*, 103, 724, 1978.
27. **McCabe, D. E., Swain, W. F., Martinell, B. J., and Christou, P.,** Stable transformation of soybean (*Glycine max*) by particle acceleration, *Biotechnology*, 6, 923, 1991.
28. **Smigocki, A. C. and Owens, L. D.,** Cytokinin gene fused with a strong promoter enhances shoot organogenesis and zeatin levels in transformed plant cell, *Proc. Natl. Acad. Sci. U.S.A.*, 85, 5131, 1988.
29. **Ooms, G. and Lenton, J. R.,** T-DNA genes to study plant development: precocious tuberisation and enhanced cytokinins in A. tumefaciens transformed potato, *Plant Mol. Biol.*, 5, 205, 1985.
30. **Estruch, J. J., Chriqui, D., Grossmann, K., Schell, J., and Spena, A.,** The plant oncogene *rolC* is responsible for the release of cytokinins from glucoside conjugates, *EMBO J.*, 10, 2889, 1991.

Chapter 22

Cytokinins and Reproductive Organogenesis in *Mercurialis*

Raymonde Durand and Bernard Durand

TABLE OF CONTENTS

I. INTRODUCTION

Phytohormones are known to play significant roles in plant development and organogenesis. Although there are general properties relating to a particular hormone, such as stimulating shoot or root formation by cytokinin and auxin, respectively,[1] the specific targets in development have not been identified. The development of a particular organ involves cell division, enlargement, differentiation, and organization and requires sequential gene expression influenced by environmental factors and plant hormones. It would be useful if individual gene actions controlling discrete developmental steps could be related to a particular hormone. Cytokinins and other hormones are known to influence the development of reproductive organs, stamens and pistil, in monoecious and dioecious species (Table 1).[2-5] Mercury plants (*Mercurialis*) have been used extensively in studying the interactions between genetic determinants controlling sex organs and hormonal effects leading to differential sex expression. This chapter summarizes key findings on sex determination, endogenous phytohormones linked to sex expression, and the results of recent molecular studies leading to cloning of genes that are differentially expressed in male and female flowers and appear to be the targets of cytokinins.

II. *MERCURIALIS*, THE PLANT

Annual mercuries are dicots of the Euphorbiaceae family. Mercuries possess "imperfect" flowers with only pistil or stamens and without any rudiment of the other sex, such as staminodes in pistillate flowers, or carpelloid structures in staminate flowers. This is in contrast to some other dioecious plants, *Asparagus*[6] or *Melandrium*,[7] where remnants of the opposite sex organs occur, but abort to give functional unisex flowers. *Mercurialis annua* is dioecious, while *M. ambigua* and *M. monoïca* are monoecious autoallopolyploids of *M. annua*.[8] The Mendelian genes controlling dioecy in *M. annua*, and consequently the stamen or pistil, have been well characterized, and the separation of male and female shoot apices facilitate the study of hormonal effects on sex expression. In monoecious plants the ratio of male to female flowers is influenced by the same hormones as in *M. annua*. Table 2 summarizes the phenotypes of mercuries. (For details on the morphology and ontogeny, the reader is referred to earlier publications.[9,10])

0-8493-6252-0/94/$0.00+$.50

Table 22-1 **Effect of Hormones on Sex Expression of some Dioecious or Monoecious Species[4,6] [a]**

Dioecious		
Cannabis sativa	Ethylene, IAA, kinetin	Masculinization of genetic females
	Ethylene	Feminization of genetic males
	GA_3	Masculinization of genetic females
	N^6-benzyladenine, IAA, ABA	Increases the % of females in populations of derooted plantlets
	GA_3	Increases the % of males in derooted plantlets
Humulus lupulus	CPTPA[b], IAA	Masculinization of genetic females
Mercurialis annua	Kinetin	Feminization of genetic males
	IAA	Masculinization of genetic female nodes
Morus nigra	Ethephon	Feminization of genetic males
Spinacia oleracea	GA_3	Increases the % of males in derooted plants
	N^6-benzyladenine, IAA, ABA[c]	Increases the % of females in derooted plants
Vitis vinifera	Kinetin	Feminization of genetic males
Asparagus officinalis	GA_3	Induces stamen development in females
	Cytokinins	Induce ovules in flowers of genetic males
Monoecious		
Begonia franconis	IAA, GA_3, N^6-benzyladenine	Modify sex ratios toward femaleness
Coriandrium sativum	GA_3	Increases female flowers
	CCC[d], Phosphon, ethephon	Inhibit feminization induced by GA_3
Cucurbita maxima	Ethephon	Increases female tendency
Cucumis sativus	GA_3	Increases male tendency
	IAA	Increases female tendency
	ABA	Increases male tendency
Luffa cylindrica	N^6-benzyladenine	Total feminization
Mercurialis ambigua	Kinetin	Total feminization
Zea mays	GA_3	Feminization of tassel and male sterility
	CCC, Phosphon, ethephon	Inhibit the female tendency induced by GA_3 in tassel

[a] Modified from Durand, B. and Durand, R., *Physiol. Plant.*, 60, 263, 1984, with permission; [b] CPTPA: 2-Chlorophenylthiopropionic acid; [c] ABA: Abscissic acid; [d] CCC: 2-Chloroethyltrimethyl ammonium chlorate.

III. PHYSIOLOGY

A. EFFECTS OF PHYTOHORMONES ON ORGANOGENESIS OF WHOLE PLANTS

At the single-leaf-pair stage (after the genetic sex of plants was recognized[11]), indolacetic acid (IAA) or naphthalene acetic acid (NAA, 10 to 1 mg/l) did not change the sex of flowers. However, the vegetative bud adjacent to the normal axial inflorescence was transformed into inflorescences, and the sex of the flowers depended on the genotype of the plant. Thus auxins that generally act as feminizing agents[4,12] in other species did not affect the reproductive organs of *Mercurialis* or the secondary sex

Table 22-2 **Phenotypes of Annual Mercuries**

Mercurialis annua Male	Female
Primary sex characteristics	
Staminate flowers with three calyx-lobes	Pistillate flowers with three calyx-lobes
Nine to ten yellow stamens in two to three groups	Two soldered green carpels, rarely three
Secondary sex characteristics	
Male inflorescence in long axillary spike with four to five sympodial clusters of male flowers	Female inflorescence in one sessile axillary cluster of three to five female flowers
Three soldered sepals in flower buds. Lateral branching poorly developed, resulting from strong apical dominance.	Three separated sepals in flower buds
	Lateral branching inducing weak apical dominance
Paler green leaf color	Darker green leaves

Mercurialis ambigua, monoica Monoecious
Monoecious inflorescence of female type (sessile cluster: one pistillate flower and three to four sympodial male flowers) or of male type (long spike), depending upon the genotype. Identical inflorescences in leaf axils of a given plant. Lateral branching generally as in female *M. annua*. Darker green leaf color.

characteristics (inflorescence length), but rather promoted flowering. Similarly, application of gibberellin (GA3) induced vegetative buds to form inflorescences and the elongation of the female sessile inflorescence,[11] but the stamens of genetic males or carpels of genetic females remained unaffected. Cytokinins, on the other hand, had dramatic effects on sex expression. Sprayings with kinetin (1 mg/l) three times daily for 2 weeks, changed the development of the staminate flowers. At the third or fourth leaf axial, male flowers bore semisterile or sterile anthers. However, flowers developed at the fifth leaf pair became pistillate. Partial male–female inflorescences never appeared. Female flowers were fertile and produced normal diploid seeds. As long as the treatment continued, all subsequent flowers of genetic males became feminized. Auxin added to kinetin solutions inhibited feminization. In contrast, genetic females were not affected. Cytokinins also had feminizing effects on monoecious mercuries, resulting in entirely female inflorescences.

Among genetic males examined, approximately one third were not feminized by kinetin, but were responsive to benzyladenine; they were designated as "resistant or strong males." The remainder of the genotypes could be divided into two categories: those very easily and quickly feminized by kinetin, "weak" males, and those requiring prolonged treatments, "intermediate or moderate" males.[13] Morphologically identical genotypes representative of these three categories of males were selected and clonally propagated for further experiments. These results suggest (1) involvement of cytokinins in female organogenesis, (2) genotypic variations in sensitivity to cytokinins, perhaps reflecting variations in cytokinin metabolism, and (3) auxin–cytokinin interactions for sex differentiation.

B. *IN VITRO* APPLICATION OF HORMONES

To recognize more precisely the effects of hormones on sex expression, male and female nodes were cultured *in vitro*.[14] Female nodes grown on a medium containing 0.1-mg/l IAA and no cytokinins developed staminate flowers, although other morphological characters remained unchanged. Other auxinic compounds, such as 2,4-D or NAA, had similar effects. Surprisingly, triiodobenzoic acid (TIBA) did not act as an antiauxin, as in *Avena* and in soybean bioassays;[15] instead, it induced male flower formation on genetic female nodes, similar to auxins (Table 3). Male nodes cultured on a medium containing auxin were not modified. The stamen-promoting effect of auxin was in contrast to its failure to induce masculinization on entire plants. It may be speculated that cytokinins are required for feminizing *in vivo*, and the level is higher in female individuals, while derooted female nodes cultured *in vitro* contain lower levels of cytokinins, thus allowing auxin in the medium to shift the

Table 22-3 **Effects of IAA and N^6-benzyladenine on Sex Expression of Flower Nodes Cultured *In Vitro***

Concentration of Benzyladenine (g/l)	0 Male	0 Female	10^{-6} Male	10^{-6} Female	10^{-4} Male	10^{-4} Female	10^{-3} Male	10^{-3} Female	10^{-2} Male	10^{-2} Female
IAA Conc. (g/l)										
0	—	—	F[a]	—	F	—	FF	—	FFF	—
10^{-6}	—	—	—	—	FF	—	FF	—	FF	—
10^{-4}	—	M[b]	ST[c]	—[d]	—	—	F	—	F	—
10^{-3}	—	M	—	—	—	—	F	—	F	—

[a] FFF, FF, and F: very strong, strong, and slight induction of female flowers; [b] M: induction of male flowers; [c] ST: induction of male sterility; [d] —: no effect.

hormone balance in favor of auxin, resulting in male flower formation. When male nodes were grown in the presence of strong cytokinins such as benzyladenine, female flowers replaced male flowers in the spike inflorescence. Isopentenyladenosine provoked the appearance of carpelloids or naked ovules. In sex-conversion experiments, only the primary sex characteristics were modified; secondary sex characteristics, principally the length of inflorescences, remained unmodified.

The balance between cytokinin and auxin in sex development was determined with a medium containing both hormones (Table 3). The masculinization effect of auxin on female nodes can be prevented at very low concentrations of cytokinins. Culturing of fertile male nodes at the minimal masculinizing concentration of auxin and low concentration of cytokinins (10^{-6} g/l) resulted in the development of a new stable and homogeneous phenotype that is male sterile. The anthers from such induced male sterile flowers exhibited characteristics typical of genetic male sterile plants, including the failure of tetrad microspores to separate, formation of dyads, and vacuolated pollen mother cells. At some concentrations there was also some semisterility.

C. FEEDER CULTURES AND REPRODUCTIVE ORGANOGENESIS

In order to determine the influence of endogenous compounds on the sex of flowers, genetically male and female nodes were cultured near feeder calluses of selected strains.[16] Strong male nodes cultivated in proximity of female calluses developed sterile male flowers, while weak male nodes developed carpelloid structures resembling pistillate flowers. Female nodes were unaffected. The sensitivity of nodes to feminizing agents was similar to that observed with external applications. Female tissues were most effective in feminization, whereas strong male tissues enhanced rhizogenesis. The inability of calluses from the strong male strain to induce male flower formation on female nodes may be related to the low level of auxin-like substances. Based on these observations, it may be concluded that auxins and cytokinins have opposite effects on sex expression in *Mercurialis*. When the balance is shifted in favor of cytokinins, a corresponding shift from maleness to femaleness occurs with the intermediate stages of semi and total male sterility. Higher levels of auxin lead to stamen development. Gibberellins promote flowering and elongation, but do not change sex expression. ABA and CCC do not have appreciable effects. Endogenous cytokinin–auxin balances are probably involved in sex differentiation of entirely male and female plants. Precise measurements and characterizations, in genotypes selected for their different sensitivities to feminizing agents, are necessary to confirm these results.

IV. GENETICS

There are no sex chromosomes in *M. annua*; both male and female karyotypes are identical with 16 chromosomes. At meiosis, eight bivalents are visible at pachytene, and all chromosomes form chiasmata. Mendelian genes controlling sex determination were identified previously.[17,18] The classification of genotypes, taking into consideration the response to feminization, is presented in Table 4. Briefly, dominant alleles of complementary genes *A* and *B* condition maleness. The sensitivity to an exogenous application of kinetin, and consequently male strength, is regulated by two forms of the *B* gene, *B1* being stronger than *B2*. For example, the genotype *AB1B2* is strongly male and resistant to kinetin, *AB1b2* is of the intermediate type, and *Ab1B2* is a weak male that is easily feminized. The recessive *a* and *b* alleles induce femaleness.

Table 22-4 **Classification of Genotypes used for Hormone Measurements According to Sex Genes, Sterility Determinants, and the Response to Exogenous Cytokinins**

Sex Series		
A/A, B1/b1, B2/b2 Strong male	N cytoplasm	Resistant to kinetin Sensitive to N^6-benzyladenine
A/A, B1/B1, b2/b2 Intermediate male	N cytoplasm	Sensitive to kinetin after prolonged treatments
A/A, b1/b1, B2/b2 Weak male	N cytoplasm	Very sensitive to kinetin, easily feminized
A/A, b1/b1, b2/b2 Female A	N cytoplasm	Fertile female
Sterile Series (Genes *A* + *B* for males; *A* + *b* for females)		
I/I, R1/R1, r2/r2 Restored fertile	S cytoplasm	Feminization by cytokinins
I/I, R1/R1, r2/r2 Semisterile	S cytoplasm	Feminization by cytokinins
I/I, r1/r1, r2/r2 Completely sterile	S cytoplasm	Feminization by cytokinins
I/I, R1/R1, R2/R2 Female 19-5	S cytoplasm	Fertile female

It is difficult to study male sterility in strict dioecious plants that do not produce pollen, but male sterile *M. annua* can be propagated as cuttings and are entirely feminized by cytokinins. They can be used in crosses to identify the determinants. Male sterility is nucleo-cytoplasmic conditioned by a nuclear inducer *I* acting in the sterile cytoplasm S. Two fertility restorers, *R1* and *R2*, have been identified; *R1* induces semirestoration, and *R2* is silent when alone. *R1* plus *R2* induce full restoration. As to sex genes, the isogenic lines differing by one, two, or three genes and by N (normal) or S cytoplasms were constructed. These lines are indispensable for studying the relationships between endogenous cytokinins and genetic factors controlling stamen and pistil development and male fertility.

V. REGULATION OF ENDOGENOUS HORMONE BALANCE

A. CYTOKININ METABOLISM AND REPRODUCTIVE ORGANOGENESIS

Cytokinin contents in apices of isogenic lines were determined by gas chromatography-mass spectrometry (GC-MS) after high-performance liquid chromatography (HPLC)[19-21] (for methods, see Chapter 24). The results are summarized in Table 5.[22] The most dramatic difference between male and female lines of the "sex series" resided in the presence of *trans*-zeatin nucleotide (t-io^{6}AMP) in male apices and *trans*-zeatin (t-io^{6}Ade) specific for females. *Trans*-ribosylzeatin (t-io^{6}Ado) and isopentenyladenosine (i^{6}Ado) and its nucleotide (i^{6}AMP) occurred in both sexes, while very small quantities of *cis*-zeatin riboside and only traces of *cis*-zeatin were identified. The quantity of *trans*-ribosylzeatin was three times higher in males than in females and correlated with the combinations of B genes. Strong males with the genotype *AB1B2* had the highest quantity. Genotypes *AB1b2* had intermediate levels and weak males (*Ab1B2*) had the least amount of ribosylzeatin. The presence of substantial levels of *trans*-ribosylzeatin in the males, without detectable levels of zeatin, suggests that the feminizing genes *b1* and *b2* are required for deribosylation. A hypothesis was proposed that in females, the active feminizing substance is *trans*-zeatin, whereas in males, zeatin production is blocked and the precursor metabolites, zeatin nucleotides, accumulate. Conversely, it is possible that male genes are related to the rapid conversion of *trans*-zeatin to the corresponding nucleoside and nucleotide. Clear interpretations would require additional information

Table 22-5 **Cytokinins and Auxin levels and IAA-Oxidase Activities in Shoot Apices of Isogenic Lines of *Mercurialis***

	Fertile Series					Sterile Series				
	Wild-type Male	Strong Male	Interm. Male	Weak Male	Female	Wild-type Female	Restored Male	Semi-fertile Male	Sterile Male	Female 19-5
Line No.		1	2	3	4		5	6	7	8
Cytokinins (ng/100 g fr.wt.)										
i^6Ado	196	170	112	103	33	66	106	51	32	14
i^6AMP	10	ND[a]	ND	ND	ND	33	ND	ND	ND	ND
c-io^6AMP	60	ND	ND	ND	ND	0	ND	ND	ND	ND
t-io^6AMP	166	ND	ND	ND	ND	0	ND	ND	ND	ND
c-io^6Ado	62	48	30	35	15	19	225	95	106	12
t-io^6Ado	186	147	105	92	43	63	404	118	133	33
c-io^6Ade	ND	TR[b]	TR	TR	TR	ND	669	178	192	18
t-io^6Ade	0	0	0	0	250	241	30	0	0	214
IAA (ng/100 g fr. wt.)	107	175	156	94	31	31	200	117	81	132
IAA Oxidase[c]	ND	309	395	456	751		370	509	627	457

[a] ND: not determined; [b] TR: trace amounts; [c] IAA oxidase: defined as methylene oxindole, ng/min/μg protein.

on the pathway of zeatin synthesis and interconversion. Preliminary results, using apex extracts, suggest the existence of a ribohydrolase activity in female apices, but the substrate of this enzyme, *cis*- or *trans*-metabolites, has not been established.

Important differences were also observed between genotypes of the fertile and sterile series, with the occurence of appreciable quantities of *cis*-zeatin and its riboside in the sterile series. In the S cytoplasm, the levels of *cis*-metabolites varied depending on the nuclear genes for fertility restoration. We speculate that two parallel pathways occur in the sterile series: one for stamen and carpel development related to *trans*-zeatin metabolites; one for anther sterility and fertility restoration, related to *cis*-zeatin metabolites. The hypothesis is supported by several observations:

1. *Cis*-zeatin and *trans*-zeatin directly induced sterility and femaleness, respectively.[20]
2. *Cis*- and *trans*-zeatin nucleotides initially occurred in shoot apices, with the *cis*-pathway becoming more pronounced as the tapetum and sporogenous tissues develop.[23]
3. Restoration of fertility did not eliminate the occurrence of *cis*-metabolites, although *trans*-zeatin compounds are restored to a limited extent.[17,24]

It is possible that gene *A*, required for stamens and pistil, also regulates *trans*-hydroxylation of i^6AMP, and that the sterility inducer *I* mediates *cis*-hydroxylation. Only by the identification of zeatin metabolic enzymes[25] and the corresponding genes can this hypothesis be confirmed.

B. CYTOKININ IN UNDIFFERENTIATED TISSUES

Analyses of cytokinins in tissue cultures of strong males, weak males, and females also confirmed the differences between sexes at the whole-plant level.[26] In addition, exogenously supplied benzyladenine accumulated in tissues of strong males, while tissues of weak males converted benzyladenine to its riboside. These observations suggest sex-determining genes affect cytokinin metabolism in organized and undifferentiated tissues.

C. AUXIN LEVEL AND REPRODUCTIVE ORGANOGENESIS

Auxin (IAA) level was determined in the same isogenic lines used for cytokinins.[27] As expected (Table 5), higher IAA content was found in genotypes expressing male characteristics. Female flowers had low IAA content, with the exception of females in the sterile series (line 8), perhaps due to an interaction between the sterility inducer gene *I* and the female determinants. In an experiment to determine the interaction between cytokinin and auxin, a weak male line (*AA/b1b1/B2B2*) was subjected to cytokinin treatment. The IAA content of treated plants decreased with a corresponding increase of IAA-oxidase activity[28,29] and coincided with complete feminization. These observations again support the balance between cytokinin and auxin, to condition female and male characteristics.

VI. MOLECULAR ANALYSES OF SEX EXPRESSION AND CYTOKININS

Differences at the molecular level, between genotypes with different sex determinants, were initially examined at the protein level, with the objective of recognizing targets of cytokinins. Total proteins were used to generate polyclonal antibodies. Using the Ouchterlony method, two antigenic proteins were identified from female flowers.[30] One of the antigens was present in males, while the other was female-specific and appeared when the weak male line was feminized by cytokinins. These preliminary results proved that genes, normally expressed in pistillate flowers, are also present in male plants, but are not expressed, while supplemental cytokinins induce their expression. Electrophoretic analyses identified male-specific anionic peroxidases.[31] Immunolocalization using antibodies specific to the male-specific isozymes indicated that the major site of the enzymes was the staminate flower primordium.[32] Isoesterases specific for female flowers were found. For both series of isozymes, when males were feminized by cytokinins, zymograms typical of females occur.

Genetic males and females have also been characterized by the array of cytokinin-containing tRNAs (tyrosine, serine, and leucine) and aminoacyl-tRNA synthetases.[33,34] In the female sex a chloroplastic leucyl-tRNA synthetase and two tyrosine-tRNAs of chloroplastic origin were unique. These qualitative differences accompanied the phenotypic transformation of males by cytokinins. Quantitative differences affecting tRNAs and aminoacyl-tRNA synthetases were also reproduced. *In vitro* translation products of poly(A)$^+$RNAs between floral types were compared.[35] Female flowers contained higher levels of translatable RNAs, and female-specific polypeptides accounted for 2% of the total.

Homologous and heterologous hybridizations between total poly(A)$^+$RNAs and cDNAs of different floral types were performed.[36] Female flowers had greater quantities of diverse poly(A)$^+$ (12,000) than did normal male flowers (8,500). Approximately 6% cDNAs was male-specific and did not hybridize with female RNAs (Figure 1). Female cDNAs readily

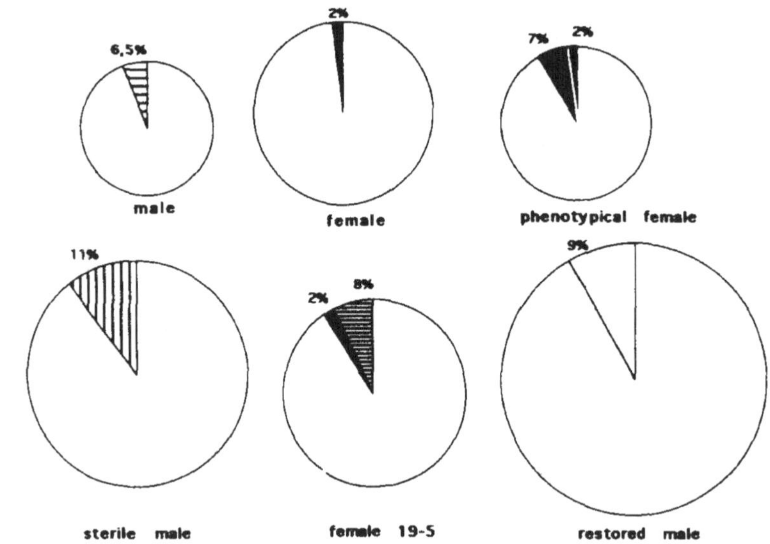

Figure 22-1 Diagrammatic representation of mRNA populations in various genotypes.

hybridized to 100% of the male RNAs, suggesting the cDNA populations of the female were composed of multiple copies of 94% male cDNAs. Pistillate flowers induced by cytokinins from genetic males expressed about 10,000 diverse sequences containing all female poly(A)$^+$RNAs plus 7% sequences specific for artificial feminization. In sterile series, restored fertile males had 10% specific poly(A)$^+$-RNAs, compared to normal fertile males, while the pistillate line (19-5) contained all female poly(A)$^+$-RNAs plus 8% unique sequence. Comparisons between normal and sterile males revealed an 11% difference (6% of the sequences were specific to fertile males compared with females).

Differential hybridizations between genetic lines resulted in the recovery of specific cDNA probes. One of the 16 clones specific for the sterile male[37] was 1550 base pairs in length, and Northern hybridization detected three RNAs from sterile genotypes. This probe did not hybridize to RNAs of other lines and appeared to be specific for sterility.

Two male-specific cDNAs were also purified.[35] One cDNA, 460 base pairs, detected a poly(A)$^+$ of 1230 bp. This message did not occur in sterile or semisterile or female flowers and may be related to normal male fertility. More importantly, the mRNA progressively disappeared from male plants treated with N^6-benzyladenine for feminization.

Three female cDNA clones were recovered from differential hybridizations between female and male cDNAs, but they detected complementary sequences in both female and male poly(A)$^+$ and were not female-specific. However, both hybridized with two RNAs of different lengths. The abundance of each particular poly(A)$^+$ was higher in females. Heterologous probes of other sources were employed with the objective of selecting female-specific cDNAs. The probes used included a cDNA from tobacco nitrate reductase, reported to be inducible by cytokinins, and several probes of alfalfa roots exposed for 48 h to *Rhizobium meliloti*. Only one of these cDNAs (Rhf 38 of alfalfa showing 85% sequence homology to a cDNA of *Phaseolus* encoding a hydroxyproline-rich glycoprotein) hybridized differentially to both sexes.[38] Four RNAs were female specific, and one was unique to males. Cytokinin treatment of male plants stimulated the progressive appearance of the four female-specific mRNAs, while the male-specific mRNA disappeared. Because Southern hybridization showed that one gene encoded these five mRNAs, we speculate that cytokinins regulate the expression of the *HRGP* gene at the posttranscriptional level in both sexes.

VII. CONCLUDING REMARKS

In *Mercurialis*, individual gene action controlling male and female flower development has been correlated to specific hormones. Sex determination genes and male sterility determinants regulate stamens or pistil formation, and anther sterility or fertility. Analyses of cytokinins in isogenic lines allowed the correlation between specific genes and metabolites. *Trans*-zeatin was correlated with gene action for femaleness, *trans*-ribosylzeatin and its nucleotide were correlated with genes controlling maleness, and *cis*-cytokinins were correlated with male sterility determinants. Characterization of cytokinin metabolic enzymes in each line should provide additional information. It is difficult to define the precise action of cytokinins on gene expression in floral organs. Changes in protein synthesis have been linked to endogenous or synthetic cytokinins, and *in vitro* translation of RNAs yielded male- and female-specific peptides. Differential hybridization kinetics between cDNAs and RNAs of various sexes, and the identification of specific cDNAs, appear to support this interpretation. However, the finding that multiple RNAs are encoded by a single gene suggest posttranscriptional control as well.

REFERENCES

1. **Skoog, F. C. and Tsui, C.,** Growth substances and the formation of buds in plant tissues, in *Plant Growth Substances*, Skoog, F., Ed., University of Wisconsin Press, Madison, 1951, 263.
2. **Frankel, R. and Galun, E.,** Pollination mechanism, in *Monographs in Theoretical and Applied Genetics, Vol. 2. Reproduction and Plant Breeding*, Springer-Verlag, Berlin, 1977, 204.
3. **Chailakhyan, M. and Kryanin, V.,** *Sexuality in Plants and Its Hormonal Regulation*, Springer-Verlag, New York, 1987.
4. **Durand, B. and Durand, R.,** Sexual differentiation in higher plants, *Physiol. Plant.*, 60, 263, 1984.
5. **Durand, B.,** Action d'une kinetine sur les caractères sexuels de *Mercurialis annua* L., *C. R. Acad. Sci.*, D, 263, 1966.
6. **Bracale, M., Caporali, E., Galli, M. G., Longo, C., Marziani-Longo, G., Rossi, G., Spada, A., Soave, C., Falavigna, A., Raffaldi, F., Maestri, E., Restivo, F. M., and Tassi, F.,** Sex determination and differentiation in *Asparagus officinalis*, *Plant Sci.*, 80, 67, 1991.
7. **Ye, D., Oliveira, M., Veuskens, J., Wu, Y., Installe, P., Hinnisdaels, S., Truong, A. T., Brown, S., Mouras, A., and Negrutiu, I.,** Sex determination in the dioecious *Melandrium*. The X/Y chromosome system allows complementary cloning strategies, *Plant Sci.*, 80, 93, 1991.
8. **Durand, B. and Durand, R.,** Dioecie, monecie et polyploidie chez les *Mercuriales annuelles*, *Bull. Soc. Bot. Fr.*, in press.
9. **Durand, B. and Durand, R.,** Sex determination and reproductive organ differentiation in *Mercurialis*, *Plant Sci.*, 80, 49, 1991.
10. **André, F., Durand-Rivieres, R., and Durand, B.,** Ontogenèse comparée des complexes inflorescentiels de *Mercurialis annua* (2× = 16) et *Mercurialis ambigua* (6× = 48). Essais préliminaires de morphogénèse expérimentale, *Bull. Soc. Bot. Fr.*, Memoires, 1967, 176.
11. **Durand, B.,** Expression du sexe chez les *Mercuriales annuelles*, *Bull. Soc. Fr. Physiol. Veg.*, 13, 195, 1967.
12. **Heslop-Harrison, J.,** The control of flower differentiation and sex expression, in *Regulateurs Naturels de la Croissance Végétale, Colloque 123*, C.N.R.S. ed., Paris, 1963, 649.
13. **Durand, B.,** Sélection de génotypes males de *Mercurialis annua* L. (2n = 16) en fonction de leur sensibilité aux cytokinines, *C. R. Acad. Sci.*, D, 263, 1309, 1966.
14. **Champault, A.,** Effets de quelques régulateurs de la croissance sur des noeuds isolés de *Mercurialis annua* (2n = 16) cultivés *in vitro*, *Bull. Soc. Bot. Fr.*, 120, 87, 1973.
15. **Galston, A. W.,** The effects of 2,3,5-triiodobenzoic acid on the growth and flowering of soybeans, *Am. J. Bot.*, 34, 356, 1947.
16. **Champault, A.,** Cytokinin contents and specific characteristics of tissue strains from three sexual genotypes of *Mercurialis annua*, evidence for sex-gene involvement at callus-tissue level, *Planta*, 166, 429, 1985.
17. **Louis, J.-P. and Durand, B.,** Studies with the dioecious angiosperm *Mercurialis annua* L. (2n = 16): correlation between genic and cytoplasmic male sterility, sex segregations and feminizing hormones (cytokinins), *Mol. Gen. Genet.*, 165, 309, 1978.
18. **Louis, J.-P.,** Genes for the regulation of sex differentiation and male fertility in *Mercurialis annua* L., *J. Hered.*, 80, 104, 1989.
19. **Dauphin-Guerin, B., Teller, G., and Durand, B.,** Different endogenous cytokinins between male and female *Mercurialis annua*, *Planta*, 148, 124, 1980.
20. **Louis, J.-P., Augur, C., and Teller, G.,** Cytokinins and differentation processes in *Mercurialis annua*, *Plant Physiol.*, 94, 1535, 1990.
21. **Letham, D. S. and Palni, L. M. S.,** The biosynthesis and metabolism of cytokinins, *Ann. Rev. Plant Physiol.*, 34, 163, 1983.
22. **Durand, B., Durand, R., Louis, J.-P., Teller, G., and Augur, C.,** Genetic regulation of cytokinin metabolism in reproductive differentiation, in *Physiology and Biochemistry of Cytokinins in Plants*, Kamínek, M., Mok, D. W. S., and Zazímalová, E., Eds., SPB Academic Publishing, The Hague, 1992, 335.

23. **Augur C.**, Etude des nucléotides de cytokinines dans les sporophytes et les tissus tumoraux de *Mercurialis annua*, *Mémoire D.E.A. Université d'Orléans*, 1985, 87.
24. **Durand, B. and Durand, R.**, Male sterility and restored fertility in annual mercuries, relations with sex differentiation, *Plant Sci.*, 80, 107, 1991.
25. **Mok, D. W. S., Mok, M. C., Martin, C., and Bassil, V.**, Zeatin metabolism in *Phaseolus*: enzymes and genes, in *Physiology and Biochemistry of Cytokinins in Plants*, Kamínek, M., Mok, D. W. S., and Zazímalová, E., Eds., SPB Academic Publishing, The Hague, 1992, 17.
26. **Besse, I., Verdeil, J.-L., Duval, Y., Sotta, B., Maldiney, R., and Miginiac, E.**, Oil palm (*Elais guineensis* Jacq) clonal fidelity: endogenous cytokinin and indolacetic acid in embryogenic callus cultures, *J. Exp. Bot.*, 43, 983, 1992.
27. **Hamdi, S., Teller, G., and Louis, J.-P.**, Master regulatory genes, auxin level and sexual organogenesis in the dioecious plant *Mercurialis annua*, *Plant Physiol.*, 85, 124, 1987.
28. **Hamdi, S. and Teller, G.**, Auxin levels, IAA-oxidase activity, cytokinins and sexual organogenesis in the dioecious plant *Mercurialis annua*, *Plant Physiol.*, 7, 105, 1988.
29. **Hamdi, S.**, Regulation of IAA-oxidase activities: correlations with sex genes, sterility determinants and IAA levels in *Mercurialis annua*, *Plant Physiol. Biochem.*, 26, 630, 1988.
30. **Durand, B. and Durand, R.**, Cytokinines et régulation de la synthèse d'une protéine spécifique du sexe femelle chez une plante dioïque *Mercurialis annua* (2n = 16), *C. R. Acad. Sci.*, D, 269, 1639, 1969.
31 **Durand, B., Durand, R., Kahlem, G., Champault, A., and Delaigue, M.**, Interactions moleculaires contrôlant la formation des organes sexuels chez une plante dioïque *Mercurialis annua* L., *Rev. Cytol. Biol. Vég.*, 37, 249, 1974.
32. **Kahlem, G.**, Isolation and localization by histoimmunology of isoperoxidases specific for male flowers of the dioecious species *Mercurialis annua* L., *Dev. Biol.*, 50, 58, 1976.
33. **Bazin, M., Chabin, A., and Durand, R.**, Comparaison des taux d'aminoacylation des tRNA leucine, valine, serine et tyrosine extraits des fleurs mâles et femelles de *Mercurialis annua* L. en présence des enzymes homologues ou hétérologues, *Plant Sci. Lett.*, 4, 207, 1975.
34. **Bazin, M., Chabin, A., and Durand, R.**, Comparison between four isoaccepting transfer ribonucleic acids and corresponding synthetases in male and female flowers of the dioecious species *Mercurialis annua* L., *Dev. Biol.*, 44, 288, 1975.
35. **Hamdi, S., Yu, L.-X., Cabré, E., and Delaigue, M.**, Gene expression in *Mercurialis annua* flowers, *in vitro* translations and sexual genotype specificity. Male specific cDNA cloning and hormonal dependence of a corresponding specific RNA, *Mol. Gen. Genet.*, 219, 545, 1989.
36. **Sagliocco, F., Yu, L.-X., and Rabouille, F.**, Hybridization kinetics and gene expression in *Mercurialis annua* flowers, sexual genotype specificity and correlation with hormonal contents, *Mol. Gen. Genet.*, 218, 168, 1989.
37. **Durand, B., Durand, R., Cabré, E., Hamdi, S., Sagliocco, F., Yu, L.-X., and Delaigue, M.**, cDNAs libraries of hormone-dependent sequences specific for normal male and sterile male flowers of *Mercurialis annua* 2n = 16, in *Signals in Plants Development*, SPB Academic Publishing, The Hague, 1989, 91.
38. **Ait-Ali, T. and Allisson, L.**, Sex specific expression of RNAs for hydroxyproline rich glycoproteins: are differential splicings also characteristic of plant sexes?, *C. R. Acad. Sci.*, D, 316, 89, 1993.

Chapter 23

Immunoanalysis of Cytokinins

Gary M. Banowetz

TABLE OF CONTENTS

ABBREVIATIONS

Abbreviations not defined in text include:
DEAE — diethylaminoethyl
HPLC — high-performance liquid chromatography
Z — zeatin
diHZ — dihydrozeatin
***trans*-Z** — *trans*-zeatin
***cis*-[9R]Z** — *cis*-zeatin riboside
(diH)[9R]Z — dihydrozeatin riboside
[9R]BAP — N^6-benzyladenosine
[9R]Z — zeatin riboside
[9R]iP — isopentenyl adenosine

I. INTRODUCTION

The past twenty years have witnessed a marked growth in the use of specific antibodies in studies of cytokinin biology. Since the first apparent report of an anticytokinin antibody in 1970,[1] anticytokinin antisera and monoclonal antibodies have been used to isolate and quantify endogenous cytokinins from a wide variety of plant tissues[2-4] and microbial sources.[5-8] Because cytokinins occur in many forms (i.e., free bases, ribosides, ribotides, *O*-, 7-, and 9-glucosides, xylosides, etc.) at relatively small concentrations ($<10^{-6}$ *M*), analytical procedures with high sensitivity and selectivity are required. Most immunoanalysis methods can detect picogram quantities of cytokinins and, combined with HPLC, provide a high level of selectivity. In addition, these techniques are well suited to handle relatively large numbers of samples. In general, quantitative estimates of cytokinin content by immunoanalysis have been in agreement with estimates provided by instrumental analyses. This chapter will discuss the preparation and use of specific antibodies in immunoaffinity and immunoassay techniques used for cytokinin analysis. Because the accuracy and reliability of a cytokinin immunoassay is highly dependent upon sample preparation methods, a section on these techniques is included. Instrumental analysis of cytokinins is discussed in this volume.[9]

II. BASIC PRINCIPLES OF CYTOKININ IMMUNOASSAY

Quantification of cytokinins by immunoassay is based upon the competition of cytokinins present in a sample with a known quantity of labeled or immobilized cytokinin for binding to an anticytokinin antibody. The degree of competition is measured using either radiolabeled cytokinins or enzyme-labeled probes. A standard curve based upon the addition of specific quantities of cytokinins to the assay is used to quantify cytokinins present in samples. Although the immunoassay may take multiple forms, all are based upon the availability of high-affinity anticytokinin antisera or monoclonal antibodies.

III. ANTIBODY PREPARATION

A. PREPARATION OF IMMUNOGENIC CONJUGATES

Because cytokinins are low-molecular-weight compounds, they are not immunogenic unless they are covalently linked to a carrier protein. The most commonly used carrier proteins are bovine serum albumin (BSA), ovalbumin (OVA), or keyhole limpet hemocyanin (KLH). Conjugation procedures are similar, regardless of the carrier protein employed. Most cytokinin-protein conjugates have been prepared by reaction of periodate-oxidized cytokinin 9-ribosides with free amino groups on the carrier protein.[10-12] Briefly, the cytokinin ribosides are converted to dialdehydes by sodium periodate oxidation and mixed with a basic (pH 9.2 to 9.4) solution of the carrier protein. The resulting conjugate is stabilized by reduction with sodium borohydride. Using conjugates prepared by this general procedure, antibodies have been prepared against *trans*-ribosylzeatin,[13-22] *cis*-ribosylzeatin,[23,24] N^6-(Δ^2-isopentenyl)adenosine,[1,12,14,18,21,25-31] dihydroribosylzeatin,[20,29,32] N^6-(*O*-hydroxybenzyl-amino)-purine,[33] and N^6-benzyladenosine.[34]

An alternative procedure that avoided the periodate oxidation in order to retain the structural integrity of the sugar moiety was described.[35] This involved a five-step procedure in which the 2′- and 3′-hydroxyl groups of the sugar ring were protected, and the cytokinin riboside was linked to amino groups on the carrier protein through the primary hydroxyl group of the sugar residue. The resulting antisera showed a high degree of specificity for the homologous cytokinin, N^6-(Δ^2-isopentenyl)adenosine, and little cross-reactivity with *trans*-zeatin or its riboside. Strnad et al. prepared a hemisuccinyl bridge through the 5′-hydroxy group of N^6-benzyladenosine,[36] which was linked to BSA, using a mixed anhydride reaction. This conjugate elicited polyclonal antibodies that cross-reacted with structurally similar cytokinins. They also prepared cytokinin-BSA conjugates, using pivalic acid to produce 2′-,3′-acetals, which were subsequently linked to BSA by a mixed anhydride reaction.

Antibodies prepared against cytokinin 9-riboside-protein conjugates usually show high cross-reactivity with cytokinin-9-glucosides, lower reactivity with 7-glucosides, variable reactivity with the free bases, and generally low cross-reactivity with cytokinin *O*-glucosides. Preparation of antibodies against cytokinins that contain *O*-substituted sugar residues also requires a conjugation procedure in which the structural integrity of the sugar residue is preserved. Brandon et al.[37] prepared a 9-(2-carboxyethyl)zeatin-*O*-glucoside-BSA conjugate, using morpholinoethylisocyanide.[38] The monoclonal antibody prepared against the conjugate showed high specificity for the *O*-glucoside.

B. ANIMAL IMMUNIZATION

Anticytokinin antibodies have been prepared in the form of antisera, mainly in rabbits, or monoclonal antibodies derived from mice or rats. A variety of immunization protocols have proven effective both in terms of the injection schedule and the adjuvant used. A general injection schedule includes primary and booster immunizations, often at 2 to 3-week intervals. Anticytokinin antibody levels in sera collected from trial bleedings are used to determine when a sufficient immune response against the cytokinin has occurred. Prior to injection, cytokinin-protein conjugates are mixed with an adjuvant to delay the release of the immunogen from the injection site and to enhance the immune response. The most widely used adjuvants for this work have been Freund's complete and incomplete adjuvants. Although Freund's complete adjuvant has been used effectively for years, certain animals that receive this adjuvant, may exhibit side effects, including fever and hypersensitivity. A recent monoclonal anti-*cis*-ribosylzeatin antibody was prepared using an adjuvant with fewer side effects than those associated with Freund's complete adjuvant.[24] Rabbit immunizations are usually given at multiple sites, subcutaneously and intramuscularly, with a primary immunizing dose of approximately 1 mg. Subsequent booster immunizations generally have contained reduced quantities of the cytokinin-protein conjugate. Intraperitoneal injections of 25 to 200 μg of conjugate have been used for most primary immunizations in mice, followed by subsequent booster injections with reduced immunizing doses. After trial bleedings have demonstrated the presence of high levels of anticytokinin antibodies, animals are either bled for polyclonal antisera or prepared for hybridoma and subsequent monoclonal antibody production.

C. MONOCLONAL ANTIBODY PRODUCTION

The choice between polyclonal antisera or monoclonal antibodies is dependent upon the facilities available and long-term research goals. The greater commitment in time, labor, and expense involved in hybridoma preparation is at least partially offset by the advantages associated with monoclonal antibodies. The capability to produce virtually unlimited quantities of identical anticytokinin antibodies provides the potential for these immunoassays to serve as standard reference assays. In addition, the capacity to provide a selection of antibodies with different affinities makes it possible to select antibodies for specific purposes in cytokinin isolation and quantification. For example, it may be desirable to select an antibody with relatively low affinity, for use in immunoaffinity isolation of cytokinins, while a high-affinity antibody may be more appropriately used in the quantitative immunoassay. Furthermore, the relative ease with which the antibodies can be purified from hybridoma culture supernatants, using commercially available adsorbents, provides a means for obtaining milligram quantities of a specific antibody. Although specific protocols in hybridoma preparation vary, all involve polyethylene glycol-mediated fusing of immune splenocytes to myeloma cells, culturing of the resulting hybrid cells in a selective medium, and subsequent cloning of antibody-secreting cells. Detailed procedures for hybridoma preparation are available.[39-40]

D. ANTIBODY CHARACTERIZATION

The specificity and sensitivity of a cytokinin immunoassay is dependent upon the cross-reactivity and relative affinity of the antibody or antiserum with structurally similar cytokinins. Consequently, cross-reactivity determinations are required of anticytokinin antibodies or antisera. Characterization studies also usually include isotype determinations as an aid in the establishment of an antibody purification protocol and may include determinations of antibody affinity constants.

1. Cross-Reactivity

Although the immunoassay format employed is not likely to affect antibody cross-reactivities, these characterization studies should be done using the protocol that will be employed in the final cytokinin determinations. Anticytokinin antisera prepared against periodate-oxidized 9-ribosides have shown remarkably similar cross-reactivities. In general, these antisera have reacted, to some extent, with the free bases and 9-derivatives of the homologous cytokinin. The antisera have shown considerable selectivity for the N^6-prenyl side chain, in most cases distinguishing *cis*- from *trans*-zeatin.[14,16-18,22,41] Similarly, anti-isopentenyl adenosine sera have generally not reacted, to any great extent, with zeatin or dihydrozeatin cytokinins. The reactivity of one antiserum prepared against the ribotide of N^6-(Δ^2-isopentenyl)adenine seemed sensitive to the presence of the phosphate group.[26] (See Badenoch-Jones et al.[29] for a comparison of cross-reactivity of selected anticytokinin antisera.)

In contrast to the similarities noted among the polyclonal antisera, monoclonal anticytokinin antibodies have been described with quite dissimilar cross-reactivities (Tables 1 and 2). The antiribosylzeatin antibodies characterized by Trione et al.[19] showed cross-reactivities with N^6-(Δ^2-isopentenyl)-adenosine, which ranged from 0 to 221%. Similar differences were seen between these antibodies in the extent of reactivity with *cis*-ribosylzeatin. Wang et al. reported marked differences in cross-reactivities of their anti-N^6-(Δ^2-isopentenyl)adenosine antibodies with certain other cytokinins.[12] These differences provide the potential for the preparation of specific "antibody cocktails" appropriate for immunoaffinity isolation of a broad (or narrow) range of cytokinins from crude extracts.

2. Isotype Analysis

Information regarding isotype (and subisotype) of monoclonal anticytokinin antibodies may be useful in the design of antibody purification procedures. Commercially available kits or antisera are used, often in an ELISA format, for isotype and subisotype determinations. A membrane-based kit that provides rapid isotype determination is also available (SangStat, Menlo Park, CA). Most monoclonal anticytokinin antibodies have been characterized as subisotypes of IgG, including IgG_1, IgG_{2a}, and IgG_{2b},[12,19,20,31] although one IgM anti-N^6-(Δ^2-isopentenyl)adenosine antibody has been described.[12]

3. Antibody Purification

Most cytokinin immunoassays are performed using antibodies in a relatively unpurified form as an antiserum or a hybridoma culture supernatant. However, some procedures (i.e., preparation of immunoaffinity matrices) require highly purified anticytokinin antibodies. Highly purified preparations of most monoclonal antibodies can be prepared rather simply using commercially available adsorbents (i.e., Protein A and Protein G) that bind

Table 23-1 **A Comparison of Cross-Reactivities of Monoclonal Anti-Ribosylzeatin Antibodies with Selected Structurally Similar Cytokinins**

Cross-reactant	#3[a]	#5	#7	#8	#10	#12	#16	J18[b] I-D1	J18 III-B5	J18 I-B3
trans-[9R]Z	100	100	100	100	100	100	100	100	100	100
trans-Z	0	57	28	30	0	0	22	36.2	52.3	47.3
cis-ZR	41	48	39	0	0	72	0	5.3	1.9	0.8
[9R]diHZ	67	52	65	52	56	0	15	2.6	2.4	1.2
iPA	0	0	0	22	28	221	0	1.8	3.1	0.5

[a] Data for #3, 5, 7, 8, 10, 12, and 16 from Trione, E. J., et al., *J. Plant Growth Regul.*, 4, 101, 1985; [b] Data for J18 I-D1, J18 III-B5, and J18 I-B3 from Eberle, J., et al., *Plant Physiol.*, 81, 516, 1986.

Table 23-2 **A Comparison of Cross-Reactivities of Monoclonal Anti-N^6-(Δ^2-isopentenyl)adenosine Antibodies with Selected Structurally Similar Cytokinins**

	Monoclonal Anti-N^6-(Δ^2-isopentenyl)adenosine Antibodies							
Cross-reactant	#3[a]	#4	#15	156[b]	157	158	159	160
iPA	100	100	100	100	100	100	100	100
iP	31	100	100	11	20	23	50	106
[9R]Z	<1	<1	<1	1.4	1.6	3.1	0.3	11
[9R]diHZ	6	<1	<1	10	4.1	36	0.8	35
diHZ	<1	<1	<1	0.7	3.2	7.1	<0.1	79
[9R]N^6-BA	100	100	100	6.8	1.9	8.4	55	>133

[a] Data for #3, 4, and 15 from Trione, E. J., et al., *Physiol. Plant.*, 70, 467, 1987; [b] Data for 156–160 from Wang, T. L., Cook, S. K., and Knox, J. P., *Phytochemistry*, 26, 2447, 1987.

the Fc portion of the immunoglobulin molecule. A specific range of isotypes and subisotypes are bound by each, with Protein G being the most broadly reactive of the two.[42,43] These adsorbents are available in a variety of forms for use in liquid chromatography, HPLC, or as part of membrane-based purification devices. Antibodies are applied to the adsorbents at a neutral or slightly basic pH, nonbinding material is removed by repeated washes, and the immunoglobulins are generally eluted by lowering the pH.[44]

Protein A or G purification methods are particularly useful when anticytokinin antibodies are the sole type of antibodies present, as is the case in hybridoma supernatant. Purification of antibodies from polyclonal antisera or ascites fluid, which likely contain immunoglobulins directed against noncytokinin antigens, can be accomplished using affinity chromatography. Periodate-oxidized cytokinin ribosides are immobilized on poly-lysine agarose, and the anticytokinin-containing fluid is passed over the matrix. Antibodies bound to the cytokinin-agarose column subsequently are eluted using a low pH buffer.[11]

4. Affinity Estimates

Meaningful immunoassay work can be performed with little direct knowledge of antibody affinity constants. Cross-reactivity characterization provides information regarding the relative affinities of the anticytokinin antibody for structurally-related cytokinins, and the determination of assay detection limits. Approximations of relative affinities of monoclonal antibodies can be obtained by simultaneous screening of hybridoma supernatants for both immunoglobulin content and anticytokinin activity. Culture fluids that show strong anticytokinin activity, but contain relatively small quantities of immunoglobulin, likely contain higher-affinity antibodies. Determinations of precise affinity constants are made by incubating a constant amount of anticytokinin antibody with a series of concentrations of radiolabeled cytokinins and preparing Scatchard[45] or Sips[46] plots. Determinations have been made for both polyclonal[13,23,34] and monoclonal anticytokinin[12,20] antibodies. Affinity constants that have been determined for the monoclonal antibodies are comparable to those noted for polyclonal antisera. A method for precise determination of an antibody affinity constant is described by Campbell.[40]

IV. ASSAY DEVELOPMENT

Immunoanalysis of cytokinins is ultimately based on the competition of a cytokinin present in a plant extract, with a known quantity of cytokinin introduced into the assay, for binding to a cytokinin-specific antibody. Because the plant extract will usually contain multiple cytokinins, and because substances that interfere with the immunoassay may be present, relatively extensive sample clean-up, including HPLC, usually precedes the immunoassay. In practice, the final immunoassay is generally performed on HPLC fractions of the original sample.

The two most common forms of cytokinin immunoassays are radioimmunoassay (RIA) and enzyme-linked immunosorbant assay (ELISA). Although the ELISA is less expensive and easier to set up and avoids the problems associated with disposal of radioactive waste, the choice between the two is somewhat dependent upon the nature of the anticytokinin antibody available. Monoclonal antibodies have been used effectively in both RIA[12,19] and ELISA.[20,31,37,47] The use of polyclonal antisera may be complicated by the presence of antibodies reactive with the epitope that represents the bridge between the carrier protein and cytokinin. These antibodies may interfere in an ELISA in which the cytokinin-protein conjugate serves as the source of competitor cytokinin. In this case RIA is preferred because the competitor cytokinin is provided by labeled free cytokinin rather than by the cytokinin-protein conjugate.

Procedures for combined HPLC-radioimmunoassay analyses of cytokinins have been published.[13,14] The RIA methods generally involve reaction of the anticytokinin antibody with a mixture that contains a known quantity of radio-labeled cytokinin and an unknown quantity of cytokinin present in a plant extract. After a brief incubation, antibody-cytokinin complexes are precipitated with ammonium sulfate, and radioactivity in the precipitate is compared to standard curve determinations prepared with known quantities of competing cytokinin. (For a detailed RIA protocol, see Weiler.[48]) A variety of ELISA protocols have been published. Eberle et al. used an immobilized antibody procedure in which rabbit anti-mouse immunoglobulin was bound to a polystyrene plate and used for subsequent immobilization of the mouse anti-cytokinin antibody.[20] (Mouse immunoglobulins do not adsorb well to the polystyrene plates typically used for ELISA analyses.) Cytokinin-containing sample or cytokinin standard solutions were added to respective wells. After appropriate incubation, wells were washed and cytokinin-alkaline phosphatase cytokinin tracer was added to each well, followed in turn by enzyme substrate. Trione et al.[31,47] used an ELISA based upon immobilized antigen (cytokinin–protein conjugate) followed by addition of either cytokinin-containing sample or cytokinin standard plus the monoclonal anticytokinin antibody. After appropriate incubation, alkaline-phosphatase-labeled goat anti-mouse immunoglobulin was used to measure the extent to which the added cytokinin competed with the immobilized cytokinin-protein conjugate for binding of the monoclonal antibody. Figure 1 provides a graphic comparison of the immobilized antibody and antigen procedures. Maldiney et al. developed a similar immobilized antigen procedure that incorporated avidin-biotin into the detection system for signal amplification.[22] They used a biotinylated goat anti-rabbit immunoglobulin to measure the amount of rabbit antibody bound by the cytokinin–protein conjugate. An avidin–alkaline phosphatase conjugate was used to provide the enzymatic detection system.

V. ASSAY VALIDATION

Whether RIA or a form of ELISA is selected, assay validation is imperative. A number of excellent discussions on assay validation are available.[48-51] The accuracy, precision, and reproducibility of the cytokinin immunoassay, along with an estimate of the sample recovery associated with a sample preparation, are factors that need to be considered. Assay accuracy may be influenced by the presence of interfering substances present in the plant extract, although their presence will be minimized by the HPLC purification that normally accompanies cytokinin quantification. Interfering substances may take the form of cross-reactive (structurally similar) compounds that copurify with the authentic cytokinin, or may represent compounds that interfere with the mechanics of the assay. Examples of the latter include inhibitors of enzymes used in an ELISA, carryover of solvents used in the extraction process that partially denature immunoglobulins, detergents that may desorb reactants from the polystyrene surface, or the presence of substances that quench radioactivity in an RIA. Radiolabeled cytokinins added to the sample provide an estimate of the percent recovery in routine analyses. Recoveries depend upon the specific cytokinin[16] and the tissue from which it is isolated.[23] Badenoch-Jones et al. showed that addition of the radiolabeled recovery markers during sample purification did not alter the shape of the RIA standard curve, as long as appropriate blanks were used.[16]

Dilution analyses of biological samples are used to determine whether interfering substances are present in a sample. Sample dilution curves of the cytokinin-containing sample should be parallel to the standard curve if no interfering compounds are present. When appropriate sample clean-up

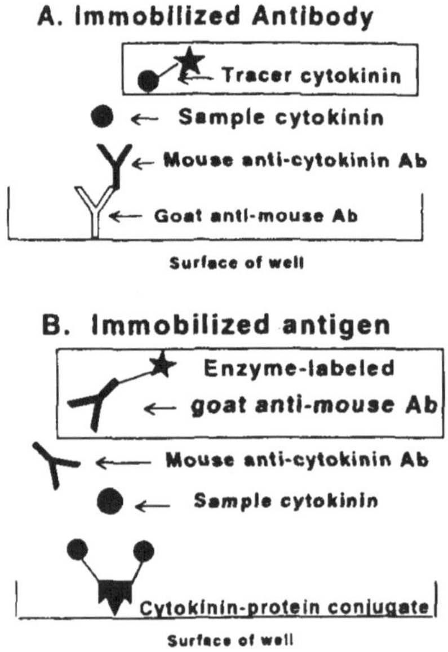

Figure 23-1 Comparison of two ELISA procedures for quantification of cytokinins. (A) Immobilized antibody technique. Tracer cytokinin (enzyme-labeled) is added after incubation of sample cytokinin with immobilized mouse anticytokinin antibody. (B) Immobilized antigen technique. Enzyme-labeled goat anti-mouse Ab is added after competition between immobilized cytokinin-protein conjugate and sample cytokinin, for binding to mouse anti-cytokinin Ab, has occurred.

procedures are used prior to a cytokinin immunoassay, the presence of interfering compounds is minimized.

Ultimately, immunoassay validation is most often confirmed by gas chromatography-mass spectrometry (GC-MS). In general, cytokinin immunoassays have been in agreement with GC-MS analyses.[4,52-56] In one notable exception, immunoassay estimates of cytokinin content did not agree with instrumental analysis and overestimated the amount of zeatin and ribosylzeatin by tenfold compared to GC-MS.[57] A number of references exist that describe methods for GC-MS analysis of cytokinins.[58-63]

VI. DATA ANALYSIS

Immunoassay standard curves, which are generated using specified quantities of cytokinin competitor, take the form of dose/response curves. A plot of the percent of competition against the log of the dose (quantity of cytokinin competitor) will generally be sigmoidal in shape. To generate a more linear standard curve from which quantities of cytokinins present in samples may be extrapolated, these data are usually represented by logit-transformed response (i.e., absorbance, radioactivity) values plotted against log dose values. Response data are transformed to respective logit values as follows: $\text{logit} = \ln[(B/B_0)/(1-B/B_0)]$ where "B_0" is the response when no competitor is present in the reaction, and "B" is the response at each respective level of competitor.[64] Using this transformed standard curve, ED_{50} values (the level of competitor that gives 50% competition) can be calculated; 50% competition occurs when logit = 0. The ED_{50} values serve as estimates of the relative affinities of anticytokinin antibodies and are useful when a selection of monoclonal antibodies are screened or anticytokinin antisera are compared. (For a comparison of nontransformed and logit-transformed standard curves, see Weiler.[13])

VII. SAMPLE PREPARATION

A. GENERAL CONSIDERATIONS

The accuracy and precision of a cytokinin immunoassay is highly dependent upon the sample preparation and degree of cytokinin purification achieved. Certain studies have used immunoassay to quantify cytokinins in crude extracts;[13,27,65-67] however, sample preparation preceding an immunoassay more often has included additional chromatographic steps,[21,37,68] HPLC,[6,8,14,18,20-24,29-31,33,69-75] or immunoaffinity purification followed by HPLC.[2,4,7,11,12,32,56,76-78]

The immunoassay of crude extracts may be useful where prior work has demonstrated the presence of only a limited number of cytokinins from the specific tissue.[13] However, most analyses require further sample cleanup. Many substances that may interfere with cytokinin immunoassay can be removed by passage of the crude extract over an ion exchange medium (i.e., DEAE cellulose) and subsequent isolation and elution of cytokinins from a C_{18} cartridge.[14,79] Cytokinins, including free bases, ribosides, ribotides, and glucosides, are retained on C_{18} cartridges, although retention is pH dependent.[79] The material eluted from the C_{18} cartridge will be highly enriched for cytokinins and adequately purified for HPLC, but will contain noncytokinin UV-absorbing substances.

B. IMMUNOAFFINITY PURIFICATION

As an alternative to C_{18} isolation of cytokinins, immunoaffinity chromatography can provide highly purified cytokinin preparations that contain minimal quantities of other UV-absorbing material. Immunoaffinity procedures involve immobilization of anticytokinin antibodies onto an insoluble matrix and application and subsequent elution of cytokinins from the matrix. Anticytokinin antibodies have been coupled to a variety of chromatographic matrices. Cyanogen bromide (CNBr)-activated cellulose served as an affinity matrix in a number of studies.[3,4,7,11,32,56] Davis et al.[32] evaluated six matrices for isolation of cytokinins, including CNBr-activated Sépharose 4B and Fast-Flow Sepharose CL-4B (Pharmacia, Piscataway, NJ), Affi-Gel 10 (Bio Rad, Richmond, CA), Ultraffinity-EP (Beckman, Berkeley, CA), and glycophase silica. Their studies indicated that the nature of the support, and the manner in which antibodies were linked to the support, affected the amount of noncytokinin UV-absorbing material retained on, and eluted from, the column. All the supports tested provided levels of purification sufficient for subsequent HPLC separation.

The useful life of anticytokinin immunoaffinity matrices is extended by prior passage of plant extracts over an anion exchange resin.[11,78] Although methanol is generally used to elute cytokinins from the immunoaffinity matrix, these columns have remarkable stability and have been reused over twenty times.[78] Binding of cytokinins to the columns may be sensitive to pH, temperature, and salt concentration, although effective binding has occurred at pH ranges from 5 to 8 at temperatures from 25 to 37°C, with salt concentrations up to 0.5 *M*.[11] Binding of the cytokinins by the antibodies occurs rapidly at 37°C, and high recoveries of cytokinins occur with a single passage of the sample over the matrix, using flow rates of 0.5 to 1.0 ml/min.

Although immunoaffinity procedures have the capability to provide highly purified cytokinins for HPLC separation, certain limitations of the method should be addressed. First of all, the technique requires relatively large (milligrams) quantities of anticytokinin antibodies that are not commercially available, except as part of immunoassay kits. Second, the range of the cytokinins bound by the immunoaffinity matrix is strictly dependent upon the reactivity of the antibody or antibodies immobilized on the matrix. Cytokinin derivatives not recognized by the antibody matrix pass through the column and, unless captured by other means, are lost to the analysis. One method used to expand the range of cytokinins bound involves the preparation of mixed-bed matrices in which multiple anticytokinin antibodies, each with defined reactivities, are used in preparation of the matrix.[7,56] Because antibodies with specificities for all metabolically important cytokinins may not be available, alternate means of isolating certain cytokinins may be required. For example, cytokinin *O*-glucosides not bound by a mixed-bed immunoaffinity matrix were collected on C_{18} cartridges.[3,4] The cytokinins eluted from C_{18} were treated with β-glucosidase, passed over the immunoaffinity matrix a second time, and analyzed as free bases and nucleosides. Two alternate strategies for cytokinin sample preparation are presented in Figure 2.

C. HIGH-PERFORMANCE LIQUID CHROMATOGRAPHY

Because plant and microbial extracts often contain multiple cytokinins, HPLC is commonly performed as a final step in sample preparation immediately preceding immunoassay. Ideally, individual cytokinins are isolated in discrete HPLC fractions that are used as the source of competing sample cytokinins in immunoassay. Comparison of ELISA results to HPLC elution profiles of authentic cytokinins may suggest the identity of the cytokinins present in the original extract, although GC-MS analysis of individual fractions is used for final confirmation.

Reversed-phase HPLC on octadecylsilica (ODS), a C_{18} medium, is the most commonly used method for cytokinin isolation.[3,7,14,18,20,30,56,70-72,80] Most protocols involve the injection of cytokinins onto the ODS column at an acidic pH (i.e., pH 3.5) in an aqueous buffer or relatively low concentration of organic solvent, and their sequential elution by a gradual increase in the amount of organic solvent in the mobile phase. Badenoch-Jones et al. used C_8 columns, in preference to ODS, for HPLC separation of cytokinins.[16,29,69]

Although elution profiles provided by HPLC UV detectors appear to indicate the separation of cytokinins into discrete fractions, immunoassay of these fractions often indicates an overlap. This is due to the difference in sensitivity of the two methods. Practical HPLC sensitivities, using a diode array detector at 254 nm, range from 4 to 15

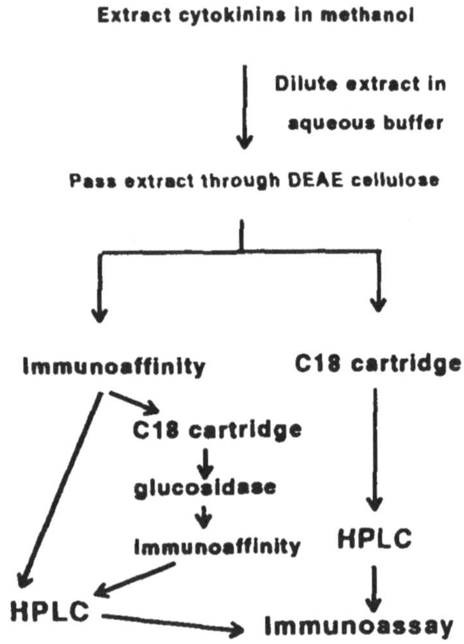

Figure 23-2 Flowchart of two procedures for isolation and quantification of cytokinins from samples. Immunoaffinity technique may require the indicated use of a C_{18} cartridge if immobilized antibodies do not bind all cytokinins (i.e., *O*-glucosides) present in sample.

ng, depending upon the cytokinin, whereas most immunoassay protocols have lower detection limits in the range of 20 pg. For this reason, a great deal of caution is required in the interpretation of the cytokinin content of a peak, based upon HPLC resolution. Immunoassays have demonstrated that HPLC peaks that appear highly purified may contain small amounts of cytokinins that represent shoulders from adjacent peaks.

VIII. CONCLUDING REMARKS

The development of immunoassay procedures for use in cytokinin research has enhanced the ease and sensitivity with which cytokinins are quantified. Previously used bioassay procedures lacked the sensitivity and specificity inherent in antibody-based assays. It should be noted, however, that a bioassay provides information regarding biological activity that is beyond the scope of an immunoassay. Cytokinin immunoassays have, in most cases, provided quantitative estimates in good agreement with instrumental analyses. GC-MS remains the method of choice for validation of immunoassays. An immunoassay is another tool available for cytokinin research. It has supplemented, but not replaced, previously used analytical methods. It is certain that new immunoassay procedures will be developed as antibodies against additional cytokinin derivatives become available. Furthermore, it seems likely that the trend toward immunoassay automation will continue and will improve accuracy and reproducibility. Ultimately, it is hoped that the improved speed and accuracy of these analytic procedures will accelerate the growth of our knowledge of cytokinin biology.

ACKNOWLEDGMENTS

I thank Drs. John Carman, and Richard Hess for their helpful comments regarding cytokinin immunoassays and the preparation of this manuscript. This research was conducted by the Agricultural Research Service, U.S. Dept. of Agriculture, Corvallis, OR, Technical Paper No. 10,047 of the Oregon Agricultural Experiment Station. Use of a company or product name by the department does not imply approval or recommendation of the product to the exclusion of others that may also be suitable.

REFERENCES

1. **Hacker, B., Vunakis, H. V., and Levine, L.,** Formation of an antibody with serological specificity for N^6-(Δ^2-isopentenyl) adenosine, *J. Immunol.*, 108, 1726, 1970.
2. **Hansen, C. E., Kopperud, C., and Heide, O. M.,** Identity of cytokinins in *Begonia* leaves and their variation in relation to photoperiod and temperature, *Physiol. Plant.*, 73, 387, 1988.
3. **Saavedra-Soto, L. A., Durley, R. C., Trione, E. J., and Morris, R. O.,** Identification of cytokinins in young wheat spikes (*Triticum aestivum* cv. Chinese Spring), *J. Plant Growth Regul.*, 7, 169, 1988.

4. **Doumas, P., Bonnet-Masimbert, M., and Zaerr, J. A.,** Evidence of cytokinin bases, ribosides and glucosides in roots of Douglas-fir, *Pseudotsuga menziesii*, *Tree Physiol.*, 5, 63, 1989.
5. **Taller, B. J. and Wong, T. Y.,** Cytokinins in *Azotobacter vinelandii* culture medium, *Appl. Environ. Microbiol.* 55, 266, 1989.
6. **Kraigher, H., Grayling, A., Wang, T. L., and Hanke, D. E.,** Cytokinin production by two ectomycorrhizal fungi in liquid culture, *Phytochemistry*, 30, 2249, 1991.
7. **Morris, R. O., Jameson, P. E., Laloue, M., and Morris, J. W.,** Rapid identification of cytokinins by an immunological method, *Plant Physiol.*, 95, 1156, 1991.
8. **Upadhyaya, N. M., Parker, C. W., Letham, D. S., Scott, K. F., and Dart, P.J.,** Evidence for cytokinin involvement in *Rhizobium* (IC3342)-induced leaf curl syndrome of pigeonpea (*Cajanus cajan* Millsp.), *Plant Physiol.*, 95, 1019, 1991.
9. **Teller, G.,** Cytokinin analysis by GC-MS, Chapter 24, this volume.
10. **Erlanger, B. F. and Beiser, S. M.,** Antibodies specific for ribonucleosides and ribonucleotides and their reaction with DNA, *Proc. Natl. Acad. Sci., U.S.A.*, 52, 68, 1964.
11. **MacDonald, E. M. S. and Morris, R. O.,** Isolation of cytokinins by immunoaffinity chromatography and analysis by high-performance liquid chromatography, *Methods Enzymol.*, 110, 347, 1985.
12. **Wang, T. L., Cook, S. K., and Knox, J. P.,** Monoclonal antibodies for the analysis and purification of isopentenyladenine cytokinins, *Phytochemistry*, 26, 2447, 1987.
13. **Weiler, E. W.,** Radioimmunoassays for *trans*-zeatin and related cytokinins, *Planta*, 149, 155, 1980.
14. **MacDonald, E. M. S. and Morris, R. O.,** Combined high-performance liquid chromatography-radioimmunoassay for cytokinins, *J. Chromatog.*, 214, 101, 1981.
15. **Vold, B. S. and Leonard, N. J.,** Production and characterization of antibodies and establishment of a radioimmunoassay for ribosylzeatin, *Plant Physiol.*, 67, 401, 1981.
16. **Badenoch-Jones, J., Letham, D. S., Parker, C. W., and Rolfe, B. G.,** Quantitation of cytokinins in biological samples using antibodies against zeatin riboside, *Plant Physiol.*, 75, 1117, 1984.
17. **Hansen, C. E., Wenzler, H., and Meins, F.,** Concentration gradients of *trans*-zeatin riboside and *trans*-zeatin in the maize stem, *Plant Physiol.*, 75, 959, 1984.
18. **Barthe, G. A. and Stewart, I.,** Enzyme immunoassay (EIA) of endogenous cytokinins in *Citrus*, *J. Agric. Food Chem.*, 33, 293, 1985.
19. **Trione, E. J., Krygier, B. B., Banowetz, G. M., and Kathrein, J. M.,** The development of monoclonal antibodies against the cytokinin zeatin riboside, *J. Plant Growth Regul.*, 4, 101, 1985.
20. **Eberle, J., Arnscheidt, A., Klix, D., and Weiler, E. W.,** Monoclonal antibodies to plant growth regulators. III. Zeatinriboside and dihydrozeatinriboside, *Plant Physiol.*, 81, 516, 1986.
21. **Hofman, P. J., Featonby-Smith, B. C., and Van Staden, J.,** The development of ELISA and IRA for cytokinin estimation and their application to a study of lunar periodicity in *Ecklonia maxima* (Osbeck) Papenf., *J. Plant Physiol.*, 122, 455, 1986.
22. **Maldiney, R., Leroux, B., Sabbagh, I., Sotta, B., Sossountzov, L., and Miginiac, E. A.,** Biotin-avidin-based enzyme immunoassay to quantify three phytohormones: auxin, abscisic acid and zeatin-riboside, *J. Immunol. Methods*, 90, 151, 1986.
23. **Parker, C. W., Badenoch-Jones, J., and Letham, D. S.,** Radioimmunoassay for quantifying the cytokinins *cis*-zeatin and *cis*-zeatin riboside and its application to xylem sap samples, *J. Plant Growth Regul.*, 8, 93, 1989.
24. **Banowetz, G. M.,** Monoclonal antibodies against the plant cytokinin, *cis*-zeatin riboside. *Hybridoma*, 12, 729, 1993.
25. **Khan, S. A., Humayun, M. Z., and Jacob, T. M.,** A sensitive radioimmunoassay for isopentenyladenosine, *Anal. Biochem.*, 83, 632, 1977.
26. **Milstone, D. S., Vold, B. S., Glitz, D. G., and Shutt, N.,** Antibodies to N^6-(Δ^2-isopentenyl) adenosine and its nucleotide: interaction with purified tRNAs and with bases, nucleosides and nucleotides of the isopentenyladenosine family, *Nucleic Acids Res.*, 5, 3439, 1978.
27. **Weiler, E. W. and Spanier, K.,** Phytohormones in the formation of crown gall tumors, *Planta*, 153, 326, 1981.
28. **Woodsworth, M. L., Latimer, L. J. P., Janzer, J. J., McLennan, B. D., and Lee, J. S.,** Characterization of monoclonal antibodies specific for isopentenyl adenosine derivatives occurring in transfer RNA, *Biochem. Biophys. Res. Commun.*, 114, 791, 1983.

29. **Badenoch-Jones, J., Parker, C. W., and Letham, D. S.,** Use of isopentenyladenosine and dihydrozeatin riboside antibodies for the quantification of cytokinins, *J. Plant Growth Regul.*, 6, 159, 1987.
30. **Sotta, B., Pilate, G., Pelese, F., Sabbagh, I., Bonnet, M., and Maldiney, R.,** An avidin-biotin solid phase ELISA for femtomole isopentenyladenine and isopentenyladenosine measurements in HPLC purified plant extracts, *Plant Physiol.*, 84, 571, 1987.
31. **Trione, E. J., Krygier, B. B., Kathrein, J. M., Banowetz, G. M., and Sayavedra-Soto, L. A.,** Monoclonal antibodies against the plant cytokinin isopentenyl adenosine, *Physiol. Plant.*, 70, 467, 1987.
32. **Davis, G. C., Hein, M. B., and Chapman, D. A.,** Evaluation of immunosorbents for the analysis of small molecules. Isolation and purification of cytokinins, *J. Chromatog.*, 366, 171, 1986.
33. **Strnad, M., Peters, W., Beck, E., and Kamínek, M.,** Immunodetection and identification of N^6-(o-hydroxybenzylamino)purine as a naturally occurring cytokinin in *Populus x canadensis* Moench cv. *Robusta* leaves, *Plant Physiol.*, 99, 74, 1992.
34. **Constantinidou, H. A., Steele, J. A., Kozlowski, T. T., and Upper, C. D.,** Binding specificity and possible analytical applications of the cytokinin-binding antibody, anti-N^6-benzyladenosine, *Plant Physiol.*, 62, 968, 1978.
35. **Papet, M.- P., Delay, D., Doumas, P., and Delmotte, F.,** New synthesis of immunogenic N^6-isopent-2-enyladenosine-protein conjugate. Preparation, purification, and specificity of related antibodies, *Bioconjugate Chem.*, 3, 14, 1992.
36. **Strnad, M., Vereš, K., Hanuš, J., and Siglerová, V.,** Immunological methods for quantification and identification of cytokinins (mini review), in *Physiology and Biochemistry of Cytokinins in Plants,* Kamínek, M., Mok, D. W. S., and Zazímalová, E., Eds., SPB Academic Publishing, The Hague, 1991, 1.
37. **Brandon, D. L., Corse, J., Higaki, P. C., and Zavala, M. E.,** Monoclonal antibodies for analysis of cytokinin O-glucosides in response to cold stress, in *Physiology and Biochemistry of Cytokinins in Plants,* Kamínek, M., Mok, D. W. S., and Zazímalová, E., Eds, SPB Academic Publishing, The Hague, 1992, 447.
38. **Aigner, H., Koch, G., and Marquarding, D.,** Isocyanides as activating reagents in peptide synthesis, in *Chemistry of Peptides and Proteins*, Vol. 1., Voelter, W., Wuensch, E., and Ovchinnikov, Y., Eds., de Gruyter, Berlin, 1982, 209.
39. **Goding, J. W.,** *Monoclonal Antibodies: Principles and Practice*, Academic Press, London, 1983.
40. **Campbell, A. M.,** *Monoclonal Antibody Technology*, Elsevier, Amsterdam, 1984.
41. **Turnbull, C. G. N. and Hanke, D. E.,** The control of bud dormancy in potato tubers. Measurement of the seasonal pattern of changing concentrations of zeatin-cytokinins, *Planta*, 165, 366, 1985.
42. **Langone, J. J.,** [^{125}I]protein A: a tracer for general use in immunoassay, *J. Immunol. Methods*, 24, 269, 1978.
43. **Björck, L. and Kronvall, G.,** Purification and some properties of streptococcal protein G, a novel IgG-binding reagent, *J. Immunol.*, 133, 969, 1984.
44. **Harlow, E. and Lane, D.,** *Antibodies: A Laboratory Manual*, Cold Spring Harbor Laboratory, Cold Spring Harbor, NY, 1988, p309.
45. **Scatchard, G.,** The attractions of proteins for small molecules and ions, *Ann. N.Y. Acad. Sci.*, 51, 660, 1949.
46. **Sips, R.,** On the structure of a catalyst surface, *J. Chem. Phys.*, 16, 490, 1948.
47. **Trione, E. J., Banowetz, G. M., Krygier, B. B., Kathrein, J. M., and Sayavedra-Soto, L.,** A quantitative fluorescence enzyme immunoassay for plant cytokinins, *Anal. Biochem.*, 162, 301, 1987.
48. **Weiler, E. W.,** Plant hormone immunoassays based on monoclonal and polyclonal antibodies, in *Immunology in Plant Sciences*, Linskens, H. F. and Jackson, J. F., Eds, Springer-Verlag, Berlin, 1986, 9.
49. **Reeve, D. R. and Crozier, A.,** Quantitative analysis of plant hormones, in *Encyclopedia of Plant Physiology*, Vol. 9, MacMillan, J., Ed, Springer-Verlag, Berlin, 1980, 203.
50. **Pengelly, W. L.,** Validation of immunoassays, in *Plant Growth Substances 1985*, Bopp, M., Ed, Springer-Verlag, Berlin, 1986, 35.
51. **Jones, H. G.,** Correction for non-specific interference in competitive immunoassays, *Physiol. Plant.*, 70, 146, 1987.
52. **Ernst, D., Schäfer, W., and Oesterhelt, D.,** Isolation and quantitation of isopentenyladenosine in an anise cell culture by single-ion monitoring, radioimmunoassay and bioassay, *Planta*, 159, 216, 1983.
53. **Wang, T. L., Futers, T. S., McGreary, F., and Cove, D. J.,** Moss mutants and the analysis of cytokinin metabolism, in *The Biosynthesis and Metabolism of Plant Hormones*, Crozier, A. and Hillman, J. R., Eds, Cambridge University Press, Cambridge, 1984, 135.

54. **Hocart, C. H., Badenoch-Jones, J., Parker, C. W., Letham, D. S., and Summons, R. E.,** Cytokinins of dry *Zea mays* seed: quantification by radioimmunoassay and gas chromatography-mass spectrometry, *J. Plant Growth Regul.*, 7, 179, 1988.
55. **Noodén, L. D., Singh, S., and Letham, D. S.,** Correlation of xylem sap cytokinin levels with monocarpic senescence in soybean, *Plant Physiol.*, 93, 33, 1990.
56. **Morris, J. W., Doumas, P., Morris, R. O., and Zaerr, J. B.,** Cytokinins in vegetative and reproductive buds of *Pseudotsuga menziesii*, *Plant Physiol.*, 93, 67, 1990.
57. **Kuraishi, S., Tasaki, K., Sakurai, N., and Sadatoku, K.,** Changes in levels of cytokinins in etiolated squash seedlings after illumination, *Plant Cell Physiol.*, 32, 585, 1991.
58. **Morris, R. O.,** Mass spectroscopic identification of cytokinins, *Plant Physiol.*, 59, 1029, 1977.
59. **Summons, R. E., Duke, C. C., Eichholzer, J. V., Entsch, B., Letham, D. S., MacLeod, J. K., and Parker, C. W.,** Mass spectrometric analysis of cytokinins in plant tissues. II. Quantitation of cytokinins in *Zea mays* kernels using deuterium labelled standards, *Biomed. Mass Spectrom.*, 6, 407, 1979.
60. **Summons, R. E., Entsch, B., Parker, C. W., and Letham, D. S.,** Mass spectrometric analysis of cytokinins in plant tissues. III. Quantitation of the cytokinin glycoside complex of lupin pods by stable isotope dilution, *FEBS Lett.*, 107, 21, 1979.
61. **Palni, L. M. S., Summons, R. E., and Letham, D. S.,** Mass spectrometric analysis of cytokinins in plant tissues. V. Identification of the cytokinin complex of *Datura innoxia* crown gall tissue, *Plant Physiol.*, 72, 858, 1983.
62. **Letham, D. S. and Singh, S.,** Quantification of cytokinin O-glucosides by negative ion mass spectrometry, *Plant Physiol.*, 89, 74, 1989.
63. **Letham, D. S., Singh, S., and Wong, O. C.,** Mass spectrometric analysis of cytokinins in plant tissue. VII. Quantification of cytokinin bases by negative ion mass spectrometry, *J. Plant Growth Regul.*, 10, 107, 1991.
64. **Berkson, J.,** A statistically precise and relatively simple method of estimating the bioassay with quantal response based on the logistic function, *J. Am. Stat. Assoc.*, 48, 565, 1953.
65. **Weiler, E. W. and Ziegler, H.,** Determination of phytohormones in phloem exudate from tree species by radioimmunoassay, *Planta*, 152, 168, 1981.
66. **Young, E.,** Cytokinin and soluble carbohydrate concentrations in xylem sap of apple during dormancy and budbreak, *J. Am. Soc. Hort. Sci.*, 114, 297, 1989.
67. **Belding, R. D. and Young, E.,** Shoot and root temperature effects on xylary cytokinin levels during budbreak in young apple trees, *Hortscience*, 24, 115, 1989.
68. **Rossi, G., Marziani, G. P., Uneddu, P., and Longo, C. P.,** Changes in endogenous levels of three cytokinins during development of excised watermelon cotyledons, *Physiol. Plant.*, 83, 647, 1991.
69. **Badenoch-Jones, J., Parker, C. W., and Letham, D. S.,** Phytohormones, *Rhizobium* mutants, and nodulation in legumes. VII. Identification and quantification of cytokinins in effective and ineffective pea root nodules using radioimmunoassay, *J. Plant Growth Regul.*, 6, 97, 1987.
70. **Vaňková, R., Kamínek, M., Eder, J., and Vaněk, T.,** Dynamics of production of *trans*-zeatin and *trans*-zeatin riboside by immobilized cytokinin-autonomous and cytokinin-dependent tobacco cells, *J. Plant Growth Regul.*, 6, 147, 1987.
71. **Hansen, C. E., Kopperud, C., and Heide, O. M.,** Identity of cytokinins in *Begonia* leaves and their variation in relation to photoperiod and temperature, *Physiol. Plant.*, 73, 387, 1988.
72. **Strnad, M., Vaněk, T., Binarová, P., Kamínek, M., and Hanuš, J.,** Enzyme immunoassays for cytokinins and their use for immuno-detection of cytokinins in alfalfa cell culture, in *Molecular Aspects of Hormonal Regulation of Plant Development*, Kutáček, M., Elliot, M. C., and Macháčková, I., Eds, SPB Academic Publishing, The Hague, 1990, 41.
73. **Upadhyaya, N. M., Letham, D. S., Parker, C. W., Hocart, C. H., and Dart, P. J.,** Do rhizobia produce cytokinins? *Biochem. Int.*, 24, 123, 1991.
74. **Vanková, R., Hsiao, K. C., Bornman, C. H., and Gaudinova, A.,** Effects of synthetic cytokinins on levels of endogenous cytokinins and respiration patterns of *Beta vulgaris* cells in suspension, *J. Plant Growth Regul.*, 10, 197, 1991.
75. **Schwartzenberg, K. V. and Hahn, H.,** The cytokinin content in needles of Norway Spruce (*Picea abies* L. Karst.) with different degrees of damage, *J. Plant Physiol.*, 139, 218, 1991.

76. **Davis, G. C., Hein, M. B., Chapman, D. A., Neely, B. C., Sharp, C. R., Durley, R. C., Biest, D. K., Heyde, B. R., and Carnes, M. G.,** Immunoaffinity columns for the isolation and analysis of plant hormones, in *Plant Growth Substances 1985*, Bopp, M., Ed., Springer-Verlag, Berlin, 1986, 44.
77. **MacDonald, E. M. S., Powell, G. K., Regier, D. A., Glass, N. L., Roberto, F., Kosuge, T., and Morris, R. O.,** Secretion of zeatin, ribosylzeatin, and ribosyl-1″-methylzeatin by *Pseudomonas savastanoi*, *Plant Physiol.*, 82, 742, 1986.
78. **Hansen, C. E., Kopperud, C., and Heide, O. M.,** Immunoaffinity chromatography for purification of cytokinins in plant extracts, *Norwegian J. Ag. Sci.*, 3, 73, 1989.
79. **Guinn, G. and Brummett, D. L.,** Solid-phase extraction of cytokinins from aqueous solutions with C_{18} cartridges and their use in a rapid purification procedure, *Plant Growth Regul.*, 9, 305, 1990.
80. **Banowetz, G. M.,** The effects of endogenous cytokinin content on benzyladenine-enhanced nitrate reductase induction, *Physiol. Plant.*, 86, 341, 1992.

Chapter 24

Gas Chromatographic-Mass Spectrometric and Related Methods for the Analysis of Cytokinins

Gerard Teller

TABLE OF CONTENTS

I. INTRODUCTION

Many cytokinin metabolites occurring in plant tissues have been characterized by mass spectrometry. Palni, Tay, and MacLeod published (in 1986) an extensive review on gas chromatographic-mass spectrometric (GC-MS) methods for the analysis of cytokinins.[1] New methods, such as immunoassays, have appeared, but mass spectrometry remains a reliable and specific means for the identification and quantification of cytokinins.

The major challenge in the application of GC-MS is the purification of the complex plant extracts to a less-complex mixture where the mass spectrometric response may specifically correlate to cytokinin content. Purification methods involve extraction, column separations, thin-layer chromatography (TLC), high-performance liquid chromatography (HPLC), and gas chromatographic methods. The methods of cytokinin extraction and purification have been described elsewhere.[2-9] In general, mass spectrometers can be directly connected to HPLC and GC, usually the final step in cytokinin purification.[10] This approach will be more useful in the future, especially when the combination of HPLC to ion-spray instruments becomes routine.

II. GAS CHROMATOGRAPHY

Cytokinins in their free forms are not amenable to GC, since they are not volatile and are thermally unstable, resulting in degradation products adsorbing on the columns, via hydrogen bonding. Fortunately, cytokinins usually can be very easily derivatized to volatile and stable compounds. The derivatization procedures are now well described: silylation (TMS) with MSTFA (*N*-methyl-n(trimethylsilyl)trifluoroacetamide) in 50% pyridine plus 1% trimethylchlorosilane for one or several hours at room temperature, or for a shorter time at 80°C, gives good results. The detection limit is about 10 ng of anhydrous cytokinins even if the sample is not completely derivatized. Other derivatives, such as permethyl derivatives,[11] t-butyldimethylsilyl (tBDMS), and trifluroacetyl compounds, can also be used. If a complete mass spectrum is desired, 1 μg of silyl derivative containing 1 to 10 ng of cytokinin is needed to get a specific and authentic response, while 1 to 10 pg of sample is sufficient for selected ion mass spectra, if there is no interference from other compounds having the same ions and retention times.

The early GC-MS runs were performed on packed columns. Currently, capillary fused-silica columns with internal diameters of 0.2 to 0.53 mm and coated with 0.1- to 2-μm films of stationary phases are used. The stationary-phase films are usually methyl (or methyl with 5% phenyl) silicone polymers, which give good separations with oven heating rates of 5°C/min, between 150 and 250°C.

Thin films of 0.1 μm are used with very pure extracts and give better chromatographic separations, but the amount of material that can be injected on thin-film coatings is very low, less than 300 ng. Therefore, when cytokinins are diluted in larger quantities of sample, it is necessary to use films of 1 or 2 μm.

0-8493-6252-0/94/$0.00+$.50

Various injectors are in use for cytokinin analysis. On-column injectors, which allow the injection of a sample into the cold column, give good results for pure extracts on small diameter columns and thin films. If the samples contain too much material, the first meter of the columns gets dirty and adsorbs the cytokinins after several injections. The derivati-zation reagents also have a degradation effect on the coating of the column.

The moving needle injector (ROS injector) gives better results with ordinary samples. The samples are deposited through a rubber septum on the end of a glass tip. The carrier gas evaporates the solvents and excess reagent to a vent. The dry sample is brought to a warm part of the injector and is evaporated near the entrance of the column, which is then heated. The unevaporated materials remain in the injector, which must be cleaned after a few days.

III. MASS SPECTROMETRY

Different types of mass spectrometers are used in combination with various kinds of inlets, analyzers, and detectors. The instruments are fitted with computers for signal detection and measurement, for control of the other electronic elements of the instruments, as well as for data manipulation.

A. GC-MS COMPONENTS

The inlet allows the sample to be brought into the ion source of the mass spectrometer. For the direct inlet, a fraction of the sample is deposited on a probe that is brought through a vacuum lock into the heated ion source if it is an electron impact or a chemical ionization source, or into a cold fast-atom bombardment source.

A flow interface is located at the end of the gas chromatographic capillary column and passes the sample through a heated interface and vacuum-tight connection to the ion source. The small gas flow of a capillary column (1 ml/min) allows the sample to pass through the ion source without degrading the vacuum of the instrument. (Formerly, packed columns required jet separators or splits to remove excess carrier gas.)

There are two main types of HPLC interfaces in common use today: the spray and flow Fast Atom Bombardment (FAB) interfaces. In the spray-type interfaces (thermospray, electrospray, or ionspray), the HPLC effluent is brought to a heated chamber and introduced into the vacuum from a fine nebulizing orifice. Ionization takes place in the fine spray of the charged droplets produced. The ions are transported through a skimmer lens into the mass spectrometer. The flow FAB interface utilizes a dynamic version of FAB for introduction of the HPLC effluent onto the probe tip for ionization by energetic argon or xenon atoms. For both methods the underivatized cytokinins undergo a very soft ionization and produce spectra rich in structural information.

Cytokinin ionization may be obtained with several modes, such as electron impact, chemical ionization, and fast-atom bombardment, or thermal, electron, or ion sprays. Electron impact uses an electron beam emitted by a hot filament, to collide with samples evaporated in the gas phase. An electron from the outer electron shell of an atom is expelled. The remaining positive molecular ion usually fragments into smaller pieces, as illustrated by the mass spectra shown below (Figures 1–3).

$$ABCD + e^- \rightarrow ABCD^+ + 2e^- \quad \text{(molecular ion species)}$$
$$\rightarrow AB^+ + CD_0 + 2e^- \quad \text{(fragment and unobserved neutral species)}$$
$$\rightarrow ABD^+ + C_0 + 2e^- \quad \text{(internal rearrangement)}$$

Electron impact ionization is commonly used, but the drawback is that molecular ions are sometimes difficult to detect. For silylated compounds, the loss of a methyl group is common $(M\text{-}15)^+$.

Chemical ionization uses a reactant gas, methane, ammonia, or isobutane, added to the ion source at a convenient pressure (about 1 torr). The electron beam produces ions, and ion-molecule reactions are observed, giving rise to such protonated species as CH_5^+, C_2H5^+, NH_4^+, or $C_4H_9^+$. These species collide with the solutes and, through an exchange process and/or ion–molecule reactions, produce protonated quasimolecular ions and specific adducts (+29, +18, or +57). This "soft" ionization method usually gives only a few fragments, if any. The spectra show quasimolecular ions that give essential information, but the low number of fragments does not lead to a complete identification.

$$ABCD + NH_4^+ \rightarrow ABCD\mathbf{H}^+ + NH_3 \quad \text{(protonated quasi-molecular ion)}$$

Fast-atom bombardment utilizes a special plasma gun, and argon or xenon ions are produced. They are sent to a pressure chamber where charge exchange takes place. With high velocity, neutral argon or xenon atoms are expelled from the gun. They impinge on a solid probe where the sample, dissolved in a liquid matrix, usually glycerol or thioglycerol, is ionized. The method produces protonated molecular ions and informative fragments for thermally labile or high-molecular-weight compounds. The usual mass range is up to 5000 Da, and

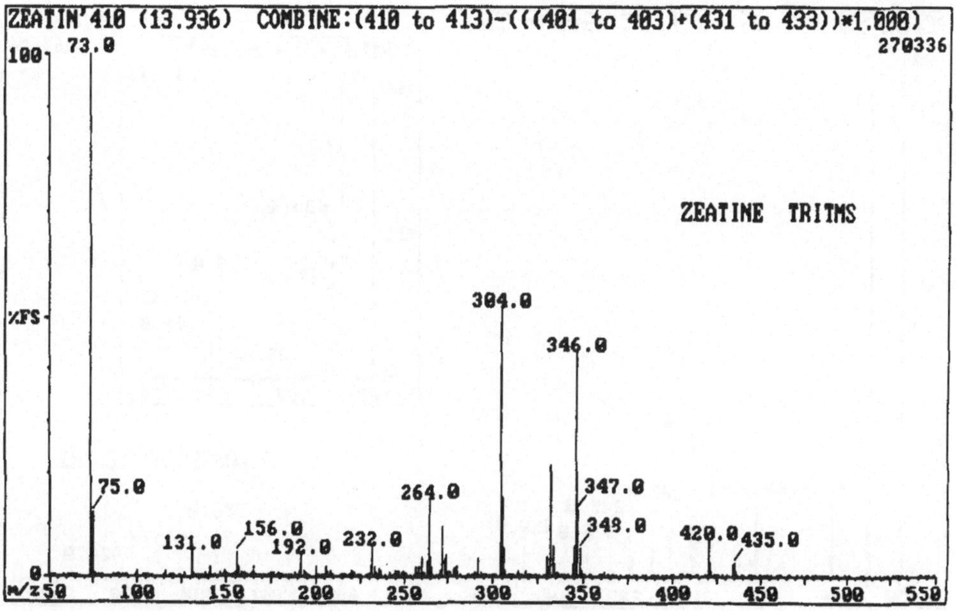

Figure 24-1 Electron impact spectrum of tri(trimethylsilyl)zeatin.

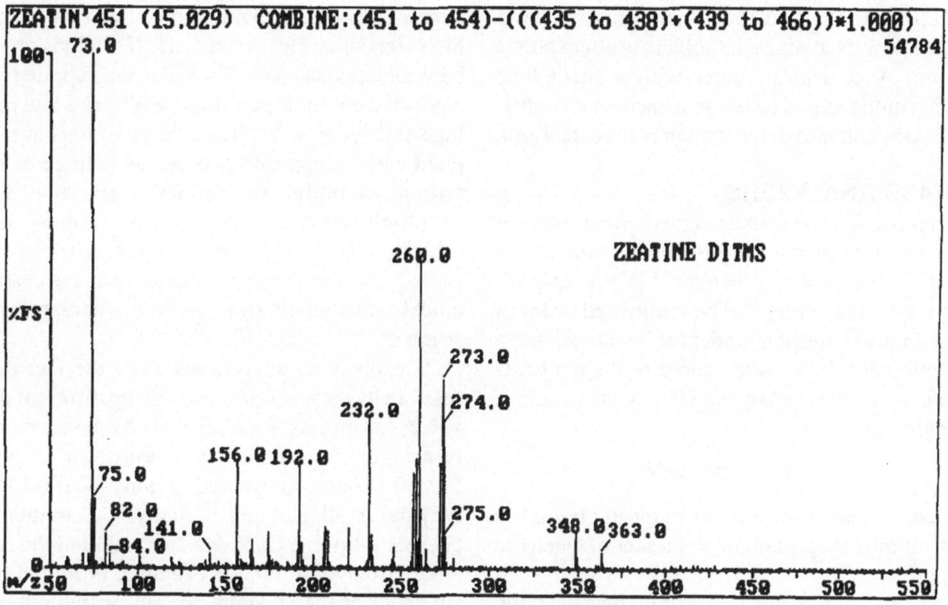

Figure 24-2 Electron impact spectrum of di(trimethylsilyl)zeatin.

for more expensive instruments a detection range of 15,000 Da is possible.

The FAB source, which uses a "soft" ionization method, produces spectra from free, underivatized cytokinins.[12] Imbault et al. were able to quantify isopentenyladenosine at the concentration of 1 ng/g fresh weight, using deuterated reference compounds.[10] For thermal or electro sprays, the effluent of an HPLC column goes to a heated evaporation chamber. As the solvent evaporates, ionization takes place. Sometimes, an electron impact device is added. Often, voltage is applied between the HPLC outlet nozzle and the entrance of the ion source. Ion spray is a powerful method of getting quasimolecular ions from an HPLC outlet, especially when the solutes are of high molecular

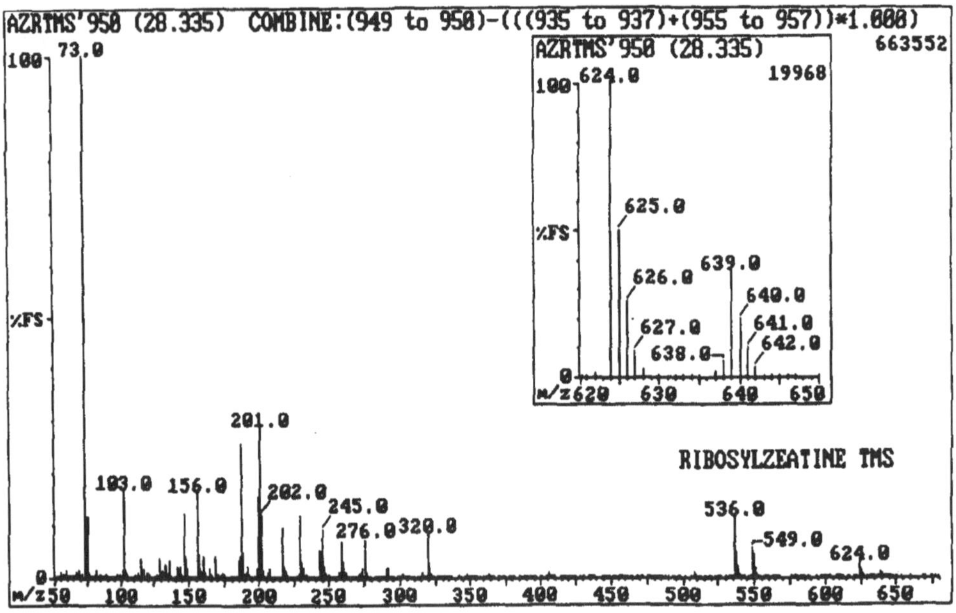

Figure 24-3 Electron impact spectrum of ribosylzeatin.

weight. High voltage (4 kV) is applied between a nebulizer and a cold (60°C) chamber, where a nitrogen flow evaporates the solvent at atmospheric pressure. A conical skimmer with a small hole extracts multicharged ions into a vacuum chamber, which is the entrance of the ion-spray mass analyzer.

B. MASS ANALYZERS

Magnetic analyzers use magnetic fields that are able to separate charged ions by momentum, analogous to the way in which an optical prism separates the wave lengths in light. The collimated beam of charged ions is injected, under high voltage, into a magnetic field. If "r" is the curve of the ion beam in the magnet, the fundamental law of the magnetic separation is

$$m/nz = B^2r^2/2V$$

where m = mass of the ion, B= magnetic field, V = acceleration voltage, and nz = number of electrical charges.

The ions are separated, measured by a detector, and recorded by a computer. The values are stored on a hard disk. If the magnetic field is changed continuously from high values to small values, all ions can be scanned in the mass spectrum. If the scan is repeated rapidly, the spectra of an effluent from a gas chromatograph or an HPLC can be recorded. It is also possible to measure selected values in a broad mass spectrum. The selected values are measured over a longer period of time, resulting in greater sensitivity (Figure 4).

For quadrupole analyzers, the ions are sent into a mass filter of four parallel metal bars. A variable high-frequency electromagnetic V^h field is applied between opposite bars. Two bars are supplied with a positive continuous voltage + V^{dc}, and a variable high-frequency + V^{rf}. The other two bars are supplied with an opposite continuous voltage of $-V^{dc}$ and a variable, in opposite phase (180°), $-V^{rf}$ high-frequency voltage. An ion is transmitted only if the frequency fits its weight (m/nz). As the frequency value can be changed quickly, fast inputs from a chromatograph can be scanned.

For ion-spray analyzers we use the fact that a mass analyzer separates ions having different m/nz values (m = mass; z = electrical charge; n= number of charges). In such a source, ions of a weight of 50,000 Da are transmitted as ions of 1000 Da if they have 50 charges. Usually, the number of charges follows a Gaussian curve around the mean value, from 40 to 60, for example. From the measurement of the apparent weight of two consecutive ions, m_1 and m_2, we may compute the value of the molecular weight (Figure 5):

$$M = (m/m_1 - m_2) \times (m_2 - 1)$$

The progress in electronics and the low cost of computers have made quadrupole analyzers linked to gas chromatographs affordable to many laboratories (range of \$100,000). A fast scanning magnet would double the price of a magnetic analyzer.

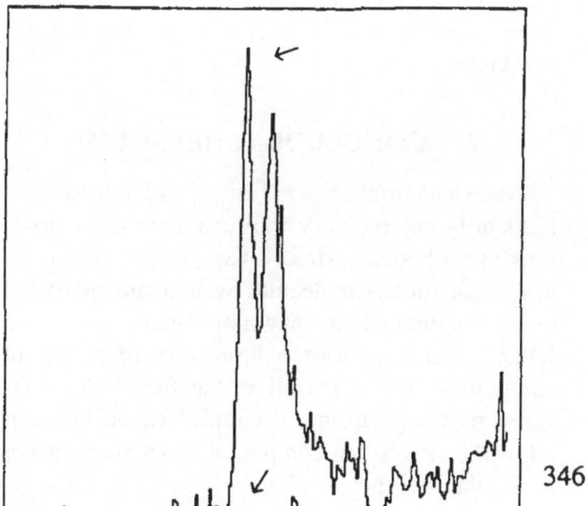

Figure 24-4 Selected ion chromatogram, ions 346, 420, and 435, showing presence of tri(trimethylsilyl)zeatin in *Mercurialis annua* L. (See simultaneous response for all selected ions, at only one location of the chromatogram. The relative intensities have the same values as in the reference spectrum.)

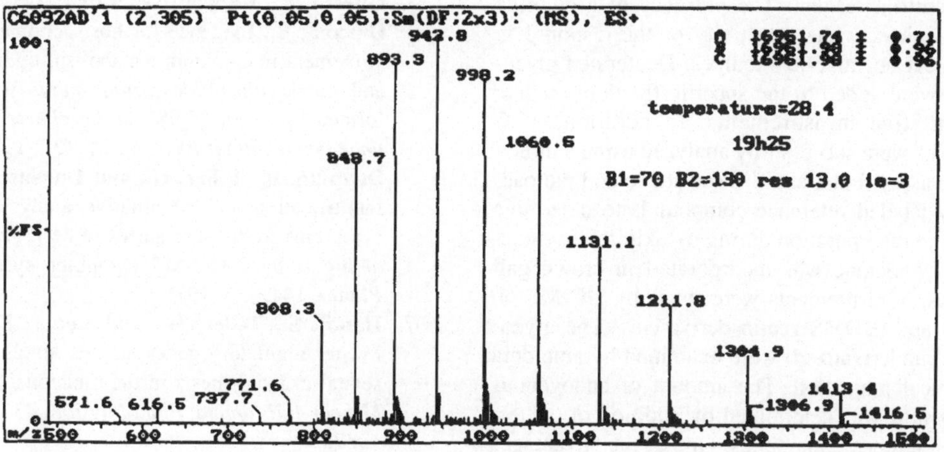

Figure 24-5 Ion-spray spectrum of myoglobin (mol. wt. = 16,951). Ion 942.8 has 18 electronic charges; ion 998.2 has 17 electronic charges. Courtesy of A. V. Dorsselaer and R. Hueber (Strasbourg).

Approximately 20 years ago, we began to collect spectra of cytokinins. The spectrum of silylated isopentenyladenosine was the first spectrum obtained and recorded on photographic paper, using an LKB integrated chromatograph-mass spectrometer. Since then we have been able, using a computerized data system, to store complete analyses of 500 scans. On the basis of selected ions from a reference compound, we can locate and determine cytokinins among a large number of eluted compounds.

Now, low-cost data systems are available, such as bench-top computerized instruments. The capillary column in the chromatograph is linked to the ion source of the mass spectrometer through a small heated interface and goes into the ion chamber, near the electron beam.

IV. SOME APPLICATIONS

GC-MS is used in combination with other techniques, to identify and quantify cytokinins. Taller et al. studied endogenous cytokinins from RNA.[13] The identification of common cytokinins involved the use of column chromatography, while unusual derivatives were authenticated by GC-MS. Louis et al. used GC-MS extensively to identify and quantify cytokinins in fertile and sterile lines of *Mercurialis annua* (see Chapter 21).[4,8] Immunoassays following HPLC separation were used by Morris et al. to isolate zeatin and isopentenyl derivatives.[14] Confirmation of the identity was made using GC-MS on permethylated derivatives. Letham and Singh quantified cytokinin *O*-glycosides by negative chemical ionization mass spectrometry.[15] They used chloro derivatives made from a chloroacrylonitrile reagent. The presence of a chlorine atom enhances the sensitivity of the response in chemical negative-ion sources. Deuterated glycosides were added to the specific fractions to give quantitative measurements. In addition, TLC samples were successfully analyzed using a direct-probe assay. Palni et al.[16] used stable and radioactively labeled reference compounds to determine adenine incorporation during cytokinin biosynthesis. ^{15}N-adenine was incorporated in crown gall tissues. Measurements were made by GC-MS on TMS and tBDMS zeatin derivatives. The glycosides (underivatized) were examined by ammonia chemical ionization. The amount of endogenous cytokinins was determined by addition of deuterated reference compounds. One of the advantages of GC-MS is the identification of new and previously unknown cytokinins. Lee et al. were able to identify *O*-xylosylzeatin and its riboside from immature seeds of *Phaseolus*.[17] The spectra of these compounds facilitate the chemical synthesis of these cytokinins.

V. CONCLUDING REMARKS

Mass spectrometric analysis is a reliable method to identify and quantify well-characterized cytokinins in biological extracts. Conjugates of cytokinins with higher molecular weight are analyzed using chemical and enzymatic degradation and HPLC separation, prior to the analysis of the parent compound. It is expected in the future that ion-spray mass spectrometry coupled to HPLC will allow the analysis of compounds with much larger molecular weights.

REFERENCES

1. **Palni, L. M. S., Tay, S. A. B., and MacLeod, J. K.,** GC-MS methods for cytokinins and Metabolites, in *Modern Methods of Plant Analysis*, Linskens, H. F. and Jackson, J. F., Eds., Springer-Verlag, Berlin, 1986, 213.
2. **Dauphin-Guerin, B., Teller, G., and Durand, B.,** Different endogenous cytokinins between male and female *Mercurialis annua*, *Planta*, 148, 124, 1980.
3. **Palni, L. M., Summons, A. E., and Letham, D. S.,** Mass spectrometric analysis of cytokinins in plant tissues, *Plant Physiol.*, 72, 853, 1983.
4. **Louis, J. P., Augur, C., and Teller, G.,** Cytokinins and differentiation processes in *Mercurialis annua*, *Plant Physiol.*, 94, 1565, 1990.
5. **Guerin, B., Kahlem, G., Teller, G., and Durand, B.,** Evidence for host genome involvement in cytokinin metabolism by male and female cells of *Mercurialis annua*, transformed by strain 15.955 of *Agrobacterium tumefaciens*, *Plant Physiol.*, 74, 139, 1984.
6. **Dauphin, B., Teller, G., and Durand, B.,** Identification and quantitative analysis of cytokinins from shoot apices of *Mercurialis ambigua* by a GC-MS computer system, *Planta*, 144, 113, 1979.
7. **Handi, S., Teller, G., and Louis, J. P.,** Master regulatory genes, auxins levels and sexual organogenesis in the dioceous plant *Mercur-ialis annua*, *Plant Physiol.*, 85, 393, 1987.
8. **Louis, J. P., Augur, C., and Teller, G.,** Cytokinins and differentiation processes in *Mercurialis annua*, *Plant Physiol.*, 94, 1535, 1990.

9. **Dauphin, B., Teller, G., and Durand, B.,** Mise au point d'une nouvelle méthode de purification, caractérisation et dosage de cytokinines endogènes extraites de bourgeons de Mercuriales annuelles, *Physiol Vég.*, 15, 747, 1977.
10. **Imbault, N., Moritz, T., Nilsson, O., Chen, H. J., Bollmark, M., and Sandberg, G.,** Separation and identification of cytokinins using combined capillary liquid chromatography/mass spectrometry, *Biol. Mass Spectrom.*, 22, 201,1993.
11. **Morris, R. O.,** Mass spectrometric identification of cytokinins, *Plant Physiol.*, 59, 1029, 1977.
12. **Greco, F., Sindona, G., Uccella, N., and Evudente, A.,** Characterization of cytokinins from plant pathogenic *Pseudomonades* by fast atom bombardment tandem mass spectrometry, *Org. Mass Spectrom.*, 27, 750, 1992.
13. **Taller, B. J., Murai, N., and Skoog, F.,** Endogeneous cytokinins in the ribosomal RNA of higher plants, *Plant Physiol.*, 83, 755, 1987.
14. **Morris, J. W., Doumas, P., Morris, R. O., and Zaerr, J. B.,** Cytokinins in vegetative and reproductive buds of *Pseudotsuga menziesii*, *Plant Physiol.*, 93, 67, 1990.
15. **Letham, D. S. and Singh, S.,** Quantification of cytokinin O-glucosides by negative ion mass spectrometry, *Plant Physiol.*, 89, 74, 1989.
16. **Palni, L. M. S., Tay, S. A. B., and MacLeod, J. K.,** Determination by GC-MS of [$^{15}N_5$] adenine incorporation into endogeneous cytokinins and the effect of tissue age on cytokinin biosynthesis in *Datura innoxia* crown gall tissue, *Plant Physiol.*, 84, 1158, 1987.
17. **Lee, Y. H., Mok, M. C., Mok, D. W. S., Griffin, D. A., and Shaw, G.,** Cytokinin metabolism in *Phaseolus* embryos: genetic difference and the occurrence of novel zeatin metabolites, *Plant Physiol.*, 77, 635, 1985.

INDEX

D

E

F

G

H

I

J

K

L

M

Q

R

S

U

V

W

X

Y

Z